Developmental Biology

A COMPREHENSIVE SYNTHESIS

Volume 3

The Cell Surface in Development and Cancer

Developmental Biology
A COMPREHENSIVE SYNTHESIS

Editor
LEON W. BROWDER
University of Calgary
Calgary, Alberta, Canada

Developmental Biology

A COMPREHENSIVE SYNTHESIS

Volume 3

The Cell Surface in Development and Cancer

Edited by

MALCOLM S. STEINBERG

Princeton University
Princeton, New Jersey

Springer Scince+Business Media, LLC

Library of Congress Cataloging in Publication Data

(Revised for vols. 3 & 4)

Developmental biology.

  Includes bibliographies and index.
  Contents: v. 1. Oogenesis —   —v. 3. The cell surface in development and cancer —
v. 4. Manipulation of mammalian development.
  1. Developmental biology—Collected works. I. Browder, Leon W.
QH491.D426 1985   574.3   85-3406
ISBN 978-1-4684-5052-1   ISBN 978-1-4684-5050-7 (eBook)
DOI 10.1007/ 978-1-4684-5050-7

Cover illustration: Whole-mount immunofluorescence histochemistry of a chicken embryo. Only the myocytes of myotome and heart show positive staining; if there are any terminally differentiated myocytes in the wing bud at this stage, they cannot be detected. (From Chapter 11 by David C. Turner.)

© 1986 Springer Science+Business Media New York
Originally published by Plenum Press, New York in 1986
Softcover reprint of the hardcover 1st edition 1986

Contributors

Kenneth E. Bassett Division of Biology, Kansas State University, Manhattan, Kansas 66506

Michael J. Bastiani Department of Biological Sciences, Stanford University, Stanford, California 94305

Paul Henry Black Department of Microbiology and Hubert H. Humphrey Cancer Research Center, Boston University School of Medicine, Boston, Massachusetts 02118

Daniel F. Bowen-Pope Department of Pathology, University of Washington, Seattle, Washington 98195

Marianne Bronner-Fraser Developmental Biology Center, University of California—Irvine, Irvine, California 92715

Peter Devreotes Department of Biological Chemistry, Johns Hopkins University School of Medicine, Baltimore, Maryland 21205

Chris Q. Doe Department of Biological Sciences, Stanford University, Stanford, California 94305

Sascha duLac Department of Biological Sciences, Stanford University, Stanford, California 94305

Edward G. Fey Department of Biology, Massachusetts Institute of Technology, Cambridge, Massachusetts 02139

Judah Folkman Department of Surgery, Children's Hospital, and Department of Surgery, Department of Anatomy and Cellular Biology, Harvard Medical School, Boston, Massachusetts 02115

Donna Fontana Department of Biological Chemistry, Johns Hopkins University School of Medicine, Baltimore, Maryland 21205. *Present address:* Department of Microbiology, University of Minnesota, Minneapolis, Minnesota 55455.

Corey S. Goodman Department of Biological Sciences, Stanford University, Stanford, California 94305

Gary Gorbsky High Voltage Electron Microscopy Laboratory and Department of Molecular Biology, University of Wisconsin—Madison, Madison, Wisconsin 53706

Russell Greig Smith Kline and French Laboratories, Philadelphia, Pennsylvania 19101.

Albert K. Harris Department of Biology, University of North Carolina, Chapel Hill, North Carolina 27514

William J. Lennarz The University of Texas System Cancer Center, M. D. Anderson Hospital and Tumor Institute, Department of Biochemistry and Molecular Biology, Houston, Texas 77030

Sheldon Penman Department of Biology, Massachusetts Institute of Technology, Cambridge, Massachusetts 02139

George Poste Smith Kline and French Laboratories, Philadelphia, Pennsylvania 19101

Jean-Paul Revel Division of Biology, California Institute of Technology, Pasadena, California 91125

Brian S. Spooner Division of Biology, Kansas State University, Manhattan, Kansas 66506

Malcolm S. Steinberg Department of Biology, Princeton University, Princeton, New Jersey 08544

Brad Stokes Division of Biology, Kansas State University, Manhattan, Kansas 66506

Douglas Dillon Taylor Department of Microbiology and Hubert H. Humphrey Cancer Research Center, Boston University School of Medicine, Boston, Massachusetts 02118

Anne Theibert Department of Biological Chemistry, Johns Hopkins University School of Medicine, Baltimore, Maryland 21205

William A. Thomas Department of Biology, Wake Forest University, Winston-Salem, North Carolina 27109

Holly A. Thompson-Pletscher Department of Chemistry, University of Montana, Missoula, Montana 59812

David C. Turner Department of Biochemistry, State University of New York Upstate Medical Center, Syracuse, New York 13210

Tit-Yee Wong Department of Biology, Johns Hopkins University, Baltimore, Maryland 21218

Preface

This series was established to create comprehensive treatises on specific topics in developmental biology. Such volumes serve a useful role in developmental biology, since it is a very diverse field that receives contributions from a wide variety of disciplines. This series is a meeting-ground for the various practitioners of this science, facilitating an integration of heterogeneous information on specific topics.

Each volume is intended to provide the conceptual basis for a comprehensive understanding of its topic as well as an analysis of the key experiments upon which that understanding is based. The specialist in any aspect of developmental biology should understand the experimental background of the field and be able to place that body of information in context to ascertain where additional research would be fruitful. At that point, the creative process generates new experiments. This series is intended to be a vital link in that ongoing process of learning and discovery.

In no other aspect of developmental biology is its multidisciplinary character more evident than in the study of growth control, which involves embryologists, cell biologists, molecular biologists, biochemists, and clinical scientists. Orderly cell division is a prerequisite to conversion of the zygote into a highly organized multicellular adult organism. Cancer, on the other hand, is characterized by faulty growth control. The cell surface plays a central role in regulating many of the cellular functions related to growth control in normal development. The cell surface is also involved in mediating cell–cell interactions, which are important in morphogenesis and which may be highly abnormal in malignancy. Clearly, the study of cell surface properties in developing cells and malignant cells is complementary; an understanding of either field improves our understanding of the other. An improved understanding of the cell biology of malignancy is the key to control of this deadly disease.

This volume is a chronicle of our current understanding of the properties of the cell surface relating to development and cancer. It is a companion to Volume 2 of this series. That book (*The Cellular Basis of Morphogenesis*, edited by Leon W. Browder) considers the role of cellular and extracellular components in the assembly of multicellular embryos. The functions of the cytoskeleton, cell surface, and extracellular matrix molecules in cellular shape changes, cellular interactions, and cell motility in development are discussed in detail. Together, these two volumes are intended to provide the basis for understanding the cellular events underlying morphogenesis and contributing to the abnormal behavior of malignant cells.

Leon W. Browder

Introduction to the Volume

Cells are the "atoms" of the metazoan organism. They are the subunits that multiply, differentiate, lay down extracellular scaffoldings, organize themselves into tissues and organs and—when the normal programs of differentiation and morphogenesis go awry—metastasize to foreign sites and invade, multiply, and destroy the tissues of the host and ultimately the host itself. Growth, differentiation, and morphogenesis must all be regulated in relationship to demand, requiring the exchange of information between the cell and its environment. All such information must be delivered either to or through the cell surface. It is not surprising, then, that as cell regulatory processes have come to be better understood, our perception of the cell envelope has changed dramatically. Whereas once it was thought of as not much more than the "bag" preventing the escape of all those interesting enzymes, today the cell membrane is recognized as a complex and heterogeneous structure with specialized domains dynamically regulating the passage of molecules of many kinds and bearing specific receptors relaying signals between the worlds without and within.

Just as the envelopes of normal cells have a crucial role in interpreting and relaying signals essential to normal cell functions, so have the envelopes of cancer cells been implicated in their refractoriness to control by those instructions to which normal cells submit. The concept of a volume bringing together accounts of contributions to our knowledge of cell-surface-related phenomena of importance to our understanding of both development and cancer arose from a symposium of like purpose sponsored by the Division of Developmental and Cell Biology of the American Society of Zoologists and generously supported by grants from NSF and NIH and by a contribution from Smith Kline and French Laboratories. This volume was assembled subsequently and addresses itself to the normal processes of controlled growth, guided cell locomotion, selective cell association and assembly, and cell communication, as well as aberrations in these processes and in the underlying mechanisms in malignant cells.

Malcolm S. Steinberg

Contents

Chapter 7 • The Growth Factor–Platelet–Transformation
Connection
Daniel F. Bowen-Pope

Chapter 8 • Intercellular Recognition and Adhesion
in Desmosomes
Gary Gorbsky

Chapter 9 • Dual Adhesive Recognition Systems in Chick
Embryonic Cells
William A. Thomas

Chapter 14 • Growth Cone Guidance and Cell Recognition in Insect
             Embryos

Corey S. Goodman, Michael J. Bastiani, Chris Q. Doe, and
Sascha duLac

Chapter 15 • Guidance of Neural Crest Migration: Latex Beads as
             Probes of Surface–Substratum Interactions

Marianne Bronner-Fraser

Chapter 16 • Cell Traction in Relationship to Morphogenesis and
             Malignancy

Albert K. Harris

Chapter 1

Cell Surfaces in the Control of Growth and Morphogenesis

MALCOLM S. STEINBERG

1. Introduction

My students and I have been concerned, over the years, with the role of cell surfaces in multicellular assembly phenomena. It pleases me greatly that two of them (Dr. Gary Gorbsky and Dr. William A. Thomas) are among the contributors to this volume. This chapter concentrates on two areas in which cell surface phenomena play a central controlling role—the control of cell proliferation and the control of multicellular assembly.

Research in my laboratory at present is concerned with bringing our understanding of morphogenetic phenomenon to the molecular level. Nevertheless, I have chosen to restrict myself in this chapter to studies at the supramolecular level that establish the foundation for molecular analyses. The observations we shall be concerned with here were largely made at the level of ordinary observation, aided in the main only by optical lenses. It is here that we discover the phenomena of central interest—that cells divide or do not divide; that certain cells move along a certain path, while others follow a different path or remain where they were; that one group of cells forms a heart while another group forms a liver. On a darker note, a cell here or there divides repeatedly; some of its progeny invade and destroy neighboring tissues, while others penetrate into the circulation, emigrating at another point to colonize, invade, and destroy an organ at a distant site. These are the central fascinations of development and of cancer. We must understand them at the level of cellular behavior before we can understand them at the molecular level.

2. Cell Surfaces in the Control of Cell Proliferation

When fibroblasts growing in tissue culture become confluent, the mitotic rate drops to a very low value, and cell density remains essentially constant in

MALCOLM S. STEINBERG • Department of Biology, Princeton University, Princeton, New Jersey 08544.

spite of periodic replacement of the culture medium. When a scratch is made through such a confluent culture, most of the cells close to the cut edge reenter the mitotic cycle. When the scraped-away cells have been replaced by migration and division of cells bordering the wound, the cells in the "wounded" area return to their original very low mitotic rate. This *wound-healing* phenomenon has been called *contact inhibition of cell division* (Golde, 1962; Stoker and Rubin, 1967; reviewed in Martz and Steinberg, 1972). We prefer the term *postconfluence inhibition of cell division*, since it is more operational and less mechanistically biased. This phenomenon implies that populations of normal cells can regulate their growth through some local parameter related to cell contact or density. In a culture of cells displaying this property, variants that have lost this control mechanism may appear and, not being subject to growth control, will eventually overrun the culture. This presents an *in vitro* model of tumorigenesis. Indeed, dense cultures of tumorigenic cells, when "wounded," do not show the local increase in mitotic rate adjacent to the wound that nontumorigenic cells show. It has been reported that *in vitro* selection for increased or decreased postconfluence inhibition correspondingly decreases or increases tumorigenesis (Aaronson and Todaro, 1968; Pollack *et al.*, 1968; Pollack and Teebor, 1969).

Postconfluence inhibition is due neither to the depletion of nutrients nor to the presence of an inhibitor produced by the cultured cells (Todaro *et al.*, 1964, 1967; Holly and Kiernan, 1968). It is shown also in the *in vitro* wound-healing phenomenon, in which cells displaying exceedingly low and near-maximal mitotic rates exist side by side in the same culture dish.

Martz and Steinberg (1972) filmed a constantly perfused and stirred culture of mouse 3T3 fibroblasts continuously for 8 days and, using time-lapse cinematomicrography, recorded, for each cell, data on its position as a function of time; the amount of its perimeter in contact with other cells; the number of cells contacting the cell being traced; local cell density; and speed of nuclear translocation (a measure of speed of cell locomotion). The cell population in the film frame showed typical postconfluence inhibition of cell division. All surface area of the coverslip was covered by cells after 3 days; saturation density was achieved at 5 days and maintained until 7.6 days, at which time filming ended. A control culture of polyoma virus-transformed 3T3 cells continued growing after confluence was achieved, never achieving a saturation density during the 6 days of observation. In the experimental 3T3 culture undergoing postconfluence inhibition of cell division, 40% of the cells with generation times of less than 50 hr (rapidly dividing) were contacted on all sides by other cells throughout the G1 period (that portion of the cell cycle during which postconfluence inhibition acts.) Thus, maximal *contact* during G1 was shown to be insufficient in itself to prevent a cell from entering a subsequent mitosis. Moreover, a superconfluent *density* greater than 1.5×10^4 cells/cm^2 was also insufficient in itself to prevent all cells experiencing it from entering a subsequent mitosis. In fact, there was no correlation whatsoever between the intermitotic durations of individual cells and their local cell density or the amount of contact they experienced with neighboring cells. Thus, such terms as "density-dependent inhibition" (Stoker and Rubin, 1967) and "contact inhibition" (Rubin, 1961; Golde, 1962) of cell division are potentially misleading; there was no correlation between the parameters of density and contact measured in this study and the generation times of the 37 intermitotic cells available for study.

One explanation consistent with the above findings is the *mechanical cell–cell contact* hypothesis, according to which postconfluence inhibition results from the mechanical or geometric effects of cell crowding, such as changes in the shapes of individual cells or changes in the cell areas exposed either to the medium or to the substratum. A number of observations have documented an effect of cell–substratum contact on cell division, discussed as *anchorage dependence* by Stoker and colleagues (House and Stoker, 1966). Normal tissue cells will not divide in suspension (Hayflick and Moorhead, 1961; MacPherson and Montagnier, 1964; Eagle and Levine, 1967) but must flatten on a solid substratum in order to divide. Tumorigenic cells, in sharp contrast, readily divide when suspended, either through agitation or by enclosure in a soft agar or other gel. The role of cell shape in growth control has been assessed directly by Folkman and Moscona (1978), using cells of several kinds, all of which were susceptible to postconfluence inhibition of cell division. They controlled cell shape in two ways: by varying cell density (cells in crowded cultures are taller and narrower), and by modifying the culture surface by application of graded amounts of a hydrophilic polymer to which cells adhere poorly. They observed that the rate of cell proliferation, as measured by [^{3}H]thymidine incorporation, was a function of cell height, regardless of whether the latter was regulated by crowding or by the degree of cell adhesiveness to the substratum (Table I).

Parallel with evidence that cell division can be controlled by cell crowding or cell shape is an impressive body of evidence that growth, including that of the very same kinds of cells, can be controlled by macromolecules and nu-

Table I. Effect of Crowding on Cell Shape and DNA Synthesis[a,b]

	Poly(HEMA)[c]			Plastic[c]	
Dilution	Cell density (cells per cm^2)	^{3}H incorporation (cpm)	Cell height (μm)	Cell density (cells per cm^2)	^{3}H incorporation (cpm)
1.00	15,000	7 ± 5	22	250,000 (confluent)	7 ± 1
4 × 100^{-2}	15,000	50 ± 5	15	100,000 (confluent)	55 ± 9
4 × 100^{-3}	15,000	250 ± 15	6	30,000 (subconfluent)	254 ± 23

[a]From Folkman and Moscona (1978).
[b]WI-38 cells were plated at the densities indicated on uncoated plastic (Falcon 3008) and also on a series of poly(HEMA) substrates of graded thickness. Cell height, number, and [^{3}H]thymidine incorporation (cpm) were measured.
[c]Poly(HEMA) plates of different adhesivity containing sparsely plated cells were selected for cells whose height most closely matched the height of cells on plastic at various densities; incorporation of [^{3}H]thymidine was more equivalent for these pairs of plates than for any other combination of plates. Similar results were obtained with endothelial cells.

trients in serum (reviewed in Holley and Kiernan, 1968), including a variety of growth factors such as insulin, epidermal growth factor, and platelet-derived growth factor. It has been found that with 3T3 cells, the saturation density can be varied by varying the serum concentration (Holley and Kiernan, 1968). Holley and co-workers combined the study of postconfluence inhibition with that of the role of serum factors in studies showing that serum factors in the medium are required for 3T3 cells to respond to "wounding" of the culture by entering DNA synthesis and that DNA synthesis is initiated in a 3T3 culture at its saturation density if the serum concentration is increased. The authors conclude that the same growth regulatory factors are involved in controlling growth in "density-dependent" or postconfluence inhibition as *in vivo* or in preconfluent cultures, but that crowded normal cells need a higher concentration of these growth factors. These workers propose that increased cell density may increase the rate of destruction of growth factors and that "cells become less responsive to a given concentration of growth factors as their surface area and movement decrease, perhaps because of a decrease in the number, availability or mobility of surface receptors." Thus, cell shape might govern cell division through its effect on the availability of cell surface receptors required for the binding of growth factors or the transport of nutrients into the cell (Folkman and Moscona, 1968; Holley and Kiernan, 1978). The reader is referred to the consideration of related topics by Bowen-Pope (Chapter 7) and by Fey and Penman (Chapter 5) in this volume.

3. Cell Surfaces and the Control of Multicellular Assembly

During the early 1960s, it was proposed (Steinberg 1962a,b,c, 1964) that the "tissue affinities" discovered by Holtfreter (1939) to govern the assembly of tissues into appropriate histological structures were based on the relative intensities of adhesion of cells of various kinds to each other. Competing explanations at that time included chemotaxis or "directed migration" (Townes and Holtfreter, 1955), and cell-type-related differences in the timing of recovery from certain effects of cell dissociation, postulated by Curtis (1961). I distinguished between these alternative possibilities through behavioral tests that were able to discriminate each of these proposed mechanisms from the others. For example, cells guided up a gradient of a diffusible substance produced within an aggregate and released at its surface should move centripetally along radii of the aggregate, accumulating at its center. On the other hand, cells grouping internally because they cohere more strongly than their neighbors and squeeze the latter out should not migrate centripetally but clump to form islands that, in turn, progressively coalesce wherever they contact each other. Through a variety of such discriminative tests, it was found that, in each case, the behavior observed was that expected if the cell rearrangements occuring during cell sorting and tissue spreading were specified by cell-type-related differences in the intensities of intercellular adhesion (Steinberg, 1962a,b,c, 1963, 1964, 1970).

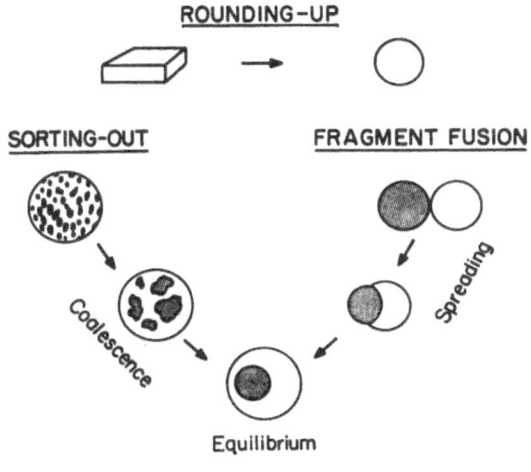

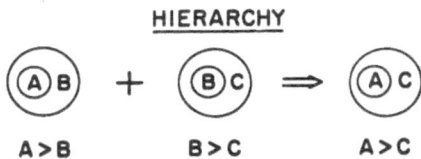

Figure 1. Many of the behaviors well-known in immiscible liquid systems are also shown by combinations of embryonic cells and tissues. (From Phillips and Steinberg, 1969.)

It became apparent early on that the sorting-out and spreading behavior observed when vertebrate embryonic cells or tissues of different kinds are combined imitated closely the well-documented behavior of immiscible liquid systems (Fig. 1), and for a very good reason. As set forth in our *differential adhesion hypothesis* (reviewed in Steinberg, 1970, 1978a,b), such reorganizing cell populations show liquidlike behavior because they share with liquids those properties that determine their behavior: (1) each consists of large numbers of subunits, (2) these subunits cohere, and (3) the subunits are mobile. Any system possessing these properties will tend to minimize its interfacial (adhesive) free energy through the regrouping of its subunits into configurations in which they are held progressively more tightly. This regrouping will continue until the system as a whole arrives at that configuration in which its subunits are most strongly bound together. This will be its "equilibrium configuration," which will be dynamically stabilized, since any rearrangement that departs from this equilibrium configuration will engender forces that oppose it.

It is worthy of particular emphasis that any system whose configuration is determined in this way will demonstrate a goal-directedness in the attainment of its equilibrium configuration, any disturbance in that configuration tending to be immediately and automatically corrected. There is no need for information to specify any particular step in the rearrangement process; instead, it is the final configuration itself that is thermodynamically specified, the system as a whole automatically and "spontaneously" approaching that configuration from any other and by any of an infinite number of morphogenetic pathways. A system operating according to this principle will automatically restore itself after an injury. In a system of living cells that are themselves undergoing differentiation even as morphogenetic movements occur, the equilibrium configuration specified by the adhesive properties of the component cells may not actually be realized, since the acquisition of new or altered adhesive properties may divert the system toward some new equilibrium configuration even as reorganization is progressing. The apparent goal directedness reflected in the "regulative" properties of embryos is undoubtedly explained at least in part in this way.

The fact that it is always the less cohesive of two immiscible liquids (the phase of lower surface free energy) that spreads over the surface of the other (Davies and Rideal, 1963) has provided the basis for three independent sets of experiments in each of which the mutual spreading tendencies of tissues have been compared with direct measurements of their relative cohesive intensities. Phillips and Steinberg (1969) devised a centrifugation procedure in which cell aggregates or tissue fragments were permitted to come to shape equilibrium while exposed to a constant centrifugal force. Such measurements constitute sessile drop measurements of the surface tension (surface free energy) of living cell aggregates. The mutual spreading tendencies of these tissues were observed to be inversely related to their surface free energies, as required. In the second such study, experimental manipulations that increased or decreased the spreading tendency of an embryonic tissue were observed to decrease or increase its surface free energy correspondingly (Wiseman et al., 1972). The third investigation (Phillips and Davis, 1978) demonstrated that the surface free energies of the rearranging components of the amphibian gastrula fall in the precise sequence predicted (Steinberg, 1964) to account for the germ layer organization generated through gastrulation.

The differential adhesion hypothesis addresses itself to the *thermodynamic* specification of structure by the adhesive properties of cells. It does not address itself to the chemical or physical bases for these adhesive properties. As was pointed out many years ago, cellular adhesiveness may be modulated through changes in either the chemical identity, the amounts, or the distributions of cell surface adhesion-mediating molecules (Steinberg, 1963, 1964). As a matter of interest, it was shown that, although two cell populations bearing different adhesion systems could obviously sort out or fail to intermix, the same should be true of two cell populations bearing identical adhesion systems but differing only in the quantities in which these molecules are expressed. Thus, the property of immiscibility—the tendency of two different tissues actively to resist intermixing—does not require the presence of a low intrinsic affinity between the adhesion molecules on the two sets of cell surfaces. Even modulation of the quantitative expression of a given adhesion system on the surfaces of a subset of cells would in principle be sufficient to cause those cells to segregate out as a unique, delimited cell population. I have proposed that the emergence of altered adhesive properties underlies the segregation of organ

primordia such as notochord, somites, pronephric rudiment, and lateral mesoderm (to which may be added gill and limb rudiments) as subdomains of what
earlier appear as smoothly contiguous areas of tissue (Steinberg and Poole,
1981, 1982). Elsewhere in this volume, a number of contributors discuss the
contribution of cell adhesion systems to morphogenetic processes (see Chapters 8, 9, 11, 12, 14, and 15.)

4. Adhesive Specification of Pronephric Duct Segregation and Migration

For some years my laboratory has been engaged in an analysis of the
mechanism guiding the primitive kidney duct from its origin where it segre-
gates from the anterior part of the dorsal lateral mesoderm to its destination at
the cloaca, into which it ultimately opens. In *Ambystoma* this duct forms by
the tailward extension of a solid stream of cells along the ventral edges of the
segregating somites (Fig. 2). Vital staining experiments (Poole and Steinberg,
1981) established that duct elongation in this species is attributable to cell
migration rather than to a wave of *in situ* delamination from the dorsal lateral

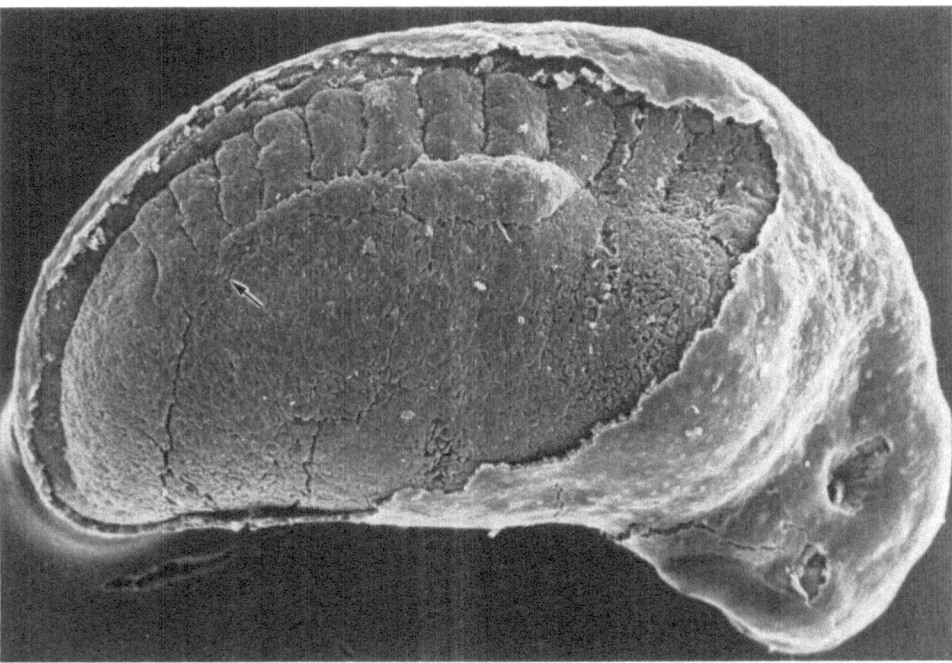

Figure 2. Scanning electron micrograph of axolotl embryo fixed before removing ectoderm from
right side. Arrow indicates migrating pronephric duct's caudal tip. (From Poole and Steinberg,
1981.)

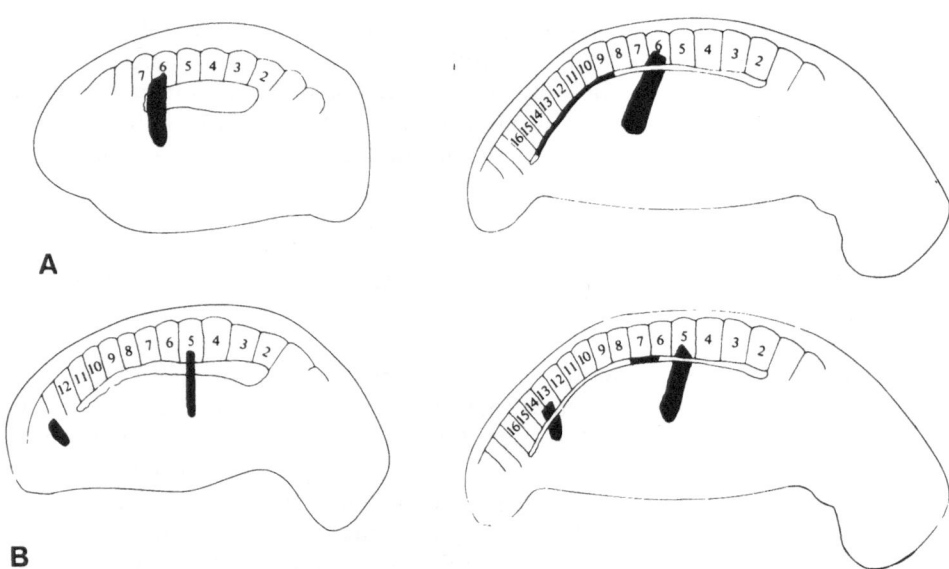

A

B

Figure 3. Camera-lucida tracings of vitally stained embryos. (A) Distal segment of pronephric duct stained with Nile blue sulfate at stage 2 has moved caudad and elongated markedly by stage 32. (B) Proximal segment of duct stained at stage 26 has moved caudad and elongated to a lesser extent by stage 32. A stained section of the duct's path is obscured as the duct passes over it. (From Poole and Steinberg, 1981.)

mesoderm (Fig. 3). As the duct tip actively migrates caudally, the cells anterior to it progressively rearrange, reducing the dorsoventral width of the duct primordium from 6–8-cell diameters to 2 or 3 by the time migration is completed. The substratum for this migration consists of the ventral margins of the forming somites and the dorsal margin of the lateral mesoderm, to both of which the duct tip cells can be seen to be attached *via* cell extensions.

Using scanning electron microscopy (SEM) to observe the results of microsurgical experiments, we sought to determine the mechanism guiding the migration of this cell population along its strictly observed path. Possibilities included chemotaxis toward some distant attractant, contact guidance along some structure such as the ventral edges of the forming somites, and adhesive differentials defining a preferred track. One might imagine that the information guiding the pronephric duct is restricted to the normal pathway of the duct, but this proved not to be the case. Pronephric duct primordia transplanted to the flank of host embryos ventrolateral to the primary duct migrated dorsocaudally across the flank to fuse with the primary duct. To determine whether these migrating duct tips were seeking out an attractant emitted by cells at a distance, tissue dorsal and caudal to the migrating duct were removed surgically. This had no effect on the migratory behavior, migrating ducts continuing their movement to the very edge of the remaining flank tissue. Transplantation experiments of several kinds revealed that the duct guidance information is not only oriented but is also directionally polarized. These experiments eliminated contact guidance as a possible mechanism, since that mechanism provides for orientation along an avenue but no clues as to which of the two possible directions should be taken.

Transplantations between embryos of different ages and at different anteroposterior levels along the flank revealed that the duct guidance information is not only directionally polarized but travels caudad as a wave synchronized with the wave of somitogenesis (Poole and Steinberg, 1982). Once the wave has passed caudad beyond a given transverse zone, that zone loses its capacity to support duct migration. Moreover, duct tip cells grafted to such regions no longer spread out and extend pseudopodia upon them, as they did earlier. Instead, such grafts rounded up, suggesting that their adhesiveness to the flank diminished as the latter lost its capacity to support their migration. These observations led to the proposal that the *Ambystoma* pronephric duct is guided by a gradient of adhesiveness to the flank mesoderm and that this gradient itself progresses caudally in synchrony with the caudally progressing differentiation of the embryonic mesoderm (Steinberg and Poole, 1982).

In the course of the above experiments, we made a number of observations concerning tissue miscibility. Pronephric ducts of host and graft invariably fuse smoothly, regardless of level, indicating a lack of regional adhesive differences that would generate immiscibility. In sharp contrast, bits of flank mesoderm accompanying pronephric duct transplants showed remarkable differences in their behavior as a function of the anteroposterior levels of origin and of implantation. When these levels were the same, the flank mesoderms of graft and host merged smoothly with no trace of a boundary (Steinberg and Poole, 1981). On the other hand, when posterior flank mesoderm was grafted to an anterior position, the graft rounded up to form a mound. These observations were pursued by Laura Gillespie, at that time a graduate student in my laboratory. Two identical grafts, both taken from the posteroventral flank mesoderm of donor embryos, were placed on the side of a single host embryo, one homotopically and the other more anteriorly, immediately ventral to the host's own pronephric primordium (Fig. 4). As before, these grafts of posterior mesoderm behaved entirely differently in these two host locations, merging smoothly with the host mesoderm when grafted homotopically but rounding up to form a mound resembling a limb bud when grafted to a more anterior position (Fig. 5).

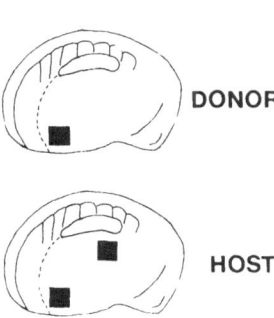

Figure 4. Operation diagram showing the posteroventral origin of flank mesoderm grafted to the two host positions indicated. See Figure 5 for results. (From Steinberg and Poole, 1982.)

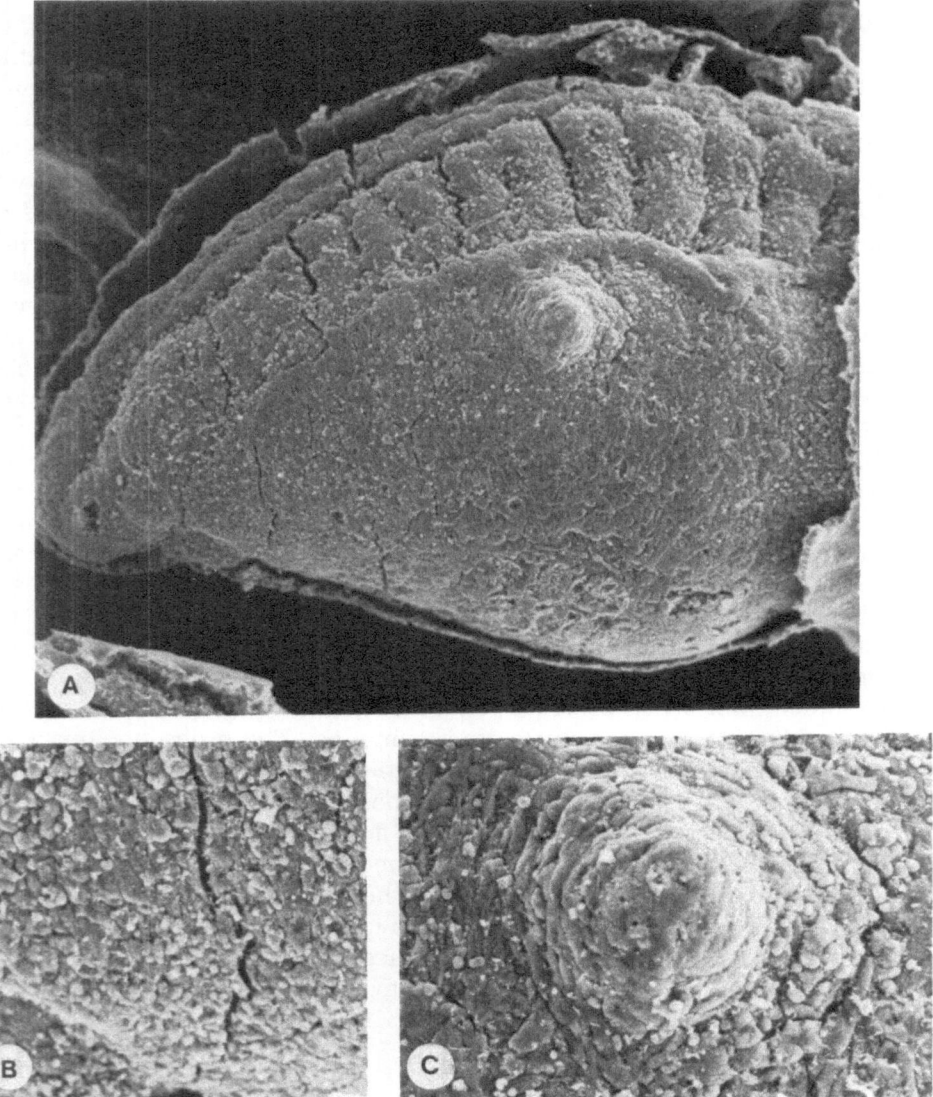

Figure 5. (A) Host embryo bearing two grafts as illustrated in Figure 4. The posterior graft has merged smoothly with the homotopic flank mesoderm of the host embryo, whereas the graft placed at the more anterior host level has rounded up into a limb-bud-like structure. Scale bar: 1 mm. (B) Enlargement of the region containing the more posteriorly placed graft. (C) Enlargement of the region containing the more anteriorly placed graft. (From Steinberg and Poole, 1982.)

This provided evidence that the flank mesoderm itself is more cohesive in its posterior than in its anterior region. Since our observations on pronephric duct migration had suggested that the adhesiveness of flank mesoderm to the pronephric duct tip was graded in this same way, the possibility arose that the cells of the flank utilize for their own mutual adhesion the same adhesion system that we had postulated was active in guiding the pronephric duct in its migration.

In an initial effort to obtain information concerning the properties of the putative adhesion system postulated to be involved in both pronephric duct guidance and flank mesoderm cohesiveness, experiments were performed to determine whether the duct guidance information is sensitive to proteolytic degradation. These experiments were designed with the knowledge that, in previously investigated cases, calcium-dependent adhesion systems have been found to be protected against proteolytic attack by the presence of calcium ions, whereas the sensitivity of calcium-independent systems to proteolytic attack is unaffected by calcium ions (reviewed in Yoshida and Takeichi, 1982). The flank mesoderm of *Ambystoma* embryos was briefly exposed to dilute trypsin solutions in the presence or absence of calcium ions. After washing and replacement of the epidermis overlying the treated mesoderm, the embryos were cultured for a period sufficient to permit substantial additional duct migration, after which the embryos were fixed and prepared for observation by SEM. It was found that, whereas somitogenesis was equally sensitive to trypsin in the presence and absence of calcium ions, pronephric duct migration was sensitive to trypsin only in the presence of calcium (Gillespie *et al.*, 1985). Thus an essential component of the pronephric duct guidance system has properties associated with calcium-dependent adhesion systems and may in fact be such a system. Efforts are under way to characterize the duct guidance system more fully.

Earlier it was suggested that adhesive differentials may function in embryogenesis not only to guide morphogenetic movements and to determine the histological structure of tissues and organs but also to bring about the segregation of organ primordia from originally smoothly contiguous expanses of tissue. The observation that a fragment of posterior flank mesoderm transposed to a more anterior level rounds up to produce a mound resembling a limb bud lends concreteness to this proposal and illustrates what this author believes to be the very real possibility that a relatively small number of molecular adhesion systems, present in many places in the embryo, are modulated to bring about many different morphogenetic events.

ACKNOWLEDGMENTS. The research carried out in the author's laboratory was supported by grants GB-2315, GB-5759X, and PCM76-84588 from the National Science Foundation and CA13605 from the National Cancer Institute, DHEW, and benefited from departmental facilities supported by the Whitehall Foundation.

References

Aaronson, S. A., and Todaro, G. J., 1968, Basis for the acquisition of malignant potential by mouse cells cultivated *in vitro, Science* **162:**1024–1026.

Curtis, A. S. G., 1961, Timing mechanisms in the specific adhesion of cells, *Exp. Cell Res. (Suppl.)* **8:**107–122.

Davies, J. T., and Rideal, E. K., 1963, *Interfacial Phenomena*, pp. 20–23, Academic Press, New York.

Eagle, H., and Levine, E. M., 1967, Growth regulatory effects of cellular interaction, *Nature (Lond.)* **213:**1102–1106.

Folkman, J., and Moscona, A., 1978, Role of cell shape in growth control, *Nature (Lond.)* **273:**345–349.

Gillespie, L. L., Armstrong, J. B., and Steinberg, M. S., 1985, Experimental evidence for a proteinaceous presegmental wave required for morphogenesis of axolotl mesoderm, *Dev. Biol.* **107:**220–226.

Golde, A., 1962, Chemical changes in chick embryo cells infected with Rous sarcoma virus *in vitro, Virology* **16:**9–20.

Hayflick, L., and Moorhead, P. S., 1961, The serial cultivation of human diploid cell strains, *Exp. Cell Res.* **25:**585–621.

Holley, R. W., and Kiernan, J. A., 1968, Contact inhibition of cell division in 3T3 cells, *Proc. Natl. Acad. Sci. USA* **60:**300–304.

Holley, R. W., and Kiernan, J. A., 1971, Studies of serum factors required by 3T3 and SV3T3 cells, in: *Growth Control in Cell Cultures: A Ciba Foundation Symposium* (G. E. W. Wolstenholme and J. Knight, eds.), pp. 3–10, Churchill Livingstone, Edinburgh.

Holtfreter, J., 1939, Gewebeaffinitat: Ein Mittel der embryonalen Formbildung, *Arch. Exp. Zellforsch Besonders Gewebezuecht.* **23:**169–209.

House, W., and Stoker, M. G. P., 1966, Structure of normal and polyoma virus-transformed hamster cell cultures, *J. Cell Sci.* **1:**169–173.

MacPherson, I., and Montagnier, L., 1964, Agar suspension culture for the selective assay of cells transformed by polyoma virus, *Virology* **23:**291–294.

Martz, E., and Steinberg, M. S., 1972, The role of cell–cell contact in "contact" inhibition of cell division: A review and new evidence, *J. Cell. Physiol.* **79:**189–210.

Phillips, H. M., 1969, Equilibrium measurements of embryonic cell adhesiveness. Physical formulation and testing of the differential adhesion hypothesis, Doctoral Dissertation, Johns Hopkins University.

Phillips, H. M., and Davis, G. S., 1978, Liquid–tissue mechanics of amphibian gastrulation: Germlayer assembly in *Rana pipiens, Am. Zool.* **18:**81–93.

Phillips, H. M., and Steinberg, M. S., 1969, Equilibrium measurements of embryonic chick cell adhesiveness. I. Shape equilibrium in centrifugal fields, *Proc. Natl. Acad. Sci. USA* **64:**121–127.

Pollack, R. E., and Teebor, G. W., 1969, Relationship of contact inhibition to tumor transplantability, morphology, and growth rate, *Cancer Res.* **29:**1770–1772.

Pollack, R. E., Green, H., and Todaro, G. J., 1968, Growth control in cultured cells: Selection of sublines with increased sensitivity to contact inhibition and decreased tumor-producing ability, *Proc. Natl. Acad. Sci. USA* **60:**126–133.

Poole, T. J., and Steinberg, M. S., 1981, Amphibian pronephric duct morphogenesis: Segregation, cell rearrangement and directed migration of the *Ambystoma* duct rudiment, *J. Embryol. Exp. Morphol.* **63:**1–16.

Poole, T. J., and Steinberg, M. S., 1982, Evidence for the guidance of pronephric duct migration by a craniocaudally traveling adhesion gradient, *Dev. Biol.* **92:**144–158.

Rubin, H., 1961, Influence of tumor virus infection on the antigenicity and behavior of cells, *Cancer Res.* **21:**1244–1253.

Steinberg, M. S., 1962a, On the mechanism of tissue reconstruction by dissociated cells. I. Population kinetics, differential adhesiveness, and the absence of directed migration, *Proc. Natl. Acad. Sci. USA* **48:**1577–1582.

Steinberg, M. S., 1962b, Mechanism of tissue reconstruction by dissociated cells. II. Time course of events, *Science* **137**:762–763.

Steinberg, M. S., 1962c, On the mechanism of tissue reconstruction by dissociated cells. III. Free energy relations and the reorganization of fused heteronomic tissue fragments. *Proc. Natl. Acad. Sci. USA* **48**:1769–1776.

Steinberg, M. S., 1963, Reconstruction of tissues by dissociated cells, *Science* **141**:401–408.

Steinberg, M. S., 1964, The problem of adhesive selectivity in cellular interactions, in: *Cellular Membranes in Development* (M. Locke, ed.), pp. 321–366, Academic Press, New York.

Steinberg, M. S., 1970, Does differential adhesion govern self-assembly processes in histogenesis? Equilibrium configurations and the emergence of a hierarchy among populations of embryonic cells, *J. Exp. Zool.* **173**:395–434.

Steinberg, M. S., 1978a, Specific cell ligands and the differential adhesion hypothesis: How do they fit together?, in: *Specificity of Embryological Interactions* (D. R. Garrod, ed.), pp. 97–130, Chapman and Hall, London.

Steinberg, M. S., 1978b, Cell–cell recognition in multicellular assembly: Levels of specificity, in *Cell–Cell Recognition*, Society for Experimental Biology Symposium No. 32 (A. S. G. Curtis, ed.), pp. 25–49, Cambridge University Press, Cambridge.

Steinberg, M. S., and Poole, T. J., 1981, Strategies for specifying form and pattern: Adhesion-guided multicellular assembly, *Philos. Trans. R. Soc. Lond. [Biol.]* **295**:451–460.

Steinberg, M. S., and Poole, T. J., 1982, Cellular adhesive differentials as determinants of morphogenetic movements and organ segregation, in: *Developmental Order: Its Origin and Regulation* (S. Subtelny and P. Green, eds.), pp. 351–378, Alan R. Liss, New York.

Stoker, M. G. P., and Rubin, H., 1967, Density dependent inhibition of cell growth in culture, *Nature (Lond.)* **215**:171–172.

Todaro, G. J., Green, H., and Goldberg, B. D., 1964, Transformation of properties of an established cell line by SV40 and polyoma virus. *Proc. Natl. Acad. Sci. USA* **51**:66–73.

Todaro, G. J., Matsuya, Y., Blom, S., Robbins, A. and Green, H., 1967, Stimulation of RNA synthesis and cell division in resting cells by a factor present in serum, in *Growth Regulating Substances for Animal Cells in Culture* (V. Defendi and M. Stoker, eds.), pp. 87–101, Wistar Institute Press, Philadelphia.

Townes, P. L., and Holtfreter, J., 1955, Directed movements and selective adhesion of embryonic amphibian cells, *J. Exp. Zool.* **128**:53–120.

Wiseman, L. L., Steinberg, M. S., and Phillips, H. M., 1972, Experimental modulation of intercellular cohesiveness: Reversal of tissue assembly patterns, *Dev. Biol.* **28**:498–517.

Yoshida, C., and Takeichi, M., 1982, Teratocarcinoma cell adhesion: Identification of a cell-surface protein involved in calcium-dependent cell aggregation, *Cell* **28**:217–224.

Chapter 2

The Cell Surface and Cancer Metastasis

GEORGE POSTE and RUSSELL GREIG

1. Introduction

The cell biology of cancer metastasis has been studied in considerable detail over the past 10 years, yielding many important new insights into the pathogenesis of this important and life-threatening disease (reviewed by Poste and Fidler, 1980; Poste, 1982; Nicolson and Milas, 1984). By contrast, efforts at understanding the underlying biochemistry and molecular biology of this process have been limited and largely unsuccessful. A much greater understanding of the molecular mechanisms responsible for the expression of the metastatic phenotype will be required if we are to develop novel and productive approaches to the therapy of metastatic disease by any approach other than random and semiempirical screening, which hitherto has made little impact in developing anticancer drugs against the more common solid malignancies of man. In the absence of any obvious biochemical target to guide the direction of such an endeavor, it is becoming increasingly clear that the process of identifying molecular properties that correlate either qualitatively or quantitatively with metastatic potential will be a prolonged and challenging task and with no guarantee that any correlation, once identified, will be causal or will offer a suitably "unique" and pharmacologically exploitable target for the design of new therapeutic agents directed against it.

The probability that alterations in the surface properties of cancer cells contribute to their aberrant behavior has long attracted the interests of cancer researchers. This attention has been productive to the extent that the modern literature contains an interminable litany of differences in cell surface properties that have been detected at one time or another (with variable consistency) in human and animal tumor cell populations of diverse tissue origins. These populations have ranged from rodent cell lines, many of which have been passaged in vitro in different laboratories for years or even decades, to human

GEORGE POSTE and RUSSELL GREIG • Smith Kline and French Laboratories, Philadelphia, Pennsylvania 19101.

tumor cells isolated from newly excised tumors. Unfortunately, the only safe conclusion that can be drawn from these studies is that we cannot yet state with certainty how any specific cell surface change contributes or is even relevant to complex multifactorial behavioral traits such as tumorigenicity, invasiveness, or the ability to metastasize. Indeed, few of the enormous number of publications on the neoplastic cell surface pertain to metastatic tumor cells. This is surprising, or perhaps disturbing, in view of the fact that metastatic disease, and not the primary tumor, is the major challenge in clinical oncology. Even fewer studies have been undertaken in which analyses of cell surface properties, conducted almost exclusively within cultured cell populations *in vitro*, have been accompanied by parallel *in vivo* studies of the behavior of the same cells. Instead, most investigators have been content to cite historical reports, or studies by others, as evidence of the *in vivo* behavior of the cells in question. This assumes that the cells in their own laboratory will not differ to any significant degree. Increasing evidence shows that this assumption is unwarranted; there is a danger that the value of the application of sophisticated molecular biological techniques to analyze tumor cells *in vitro* will be undermined by failure to assay the *in vivo* behavior of the cell preparations being studied.

There is compelling circumstantial evidence to implicate changes in the structural and functional organization of the cell surface in the etiology of the altered behavior of malignant cells. Within the primary tumor, alterations in the surface properties of tumor cells are presumed to contribute to their escape from many of the regulatory controls to which normal cells are subject. The proliferation of tumor cells is no longer effectively regulated by cell—cell contact interactions, and surface changes are reasonably viewed as contributing to the altered responsiveness of tumor cells to various autocrine, paracrine, and endocrine signals.

Alterations in plasma membrane transport processes are common in tumor cells. Such changes may facilitate improved nutrient utilization by tumor cells, thereby conferring an adaptive growth advantage on these cells under conditions in which the nutrient supply is limiting. Changes in membrane transport functions can also render tumor cells resistant to diverse classes of chemotherapeutic agents, either by frustrating drug uptake or by promoting active drug efflux. The aberrant regulation of positional control in invasive tumor cells that enables them to infiltrate surrounding host tissues probably involves lesions in surface properties that alter cellular cognitive responses to homologous and heterologous cells. Similarly, progression of the invasion process, culminating in the entry of malignant tumor cells into lymphatics and blood vessels, involves a complex set of events in which changes in surface properties almost certainly contribute to the defective control of cell locomotion. Thus, tumor cells are able to infiltrate normal tissues and to destroy host tissues *via* the presumed release of tissue degradative enzymes. In malignant tumors, these processes are augmented by the metastatic cascade in which the surface properties of tumor cells are believed to play a major role in the dissemination of tumor cells to distant organs, including, in certain tumors, the apparent selective, nonrandom localization of tumor cells in particular "target" organs. Finally, at all stages in tumor progression, the surface properties of neoplastic cells affect their susceptibility to recognition and destruction by host defense mechanisms.

Surface alterations in tumor cells are also viewed logically as potential targets for antineoplastic agents. Unfortunately, with the exception of determinants expressed by tumor cells in a temporally inappropriate fashion and absent from non-neoplastic cells of the same lineage, most surface changes detected in tumor cells to date are of a quantitative rather than qualitative nature, thereby limiting their value as targets for achieving selective therapeutic destruction of neoplastic cells. However, with the rapid progress in developing monoclonal antibodies to human tumor cells have come increasing number of reports of epitopes expressed by tumor cells that do not appear to be present on normal cells of the same tissue; these may be of considerable value in cancer diagnosis and therapy. These observations have led to proposals to use the shedding of such determinants as diagnostic markers and/or their exploitation in engineering antibody-mediated "targeting" of imaging and therapeutic agents to tumor cells. Although these approaches are in their infancy and have perhaps generated more publicity than their clinical success to date warrants, future refinements in these approaches will almost certainly generate worthwhile improvements in the diagnosis, staging, and/or therapy of malignant disease.

As the diversity of the surface changes detected in cancer cells, their potential functional importance, and their value as possible targets for therapeutic assault have been discussed at length in several recent reviews a further survey would be redundant (e.g., see Bruchovsky and Goldie, 1982; Davies and Crumpton, 1982; Fidler and Poste, 1982; Hawkes and Wang, 1982; Owens *et al.*, 1983; Poste and Nicolson, 1983; Bresciani *et al.*, 1984; Nicolson and Milas, 1984). This chapter focuses instead on the type of experimental approaches needed to translate our current descriptive perspectives on cell surface changes in cancer into a mechanistic framework in which specific surface changes can be shown to be causally correlated with particular aspects of tumor cell behavior. The long-term goal of this exercise is to define the particular subsets of tumor cell properties responsible for clinically important aspects of tumor cell behavior, such as metastasis and resistance to killing by therapeutic agents or host defenses. With this knowledge, it might then be possible to embark on the rational design of novel therapeutic agents directed against tumor cells that display these traits.

2. Tumor Models

The use of tumor models has been of overwhelming importance in research on the pathogenesis of neoplastic disease and the search for new antitumor agents. This will not change in the foreseeable future. It is essential, however, that the criteria used in choosing a tumor model be reexamined at regular intervals to determine whether the model still fulfills the original intention and whether it should be discarded or refined to accommodate new advances in cancer biology that are occurring at a rapid pace as a result of dramatic progress in cell biology, immunology, and molecular genetics.

Animal tumor models exhibit many important differences from human neoplasms. These have been documented in numerous publications (see reviews by Hewitt, 1982; Poste, 1982; Goldin, 1983; Donelli et al., 1984; Poste and Greig, 1985). Prominent among these are differences in body size and life span; species variation in the anatomy and physiology of major organ systems; variation in the contribution of different elements of host-immune and nonimmune reactions in combatting neoplasia; and significant species variation in the absorption, distribution, metabolism, and excretion of therapeutic agents.

The extensive reliance on transplantable established tumor cell lines (usually of rodent origin) as models for human tumors represents a major source of potential irrelevancies and artifacts. Most solid malignancies in humans grow relatively slowly and contain cells that retain many of the properties of the cell type from which they originated. By contrast, many of the more widely used tumor cells represent established lines of human and animal origin that typically contain rapidly growing anaplastic cells bearing little or no phenotypic relationship even to the cell type from which they arose originally, yet alone tumors originating in entirely different cell types. There is an urgent need to pay greater attention to the biological uniqueness of neoplasms arising in different cell lineages (reviewed by Poste and Greig, 1985). The increasing attention given by many laboratories to the development of model systems using freshly isolated human tumor cells is particularly relevant and represents a major initiative in increasing the technical sophistication of current experimental approaches.

Irrespective of whether human or animal tumor cells are used, we must begin to give more attention to the nature and extent of phenotypic change that results from the selection and serial propagation of tumor cells in vitro or in vivo. We must also investigate how far such changes render the cells unsuitable, hence irrelevant, for detecting the cellular alterations responsible for tumorigenicity and metastasis and for evaluating cellular responses to antitumor agents (reviewed by Poste, 1982; Poste and Greig, 1985).

No attempt is made in this chapter to debate the merits of one tumor model versus another. Each different application of metastatic tumor models will almost certainly impose unique experimental requirements that must be fulfilled by the model selected. The strengths and weaknesses of the transplantable animal tumors commonly used in research on metastasis have been discussed at length by Poste (1982), Nicolson and Poste (1983), Poste and Nicolson (1983), and Poste and Greig (1985). Each may have merits for the purpose for which it was chosen. Similarly, every model will have shortcomings, and personal views concerning the balance between the strengths and weaknesses of a model will vary enormously. Some of the more obvious examples are reflected in (1) the longstanding, yet unresolved, debates regarding the value of using induced versus spontaneous tumors; (2) the merits of autochthonous tumors versus transplantable tumors (whether of spontaneous origin or induced); and

(3) the use of highly antigenic immunogenic tumors versus weakly immunogenic or nonimmunogenic (?) tumors in evaluating host immune responses to tumors. Similarly, the incidence of spontaneous tumors arising in specific cell types varies significantly in humans and animals. Also, the risk of tumor formation occurring in the same target organ after exposure to carcinogens varies markedly between species and is a major limiting factor in using animals to predict the carcinogenic liability of food additives, chemicals, or drugs (see reviews by Efron, 1984; National Toxicology Program, 1984).

Notwithstanding the complexity of these unresolved issues, several important conceptual and technical advances have been made in the development of metastatic tumor models that must be accommodated in experimental strategies that seek to correlate changes in tumor cell surface properties with the ability to metastasize. First, it is necessary to study metastatic tumors. This may seem to be stating the obvious to the point of absurdity, yet it is still not uncommon, although far less frequent than a few years ago, to read papers in which the stated rationale is to study malignancy or metastatic disease—but using a nonmetastasizing tumor. In addition to this fundamental requirement, it is also helpful if the tumor selected for study displays all or most of the following properties: (1) its origin and passage history is known in detail; (2) it can be transplanted in syngeneic hosts or immune-deficient animals to produce a reproducible and consistent pattern of metastatic disease in defined organ(s); (3) the plating efficiency of the cells is sufficient to permit routine recovery of tumor cells from tumors *in vivo*; (4) the phenotypic properties expressed by defined subpopulations of tumor cells *in vitro* persist *in vivo* and vice versa; (5) assay procedures are available to quantify the metastatic burden; and (6) for the development of therapeutic agents, it is useful if the tumor exhibits a similar response profile to human tumors of comparable histological origin.

Second, with the demonstration that the cellular composition of most, if not all, malignant tumors is heterogeneous, and that only certain subpopulations of tumor cells express metastatic properties, perhaps transiently, it is now almost obligatory to isolate representative examples of the constituent cellular subpopulations for detailed comparison (Poste and Fidler, 1980; Fidler and Poste, 1982; Poste, 1982, 1984; Poste and Greig, 1982; Owens *et al.*, 1983; Heppner, 1984; Nicolson and Poste, 1984; Poste *et al.*, 1984).

3. Tumor Cell Heterogeneity and Correlation of Cell Surface Changes with Metastatic Behavior

The knowledge that all human and animal tumor cell populations studied to date have been found to contain phenotypically different subpopulations of cells has important implications for experimental efforts to define the cellular and subcellular properties that correlate with complex multifactorial behavioral traits such as metastasis. The coexistence within the same tumor (or tumor cell line) of subpopulations of cells with different metastatic properties dictates that the use of heterogeneous populations containing both nonmetastatic and metastatic cells cannot be interpreted as yielding information about the metastatic subpopulations *per se* (see Poste, 1982; Poste and Greig, 1983; Nicolson and Poste, 1984; Fidler and Poste, 1985). If, for example, the metastatic cell subpopulations are in the minority, analysis of the entire cell population may provide little or no insight into the cell surface properties that are unique to metastatic cells, since the probability is high that these would be obscured or lost in the "background noise" imposed by the majority fraction of nonmetastatic cells (Poste, 1982). Consequently, characterization of the surface properties of metastatic tumor cell subpopulations requires either that such subpopulations are present as the overwhelming majority in a heterogeneous cell population or that individual subpopulations be isolated by cloning to generate homogeneous cell preparations for comparison with nonmetastatic clones isolated from the same parent cell population. In the former strategy, selection pressures are imposed to favor the metastatic subpopulations, with resulting enrichment of their contribution to the total cell population (Poste, 1982). Conversely, negative selection pressures can be applied to deplete the fraction of metastatic cells (see Poste, 1982 and references cited therein). It is important to emphasize, however, that tumor cell preparations subjected to selection pressures to enrich or deplete specific subpopulations of cells are still heterogeneous and that nonmetastatic subpopulations are still likely to be present.

A more satisfactory approach for addressing the problems posed by tumor cell heterogeneity involves cloning of tumor cells from a heterogeneous cell population to create a panel of clonal populations whose phenotypes can be compared (Poste, 1982). The aim of this strategy is to isolate a sufficient number of clones to permit comparison of clones with the following properties in order to identify surface properties that are invariant features of metastatic clones but that are absent from nonmetastatic clones:

Tumorigenic, noninvasive, nonmetastatic clones ($T^+I^-M^-$)
VERSUS
Tumorigenic, invasive, nonmetastatic clones ($T^+I^+M^-$)
VERSUS
Tumorigenic, invasive, metastatic ($T^+I^+M^+$)

Although this approach is conceptually straightforward, it is technically demanding and labor intensive, necessitating the isolation and examination of a large number of clones from the same parent tumor cell population. Furthermore, reliable correlation of a specific cell surface change assayed *in vitro* with the metastatic properties of cells *in vivo* will require additional experiments to confirm that the same surface change is expressed by the metastatic cells *in vivo* and in cells recovered from metastatic lesions and cultivated *in vitro* surface properties (see Poste, 1982; Poste *et al.*, 1982a,b).

Since cell cloning offers the most direct and effective method for studying variation in the surface properties of different cells residing in the same population, it is helpful if the tumor selected for study has a reasonably high cloning efficiency. Furthermore, the use of defined media to facilitate the cloning and the recovery of the more fastidious tumor cell clones merits attention. These requirements are of particular relevance to studies with human tumor cells, and a substantial effort will be required to solve these problems before detailed analyses of the clonal composition of tumors of the kind already completed for a few animal tumors can become routine practice for human tumors. Identification of clonal variation among cells present in a specific tumor also demands that a panel of stable phenotypic markers, including surface markers suitable for both *in vitro* and *in vivo* analyses, be available to facilitate reliable identification of individual clones within heterogeneous populations and to follow clonal lineages (Poste *et al.*, 1982a,b). Ideally, the panel should include a series of karyotypic, immunological, and biochemical markers that can be assayed in reproducible fashion in tumor cells both *in vitro* and *in vivo* (Poste and Greig, 1985). As emphasized earlier, few panels of this kind are currently available. Also, for experimental purposes, it is acceptable to engineer the introduction of specific markers into individual subpopulations using gene transfer methods, mutagenesis, or treatments that induce significant phenotypic change such as treatment with drugs that alter DNA methylation patterns (Poste and Greig, 1985). In creating a panel of phenotypic markers, it is essential, however, to determine the (in)stability of such markers in clonal tumor cell populations when maintained under different conditions *in vitro* and *in vivo*.

In comparing the surface properties of metastatic and nonmetastatic clones isolated from the same parent cell population, it is necessary to confirm that the metastatic properties of the clones are stable. This is assessed by repeated subcloning at intervals of only a few weeks to examine whether the metastatic behavior of the original clone and progeny subclones are maintained or if variant subclones whose metastatic properties differ significantly from the original clone emerge to create a heterogeneous cell population. Studies on several animal tumors have revealed that the metastatic properties of certain clones may be highly unstable and that subclones with altered metastatic behavior emerge rapidly, generating a heterogeneous cell population (see Poste, 1982; Heppner, 1984; Nicolson and Poste, 1984; and references cited therein).

If clones from the tumor being studied undergo rapid phenotypic drift, frequent subcloning will be needed to ensure that the clonal populations being compared are homogeneous and that repeat experiments are done using replicate subclones that display identical metastatic properties to the original clones. It is also informative to obtain data on the rate of phenotypic "drift." If no significant changes in metastatic behavior or other phenotypic properties, occur, or take place only over a period of several months rather than weeks, *in vitro* analyses of specific tumor cell properties can be undertaken in the reasonable belief that the metastatic properties of the cells being assayed are still comparable to samples of the same clone assayed a few weeks before or after the experiment. If, however, the rate of drift in metastatic properties is rapid due to the emergence of variant subclones with altered metastatic abilities, the experimental requirements for correlating cell surface alterations assayed *in vitro* with metastatic behavior *in vivo* became far more stringent. For clones that show rapid drift in metastatic behavior (or other properties), the correlation of a specific surface property assayed *in vitro* with metastatic capacity *in* vivo will require that the in vitro and in vivo assays be conducted synchronously using replicate cell preparations assayed at the shortest interval after cloning consistent with generating sufficient cells for the assays (Greig et al., 1983). Unless such precautions are taken, the in vitro assay of a cell surface property today cannot be correlated with the metastatic behavior of the same "clone" measured a few weeks earlier or later, because rapid formation of variant subclones will have rendered the "clone" heterogeneous. Consequently, for tumor cell clones that exhibit unstable metastatic properties during serial passage, these more demanding synchronous assay protocols must be adopted as routine practice, unless experimental conditions can be developed to eliminate the drift in metastatic properties (Poste, 1982).

The lability of metastatic properties and many other phenotypic traits in tumor cell clones during serial passage in vitro and in vivo poses formidable logistical problems in designing experiments to correlate cell surface changes with metastatic competence, or at the next level of analytical detail, correlation with specific steps in the metastatic process, such as invasion, tumor cell survival in the curculation, and arrest. Although polyclonal cell lines are relatively stable under similar conditions, the disadvantages of employing these heterogeneous populations have been cited. Consequently, to ensure that in vitro assays of biochemical or molecular traits can be related to the in vivo behavior of the cells being studied, correlative experiments are performed synchronously using replicate tumor cell populations on the same day with an additional flask harvested for storage as a reference cell stock (Greig et al., 1983).

By far the most perplexing problem in performing experiments of this type is the time required for expanding any given clone to generate sufficient cells for biochemical and biological analysis. There is little guarantee during this relatively lengthy procedure (6–10 weeks) that the "clones" will not be rendered heterogeneous by rapid formation of new variant subclones with altered biological and biochemical phenotypic traits. To minimize complications attributable to phenotypic destabilization, several precautions can be taken: (1) all tumor cell clones should be used as soon as possible after their intial isolation; (2) biochemical or other in vitro analyses of cell surface properties are performed on a panel of clones so that atypical information derived from a single clone that has undergone rapid phenotypic "destabilization" will not have a disproportionate effect on the interpretation of the data; and (3) at the time of the final in vivo and in vitro analyses, clones should be subcloned to evaluate the homogeneity of the parent clone, with approximately 10 to 15 subclones being tested from each parent clone. Although the last procedure is the most informative in evaluating the phenotypic stability of the original clone, it is too tedious and time consuming to be performed on a routine basis; consequently, only a minority of clones in a given screen can be investigated in this way.

The dilemma of phenotypic drift can never be totally circumvented, but the above precautions are at least sufficient to ensure that most tumor cell clones being studied represent homogeneous cell populations that can be legitimately compared with each other. This set of precautions is also adequate to detect phenotypic drift in clones (Poste *et al.*, 1981). For example, the ability of certain B16 melanoma clones to accumulate cAMP when challenged with activators of adenylate cyclase (forskolin and MSH) has been shown to correlate positively with their metastatic performance (Sheppard *et al.*, 1984). This conclusion was based on a study of more than 30 B16 melanoma clones. In pursuing this observation, two clones, F1-C14 and F1-C16, were selected for detailed biochemical investigation. During cultivation *in vitro*, the response of these clones to forskolin and MSH challenge was found to alter radically and was accompanied by a concomitant change in metastatic properties (Sheppard *et al.*, 1984). Phenotypic drift in these clones was also confirmed by subcloning experiments that demonstrated the emergence of phenotypically heterogeneous subclones in both "clones" (Sheppard *et al.*, 1984). By contrast, it should also be pointed out that a number of other B16 clones isolated and expanded in culture at the same time as F1-C14 and F1-C16 have displayed remarkable phenotypic stability in that their metastatic capacity and hormonal responsiveness have remained unaltered after continuous propagation *in vitro* (and *in vivo*) for several months (B. Lester *et al.*, unpublished).

If the cell surface change being studied is relevant to the expression of the metastatic phenotype, it would be reasonable to predict that metastases formed *in vivo* should express the same alteration. It is therefore informative to examine expression of the property in question in biopsies both from the primary tumor implant and, more important, individual lung metastases, either assayed directly after excision from the animal or after recovery of cells from the lesion and establishment in culture (Poste *et al.*, 1982a,b).

Demonstration of an association between a specific cell surface alteration and metastatic performance in other tumor systems of similar or dissimilar histological origins would then further strengthen the proposed association.

4. Obligatory and Nonobligatory Phenotypes: Demonstrating Causality between Phenotypic Change and Metastatic Behavior

Even when the rigorous methods outlined in Section 3 are used, further technical complexity is encountered in attempting to define which of the many altered properties exhibited by tumor cells are obligatory for the expression of tumorigenicity or metastasis (obligatory phenotypes). It is also difficult to define those traits that may confer valuable adaptive advantage on tumor cells and that facilitate their survival in the face of potentially destructive selection pressures imposed by the host or therapy but that are not absolutely necessary for expression of tumorigenic or metastatic behavior (nonobligatory phenotypes). Stated another way, the task facing investigators interested in defining the molecular basis of metastatic behavior is how to distinguish the various obligatory phenotypic traits presumably expressed by all metastatic cells (i.e., homogeneous traits) against the "background noise" of extensive phenotypic heterogeneity created by variable expression of nonobligatory traits that are, in the strictest sense, irrelevant to the expression of tumorigenic of metastatic behavior. This assumes, however, that all the metastatic tumor cell subpopulations present in different tumors of common histological origin will exhibit the same panel of obligatory phenotypic changes. It is perhaps not unreasonable to assume that at least some obligatory phenotypes will be shared and should this be detected as an invariant feature of metastatic cells, and some of these might be shared with metastatic tumor cell subpopulations in neoplasms of different histologic origins arising in tissues with the same germ cell layer origins. However, metastatic behavior might also arise *via* the expression of a set of phenotypic changes that collectively confer metastatic competence (i.e., complementing phenotypes) but that cannot induce this behavioral change in a cell if expressed individually. Such phenotypes will only be detected by analysis of the expression of multiple phenotypes in a large number of tumor cell clones. The technical demands of this exercise do not require emphasis, and it is probably impossible using today's methods. Also, the identification of complementing phenotypes requires that all the phenotypes involved are already identified. If complementation involves a trait that has yet to be detected, the correlation of the other complementing traits to metastatic behavior will be impossible. Thus, "known" traits could be coexpressed in clones that are nonmetastatic because they lack the yet "undiscovered" complementing trait. Coexpression of the "known" traits in metastatic clones that express the yet "undiscovered" trait would then create the risk of drawing the erroneous conclusion that the known traits have no relevance to metastasis. The amount of analysis needed to identify complementing phenotypes will increase on a daunting scale as the number of traits involved increases.

Theoretically, cell surface changes and other phenotypic alterations found in tumor cells can be assigned to a number of categories on the basis of their functional relevance.

4.1. Obligatory Phenotypes

Obligatory phenotypes these represent changes in cell surface properties or other cellular traits essential to full expression of a multifactorial trait, such as tumorigenicity or metastasis. Cells lacking the ability to express an obligatory metastatic phenotype would not be metastatic. However, expression of an obligatory phenotype need not be a permanent feature and might only be detectable at certain specific stages of the metastatic process (Poste, 1982). The concept of temporal variation in the expression of metastatic competence is consistent with the idea of "transient metastatic compartments" as proposed by Weiss (1980). According to this hypothesis, all tumor cells within the primary neoplasm have the potential to metastasize but the expression of this behavior in an individual tumor cell is only temporary. Tumor cells are envisaged as moving asynchronously through a series of functional states, referred to by

Weiss (1980) as compartments, and thus constantly acquiring, expressing and losing the capacity to metastasize under these circumstances. The expression of obligatory molecular traits associated with the metastatic phenotype would also be transient thus enormously complicating their identification and characterization.

Obligatory properties can be subdivided into two groups: "threshold" phenotypes and "incremental" phenotypes. A "threshold" property is one that is expressed by all metastatic cells (for a given tumor type) but that displays little or no quantitative change with variation in metastatic proficiency. In other words, the successfully metastatic cell need not excel at completing all of the steps in the metastatic process. It need only display "minimal competence." For example, the most invasive tumor cells (as judged, for example, by secretion of degradative enzymes) need not necessarily be the most metastatic. A metastatic cell needs only to produce sufficient proteolytic enzymes to achieve breakdown of the extracellular matrix to provide access to the vasculature. The concept of "minimal competence" cautions against expecting a linear or simplistic relationship between a given biochemical trait and metastatic performance. Rather, metastatic tumor cells may require only a critical "threshold" of functional competence in the properties required to complete any particular step in the metastatic process. In contrast to threshold phenotypes, "incremental" phenotypes are also expressed by all metastatic cells but show significant quantitative variation (positive or negative) in association with gradations in metastatic proficiency. For example, the ability of circulating metastatic tumor cells to respond to local growth factors may be related directly to the number of cell surface receptors for the mitogen and the rate of proliferation of individual metastases will be a direct reflection of receptor density on the cell surface.

4.2. Nonobligatory Phenotypes

4.2.1. Secondary (Advantageous) Phenotypes

Some of the changes detected in tumor cells may not be obligatory for metastasis and can thus be classified as secondary or advantageous phenotypes. Such properties are not required for metastasis, but their presence confers significant adaptive advantage upon the metastatic tumor cell in the face of potentially destructive selection pressures mounted by host defenses and competition by other clones within the tumor. Examples of this kind might include a greater tolerance of the tumor cell to stressful conditions (e.g., hypoxia, pH fluctuations, nutrient deprivation) and reduced sensitivity to recognition and destruction by host defense mechanisms and therapeutic agents.

4.2.2. Predisposing Phenotypes

Certain phenotypic changes detected in tumor cells may not be essential for metastasis but their presence may enhance the risk and/or frequency with which obligatory metastatic properties are generated. Error-prone DNA polymerases and gene transpositions might represent examples of "predisposing" phenotypes.

4.2.3. Complementing Phenotypes

The final category of phenotypic changes that may occur in tumor cells represent traits that if expressed singly do not confer metastatic competence but when expressed concomitantly generate metastatic cells. With our current limited level of analytical sophistication, the identification of complementing phenotypes will be extremely difficult to achieve. Nevertheless, their existence, although theoretical, would not be unexpected. For example, the ability of a tumor cell to synthesize and secrete its own growth factors (phenotype 1) confers no advantage unless the cell is also capable of synthesizing and expressing the specific cell surface receptor (phenotype 2), with aberrant autocrine regulation occurring as a consequence of these two "complementing" phenotypes. Complementing interactions may also occur between cells. Noninvasive tumor cells may gain access to the circulation, not through their own efforts, but by exploiting the invasive capacity of neighboring tumor cells. However, even though they may lack the properties to invade the circulation in their own right, once intravasation has occurred they may possess all of the traits needed to complete the remaining steps in the metastatic process and thus be able to form metastases. Similarly, a tumor cell displaying surface determinants recognized by host defense mechanisms could be passively carried through the vasculature in a protected manner by passive entrapment within a clump of tumor cells that are not recognized by circulating NK cells. The existence of complementing phenotypes also implies that the isolation of tumor cell populations from metastases which are devoid of a biochemical trait previously considered to be obligatory for metastatic competence does not necessarily negate the importance of the trait in the expression of the malignant phenotype, since cellular complementation may have enabled the cells recovered from the lesion to give rise to a metastasis. However, in such cases, the isolated tumor cell population would not be expected to metastasize when reinjected into animals since, by itself, it lacks one or more phenotypic traits essential for successful completion of the metastatic process.

It could be argued that since tumorigenicity and metastasis are complex composite traits that require alterations in a wide variety of cell properties, each of these alterations, including obligatory phenotypes, represents complementing phenotypes. This is true only if metastasis is viewed as a single event. It clearly is not and comprises a series of sequential steps. The ability of a tumor cell to complete any particular step could presumably be shown to result from specific tumor cell properties if suitable assays were available to allow experimental dissection of metastasis into discrete stages. In short, the functional importance of changes in cell surface properties, or other cellular functions, will eventually need to be defined in relationship to specific steps in metastasis, rather than the entire metastatic cascade. This requires, of course, that assays are available to assay the competence of tumor cells to complete each step in the metastatic process. Experimental efforts to develop such assays are only just beginning.

When defined in relation to a specific stage in the metastatic process rather than the entire process, the distinction between obligatory and complementing phenotypes is clearer. Expression of an obligatory phenotype by a tumor cell will automatically allow it to complete the step in question. By contrast, the ability of complementing phenotypes to substitute for an obligatory phenotype in allowing a tumor cell to complete the same step will require that all the interacting phenotypes be expressed. None of these traits is able to render the cell competent to complete the same step if expressed in anything other than the full complement and thus cannot be viewed as obligatory phenotypes in their own right.

Finally, irrespective of the type of biochemical trait being sought, a long recognized concern, still unresolved, is the issue of how far tumor cell properties examined *in vitro* faithfully reflect those exhibited by the same cells *in vivo*. Past failures to reveal biochemical correlates of the metastatic phenotype may be explained, in part, by the artificial environment of *in vitro* culture conditions. Under the relatively quiescent conditions of *in vitro* culture, and in the absence of host selection pressures, such phenotypes may not be activated and thus go undetected. Biochemically, tumor cell populations with widely differing metastatic capacities may be indistinguishable *in vitro*. However, after inoculation into animals and exposure to a multiplicity of host stimuli, they may display a spectrum of biochemical responses that account for their distinct metastatic capacities. Investigations of metastatic clones derived from the B16 melanoma have confirmed this suspicion and revealed that some biochemical differences between clones with different metastatic capacities are more apparent in cultures that have been exposed to metabolic challenges such as exogenous hormones, growth factors, and stressing agents than unchallenged cultures (Sheppard *et al.*, 1984).

5. Experimental Manipulation of the Metastatic Properties of Tumor Cells: Fulfilling Koch's Postulates at the Molecular Level

To establish a causal relationship between infection by specific microorganisms and development of disease, the nineteenth century microbiologist Robert Koch formulated a set of experimental requirements that are equally applicable to contemporary efforts to demonstrate causal correlations between specific changes in tumor cells and their behavior *in vivo*. Koch's postulates required that the microorganism suspected of causing disease should be found consistently in all cases of the disease, should be recovered consistently from disease lesions and produce the disease when introduced into susceptible hosts. In analyzing causal events in tumor cell behavior, the equivalent of the first two of Koch's postulates would be satisfied by the experimental approaches described in previous sections in which a specific cell surface change is shown to be a consistent feature of (1) metastatic cells *in vitro*; (2) cells in metastatic lesions *in vivo*; and (3) cells recovered from metastases and re-cultured *in vitro*. The equivalent of the third of Koch's postulates of causality would be to demonstrate that experimental manipulations that induce loss or acquisition of the surface property of interest are accompanied by loss or gain of metastatic capacity.

The study of microbial disease has the substantial advantage that disease results from the action of a single agent (a pathogenic microorganism) and the demonstration of causality is relatively straightforward (at least to the twentieth century pathologist). Disease can be induced by introduction of the organism into a susceptible host, and progression of the disease can be arrested by agents that impair microbial function (host immunity, vaccination, antibiotics). By contrast, the suspected multiplicity of phenotypic changes needed to confer metastatic behavior on a cell dictate that it is far more complicated to show whether loss or gain of any specific cellular characteristic affects metastatic behavior. It may be easier, for example, to induce loss of metastatic ability than acquisition of this behavior. Experimentally induced loss of a single obligatory phenotypic trait may be sufficient to abrogate the entire metastatic process by rendering a tumor cell incapable of completing just one step in the metastatic process. It is important to emphasize, however, that depending on the mechanism of action of the manipulation responsible for loss of metastatic competence, other phenotypic traits essential for metastasis may continue to be expressed. It would be erroneous to conclude that their expression in cells that have been rendered nonmetastatic invalidates any previously established causal correlations for these traits.

To confer metastatic ability on a tumorigenic but nonmetastatic cell by experimental alteration of a single phenotypic trait is likely to be far more difficult. Success would probably require that the change be imposed on a cell that possesses all of the other phenotypic changes needed to metastasize except the property being investigated. A possible experimental strategy in this regard would be to show that experimentally induced loss of a single (obligatory) phenotype from a metastatic cell eliminates its metastatic capacity and that induction of the same trait in the modified cell restores metastatic behavior. By contrast, conversion of non-neoplastic cells and many tumorigenic, non-metastatic cells into metastatic cells may require simultaneous alterations in a variety of cell properties that are not technically feasible using current techniques.

Our present ignorance of the extent and nature of the phenotypic changes needed to confer metastatic properties on a malignant tumor cell suggests that a more productive approach is to focus experimental questions on how changes in specific surface properties affect the ability of tumor cells to complete specific steps in the metastatic process. By defining the functional importance of a specific property at different stages in the metastatic cascade, a catalog of phenotypic changes needed to complete the entire process might eventually be assembled. For example, Poste and Nicolson (1980) showed that when plasma membrane vesicles isolated from B16 melanoma cells that metastasized preferentially to the lung were fused with B16 cells that were less efficient in metastasizing to this organ, the "membrane-modified" cells displayed a significantly higher arrest in the lung microcirculation. This experiment suggests that plasma membrane components introduced from the vesicles could alter the organ distribution and arrest patterns of circulating B16 cells.

The powerful new methods for altering cell function that are now emerging as a result of rapid advances in molecular genetics offer significant opportunities for exploring the functional importance of specific surface determinants at different steps in the metastatic process. For example, although the evidence remains circumstantial, there is a strong belief that the secretion of proteases by malignant cells is essential for successful metastasis. Two ways to test this relationship for causality are (1) to abrogate protease activity by using selective enzyme inhibitors, or (2) to endow a tumorigenic nonmetastatic cell with protease activity. Theoretically, the latter can now be achieved using molecular genetic techniques. Once cloned, the gene for a particular tissue degradative enzyme (e.g., collagenase, plasminogen activator) can be introduced into the target tumor cell and its effect on metastatic properties assayed. In the future, similar experiments should be possible for a whole range of tumor cell gene products implicated in the metastatic process; for example laminin receptors, adenylate cyclase, enzymes regulating arachadonic acid metabolism and the various surface determinants whose expression has been shown to vary with changes in metastatic proficiency (Nicolson and Poste, 1984). These techniques offer powerful new approaches for the identification of specific biochemical properties that may contribute to the expression of metastatic behavior. Advances in DNA technology also raise the distant nevertheless intriguing prospect of targeting drugs not only to the protein products of genes but also to the genes themselves. This represents an exciting new area in fundamental research with direct implications for the treatment of metastasis.

Finally, gene transfer techniques are beginning to be used to study whether transfection of nonmetastatic cells with genomic material from metastatic cells might render them metastatic in a fashion analogous to the tumorigenic transformation of cells by oncogenes. Once again, however, the multifactorial nature of the metastatic phenotype may be an obstacle to detection of "metastatic genes." Even if such genes exist, expression of metastatic properties may occur only if these elements are introduced into tumor cells that already possess many of the phenotypic alterations needed to metastasize and that have accumulated as a result of multiple changes in geonomic expression during serial cultivation of these cells *in vitro* or progressive growth *in vivo*. This may well be the case, since metastatic cells do not appear to be present from the outset in host tumors and emerge at a later stage in tumor progression; the precise stage differs between neoplasms arising in different cell types (Sugarbaker, 1979).

Even if transfection of metastatic properties does not prove feasible, it is clear that the new tools of molecular genetics, coupled with equally powerful improvements in techniques and analytical methods in cell biology and immunology, offer substantial promise for altering the expression and regulation of specific gene products in normal and neoplastic cells. They also herald exciting opportunities for manipulating a wide range of cell properties that can be confidently expected to be of great value in defining the functional importance of various alterations in cell surface properties in determining the aberrant behavior of neoplastic cells.

References

Bresciani, F., King, R. J. B., Lippman, M. E., Namer, M., and Raynaud, J. P., (eds.), 1984, *Hormones and Cancer, Part 2*, Vol. 31, *Progress in Cancer Research and Therapy*, Raven Press, New York.

Bruchovsky, N., and Goldie, J. H., (eds.), 1982, *Drug and Hormone Resistance in Neoplasia*, Vols. I and II, CRC Press, Boca Raton, Florida.

Davies, A. J. S., and Crumpton, M. J., (eds.), 1982, *Experimental Approaches to Drug Targeting, Cancer Surveys*, Vol. 1, pp. 349–559, Oxford University Press, Oxford.

Donelli, M. G., D'Incalci, M., and Grattini, S., 1984, Pharmacokinetic studies of anticancer drugs in tumor-bearing animals, *Cancer Treatm. Rep.* **68**:381–400.

Efron, E., 1984, *The Apocalyptics*, Simon and Schuster, New York.

Fidler, I. J., and Poste, G., 1982, The heterogeneity of metastatic properties in malignant tumor cells and regulation of the metastatic phenotype, in: *Tumor Cell Heterogeneity* (A. Owens, D. S. Coffey, and S. B. Baylin, eds), pp. 127–145, Academic Press, New York.

Fidler, I. J., and Poste, G., 1985, The cellular heterogeneity of malignant neoplasms: Implications for adjuvant chemotherapy, *Semin. Oncol.* (in press.)

Goldin, A., 1983, Animal models for cancer chemotherapy, in: *Cancer Chemotherapy* (F. M. Muggia, ed.), pp. 65–102, Martinus Nijhoff, The Hague.

Greig, R. G., Caltabiano, L., Reid, R., Jr., Field, J., and Poste, G., 1983, Heterogeneity of protein phosphorylation in metastatic variants of B16 melanoma, *Cancer Res.* **42**:6057–6065.

Hawkes, S., and Wang, J. L., (eds.), 1982, *Extracellular Matrix*, Academic Press, New York.

Heppner, G. H., 1984, Tumor heterogeneity, *Cancer Res.* **44**:2259–2265.

Hewitt, H. B., 1982, Counterpoint: Animal tumor models and their relevance to human tumor immunology, *J. Biol. Respir. Mod.* **1**:107–119.

National Toxicology Program, 1984, *Report of the NTP Ad Hoc Panel on Chemical Carcinogenesis Testing and Evaluation*, U.S. Department of Health and Human Services, Washington, D.C.

Nicolson, G. L., and Milas, L., (eds.), 1984, *Cancer Invasion and Metastasis: Biologic and Therapeutic Aspects*, Raven Press, New York.

Nicolson, G. L., and Poste, G., 1983, Tumor implantation and invasion of metastatic sites, *Int. Rev. Exp. Pathol.* **25**:77–181.

Owens, A. H., Jr., Coffey, D. S., and Baylin, S. B., (eds.), 1983, *Tumor Cell Heterogeneity: Origins and Implications*, Academic Press, New York.

Poste, G., 1982, Experimental systems for analysis of the malignant phenotype, *Cancer Metast. Rev.* **1**:141–199.

Poste, G., 1983, Tumor cell heterogeneity and the metastatic process, in: *Understanding Breast Cancer: Clinical and Laboratory Concepts* (M. A. Rich, J. C. Hager, and P. Furmanski, eds.), pp. 119–143, Marcel Dekker, New York.

Poste, G., and Fidler, I. J., 1980, The pathogenesis of cancer metastasis, *Nature (Lond.)* **283**:139–146.

Poste, G., and Greig, R., 1982, On the genesis and regulation of cellular heterogeneity in malignant tumors, *Invasion Metast.* **2**:137–176.

Poste, G., and Greig, R., 1983, The experimental and clinical implications of cellular heterogeneity in malignant tumors, *J. Cancer Res. Clin. Oncol.* **106**:159–170.

Poste, G., and Greig, R., 1985, Experimental models for studying the pathogenesis and therapy of metastatic disease, in: *Mechanisms of Metastatasis: Potential Therapeutic Implications* (K. V. Honn, J. D. Crissman, W. E. Powers, and B. F. Sloane, eds.), Martinus Nijhoff, The Hague.

Poste, G., and Nicolson, G. L., 1980, Arrest and metastasis of blood-borne tumor cells are modified by fusion of plasma membrane vesicles from highly metastatic cells, *Proc. Natl. Acad. Sci. USA* **77**:399–403.

Poste, G., and Nicolson, G. L., 1983, Experimental systems for analysis of the surface properties of metastatic tumor cells, in: *Biomembranes*, Vol. 11 (A. Nowotny, ed.), pp. 341–364, Plenum Press, New York.

Poste, G., Doll, J., and Fidler, I. J., 1981, Interactions among clonal subpopulations affect the stability of the metastatic phenotype in polyclonal populations of B16 melanoma cells, *Proc. Natl. Acad. Sci. USA* **78**:6226–6230.

Poste, G., Doll, J., Brown, A. E., Tzeng, J., and Ziedman, I., 1982a, Comparison of the metastatic properties of B16 melanoma clones isolated from cultured cells lines, subcutaneous tumors and individual lung metastases, *Cancer Res.* **42**:2770–2778.

Poste, G., Tzeng, J., Doll, J., Greig, R., Rieman, D., and Ziedman, I., 1982b, Evolution of tumor cell heterogeneity during progressive growth of individual lung metastasis, *Proc. Natl. Acad. Sci. USA* **79**:6574–6578.

Poste, G., Greig, R., Tzeng, J., Koestler, T., and Corwin, S., 1984, Interactions between tumor cell subpopulations in malignant tumor cell subpopulations in malignant tumors, in: *Cancer Invasion and Metastasis: Biologic and Therapeutic Aspects* (G. L. Nicolson and L. Milas, eds.), pp. 223–243, Raven Press, New York.

Sheppard, J. R., Koestler, T. P., Corwin, S. P., Buscarino, C., Doll, J., Lester, B., Greig, R. G., and Poste, G., 1984, Experimental metastasis correlates with cyclic AMP accumulation in B16 melanoma clones, *Nature (Lond.)* **308**:544–547.

Sugarbaker, E. V., 1979, Cancer metastasis: A product of tumor–host interactions, *Curr. Probl. Cancer* **3**:3–59.

Weiss, L., 1980, Metastases: Differences between cancer cells in primary and secondary tumors, in: *Pathobiology Annual*, Vol. 10 (H. Loachin, ed.), pp. 51–81, Raven Press, New York.

Chapter 3

Shedding of Plasma Membrane Fragments
Neoplastic and Developmental Importance

DOUGLAS DILLON TAYLOR and PAUL HENRY BLACK

1. Introduction

It has been hypothesized that neoplasia may result from some aberration of cell differentiation. The evidence supporting this viewpoint includes production by tumors of substances normally associated exclusively with fetal tissues, such as placental-type alkaline phosphatase (Doellgast and Fishman, 1975; Fishman *et al.*, 1975); formation by tumors of materials normally produced by other cell types, such as the ectopic production of hormones (Stolbach *et al.*, 1976; Nishiyama *et al.*, 1980; Ahluwalia *et al.*, 1981); and changes in the histo-logical composition of tumor cells, such as different levels of differentiation present in some teratocarcinomas (Goodfellow and Andrews, 1982).

Of the similarities between neoplastic and embryonic cells, one of the most crucial is the presence of shared cell surface antigens. These antigens, if recognized by the host as foreign, could initiate immune reactions (both cellular and humoral), which would result in their elimination. One of the main questions that arises concerns the development of a tumor in a host with immune surveillance. Since tumors and fetal tissues share many surface antigens (Rosenberg *et al.*, 1978; McIntyre and Faulk, 1980), tumor cells, like fetal cells, must develop mechanisms to escape rejection by the host's immune system. The general concept of antigen shedding as an immunosuppressive mechanism was expressed by Alexander and Currie (Alexander, 1974; Currie and Alexander, 1974) who noted that antigen shedding may also be relevant in the escape of embryonic cells from rejection by the maternal immune system (Alexander, 1974).

Shedding of antigens from the cell surface occurs to a significant extent in

DOUGLAS DILLON TAYLOR and PAUL HENRY BLACK • Department of Microbiology and Hubert H. Humphrey Cancer Research Center, Boston University School of Medicine, Boston, Massachusetts 02118.

tumors; shed antigens alone or combined with antibody may form circulating blocking factors (Hellstrom et al., 1969; Currie and Basham, 1972; Baldwin et al., 1973; Doljanski and Kapeller, 1976; see Black, 1980, for review). Recent work has indicated that blocking factors may also include tumor-derived membrane fragments, alone or associated with antibody (Doellgast et al., 1982; Taylor et al., 1983b). The significance of these membrane fragments and soluble tumor antigens was described by Currie (1973, 1976), who attributed the immunological escape of tumors to the shedding of antigens. During the development of a primary tumor, the local tissue fluid may become saturated by shed tumor antigens, which could combine with and neutralize both specific effector cells and humoral antibody. Current evidence indicates that the shedding of antigens provides protection against immune surveillance by several mechanisms: (1) competition with the tumor for the effector processes of the immune response, (2) prevention of immune effector response formation, and (3) depletion of tumor or "foreign" antigens from the cell surface.

This review analyzes shedding of tumor antigens, primarily their release as discrete particles (termed membrane fragments) from the plasma membrane, their characteristics, and possible mechanisms of release of the fragments. In addition, this chapter examines both the immunologic and nonimmunological consequences of the shedding of this material. Finally, as shedding from tumor cells represents an uncontrolled activated state, shedding is described as it occurs during normal development and proliferation.

2. Historical Background

While the phenomenon of tumor antigen shedding, both in vivo and in vitro, has been recognized for more than two decades, the distinction between soluble and particulate forms of these antigens has not always been made, and the significance of the particulate form (membrane fragments) to tumor immunology is only beginning to be understood.

Initially, studies on sera from cancer patients, using gel filtration on Sephadex G-200, demonstrated two separate peaks of alkaline phosphatase activity, designated 7S and 19S (Dunne et al., 1967). This designation was based on their elution relative to IgG (7S) and IgM (19S). The 19S fraction, when subjected to starch gel electrophoresis, remained at the origin; however, after the 19S fraction was treated with butanol, the alkaline phosphatase activity associated with it comigrated with the 7S fraction (Jennings et al., 1970). This butanol extractability was believed to be a property of lipid–protein complexes, which was further supported by disruption of the shed membrane fragments by Triton X-100 (Taylor et al., 1983a) (Fig. 1). Further studies using high-exclusion limit agarose-based gels (Sepharose 2B, Bio-Gel A-50 m, and Ultrogel A2), indicated the molecular weight of these lipid–protein complexes to be in excess of 50 million (Taylor et al., 1983a).

Singer et al. (1975) examined the Sepharose-excluded high-molecular-weight (HMW) complexes, which exhibited characteristics of plasma mem-

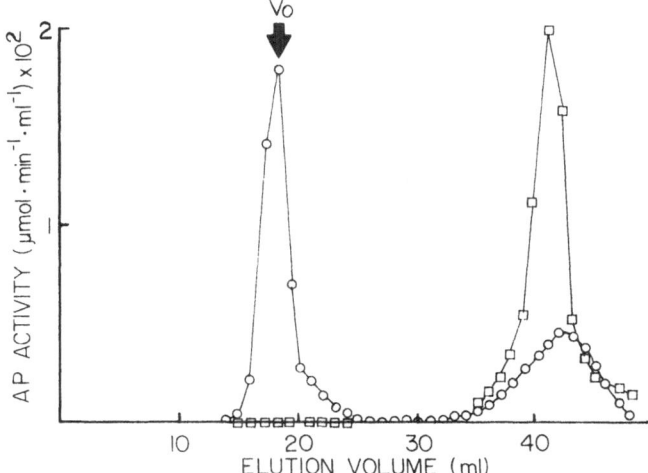

Figure 1. Elution profile of the ascites fluid from an ovarian cancer patient chomatographed on Bio-Gel A-50 m, in the presence (□————□) or absence (○————○) of 5% Triton X-100. V_o represents column void volume. The fractions were assayed for alkaline phosphatase activity.

branes, found in the media of cultured human fetal cells and HeLa cells. They also demonstrated similarities between HMW fragments in sera and fluid specimens (cyst and ascites fluids) of ovarian cancer patients and those present in the media of cultured cells. These studies have been expanded to include cultured human melanoma, murine melanoma, and murine thymoma (Taylor *et al.*, 1983*a*, 1985).

Raz *et al.* (1978) examined various marker enzymes, such as 5′-nucleotidase (a plasma membrane marker), acid phosphatase (a lysosomal marker), and succinic dehydrogenase (a mitochondrial marker) of membrane fragments released by YAC lymphoma cells. These workers found high levels of 5′-nucleotidase, low levels of acid phosphatase, and no detectable succinic dehydrogenase. In addition to these properties, the presence of membrane glycoprotein receptors on the membrane fragments suggested that the ascites fluid-derived membrane fragments were from the plasma membranes of viable tumor cells. Similar conclusions and findings have been made regarding membrane fragments from the culture medium of the murine melanoma, B16-F10 (Taylor *et al.*, 1983*a*). Shinkai and Akeido (1972) also observed, utilizing hepatoma patients' sera, that the Sepharose-excluded alkaline phosphatase was associated with other plasma membrane markers.

Morphologically, the released membrane fragments appear to be predominantly vesicles both *in vivo* and *in vitro* (Fig. 2). However, the vesicular nature can be lost, or at least altered, depending on the isolation procedure. Detailed studies on various techniques for membrane fragment isolation have demonstrated that separation by column chromatography results in a highly purified preparation, whereas the vesicular form remains unaltered (Taylor *et al.*,

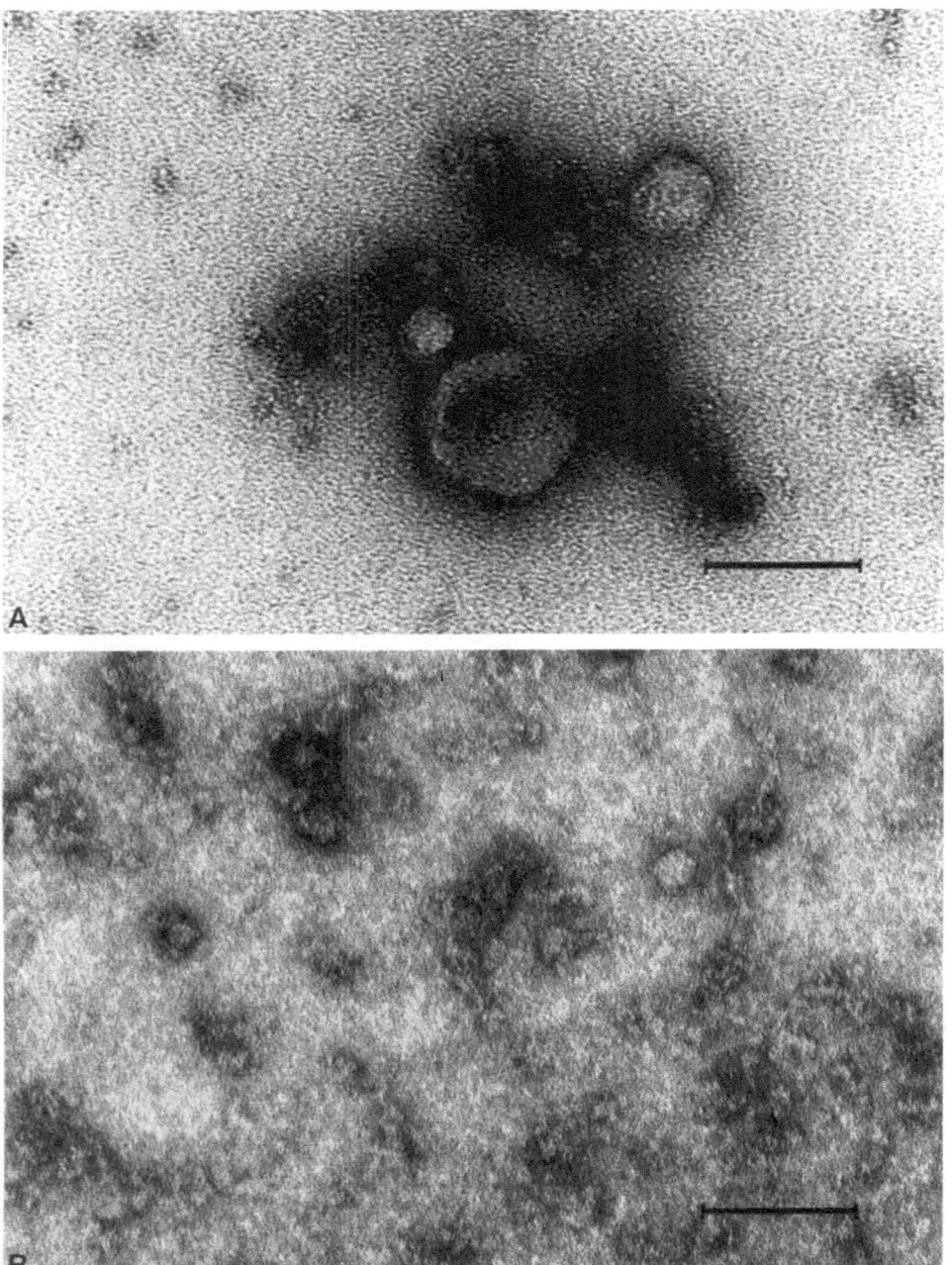

Figure 2. Electron micrographs of shed membrane fragment preparations isolated by (A) column chromatography as described by Taylor *et al.* (1983a) or (B) centrifugation at 150,000 × g as described by Taylor *et al.* (1980). Scale bar: 0.1 μm.

1983a). By contrast, isolation by ultracentrifugation yields a heterogeneous preparation composed of smaller vesicles, broken membranous fragments, and nonspecific lower-molecular-weight (LMW) material (less than 20 million daltons).

Comparisons between isolated tumor cell plasma membranes and their shed membrane fragments suggest that the fragments represent "micromaps" of selected portions of the plasma membranes (van Blitterswijk *et al.*, 1979, 1982; Lerner *et al.*, 1983). This conclusion was reached by two groups: one using primarily lipid analysis (van Blitterswijk *et al.*, 1979, 1982) and the other using enzyme analysis (Lerner *et al.*, 1983). In the initial study by van Blitterswijk *et al.* (1979), the membrane fragments in the ascites fluids from murine leukemia-bearing mice were examined, and their composition was compared with the plasma membrane of the tumor. The shed plasma membrane fragments, while containing the same antigens as the original tumor plasma membranes, were enriched in protein but exhibited less phospholipid. The membrane fragments also contained more cholesterol and sphingomyelin than did the tumor cell membranes. All these factors result in a more rigid membrane, as confirmed by fluorescence polarization. Lerner *et al.* (1983) compared the activities of al-kaline phosphatase, γ-glutamyltranspeptidase, and protein kinase, in addition to cytotoxic antibody binding. Alkaline phosphatase activity and cytotoxic antibody binding were both increased in membrane fragments as compared with the plasma membranes of the cell; however, the activities of γ-glutamyl-transpeptidase and protein kinase were high in the plasma membranes and absent, or decreased, in shed membrane fragments. The differences between the plasma membrane and shed membrane fragments caused both groups to conclude that fragments represent selected domains of the tumor plasma membrane.

3. Membrane Structure

In order to delineate mechanisms of shedding and its possible conse-quences, it is necessary to review the basic plasma membrane structure. The plasma membrane (in its most inclusive definitions) is a complex structure, whose outer portion is poorly demarcated from the extracellular matrix and whose inner portion is poorly demarcated from the cytoplasm (Fig. 3).

The outer portion consists of proteoglycans (glycosaminoglycans) associ-ated with proteins and glycoproteins by ionic or covalent interactions (Burger, 1979). They are loosely associated with extracellular networks containing a variety of extracellular materials, including collagen and sometimes elastin. These complexes can be readily released from the cell, spontaneously or by ionic treatment, and can range in molecular weight up to 20 million daltons. Fibronectin is also found on the cell surface (as well as in serum) in several molecular forms (predominantly dimers of the 220,000-M_r subunit) and, al-though it may form complexes with collagen and glycosaminoglycans, it does

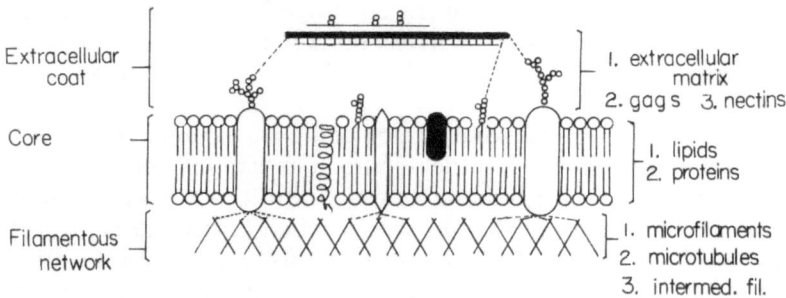

Figure 3. General schematic structure of the plasma membrane broadly defined, showing the three major portions.

not appear to be directly anchored in the lipid core of the plasma membrane (Vaheri and Mosher, 1978).

The central portion of the plasma membrane consists of the classic components of lipids and proteins. This portion exhibits a general asymmetry, not only of phospholipids (i.e., phosphatidylserine on the cytoplasmic side and phosphatidylcholine on the outer portion), but on the carbohydrate-bearing moieties (such as glycoproteins and glycolipids). These glycosylated components are oriented such that their oligosaccharide portions are exposed to the extracellular environment. The importance of this asymmetry may be of a functional nature, such as receptors for hormones, cell–cell communication, and adhesion to extracellular matrices or other cells.

This central region is also responsible for the "fluidity" of the plasma membrane, since the phospholipids are generally free to rotate and diffuse laterally. The three main characteristics that determine the fluidity of membranes (Shinitzky and Inbar, 1976) are (1) the molar ratio of cholesterol to phospholipids, (2) the relative composition of the different phospholipids (in particular the molar ratio of sphingomyelin to phosphatidylcholine), and (3) the degree of saturation of the phospholipid acyl chains. An increased sphingomyelin : phosphatidylcholine ratio produces increased rigidity of the membrane, while increased unsaturation of the acyl chain will enhance fluidity. The most prominent factor is the cholesterol level. Using a constant composition of phospholipids, at a constant temperature, an increase in the cholesterol : phospholipid ratio (M/M) will result in an increased rigidity of the membrane.

The inner portion of the plasma membrane is contiguous with the cytoskeletal network, composed of microtubules (25 nm), intermediate filaments (10 nm), and microfilaments (6 nm) interlinking cellular regions (McNutt *et al.*, 1971, 1973; Snyder and McIntosh, 1976; Osborn *et al.*, 1980). This submembranous network appears to influence cell shape and to mediate shape changes and cell movement. Direct evidence has demonstrated that microtubules in conjunction with microfilaments are important in plasma membrane receptor movement.

The major protein of microfilaments is actin (43,000 M_r) with the additional presence of α-actinin (110,000 M_r), myosin (220,000 M_r), vinculin (130,000 M_r), and filamin (250,000 M_r). The microfilament bundles transverse the cell and are anchored in the adhesion plaques (focal contacts) on the ventral cell surface (Heath and Dunn, 1978). The nature of the attachment of microfilaments to the plasma membrane is not clear; however, α-actinin and vinculin have been implicated as linkage proteins between microfilaments and an unidentified intrinsic membrane protein (Lazarides and Burridge, 1975; Geiger *et al.*, 1980). Microfilaments alone have been generally implicated as the primary cytoskeletal alteration in cell transformation (Boschek, 1982). There are significant differences in the microfilament networks of metastatic variants. Poorly metastatic cells have been shown to possess prominent actin cables that terminate in vinculin-containing focal contacts, whereas highly metastatic cells were demonstrated to contain low numbers of disorganized actin bundles and reduced focal contacts (Raz and Geiger, 1982).

4. Characteristics of Shedding

Shed plasma membrane materials has been demonstrated to be a heterogeneous population of lipids, glycolipids, proteoglycans, proteins, and glycoproteins that vary in size from small individual molecules to vesicles composed of membrane components. As compared with the plasma membrane of tumor cells, shed membrane fragments contain increased protein, decreased phospholipids, and increased cholesterol and sphingomyelin (van Blitterswijk *et al.*, 1979). Such a composition results in more rigid fragments. In addition to these alterations, these shed fragments contain elevated levels of certain enzymes (see Section 2) (Lerner *et al.*, 1983). The proteins present on the membrane fragments appear to be more exposed (Taylor *et al.*, 1980; Lerner *et al.*, 1983), thus competing with the original tumor for the binding of anti-tumor antibody.

Characterization of the shed material indicates that it is not liberated as a consequence of cell death, since there is no evidence that soluble cytosol proteins (such as lactic dehydrogenase) (Doljanski and Kapeller, 1976) or other organelle markers (Raz *et al.*, 1978; Taylor *et al.*, 1983a) are released. Although shedding of soluble molecules (such as proteins and glycoproteins) has been demonstrated in growing normal and activated cells, membrane fragments of greater than 50 million M_r are not released from either growing normal human fibroblasts or nontransformed mouse 3T3 cells (Taylor and Black, 1985).

In studies of shedding of membrane material (both soluble and vesicular) from tumors, the importance of several factors in evaluating the data must be considered.

1. *Cell types:* Studies on shedding have been performed on monolayer cultures (both human and murine melanomas) and suspension cultures (mastocytomas). These different cell types possess different cytoskeletal

systems, which may alter membrane shedding dramatically (see Section 5).

2. *Physiologic state of the culture:* Quantitation of shed membrane molecules may be misinterpreted by the presence of material in the culture medium due to secretion by tumor cells and the presence of end products as a result of cell lysis.

3. *Labeling:* The determination of shed molecules can be influenced by the labeling procedure (i.e., metabolic or external labeling).

4. *Form of the shed material:* Since tumors release materials in various forms, particulate and soluble material must be distinguished.

Shedding has been demonstrated to be a selective process in that only some membrane proteins are shown to be released (Cone et al., 1971; Huang et al., 1974; Lerner et al., 1983). The shedding of membrane material is a slow process, with the release of 36% of the newly synthesized macromolecules within 48 hr (mouse melanoma cells) (Bystryn et al., 1974; Bystryn, 1976). The amount of shed materials appears to increase linearly with time, indicating the coupling of synthesis, shedding, and the presence of intracellular pools (Siekevitz, 1972; Black, 1980). Shedding of at least soluble molecules requires respiration, protein synthesis, and energy (Doljanski and Kapeller, 1976). Whether these metabolic processes are required for the shedding of membrane fragments has not been determined.

5. Mechanisms of Shedding

The mechanisms involved in shedding are only partially understood; however, several hypotheses have been proposed: (1) enhanced proteolysis at the cell surface, (2) lipolysis at the cell surface mediated by phospholipases, (3) disruption of cytoskeletal components (microfilaments and microtubules), (4) destabilization of membranes by ion fluxes, and (5) mechanical shearing of rigid portions of fluid membranes.

There is suggestive evidence for the involvement of cell surface proteases in the phenotypic conversion of transformed cells and shedding. It has been shown that molecules released by protease treatment are similar to those that are spontaneously shed (Roblin et al., 1975). Exposure of certain normal cells to proteases can alter many of their properties to resemble the transformed phenotype and even stimulate cell growth (Blumberg and Robbins, 1975; Zetter et al., 1976). Proteolysis at the cell surface can generate cell surface enzymatic activity (Neurath and Walsh, 1976; Richert and Ryan, 1977), certain receptors during development (Rutishauser et al., 1976), and viral proteins by cleavage of precursor molecules (Shapiro and August, 1976; Witte et al., 1977). Although the data on proteases in general have not been conclusive, interesting results have been obtained with plasminogen activator (PA). Quigley and Goldfarb (1978) reported that the changes in morphology of chick cells induced by the tumor promoter phorbol myristate acetate (PMA) were accompanied by a large increase in synthesis and shedding of PA. The alterations attributable to PMA could be prevented by inhibitors of PA such as diisopropylfluorophosphate (DFP), leupeptin, antipain, and benzamidine, but not by inhibitors of plasmin (aprotinin, ϵ-aminocaproic acid, and soybean trypsin inhibitor). These results suggested that an arginine-specific DFP-sensitive serine protease may be responsible for the PMA-induced morphological and cellular alterations. However, experiments utilizing protease inhibitors must be cautiously interpreted until the toxicity and permeability of these compounds are carefully studied.

The role of phospholipases in shedding is unclear; it is well established, however, that they are necessary in the fusion event (Peretz *et al.*, 1974; van der Bosch, 1982). After the formation of blebs and before this material is released into the environment, it is necessary to initiate fusion at the base of the bleb between the two apposing sections of membrane. Phospholipases may mediate fusion by the formation of lysophospholipids, which exhibit detergentlike activity and thus permit free interactions between membrane structures.

Of all the proposed mechanisms for shedding, disruption of cytoskeletal components has received the most extensive study. There is good evidence that the cytoskeleton is involved in the shedding of histocompatibility antigens and immunoglobulins from murine lymphocytes (Emerson and Cone, 1979a,b, 1981). Certain chemicals or drugs can induce tumor cells to shed cell surface membranes that share antigenic properties with the tumor cell. Liepins and Hillman (1981; Liepins, 1983) demonstrated that blebbing and shedding of membrane fragments (vesicles) by P815 tumor cells could be induced by low temperature, colchicine, and vinblastine, all of which disrupt microtubules. Microtubule stabilization with deuterium oxide and hexylene glycol was found to inhibit the low-temperature-induced membrane fragment shedding process. These results indicate that disruption of cytoplasmic microtubules may be essential in membrane fragment shedding. However, this induced shedding of membrane fragments differs from spontaneous shedding of tumor cells in that it results in a significant reduction in cell size. It is important to note that whereas these agents or conditions can result in the induction or inhibition of shedding, there is no evidence that spontaneous shedding can be inhibited in this manner. Thus, the involvement of disruption of cytoskeleton in the process of shedding of vesicles by tumor cells is unclear.

It is of interest that other chemicals (such as dicoumarol and menadione) can induce blebbing, and potentially shedding, without interacting directly with the cytoskeleton. The formation of blebs may reflect a change in cytoskeletal structure caused by altering intracellular calcium homeostasis. Intracellular calcium and its binding proteins have a crucial role in regulating cytoskeletal structure and function (Albrecht-Buehler, 1981; Flores and Galston, 1982). Two mechanisms have been suggested for the altered calcium levels—a change in the intracellular distribution of calcium, and the influx of extracellular calcium (Flores and Galston, 1982). Evidence indicates blebbing is due not to the influx of calcium but to changes in its intracellular distribution (Flores and Galston, 1982).

Although several studies (van Blitterswijk *et al.*, 1979, 1982) have demonstrated that membrane fragments represent rigid domains in a fluid membrane, the possible role of potential mechanical shearing has not been addressed. However, the presence of these selected rigid domains, combined with the reduced cytoskeletal organization (Raz and Geiger, 1982) seen in tumor cells, could conceivably produce shearing during normal events such as cellular motion.

6. Nonimmunological Consequences of Shedding

Shedding of membrane-associated material may account for certain aspects of the pathogenesis of cancer. The following consequences have been demonstrated to be due, at least in part, to membrane fragment-associated material.

In cancer, the hemostatic mechanism is poised between hyperactivity of the coagulation process, with excessive fibrin deposition and thrombosis, and hyperactivity of the fibrinolytic system (Sack et al., 1977). The shedding of membrane material appears to be important in the excessive stimulation of the coagulation and fibrinolytic pathways. Coagulation normally occurs at the site of vascular injury by the release of thromboplastin or tissue factor (TF). TF is sequestered on the surface of a number of cells, such as fibroblasts (Zacharski and McIntyre, 1972; Maynard et al., 1975), endothelial cells, and muscle cells. TF-like activity (or procoagulant activity) is associated with tumor cells, and evidence indicates tumor cells are capable of shedding this activity in membrane fragments (Dvorak et al., 1981). Local fibrin formation (due to shedding of TF-like activity) may be important in the growth of primary and metastatic tumors, since these cells may use a fibrin network as a matrix for growth and movement.

The fibrinolysis associated with cancer is generally considered secondary to excessive coagulation (Davidson et al., 1969; Harker and Slichter, 1972). Many tumor cells express plasminogen activator (PA) on their surfaces, which can be released. Whereas the form in which PA is released is unknown, the predominant form shed from B16-derived lines (murine melanoma) appears to be as a membrane vesicle (D. Dillon Taylor and P. H. Black, unpublished observation). PA converts plasminogen to plasmin, which hydrolyzes fibrin. Fibrinolysis is normally confined locally to a clot; however, in cancer, plasmin may overwhelm the fibrinolytic system. Increased shedding of PA by tumor cells with subsequent generation of plasmin is important clinically with respect to the fibrinolytic syndromes. The generation of plasmin has also been suggested to have a pivotal role in tumor invasion and metastasis by enabling tumor cells to separate from the primary tumor (Svanberg and Astedt, 1975).

Of the nonimmunological consequences of shedding, one of the most important relates to their involvement in the steps of the metastatic cascade (Fig. 4). Preliminary evidence in our laboratory indicates that variants of high metastatic (or colonizing) potential shed more membrane fragment material than do less or nonmetastasizing variants (at least for highly metastatic vari-

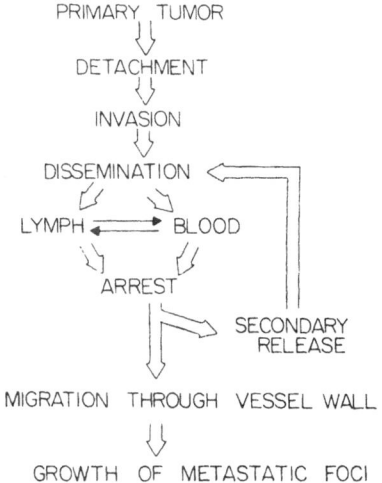

Figure 4. Sequence of the metastatic cascade.

ants, B16-F10 and MDAY-D2 as compared with their poorly metastatic variants, B16-F1 and MDW1, respectively).

In order for metastases to occur, the tumor cells must be released from the primary tumor. During metastasis formation, tumor cells must gain entry into the vascular or lymphatic systems, disseminate through the host, arrest in the target organ, extravasate, and grow in the extravascular space. Factors that may be operative in the release of tumor cells from the primary site include increased tissue pressure, motility of tumor cells, a decrease in cell–cell or cell–extracellular matrix adhesion, and an increase in proteolytic activity. The loss of adhesion may result from changes in the physicochemical bonds on the surface of adjacent cells or between the cell and extracellular matrix. The decrease of adhesion proteins (such as fibronectin or laminin) is also likely to be a factor operative in the loss of adhesion of tumor cells.

Several studies have indicated a relationship between metastatic potential and collagenase activity. Metastatic tumor cells from the venous effluent draining a tumor were shown to possess higher proteolytic activity than the primary tumor (Liotta *et al.*, 1977). These cells have an enhanced ability to degrade collagen and isolated basement membrane material. This proteolytic activity of tumor cells has been thought to be responsible for breaks in basement membranes, seen in areas adjacent to invading tumor cells (Kellner and Sugar, 1967; Hashimoto *et al.*, 1973). In areas surrounding infiltrating tumor cells, in which collagen breakdown occurs, it is unclear whether the collagenolytic activity is cell associated or shed. It has been suggested that procollagenase on the cell surface is activated and subsequently released by shedding (Rose and Robertson, 1977). Plasmin can convert this procollagenase to the active collagenase. Evidence that both PA and procollagenase are expressed on the same cell suggests that the generation of collagenase activity may result from the activation of plasmin (Werb *et al.*, 1977). Activation of procollagenase may also be produced by other neutral proteases present on or shed from the tumor cell (Horwitz *et al.*, 1976).

In addition to degradative enzymes, tumor cells contain material causing endothelial cell retraction. Senger *et al.* (1983) demonstrated the presence of a component on the shed tumor membrane fragments from ascites fluids of tumor-bearing guinea pigs, hamsters, and mice, enhancing vascular permeability. These workers proposed that this increased permeability may be due to endothelial cell retraction. Also, shed membrane fragments can produce platelet aggregation as well as the concomitant release of platelet derived factors, such as serotonin (which can cause endothelial cell retraction) (Gasic *et al.*, 1978). This retraction can also be induced by thrombin (Laposata *et al.*, 1983) or histamine (Majno *et al.*, 1969).

In addition to the increase in vascular permeability, platelet aggregation may play a central role in tumor embolus formation. Single cells do not generally adhere to vascular endothelium and are swept into the vessels. The tumor emboli, which consist of tumor cells and platelets, readily arrest on vessel walls (Fidler, 1973). It appears likely that fibrin is associated with the tumor emboli and that adherent tumor emboli are enmeshed in a fibrin network (Chew and Wallace, 1976). Surface and shed tissue factor-like activity can promote fibrin formation associated with the primary tumor and the tumor emboli. Some studies have indicated the beneficial effects of anticoagulants in limiting metastatic spread of cancer cells, indicating the possible importance of the coagulation process in tumor cell arrest and entrapment.

Cryofibrinogenemia is associated with thromboembolic conditions, particularly the migratory thrombophlebitis accompanying cancer (Sack *et al.*, 1977) or during disseminated intravascular coagulation (Mosesson *et al.*, 1968). It can result from adding thrombin to normal plasma or to plasma treated with drugs that enhance fibrin formation. Fibrinogen, fibrin, and cold-insoluble globulin (CIG; fibronectin), suggested to result from released cell associated material, are the components of cryofibrinogen. Increased levels of fibrin were suggested previously to result from enhanced shedding of membrane fragment associated procoagulant activity. CIG apparently acts as the nucleus with multiple binding sites for the fibrin and fibrinogen complexes (Stathakis *et al.*, 1978). The increased shedding of cell surface fibronectin, combined with increased coagulation seen in some cancers, may promote cryofibrinogenemia.

7. Immunological Consequences of Shedding

The conversion of a normal cell to a cancer cell may be accompanied by the appearance of new antigens on the cell surface. These antigens generally represent either developmental or altered self, or, in the case of viral transformation, virus-associated antigens. Immunocompetent hosts can use these new components as "transplantation antigens" and mount an immune response against the tumor. The cell-mediated cytotoxic immune response by T cells, natural killer (NK) and "null"(K) cells and macrophages are generally considered the main factors in the inhibition of tumor growth. K cells (and probably certain populations of NK cells) possessing Fc receptors can mediate antibody-dependent cellular cytotoxicity, and NK cells may be involved in direct tumor cell destruction. Macrophages have been shown to consist of subpopulations of cells with stimulatory and inhibitory activities toward various immune responses. Macrophages, activated specifically or nonspecifically, are capable of both tumor cell cytolytic and cytostatic activities (Fidler, 1978; Loveless and Munson, 1981), in addition to various immunoregulatory functions (Unanue, 1972; Boraschi and Niederhuber, 1982). In many instances in which cancer develops, the tumor presumably escapes the immune constraints of the host, proliferates, and ultimately may kill the host.

Several possibilities have been suggested to account for the escape of tumor cells from immune destruction.

1. *Antigenicity:* The antigenic properties of certain spontaneous tumors are not always detectable (that is, many spontaneous tumors do not manifest "perceptible" antigenicity) (Sjogren, 1965; Hewitt et al., 1976; Herberman, 1977).
2. *Immunological paralysis:* Highly immunogenic antigens may tolerize B lymphocytes when present in high doses, by reducing antibody secretion rate through direct attachment to the cell surface (Stutman, 1975). Antibody–antigen complexes are capable of tolerizing both T and B lymphocytes. It is important to point out that concepts of immunological tolerance are based on the lack of a detectable immune response to tumor antigens, which does not necessarily mean a complete lack of response. A specific immune deficiency rather than tolerance may be responsible for tumor escape from destruction (Stutman, 1975). An immune deficiency due to altered levels of helper cells or suppressor cells, whether resulting from the action of tumor-derived material or from genetic defects, could be responsible for a tolerancelike condition.
3. *Immunological enhancement:* Two mechanisms have been proposed for immunological enhancement: attachment of antibodies to antigens on the tumor cell surface can (a) hinder the metabolic release of these antigens into the circulation, and (b) interfere with the establishment of direct contact between lymphoid cells and tumor cells (Feldman, 1972; Ting and Herbermann, 1975).
4. *Blocking factors:* By definition, blocking factors are efferent inhibitors acting on the expression of immunoreactivity (Hellstrom et al., 1977). Sera from hosts bearing progressively growing tumors can block the cytolytic interaction of specifically sensitized lymphocytes with the tumor cells. This blocking may be caused by tumor antigens and/or antigen–antibody complexes shed into the circulation (Halliday et al., 1974, 1980). Tumor ascites fluids have been shown to possess immunosuppressive properties. These ascites fluids have been demonstrated to contain factors that interfere with the receptors on sensitized lymphocytes that bind the antigen (Mongini and Rosenberg, 1975). In addi-

tion, *in vitro* components of ascites fluids have been shown to suppress the mixed lymphocyte reaction and the response of peripheral blood lymphocytes to keyhold limpet hemocyanin (Badger *et al.*, 1977). In addition to the blocking activities of antigen–antibody complexes, they may also activate suppressor T lymphocytes (Thomson, 1975; Hellstrom *et al.*, 1982).

5. *Immunoresistance:* It is possible for variant cell clones to arise with altered antigenicity such that they are rendered insusceptible to immune destruction. In addition to altered antigens, some tumor cells exhibit altered lipid composition of the plasma membrane such that they are resistant to destruction by the humoral immune response (Friedberg and Halpert, 1978; Schlager *et al.*, 1978).

6. *Sneaking through:* Sneaking through can be described as a discrepancy in timing between growth of the tumor and immune recognition by the host that favors the tumor (Gatenby *et al.*, 1981). It is conceivable that when a tumor is extremely small, an insufficient amount of the tumor antigen reaches the immune system to initiate stimulation, and that by the time the immune system is triggered, a threshold level of growth has been achieved such that the tumor is insensitive to immune attack. The host does not appear to be tolerant to the tumor antigens; rather the immune system is apparently too inefficient to detect and respond to the extremely small antigen doses (Grossman and Berke, 1980).

7. *Antigen masking:* In some cases, potentially antigenic components on tumor cell surfaces are "masked" or are in such close association with other surface proteins as to prevent their recognition. Transplantation of cultured tumor cells into syngeneic hosts may result in the alterations of the expression of cell surface antigens (Baldwin and Robins, 1977; Hellstrom and Brown, 1979). Generally, a decreased expression is seen that can be reversed in some cases by exposure to proteolytic enzymes, indicating the presence of protein material masking the antigens.

8. *Antigen shedding:* It has been suggested that shedding of antigens may provide protection for the tumor by competing for the effector processes of the immune system (Raz *et al.*, 1978) or by suppressing the formation of the immune effector response (Bast *et al.*, 1983).

There is evidence to indicate that the release of plasma membrane antigens from tumor cells, potentially followed by activation of suppressive responses, provides the most crucial of these escape mechanisms (Hellstrom *et al.*, 1982). Tumor cells release components from the cell surface that can inhibit nonspecific immune responses such as the lymphocyte proliferation response to mitogens and antigens (von Eyben and Arends, 1983; Whisler and Yates, 1980). The mitogenic response of lymphocytes can be inhibited by tumor-derived proteins and gangliosides. Recently, we demonstrated that shed membrane fragments could inhibit T cell mitogenesis in a dose-dependent manner and that this inhibition may be mediated through adherent accessory cells (presumably macrophages) (Taylor et al., 1985).

One of the earliest events in the induction of an immune response consists of the processing of antigens by macrophages. Macrophages take up antigens, primarily if they are large polymers or if the antigens are coated with antibody and complement. The macrophage serves an essential role not only by processing the bulk of the antigens but by presenting these antigens to the lymphocyte after reexpression at the cell surface. Macrophages must also express Ia (in mice or HLA-DR in humans) which, in addition to the processed antigen, can induce the lymphocytes (both B and T cells) to respond to the antigens. In addition to the presentation of antigen, macrophages can secrete products that modify lymphocyte response. Thus, macrophage function may result from a combination of three properties: (1) removal of excess antigens, (2) antigen processing and presentation, and (3) secretion of stimulatory molecules (such as interleukin 1).

There is evidence to indicate that released membrane components from tumor cells interfere with tumoricidal activity of normal macrophages against tumor cells (Evans and Alexander, 1970, 1971; Hibbs, 1973; Keller, 1973; Raz et al., 1977). In general, the elimination of tumors by macrophages results from their adherence to the tumor cell (Currie, 1976). Although extensive and efficient tumor elimination by macrophages occurs in vitro under defined conditions, this potential killing capacity of macrophages does not always succeed in the elimination of tumor cells in vivo. The nature of the escape mechanism has been suggested to be the result of antigen modulation (Calafat et al., 1976) or active shedding of tumor plasma membrane fragments (Nowotny et al., 1974; van Blitterswijk et al., 1979; Raz et al., 1978). Raz et al. (1978) demonstrated that shed tumor membrane fragments could specifically inhibit the association of normal macrophages and lymphoma cells, indicating that in vivo shedding of membrane fragments from tumor cells may induce protection of tumor from immune rejection.

Recent work in our laboratory (Taylor and Black, 1985; Taylor et al., 1985) has demonstrated that the interaction of macrophages with membrane fragments released from murine and human cultured melanoma cells can inhibit the expression of Ia by macrophages. The importance of this observation is that Ia-positive macrophages are necessary for antigen presentation and subsequent helper cell functions (and ultimately, the formation of cytotoxic immune cells) (Emerson and Cone, 1979a).

In addition to processing and presenting antigens, macrophages can directly suppress immune processes in tumor-bearing animals. This suppression is seen only when the tumor load is relatively high and appears to be antigenically nonspecific (Herberman et al., 1979; Garrigues et al., 1981). Although interactions have been postulated between T-cell- and macrophage-mediated suppression of tumor immunity, the nature of this interaction is unclear.

Several model suppressor systems involving both humoral and cell-mediated immune responses have been proposed. Two components have emerged: (1) the requirement for interactions between multiple, distinct T-cell subsets for the efficient action of the suppressor regulatory pathway; and (2) the importance of genes of the major histocompatibility complex in the control of T-cell subset interaction.

Antigens, administered in tolerizing regimens, can elicit idiotypic Ts_1 cells. The activity of the Ts_1 cell is to trigger anti-idiotypic, antigen-primed Ts_2 cells, by the release of TsF_1 (a soluble immunoregulatory factor). Ts_2 cells may suppress idiotypic B cells directly or act with a third population of cells (an idiotypic antigen-primed Ts_3 cell) to generate Ts_2 suppressor activity for cell-mediated response. Ts_1 cells are considered the equivalent to the "inducer" of suppression (afferent suppressors), while Ts_2 cells are considered efferent suppressors (Dietz et al., 1981). The Ts_1 cell is a $Ly1^+2^-$, $I-J^+$, and $Thy1^+$ cell and can be functionally identified in the spleens of primed animals 3–7 days after antigen administration (Germain et al., 1980) Ts_1 cells are suppressive only when given at the time of antigen priming, thereby characterizing them as afferent suppressors (Dietz et al., 1981). It is this population of suppressor cells (Ts_1) that is suggested to be responsible for immunosuppression which is induced by shed tumor membrane fragments in tumor-bearing mice (Bast et al., 1983). This conclusion was based on the timing of suppression, in addition to its removal by anti I-J or anti-Thy 1^+ and complement.

In addition to the role of shed tumor membrane fragments in immune escape mechanisms, it is important to note that shed membrane fragments from normal cells play a central role in the normal differentiation and functioning of cells (such as lymphocytes). It is well established that immune cells (T cells, B cells, and macrophages) interact in the generation of an immune response. These interactions occur at the cell surface and evidence indicates that plasma membrane molecules mediate many of these interactions. It has been suggested that shedding of cell surface alloantigens may be central to the function of these molecules in regulating immune responses (Emerson and Cone, 1981). The key role for membrane Ia turnover in immune cell interactions appears to involve soluble proteins bearing Ia antigenic determinants in mediating helper and suppressive interactions by macrophages, as well as T and B lymphocytes (Tada et al., 1975). Ia antigens, detected primarily on splenic B lymphocytes, were shown to decrease during the culturing of lymphocytes. This decline in Ia activity appeared to be due to shedding of intact Ia antigens (predominantly the release of I-A subregion coded proteins) (Emerson and Cone, 1979a). It has been demonstrated that cell surface I-A, H-2K, and immunoglobulins are rapidly turned over by shedding of these molecules, whereas I-E and H-2D present on the same cells, are not (Emerson and Cone, 1979b, 1981). The I-A and H-2K material appeared to be released as particles associated with lipid and appeared as small membrane vesicles. Although activated lymphocytes release membrane vesicles similar to those released by tumor cells, the composition of these vesicles is very different. van Blitterswijk et al. (1982) examining levels of proteins, cholesterol, sphingomyelin, and phospholipids (and their fatty acid composition) concluded that membrane vesicles from tumor cells (lymphoma) represented selected domains from the tumor cell surface, while membrane vesicles from activated lymphocytes were representative of the general cell surface.

8. Embryological and Developmental Importance of Shedding

Shedding has been indicated to occur in a variety of cells subsequent to cell activation. In this section, shedding is considered in hormone-activated ovarian granulosa cells, certain embryonic cells, and erythroid stem cells. It is important to note, however, that the molecular nature of this shed material is unclear in most instances.

The degradation of proteins accompanying tissue remodeling during embryogenesis led to the suggestion that proteases may play an important role in development. This concept was initially proposed by the work of Gross (1974), who demonstrated the correlation between production of collagenase and tail resorption in tadpoles during metamorphosis. Plasminogen activator has been proposed as a candidate for embryonic proteolysis (Wilson and Reich, 1978; Strickland, 1980). PA is associated with the cell surface of many cell types, including tumor (Wang et al., 1980) and embryonic cells (Marotti et al., 1982) and may be released in both soluble and particulate forms.

In mammals, PA has been demonstrated to be involved in various processes requiring tissue degradation. These include the rupture of the ovarian follicle during ovulation (Beers et al., 1975; Strickland and Beers, 1976), the implantation of the embryo (Strickland et al., 1976), and the involution of the mammary gland after lactation (Ossowski et al., 1979).

During the rupture of the egg from the follicle at the time of ovulation, PA production and release appear to increase. PA was detectable in rat ovarian follicles destined to ovulate and it increased in granulosa cells as ovulation approached (Beers et al., 1975). Exposure of inactive cells to follicle-stimulating hormone (FSH) enhanced the production of PA (Strickland and Beers, 1976). During this activation process, there is an increase in cellular PA as well as in the shedding of PA into the follicle fluid. The release of PA peaked 6–10 hr after exposure to FSH. The shed PA, in association with plasminogen present in the follicle fluid, may be responsible for the weakening and eventual disruption of the follicle wall during ovulation (Strickland and Beers, 1976).

In trophoblast cultures, PA was demonstrated to be released 6–10 days after fertilization, which represents the time of invasion of the uterine wall by the trophoblast cells (Sherman et al., 1976; Strickland et al., 1976). In a second cell type, the parietal endoderm cells, which develop at the junction of the inner cell mass and the blastocoel cavity, PA appears to be released continuously. Although all these cell types produced PA, the enzyme released by parietal endoderm cells was demonstrated to be predominantly 79,000 M_r, whereas PA associated with other embryonic cell types (such as visceral endoderm and extraembryonic mesoderm) was shown to be 38,000 M_r. The parietal endoderm cells migrate outside the yolk sac cavity and along the trophectoderm, until they form a continuous layer apposed to the trophectoderm. At this point, they secrete and adhere to a thick basement membrane (Reichert's membrane). During events of embryogenesis, modification of tissue must be induced by both the invading trophoblastic cells and the migrating parietal endodermal cells, as they separate from Reichert's membrane. The precise function of PA in the embryo remains unclear; however, parietal endoderm PA was envisioned as being responsible for the degradation of Reichert's membrane. Incubation *in vitro* of parietal endoderm cells with Reichert's membrane failed to demonstrate degradation of this basement membrane (Smith and Strickland, 1981). PA has been demonstrated in some interactions between embryonic cells and their surrounding matrices (Strickland and Beers, 1976; Strickland *et al.*, 1976), such as facilitating the movement of mesodermal cells (which wedge between ectoderm and endoderm) in a manner analogous to the invasion process of metastases (Wilson and Reich, 1978; Ossowski and Reich, 1980). However, it is apparent that PA production and shedding from embryonic cells are distinct from similar events in tumor cells, in that they are temporally controlled in embryonic cells.

The control of cell proliferation by cell–cell interactions has been previously demonstrated, although the nature of this regulation is uncertain (Bunge *et al.*, 1979). It has been suggested that differentiation of hematopoietic stem cells may be regulated by cell–cell interactions. The material responsible for the regulation is expressed on the cell surface and appears to be shed (Price and McCulloch, 1978). Peripheral blood leukocytes were demonstrated to shed plasma membrane components that possess granulopoietic colony-stimulating activity. Dainiak and Cohen (1982) examined the erythroid burst-promoting activity (BPA) released from peripheral blood mononuclear cells, erythrocytes, bone marrow cells, spleen cells, and T lymphoblasts. The BPA was apparently derived predominantly from lymphocyte-rich mononuclear cells (but not from monocyte preparations). BPA was shown to be associated with membrane vesicles, the activity of which appears to be the result of a heat-labile membrane protein (Dainiak and Cohen, 1982). In addition to the modulation of immune responses by lymphocyte-derived membrane material (described in Section 7), membrane vesicles spontaneously shed from mononuclear cell plasma membranes appear to be important in the regulation of erythroid burst proliferation.

9. Summary

The phenomenon of shedding of cell surface macromolecules and their importance in the cancer process has been reviewed with particular emphasis on tumor membrane fragments. With cell activation (during growth or stimulation of normal cells), there is an increase in synthesis, processing, insertion, and eventual, intact release of certain membrane proteins, some of which are proteases. In cancer, these events occur spontaneously and without the temporal, physiological, or hormonal control apparent in normal cells. In a previous review (Black, 1980), many of the consequences of shedding tumor products were described, but the nature of the shed material was not clear. It now seems likely that some proteolytic, procoagulant, and immunosuppressive activities of shed material are contained within membrane particulate material (vesicles).

Under normal conditions, shed membrane material (particularly pro-

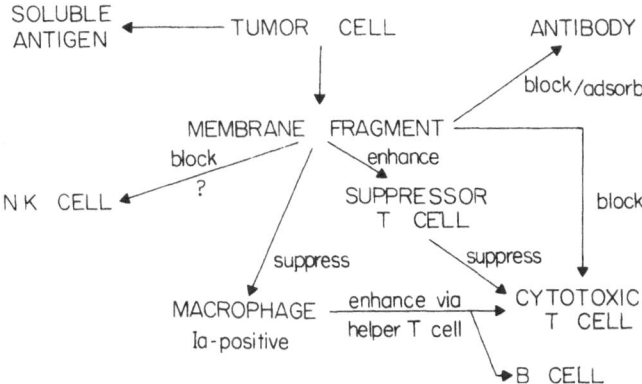

Figure 5. Interaction of shed tumor plasma membrane fragments with immune pathways, resulting in either the enhancement or blocking of effector pathways or the suppression of the formation of effector mechanisms.

teolytic activity) may be necessary for cell movement and tissue remodeling which occur during embryogenesis. In cancer, shedding of plasma membrane fragments may be responsible for the key features of the malignant phenotype by the presence and release of proteolytic activity producing the separation of tumor cells from the primary site, invasion of the surrounding tissues by tumor cells, and formation of distinct metastases. Shed plasma membrane fragments may play a central role in tumor progression by enhancing the steps of the metastatic cascade, in particular by increasing tumor embolus formation (by enhanced fibrin deposition and platelet aggregation) and vascular permeability, as well as increasing basement membrane degradation. Shed membrane fragments (containing tumor antigens) either alone or complexed with antibody, may be responsible for blocking the cell-mediated immune reaction by the formation of "blocking factors" or by suppressing the formation of cytotoxic immune pathways. The suppression of immune response formation may be due to blocking of antigen presentation by macrophages (due to inhibition of Ia) or by the induction of Ts_1 cells (Fig. 5).

References

Ahluwalia, B., Jackson, M. A., Jones, G. W., Williams, A. O., Rao, M. S., and Rajgunu, S., 1981, Blood hormone profiles in prostate cancer patients in high risk and low risk populations, *Cancer* **48:**2267–2273.

Albrecht-Buehler, G., 1981, Does blebbing reveal the convulsive flow of liquid and solutes through the cytoplasmic meshwork, *Cold Spring Harbor Symp. Quant. Biol.* **44:**45–49.

Alexander, P., 1974, Escape from immune destruction by the host through shedding of surface antigens: Is this a characteristic shared by malignant and embryonic cells?, *Cancer Res.* **34:**2077–2082.

Badger, A. M., Cooperband, S. R., Merluzzi, V. J., and Glasgow, A. H., 1977, Immunosuppressive activity of ascitic fluid from patients with cancer metastatic to the peritoneum, *Cancer Res.* **37:**1220–1226.

Baldwin, R. W., and Robins, R. A. 1977, Induction of tumor-immune responses and their interactions with the developing tumor, in: *Contemporary Topics in Molecular Immunology.* Vol. 6 (R. R. Porter and G. L. Ada eds.), pp. 177–207, Plenum Press, New York.

Baldwin, R. W., Bowen, J. G., and Price, M. R., 1973, Detection of circulating hepatoma D23 antigen and immune complexes in tumor bearer serum, *Br. J. Cancer* **28:**16–24.

Bast, R. C., Klein, S., Inbar, M., Karnovsky, M., Knapp, R. C., and Greene, M. I., 1983, Tumor cell membrane vesicles block T cell mediated cytotoxicity and induce Ts$_1$ suppressor cells, *Fed. Proc.* **42:**1081.

Beers, W. H., Strickland, S., and Reich, E., 1975, Ovarian plasminogen activator: Relationship to ovulation and hormonal regulation, *Cell* **6:**387–394.

Black, P. H., 1980, Shedding from the surface of normal and cancer cells, *Adv. Cancer Res.* **32:**75–199.

Blumberg, P. M., and Robbins, P. W., 1975, Effect of proteases on activation of resting chick embryo fibroblasts and on cell surface proteins, *Cell* **6:**137–147.

Boraschi, D., and Niederhuber, J. E., 1982, Regulation of macrophage suppression and cytotoxicity by interferon: Role of Ia bearing macrophages, *J. Immunol.* **129:**1854–1858.

Boschek, C. B., 1982, Organizational changes of cytoskeletal proteins during cell transformation, in: *Advances in Viral Oncolgoy,* Vol. 1 (G. Klein, ed.), pp. 173–187, Raven Press, New York.

Bunge, R., Glaser, L., Lieberman, M., Raben, D., Salzer, J., Whittenberger, B., and Woolsey, T., 1979, Growth control by cell to cell contact, *J. Supramol. Struct.* **11:**175–187.

Burger, M. M., 1979, The cell surface and metastasis, in: *Biology of the Cancer Cell* (K. Letnansky, ed.), pp. 193–208, Kugler Publications, Amsterdam.

Bystryn, J. C., 1976, Release of tumor associated antigens by murine melanoma cells, *J. Immunol.* **116:**1302–1305.

Bystryn, J. C., Schenkein, I., and Baur, S., 1974, Partial isolation and characterization of antigen(s) associated with murine melanoma, *J. Natl. Cancer Inst.* **52:**1263–1269.

Calafat, J., Hilgers, J., van Blitterswijk, W. J., Verbeet, M., and Hageman, P. C., 1976, Antibody-induced modulation and shedding of mammary tumor virus antigens on the surfaces of 6R ascites leukemia cells as compared with normal antigens, *J. Natl. Cancer Inst.* **56:**1019–1023.

Chew, E. C., and Wallace, A. C., 1976, Demonstration of fibrin in early state of experimental metastases, *Cancer Res.* **36:**1904–1909.

Cone, R. E., Marchalonis, J. J., and Rolley, R. T., 1971, Lymphocyte membrane dynamics: Metabolic release of cell surface proteins, *J. Exp. Med.* **134:**1373–1384.

Currie, G., 1973, The role of circulating antigen as an inhibitor of tumor immunity in man, *Br. J. Cancer* (Suppl. 1) **28:**153–161.

Currie, G., 1976, Immunological aspects of host resistance to the development and growth of cancer, *Biochim. Biophys. Acta* **458:**135–165.

Currie, G. A., and Alexander, P., 1974, Spontaneous shedding of TSTA by sarcoma cells: Its possible role in facilitating metastatic spread, *Br. J. Cancer* **29:**72–75.

Currie, G. A., and Basham, C., 1972, Serum mediated inhibition of the immunological reactions of the patient to his own tumor: A possible role for circulating antigen, *Br. J. Cancer* **26:**427–438.

Dainiak, N., and Cohen, C. M., 1982, Surface membrane vesicles from mononuclear cells stimulate erythroid stem cells to proliferate in culture, *Blood* **60:**583–594.

Davidson, J. F., McNicol, G. P., Frank, G. L., Anderson, T. J., and Douglas, A. S., 1969, Plasminogen-activator-producing tumour, *Br. Med. J.* **1:**88–91.

Dietz, M. H., Sy, M. S., Benacerraf, B., Nisonoff, A., Greene, M. I., and Germain, R. N., 1981, Antigen and receptor driven regulatory mechanisms. VII. H-2-restricted anti-idiotypic suppressor factor from efferent suppressor T cells, *J. Exp. Med.* **153:**450–463.

Doellgast, G. J., and Fishman, W. H., 1975, New chromatographic approaches to the separation of human alkaline phosphatase isozymes, in: *Isozymes. I. Molecular Structure* (C. L. Markert, ed.), pp. 293–314, Academic Press, New York.

Doellgast, G. J., Taylor, D. D., and Roh, B. H., 1982, "High molecular weight" immunoglobulin in ovarian cancer fluids, *Am. J. Reprod. Immunol.* **2:**185.

Doljanski, F., and Kapeller, M., 1976, Cell surface shedding: The phenomenon and its possible significance, *J. Theor. Biol.* **62**:253–270.

Dunne, J., Fennelly, J. J., and McGeeney, K., 1967, Separation of alkaline phosphatase enzymes in human serum using gel filtration techniques, *Cancer* **20**:71–76.

Dvorak, H. F., Quay, S. C., Orenstein, N. S., Dvorak, A. M., Hahn, P., and Bitzer, A. M., 1981, Tumor shedding and coagulation, *Science* **212**:923–924.

Emerson, S. G., and Cone, R. E., 1979a, Turnover and shedding of Ia antigens by murine spleen cells in culture, *J. Immunol.* **122**:892–899.

Emerson, S. G., and Cone, R. E., 1979b, Differential effects of colchicine and cytochalasins on the shedding of murine B cell membrane IgM and IgD, *Proc. Natl. Acad. Sci. USA* **76**:6582–6586.

Emerson, S. G., and Cone, R. E., 1981, I-K^k and H-2K^k antigens are shed as supramolecular particles in association with membrane lipids, *J. Immunol.* **127**:482–486.

Evans, R., and Alexander, P., 1970, Cooperation of immune lymphoid with macrophages in tumor immunity, *Nature (Lond.)* **228**:620–622.

Evans, R., and Alexander, P., 1971, Rendering macrophages specifically cytotoxic by a factor released from immune lymphoid cells, *Transplantation* **12**:227–229.

Feldman, J. D., 1972, Immunological enhancement: A study of blocking antibodies, *Adv. Immunol.* **15**:167–214.

Fidler, I. J., 1973, The relationship of embolic homogeneity, number, size and viability to the incidence of experimental metastasis, *Eur. J. Cancer* **9**:223–227.

Fidler, I. J., 1978, Recognition and destruction of target cells by tumoricidal macrophages, *Isr. J. Med. Sci.* **14**:177–191.

Fishman, W. H., Inglis, N. R., Vaitukaitis, J., and Stolbach, L. L., 1975, Regan isoenzyme and human chorionic gonadotrophin in ovarian cancer, *Natl. Cancer Inst. Monogr.* **42**:63–73.

Flores, H. E., and Galston, A. W., 1982, Bleb formation in hepatocytes during drug metabolism is caused by disturbances in thiol and calcium ion homeostasis, *Science* **217**:1257–1261.

Friedberg, S. J., and Halpert, M., 1978, Ehrlich ascites tumor cell surface membranes: An abnormality in ether lipid content, *J. Lipid Res.* **19**:57–64.

Garrigues, H. J., Romero, P., Hellstrom, I., and Hellstrom, K. E., 1981, Adherent cells (Macrophages?) in tumor-bearing mice suppress MLC responses, *Cell. Immunol.* **60**:109–118.

Gasic, G. J., Boettiger, D., Catalfamo, J. L., Gasic, T. B., and Stewart, G. J., 1978, Aggregation of platelets and cell membrane vesiculation by rat cells transformed *in vitro* by RSV, *Cancer Res.* **38**:2950–2955.

Gatenby, P. A., Basten, A., and Creswick, P., 1981, "Sneaking through": A T-cell-dependent phenomenon, *Br. J. Cancer* **44**:753–756.

Geiger, B., Tokuyasu, K. T., Dutton, A. H., and Singer, S. J., 1980, Vinculin, an intracellular protein localized at specialized sites where microfilament bundles terminate at cell membranes, *Proc. Natl. Acad. Sci. USA* **77**:4127–4131.

Germain, R. N., Waltenbauch, C., and Benacerraf, B., 1980, Antigen specific T cell mediated suppression. V. H-2 linked genetic control of distinct antigen defects in the production and activity of L-glutamic acid-L-tyrosine suppressor factor, *J. Exp. Med.* **151**:1245–1259.

Goodfellow, P. N., and Andrews, P. W., 1982, The biology of teratocarcinomas, *Nature (Lond.)* **300**:107–108.

Gross, J., 1974, Collagen biology: Structure, degradation, and disease. *The Harvey Lectures*, Series 68, pp. 351–432, Academic Press, New York.

Grossman, Z., and Berke, G., 1980, Tumor escape from immune elimination, *J. Theoret. Biol.* **83**:267–296.

Halliday, W. J., Maluish, A. E., and Isbister, W. H., 1974, Detection of anti-tumor cell mediated immunity and serum-blocking factors in cancer patients by the leucocyte adherence inhibition test, *Br. J. Cancer* **29**:31–35.

Halliday, W. J., Koppi, T. A., Kahn, J., and Davis, N. C., 1980, Leukocyte adherence inhibition: Tumor specificity of cellular and serum-blocking reactions in human melanoma, breast cancer, and colorectal cancer, *J. Natl. Cancer Inst.* **65**:327–335.

Harker, L. A., and Slichter, S. J., 1972, Platelet and fibrinogen consumption in man, *N. Engl. J. Med.* **287**:999–1005.

Hashimoto, K., Yamanishi, Y., Maeyens, E., Dabbous, M. K., and Kanzaki, T., 1973, Collagenolytic activities of squamous cell carcinoma of the skin, *Cancer Res.* **33**:2790–2801.

Heath, J. P., and Dunn, G. A., 1978, Cell to substratum contacts of chick fibroblasts and their relation to the microfilament system, *J. Cell Sci.* **29**:197–212.

Hellstrom, K. E., and Brown, J. P., 1979, Tumor antigens, in: *The Antigens* (M. Sela, ed.), Vol. 5, pp. 1–82, Academic Press, New York.

Hellstrom, I., Hellstrom, K. E., Evans, C. A., Huppner, G. H., Pierce, G. E., and Yang, P. S., 1969, Serum-mediated protection of neoplastic cells from inhibition by lymphocytes immune to tumor-specific antigens, *Proc. Natl. Acad. Sci. USA* **62**:362–368.

Hellstrom, K. E., Hellstrom, I., and Nepom, J. T., 1977, Specific blocking factors—Are they important? *Biochim. Biophys. Acta* **473**:121–148.

Hellstrom, K. E., Hellstrom, I., and Nelson, K., 1982, Antigen specific suppressor ("blocking") factors in tumor immunity, in: *Pathological Membranes* (A. Nowotny, ed.), pp. 1–39, Plenum Press, New York.

Herberman, R. B., 1977, Tumor immunology and immunotherapy, *Laryngoscope* **87**:722–725.

Herberman, R. B., Holden, H. T., Djeu, J. Y., Jerrel, T. R., Varesi, L., Tagliabue, A., White, S. L., Ohelen, J. R., and Dean, J. H., 1979, Macrophages as regulators of immune responses against tumors, *Adv. Exp. Med. Biol.* **121B**:361–379.

Hewitt, H. B., Blake, E. R., and Walder, A. S., 1976, A critique of the evidence for active host defense against cancer based on personal studies of 27 murine tumors of spontaneous origin, *Br. J. Cancer* **33**:241–259.

Hibbs, B. J., Jr., 1973, Macrophage non-immunologic recognition: Target cell factors related to contact inhibition, *Science* **180**:868–870.

Horwitz, A. L., Kelman, J. A., and Crystal, R. G., 1976, Activation of alveolar macrophage collagenase by a neutral protease secreted by the same cell, *Nature (Lond.)* **264**:772–774.

Huang, C. C., Tsai, C. M., and Canellakis, E. S., 1974, Iodination of cell membranes. I. Characterization of HeLa cell membrane surface proteins, *Biochim. Biophys. Acta* **332**:59–68.

Jennings, R. C., Brocklehurst, D., and Hirst, M., 1970, A comparative study of alkaline phosphatase enzymes using starch gel electrophoresis with special reference to high molecular weight enzymes, *Clin. Chim. Acta* **30**:509–517.

Keller, R., 1973, Cytostatic elimination of syngeneic rat tumor cells *in vitro* by non-specifically activated macrophages, *J. Exp. Med.* **138**:625–644.

Kellner, B., and Sugar, J., 1967, Morphological factors accompanying growth and invasion, in: *Endogenous Factors Influencing Host–Tumor Balance* (R. W. Wissler, T. L. Doa, and S. Wood, eds.), pp. 239–273, Chicago University Press, Chicago.

Laposata, M., Dovnarsky, D. K., and Shin, H. S., 1983, Thrombin-induced gap formation in confluent endothelial cell monolayers *in vitro*, *Blood* **62**:549–556.

Lazarides, E., and Burridge, K., 1975, α-Actinin: Immunofluorescent localization of a muscle structural protein in nonmuscle cells, *Cell,* **6**:289–298.

Lerner, M. P., Lucid, S. W., Wen, G. J., and Nordquist, R. E., 1983, Selected area membrane shedding by tumor cells, *Cancer Lett.* **20**:125–130.

Liepins, A., 1983, Possible role of microtubules in tumor cell surface membrane shedding, permeability, and lympholysis, *Cell. Immunol.* **76**:120–128.

Liepins, A., and Hillman, A. J., 1981, Shedding of tumor cell surface membranes, *Cell Biol. Int. Rep.* **5**:15–26.

Liotta, L. A., Kleinerman, J., Catanzaro, P., and Rynbrandt, D., 1977, Degradation of basement membrane by murine tumor cells, *J. Natl. Cancer Inst.* **58**:1427–1431.

Loveless, S. E., and Munson, A. E., 1981, Maleic vinyl ether activation of murine macrophages against lung metastasizing tumor, *Cancer Res.* **41**:3901–3906.

Majno, G., Shea, S. M., and Leventhal, M., 1969, Endothelial contraction induced by histamine-type mediators, *J. Cell Biol.* **42**:647–672.

Marotti, K. R., Berlin, D., and Strickland, S., 1982, The production of distinct forms of plasminogen activator by mouse embryonic cells, *Dev. Biol.* **90**:154–159.

Maynard, J. R., Heckman, C. A., Pitlick, F. A., and Nemerson, Y., 1975, Association of tissue factor activity with the surface of cultured cells, *J. Clin. Invest.* **55**:814–824.

McIntyre, J. A., and Faulk, W. P., 1980, Cross-reactions between cell surface membrane antigens of human trophoblast and cancer cells, *Placenta* 1:197–207.

McNutt, N. S., Culp, L. A., and Black, P. H., 1971, Contact-inhibited revertant cell lines isolated from SV40-transformed cells. II. Ultrastructural study, *J. Cell Biol.* 50:691–708.

McNutt, N. S., Culp, L. A., and Black, P. H., 1973, Contact inhibited revertant cell lines isolated from SV40-transformed cells. IV. Microfilament distribution and cell shape in untransformed, transformed, and revertant Balb/c 3T3 cells, *J. Cell Biol.* 56:412–428.

Mongini, P. K. A., and Rosenberg, L. T., 1975, Inhibition of lymphocyte trapping by cell-free ascitic fluids cultivated in syngeneic mice, *J. Immunol.* 114:650–654.

Mosesson, M. W., Colman, R. W., and Sherry, S., 1968, Chronic intravascular coagulation syndrome. Report of a case with special studies of an associated plasma cryoprecipitate, *N. Engl. J. Med.* 278:815–821.

Neurath, H., and Walsh, K. A., 1976, Role of proteolytic enzymes in biological regulation (a review), *Proc. Natl. Acad. Sci. USA* 73:3825–3832.

Nishiyama, T., Stolbach, L. L., Rule, A. H., DeLellis, R. A., Inglis, N. R., and Fishman, W. H., 1980, Expression of oncodevelopmental markers in tumor tissues and uninvolved bronchial mucosa, an immunohistochemical study, *Acta Histochem. Cytochem.* 13:245–253.

Nowotny, A., Groshman, J., Abdelnoor, A., Note, N., Jang, C., and Waltersdroff, R., 1974, Escape of TA3 tumors from allogenic immune rejection. Theory and experiments, *Eur. J. Immunol.* 4: 73–78.

Osborn, M., Franke, W., and Weber, K., 1980, Direct demonstration of the presence of two immunologically distinct intermediate-sized filament systems in the same cell by double immunofluorescence microscopy, *Exp. Cell Res.* 125:37–46.

Ossowski, L., and Reich, E., 1980, Experimental model for quantitative study of metastasis, *Cancer Res.* 40:2300–2309.

Peretz, H., Toister, Z., Laster, V., and Loyter, A., 1974, Fusion of intact human erythrocytes and erythrocyte ghosts, *J. Cell Biol.* 63:1–11.

Price, G. B., and McCulloch, E. A., 1978, Cell surfaces and the regulation of hemopoiesis, *Semin. Hematol.* 15:283–300.

Raz, A., and Geiger, B., 1982, Altered organization of cell substrate contacts and membrane-associated cytoskeleton in tumor cell variants exhibiting different metastatic capabilities, *Cancer Res.* 42:5183–5190.

Raz, A., Inbar, M., and Goldman, R., 1977, A differential interaction *in vitro* of mouse macrophages with normal lymphocytes and malignant lymphoma cells, *Eur. J. Cancer* 13:605–615.

Raz, A., Goldman, R., Yuli, I., and Inbar, M., 1978, Isolation of plasma membrane fragments and vesicules from ascites fluids of lymphoma bearing mice and their possible role in the escape mechanism of tumors from host immune rejection, *Cancer Immunol. Immunother.* 4:53–59.

Richert, N. D., and Ryan, R. J., 1977, Proteolytic enzyme activation of rat ovarian adenylate cyclase, *Proc. Natl. Acad. Sci. USA* 74:4857–4861.

Roblin, R., Chou, I. H., and Black, P. H., 1975, Proteolytic enzymes, cell surface changes, and viral transformation. *Adv. Cancer Res.* 22:203–260.

Rose, G. G., and Robertson, P. B., 1977, Collagenolysis by human gingival fibroblast cell lines, *J. Dent. Res.* 56:416–424.

Rosenberg, S. A., Parker, G. A., and Thorpe, W. P., 1978, Expression of oncofetal antigens by murine and human cells in tissue culture, *Isr. J. Med. Sci.* 14:98–104.

Rutishauser, U., Thiery, J. P., Brackenbury, R., Sela, B.-A., and Edelman, G. M., 1976, Mechanisms of adhesion among cells from neural tissues of the chick embryo, *Proc. Natl. Acad. Sci. USA* 73:577–581.

Sack, G. H., Jr., Levin, J., and Bell, W. R., 1977, Trousseau's syndrome and other manifestations of chronic disseminated coagulopathy in patients with neoplasms, *Medicine (Baltimore)* 56:1–37.

Schlager, S. I., Ohanian, S. H., and Borsos, T., 1978, Identification of lipids associated with the ability of tumor cells to resist humoral immune attack, *J. Immunol.* 120:472–480.

Senger, D. R., Galli, S. J., Dvorak, A. M., Perruzzi, C. A., Harvey, V. S., and Dvorak, H. F., 1983,

Tumor cells secrete a vascular permeability factor that promotes accumulation of ascites fluid, *Science* **219**:983–985.

Shapiro, S. Z., and August, J. T., 1976, Proteolytic cleavage events in oncornavirus protein synthesis, *Biochim. Biophys. Acta* **458**:375–396.

Sherman, M. I., Strickland, S., and Reich, E., 1976, Differentiation of early mouse embryonic and teratocarcinoma cells *in vitro*: Plasminogen activator production, *Cancer Res.* **36**:4208–4216.

Shinitzky, M., and Inbar, M., 1976, Microviscosity parameters and protein mobility in biological membranes, *Biochim. Biophys. Acta* **433**:133–149.

Shinkai, K., and Akeido, H., 1972, A multienzyme complex in serum of hepatic cancer, *Cancer Res.* **32**:2307–2313.

Siekevitz, P., 1972, The turnover of proteins and the usage of information, *J. Theor. Biol.* **37**:321–334.

Singer, R. M., White, L. J., Perry, J. E., and Doellgast, G. J., 1975, The release of high molecular-weight alkaline phosphatase and leucine aminopeptidase into the media of cultured human cells, *Cancer Res.* **35**:3048–3050.

Sjogren, H. O., 1965, Transplantation methods as a tool for detection of tumor-specific antigens, *Prog. Exp. Tumor Res.* **6**:289–322.

Smith, K. K., and Strickland, S., 1981, Structural components and characteristics of Reichert's membrane, an extra-embryonic basement membrane, *J. Biol. Chem.* **256**:4654–4661.

Snyder, J., and McIntosh, J. R., 1976, Biochemistry and physiology of microtubules, *Annu. Rev. Biochem.* **45**:699–727.

Stathakis, N. E., Mosesson, M. W., Chen, A. B., and Galanakis, D. K., 1978, Cryoprecipitation of fibrin-fibrinogen complexes induced by the cold-insoluble globulin of plasma, *Blood*, **51**:1211–1222.

Stolbach, L. L., Inglis, N. R., Lin, C. W., Turksoy, R., Fishman, W. H., Marchant, D., and Rule, A., 1976, Measurement of Regan isoenzyme, HCG, CEA, and histaminase in the serum and effusion fluids of patients with cancer of the breast, ovary, or lung, in: *Oncodevelopmental Gene Expression* (W. H. Fishman and S. Sell, eds.), pp. 433–444, Academic Press, New York.

Strickland, S., 1980, Plasminogen activator in early development, in: *Development in Mammals* (M. H. Johnson, ed.), pp. 81–100, Elsevier/North-Holland, Amsterdam.

Strickland, S., and Beers, W. H., 1976, Studies on the role of plasminogen activator in ovulation. *In vitro* response of granulosa cells to gonadotropins, cyclic nucleotides and prostaglandins, *J. Biol. Chem* **251**:5694–5702.

Strickland, S., Reich, E., and Sherman, M. I., 1976, Plasminogen activator in early embryogenesis. Enzyme production by trophoblast and parietal endoderm, *Cell* **9**:231–240.

Stutman, O., 1975, Immunodepression and malignancy, *Adv. Cancer Res.* **22**:261–422.

Svanberg, L., and Astedt, B., 1975, Coagulative and fibrinolytic properties of ascitic fluid associated with ovarian tumors, *Cancer* **35**:1382–1387.

Tada, T., Taniguchi, M., and Takemori, T., 1975, Properties of primed suppressor T cells and their products, *Transpl. Rev.* **26**:106–129.

Taylor, D. D., and Black, P. H., 1985, Inhibition of Macrophage Ia antigen expression by shed plasma membrane vesicles from metastatic murine melanoma lines, *J. Natl. Cancer Inst.* **74**:859–867.

Taylor, D. D., Homesley, H. D., and Doellgast, G. J., 1980, Binding of specific peroxidase-labeled antibody to placental-type phosphatase on tumor-derived membrane fragments, *Cancer Res.* **40**:4064–4069.

Taylor, D. D., Chou, I. N., and Black, P. H., 1983a, Isolation of plasma membrane fragments from cultured murine melanoma cells, *Biochem. Biophys. Res. Commun.* **113**:470–476.

Taylor, D. D., Homesley, H. D., and Doellgast, G. J., 1983b, "Membrane associated" immunoglobulins in cyst and ascites fluids of ovarian cancer patients, *Am. J. Reprod. Immunol.* **3**:7–11.

Taylor, D. D., Levy, E. M., and Black, P. H., 1985, Shed membrane vesicles: A mechanism for tumor-induced immunosuppression, in: *Immunity to Cancer* (A. E. Reif, and M. Mitchell, eds), pp. 369–373, Academic Press, New York.

Thomson, D. M. P., 1975, Soluble tumour-specific antigen and its relationship to tumour growth, *Int. J. Cancer* **15**:1016–1029.

Ting, C. C., and Herberman, R. B., 1975, Specific afferent interference by antiserum of *in vivo* immunity, *Nature (Lond.)* **257**:801–802.

Unanue, E. R., 1972, The regulatory role of macrophage in antigenic stimulation, *Adv. Immunol.* **15**:95–165.

Vaheri, A., and Mosher, D. F., 1978, High molecular weight, cell surface-associated glycoprotein (fibronectin) lost in malignant transformation, *Biochim. Biophys. Acta* **516**:1–25.

van Blitterswijk, W. J., Emmelot, P., Hilkmann, H. A., Hilgers, J., and Feltkamp, C. A., 1979, Rigid plasma-membrane derived vesicles, enriched in tumour-associated surface antigens (MLr) occurring in the ascites fluid of a murine leukaemia (GRSL), *Int. J. Cancer* **23**:62–70.

van Blitterswijk, W. J., DeVeer, G., Krol, J. H., and Emmelot, P., 1982, Comparative lipid analysis of purified plasma membranes and shed extracellular membrane vesicles from normal murine thymocytes and leukemia GRSL cells, *Biochim. Biophys. Acta* **688**:495–504.

van der Bosch, H., 1982, Phospholipase, in: *Phospholipids* (J. N. Hawthorne and G. B. Ansell, eds.), pp. 313–358, Elsevier Biomedical Press, Amsterdam.

von Eyben, F. E., and Arends, J., 1983, Suppression of the mitogen response of normal lymphocytes by serum from patients with testicular germ cell tumors, *Am. J. Reprod. Immunol.* **4**:5–10.

Wang, B. S., McLoughlin, G. A., Richie, J. P., and Mannick, J. A., 1980, Correlation of the production of plasminogen activator with tumor metastasis in B16 mouse melanoma cell lines, *Cancer Res.* **40**:288–292.

Werb, Z., Mainardi, C. L., and Vater, C. A., 1977, Endogenous activation of latent collagenase by rheumatoid synovial cells. Evidence for a role of plasminogen activator, *N. Engl. J. Med.* **296**:1017–1023.

Whisler, R. L., and Yates, A. J., 1980, Regulation of lymphocyte responses by human gangliosides, *J. Immunol.* **125**:2106–2111.

Wilson, E. L., and Reich, E., 1978, Plasminogen activator in chick fibroblasts: Induction of synthesis by retinoic acid; synergism with viral transformation and phorbol ester, *Cell* **15**:385–392.

Witte, O. N., Tsukamoto-Adey, A., and Weissman, I. L., 1977, Cellular maturation of oncornavirus glycoproteins: Topological arrangement of precursor and product forms in cellular membranes, *Virology* **76**:539–553.

Zacharski, L. R., and McIntyre, O. R., 1972, Physical stability of cell culture procoagulants detectable in first stage coagulation factor assays, *Proc. Soc. Exp. Biol. Med.* **139**:713–715.

Zetter, B. R., Chen, L. B., and Buchanan, J. M., 1976, Effects of protease treatment on growth, morphology, adhesion, and cell surface proteins of secondary chick embryo fibroblasts, *Cell* **7**:407–412.

Chapter 4

Regulation of Glycoprotein Synthesis during Development of the Sea Urchin Embryo

WILLIAM J. LENNARZ

1. Introduction

In recent years there has been much progress in our understanding of the assembly mechanism for the oligosaccharide chain of N-linked glycoproteins (Struck and Lennarz, 1980). N-linked glycoproteins fall into two classes—one in which the chains terminate with polymannose residues and a second in which the chains terminate with the trisaccharide consisting of sialic acid, galactose, and N-acetylglucosamine (see Fig. 1). Despite the structural differences in these two types of chains, both have the common feature of an N-glycosidic bond between the innermost N-acetylglucosamine unit and the amino side chain of an asparagine residue in the polypeptide chain. Furthermore, both types of chain are formed *via* a common precursor, an oligosaccharide that is preassembled in the rough endoplasmic reticulum.

Figure 2 shows this preassembly process in abbreviated form. The novel features are that (1) dolichylphosphate, a long-chain polyisoprenoid, acts as an acceptor of sugar units from sugar nucleotides, and (2) the complete oligosaccharide chain is built up on this lipid molecule. Transfer of the oligosaccharide chain to the asparagine residue in a protein destined to be glycosylated is, as shown, a cotranslation process that occurs while the polypeptide is being elongated. After transfer is completed, glucosyl residues are excised, and the molecule is then moved *via* transfer vesicles to the Golgi complex. At this stage, if the glycoprotein is destined to be a polymannose type, no further processing occurs. On the other hand, if it is destined to be a complex-type chain, a series of glycosidases and glycosyltransferases act on it as shown to convert it to a complex-type chain. The process illustrated in Figure 2 is for a secretory pro-

WILLIAM J. LENNARZ • The University of Texas System Cancer Center, M. D. Anderson Hospital and Tumor Institute, Department of Biochemistry and Molecular Biology, Houston, Texas 77030.

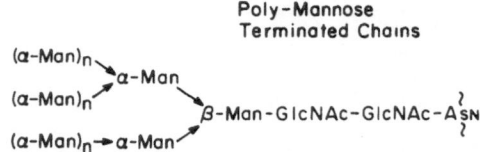

Poly-Mannose
Terminated Chains

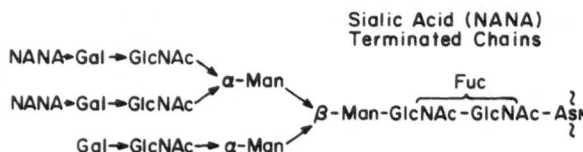

Sialic Acid (NANA)
Terminated Chains

Figure 1. Two major classes of N-linked glycoproteins. In one class the oligosaccharide chains terminate with polymannose units, whereas in the other the chains terminate in sialic acid residues. (From Lennarz, 1983.)

tein, but there is strong evidence that this process is also operative in the case of membrane glycoproteins. In either case, the luminally oriented membrane glycoprotein or the protein destined for secretion is moved from the Golgi complex *via* secretory vesicles to the cell surface, where it becomes externalized.

Some years ago we became interested in investigating the mechanism of regulation of glycoprotein synthesis in a developmentally regulated system. The system we chose was the sea urchin embryo because large numbers of embryos can be cultivated in synchrony and there is a wealth of information about the origins and the fates of various cell types in this well-studied system. The key features of the early stages of development of the sea urchin embryo are shown in Figure 3. After initial cleavage stages, the solid ball of cells gradually becomes hollow, forming the blastula. At about 18 hr the blastula hatches; primary mesenchyme cells emerge from the presumptive ectoderm at the vegetal pole and enter the blastocoel. Shortly afterward, the tissue at the vegetal pole invaginates, signaling the onset of gastrulation. This is accompanied by the appearance of secondary mesenchyme cells at the tip of the advancing gut tube. At the later stages of gastrulation, the primary mesenchymal cells fuse to form a syncytium, and calcium carbonate is deposited to produce primitive skeletal elements called spicules. Further development leads to the prism and then the pluteus, at which point the organism for the first time depends on organic material from the outside world.

2. The Effect of Tunicamycin on Development

Our initial studies on regulation of glycoprotein synthesis in the embryos were based on the observation that tunicamycin blocks protein glycosylation in a wide variety of systems (Struck and Lennarz, 1980). We sought to determine whether this drug was effective in the sea urchin embryo and, if so, what the

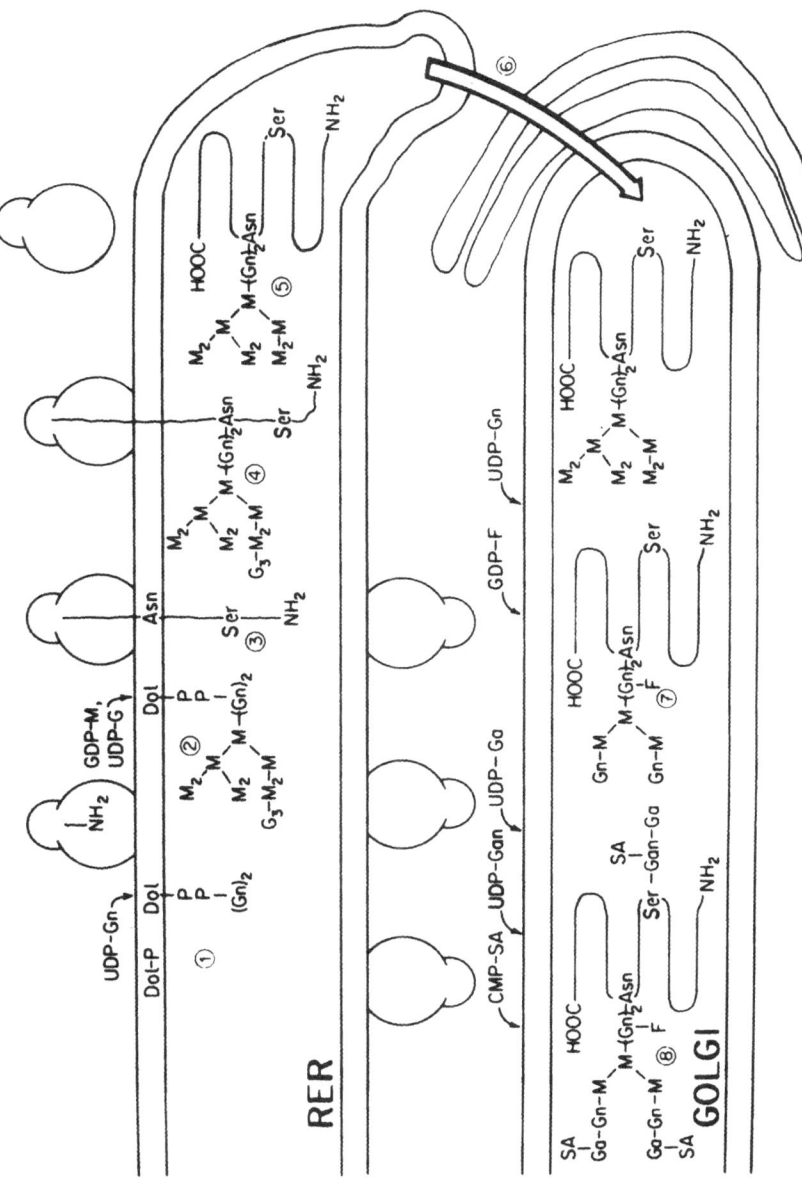

Figure 2. Schematic diagram of the assembly of oligosaccharide chains attached to dolichylphosphate and transfer to an Asn residue in a nascent polypeptide chain in the rough endoplasmic reticulum (RER). Processing of the transferred chain is initiated in the RER and completed in the Golgi complex in the case of complex-type chains. Gn, N-acetylglucosamine; M, mannose, G, glucose; Ga, galactose; SA, sialic acid; Gan, N-acetylgalactosamine. (Reprinted with permission from Hanover and Lennarz, 1981.)

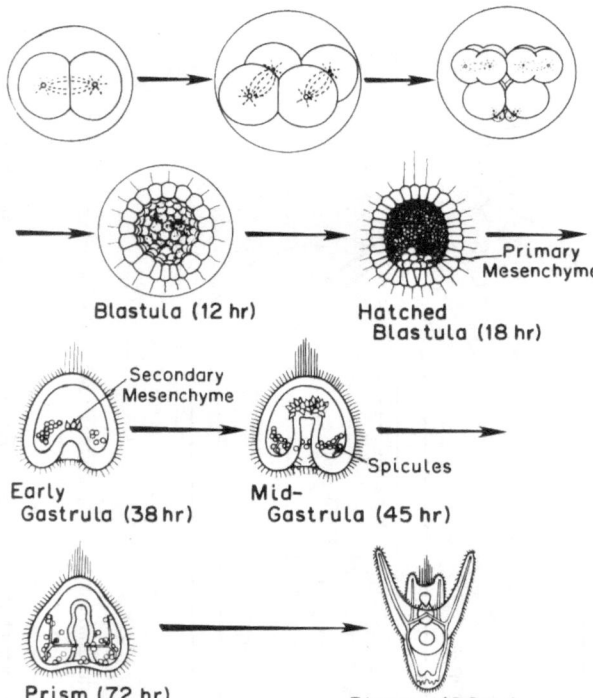

Figure 3. Key features of the early stages in development of sea urchin embryos. The times shown are for *Strongylocentrotus purpuratus.* (From Lennarz, 1983.)

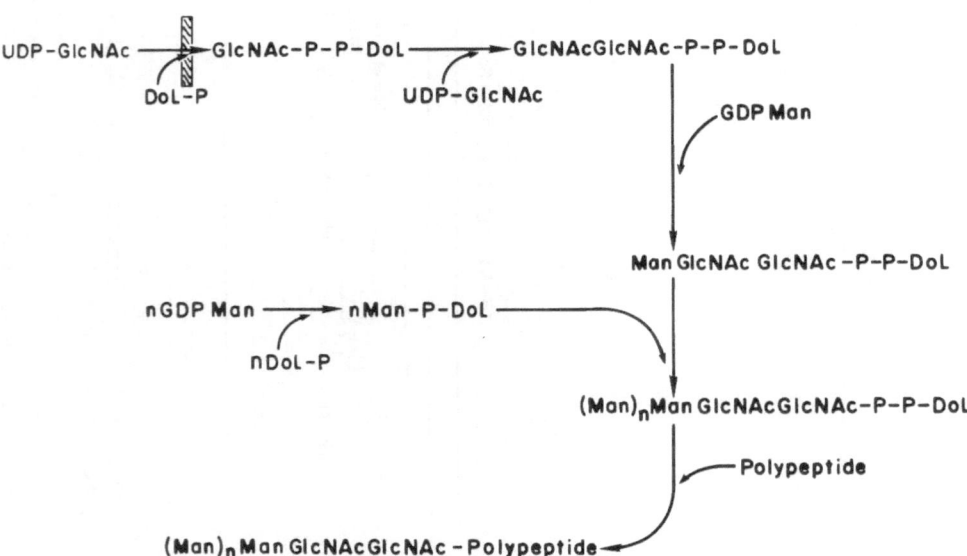

Figure 4. Tunicamycin blocks assembly of oligosaccharide chains by blocking transfer of the first N-acetylglucosamine unit to dolichylphosphate.

developmental consequences of this block might be. As shown in Figure 4, work in other systems (Struck and Lennarz, 1980) has established that tunicamycin blocks the very first step in the assembly pathway by inhibiting the first transferase. As a consequence, the oligosaccharide is not assembled on dolichylphosphate and polypeptide chains are not glycosylated. Initially in *Arbacia punctulata* (Schneider *et al.*, 1978) and then in *Strongylocentrotus purpuratus* (Heifetz and Lennarz, 1979) we carried out a series of simple experiments in which embryos were cultivated from an early stage in the presence of tunicamycin. Much to our surprise, neither fertilization nor development up to the blastula stage was affected by the presence of tunicamycin. Furthermore, the blastula stage embryo hatched and primary mesenchyme cells emerged in the blastocoel. However, at this point further development was arrested; the organism did not gastrulate. Suitable control experiments showed that this block to gastrulation was not a result of toxicity of the drug since the embryos respired normally and incorporated DNA and RNA precursors at levels comparable to control embryos.

On the basis of these observations, we postulated that development was blocked at the mesenchyme blastula stage, because glycoprotein synthesis was a requirement for gastrulation and this process could not occur in the presence of tunicamycin. To test this idea, we carried out *in vivo* labeling experiments.

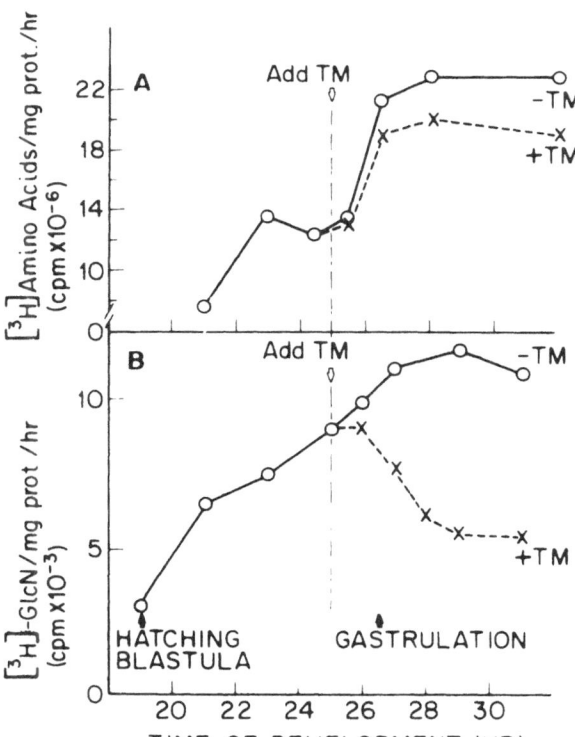

Figure 5. Tunicamycin (TM, added at the point indicated by the arrow) markedly decreases the rate of incorporation of glucosamine (bottom) into the N-linked oligosaccharide chains synthesized by the embryos. By contrast, this drug has little effect on the rate of amino acid incorporated in proteins (top). (From Heifetz and Lennarz, 1979.)

Glucosamine was added to cultures of embryos cultivated in artificial seawater, and incorporation of these labeled sugars into glycoproteins was measured. As shown in Figure 5, when tumicamycin was added, the rate of glucosamine incorporation into the glycoproteins of developing embryos was strikingly inhibited. By contrast, tunicamycin had only a very modest effect on the rate of amino acid incorporation into proteins of the embryo. As a more direct test of the idea that glycosylation was being blocked in sea urchin embryos by tunicamycin, we measured the immediate end product of the dolichol-linked pathway, oligosaccharide lipid. Figure 6 shows incorporation of both glucosamine and mannose into the oligosaccharide lipid of developing embryos to be dramatically inhibited by the addition of tunicamycin. From these results, we concluded that gastrulation requires glycoprotein synthesis. We began to wonder why earlier stages of development were not affected by tunicamycin. The answer became apparent when we did a more complete developmental study of glycoprotein synthesis, again using labeled sugars as precursors (Lennarz, 1983). As shown in Figure 7, there is only a very low rate of N-linked glycoprotein synthesis in the embryos until the mesenchyme blastula stage. Thus, the effect of tunicamycin is seen only at gastrulation because before this stage little glycoprotein synthesis occurs. At gastrulation there is a striking increase in N-

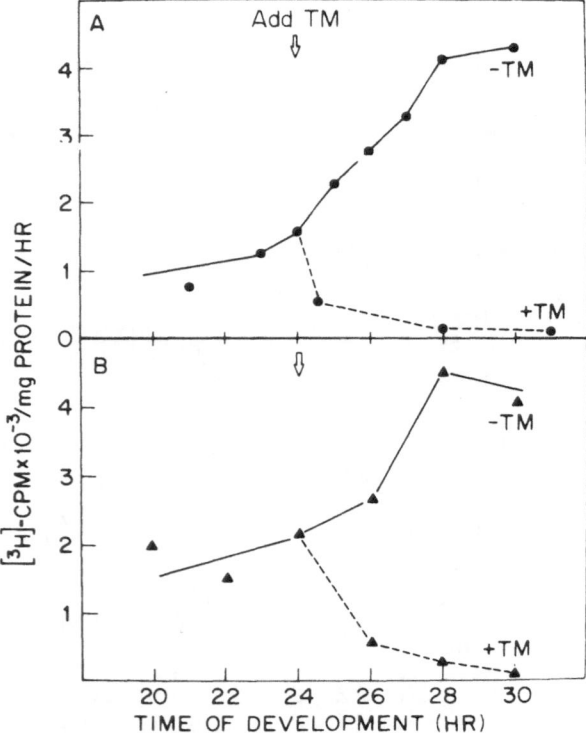

Figure 6. Tunicamycin (TM, added at the point indicated by the arrow) markedly decreases the rate of glucosamine (top) and mannose (bottom) incorporation into oligosaccharide lipid. (From Heifetz and Lennarz, 1979.)

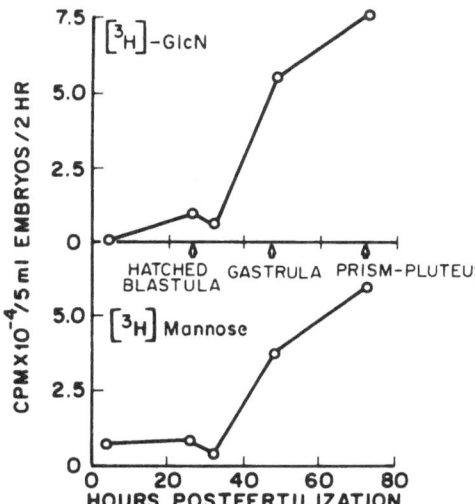

Figure 7. Rate of synthesis of polymannose type N-linked glycoproteins measured over the course of development by glucosamine (top) and mannose (bottom) incorporation into concanavalin A-bindable glycopeptides. (From Lennarz, 1983.)

linked glycoprotein synthesis, and this increase apparently is a prerequisite to this morphogenetic process.

3. Requirement of Dolichol Synthesis for Gastrulation

Unlike the embryos of higher organisms, the sea urchin embryo does not require any organic nutrients until the pluteus stage. That is, it develops normally in a simple medium containing only seawater and oxygen. In view of this, it became evident that the hypothesis that N-linked glycoprotein synthesis is essential for gastrulation could be tested by a completely different approach. The pathway of biosynthesis of dolichol shares a number of steps with the pathway for cholesterol and coenzyme Q biosynthesis (see Fig. 8). A potent inhibitor of this pathway is the drug compactin, which inhibits HMG CoA reductase. If, indeed, the formation of N-linked glycoprotein is necessary for gastrulation, and if the early embryo has little or no stored dolichylphosphate made during oogenesis, compactin, like tunicamycin, should arrest gastrulation. In fact, we found this to be the case—treatment of early embryos with compactin had no effect on development until the mesenchyme blastula stage, at which point, the embryo was arrested and normal gastrulation did not occur (Carson and Lennarz, 1979, 1981). However, because the site of action of com-

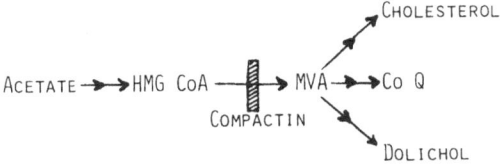

Figure 8. Site of action of compactin, an inhibitor of HmG CoA reductase. (From Lennarz 1983.)

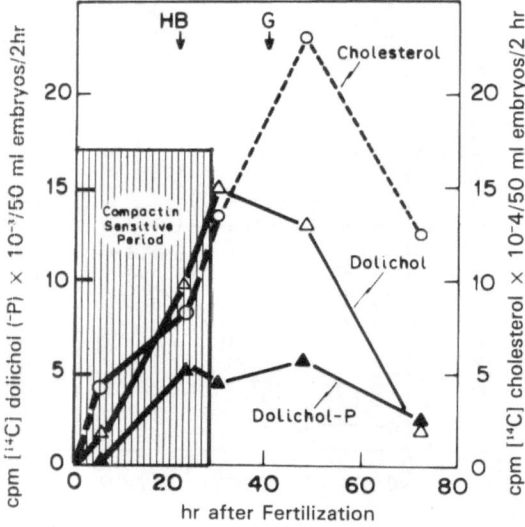

Figure 9. Rate of incorporation of labeled acetate into dolichol, dolichylphosphate, and cholesterol over the course of development. Embryos become insensitive to compactin at the mesenchyme blastula stage, as discussed in the text. (From Carson and Lennarz, 1981.)

pactin is early in the polyisoprenoid pathway, it was impossible to conclude that the defect in development was solely due to a deficiency in dolichol, since coenzyme Q synthesis and cholesterol synthesis would also be expected to be blocked by compactin. Therefore, we carried out labeling experiments in which embryos were cultivated in the presence of radioactive acetate in order to measure cholesterol, dolichol, and dolichylphosphate synthesis. A marked increase in the rate of incorporation of acetate into these three isoprenoid-derived lipids occurs immediately after fertilization (Fig. 9). Furthermore, in experiments in which compactin was added at various stages of development, the embryos became insensitive to the drug compactin shortly after hatching, just before gastrulation. While there are many possible interpretations of this observation, one explanation was that this insensitivity was a reflection of the fact that by that time in development the embryos had synthesized sufficient dolichylphosphate for protein glycosylation; the drug therefore no longer affected gastrulation when added at this late stage.

The effect of compactin on synthesis of both dolichol and cholesterol is shown in Table I. It is clear that under conditions in which compactin prevents 75% of the embryos from gastrulating, virtually complete inhibition of choles-

Table I. Compactin Inhibition of Both Sterol and Dolichol Synthesis

Addition	% Normal gastrula	cpm [^{14}C]acetate incorporated into:	
		Sterol	Dolichol
None	96	33,200 (100%)	3,073 (100%)
Compactin	25	456 (1.4%)	548 (18%)

Table II. Effect of Supplementation with Isoprenoid-Derived Lipids on the Development of Embryos Cultured in the Presence of Compactin

Supplements	Compactin (20 μM)	Normal gastrulae observed (%)
None	–	89–98
None	+	4–18
Cholesterol, coenzyme Q_{10}, Dolichol	+	75–89
Cholesterol, Coenzyme Q_{10}	+	2–12
Dolichol	+	61–70
Cholesterol, Dolichol	+	73
Coenzyme Q_{10}, Dolichol	+	75
Dolichol Phosphate	+	84

terol synthesis and about 80% inhibition of dolichol synthesis takes place. The question then arose of whether the lesion in development was due to a deficiency in the level of dolichol, cholesterol, or coenzyme Q, or in all three of these compounds. To answer this question, we carried out nutritional supplementation experiments. As shown in Table II, when embryos were cultivated in the presence of compactin, gastrulation was nearly totally inhibited. On the other hand, if not only compactin but also cholesterol, dolichol, and coenzyme Q were present in the cultivation medium, the gastrulation block was obviated. However, it is clear that not all three of these isoprenoid-derived lipids are required, because dolichol or, even better, dolichylphosphate added to the seawater restores the percentage of embryos that gastrulate to near-control values. On the basis of these observations, we conclude that although cholesterol synthesis is blocked by compactin, the embryo apparently has adequate stores of this lipid to develop normally in the absence of de novo synthesis. By contrast, the stores of dolichol apparently are inadequate, and a block in its de novo synthesis impedes development. The implication of these findings is that the deficiency in dolichol causes a deficiency in protein glycosylation. To test this idea, in vivo labeling experiments with mannose were carried out (see Table III). In this series of experiments, mannose incorporation into mannose

Table III. Effects of Compactin and Dolichol Supplementation on the Synthesis of Mannose-Labeled Glycoproteins and Lipids

Supplement to culture medium	Incorporation of [³H]mannose into:		
	Mannosyl-lipid (%)	Oligosaccharide-lipid (%)	N-linked glycoproteins (%)
None	100	100	100
Compactin	41–46	9–11	56
Compactin and dolichol	96–129	168–230	177

lipid, oligosaccharide lipid, and glycoproteins was measured in control embryos, in compactin-treated embryos, and in compactin-treated embryos supplemented with dolichol. It is clear from the results that compactin markedly
inhibits synthesis of the lipid-linked intermediates and glycoproteins and that
supplementation with dolichol obviates these blocks in the glycosylation
process.

4. Metabolic Interconversion of Dolichol and Dolichylphosphate during Development

The major product synthesized by the sea urchin embryo is dolichol, not
dolichylphosphate, the phosphorylated form that serves as the intermediate in
glycoprotein synthesis (see Fig. 9). Furthermore, we have found that although
the egg contains some dolichol, none of it is present in the phosphorylated form
(Carson and Lennarz, 1981). On the basis of these observations and the knowledge that other biological systems have a two-component enzyme system involved in interconverting dolichol and dolichylphosphate, i.e., a CTP-depen-
dent kinase and a phosphatase (Fig. 10), we investigated the possibility that
this pair of enzymes was important in development in the sea urchin embryo.
More specifically, the question we sought to answer was: Are these enzymes
present in the sea urchin embryo and, if so, are they regulated in this developing system in which there is a marked increase in glycosylation at gastrulation?
These studies, carried out in collaboration with Dr. C. J. Waechter and Dr.
Malka Scher, then at the University of Maryland, were initiated by assaying
cell-free membrane preparations from embryos at various stages of development for dolichol kinase activity (Rossignol et al., 1981).

The egg has detectable kinase activity and, upon fertilization, the level of
this enzyme progressively increases up to the mesenchyme blastula stage and
then begins to decline, reaching low levels by the time the organism has
reached the prism stage (see Fig. 11). The activation of dolichol kinase suggested the possibility that the principal pathway for formation of dolichylphosphate might be via phosphorylation of dolichol either synthesized after fertilization de novo or stored after synthesis during oogenesis. To investigate the
former possibility, in vivo labeling experiments were carried out using labeled
acetate and measuring formation of dolichylphosphate and dolichol (Rossignol
et al., 1983). These short-term labeling experiments revealed that the product

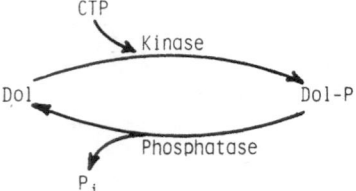

Figure 10. Dolichol kinase and dolichylphosphate phosphatase are two enzymes that interconvert dolichol and
dolichylphosphate.

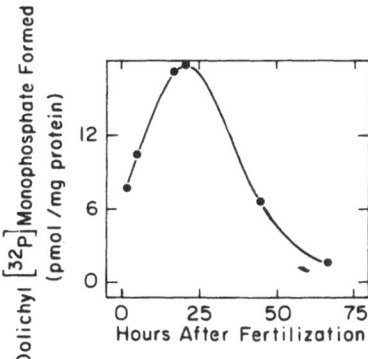

Figure 11. Dolichol kinase activity over the course of development. (From Rossignol *et al.*, 1981.)

first labeled from acetate was not dolichylphosphate but dolichol, even at the earliest times that the formation of these two compounds could be detected (Fig. 12).

In view of these findings, pulse-chase experiments were carried out in which the embryos were first labeled with acetate to permit formation of both labeled dolichol and dolichylphosphate. The embryos were then switched to unlabeled medium, and the potential precursor–product relationship between these two compounds was investigated. Dolichol undergoes turnover during the chase period, and the extent of its decrease roughly approximates the increase in the amount of dolichylphosphate (Fig. 13). The sum of the two compounds remains unchanged, indicating that there is little net breakdown of the isoprenoid backbone (Fig. 13, inset). From these results, it is clear that dolichol synthesized *de novo* serves as a precursor of dolichylphosphate. In other experiments we have also found that even after blockage of *de novo* synthesis of dolichol with compactin, dolichylphosphate can be formed (Rossignol *et al.*, 1983). These results indicate that the kinase acts not only on newly synthesized dolichol, but on stored dolichol as well.

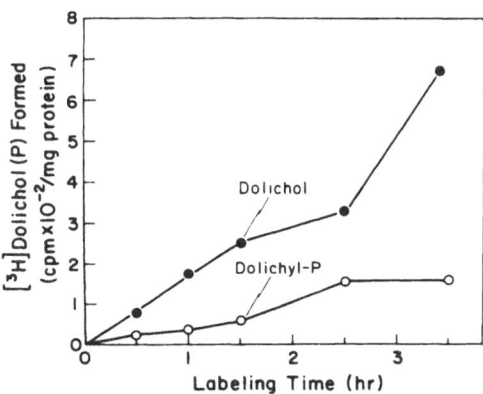

Figure 12. Kinetics of dolichylphosphate and dolichol synthesis from labeled acetate. (From Rossignol *et al.*, 1983.)

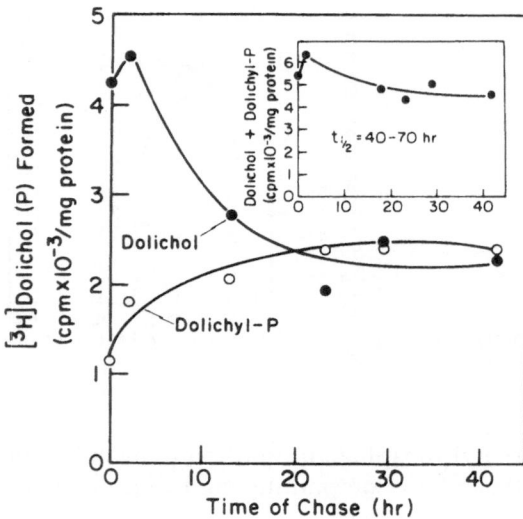

Figure 13. Precursor–product relationship between dolichol and dolichylphosphate. Both compounds were labeled with a pulse of labeled acetate and then chased at zero time by dilution of the labeled acetate. The inset indicates that there is little net breakdown of the isoprenoid backbone. (From Rossignol *et al.*, 1983.)

Next we studied the other process known to affect interconversion of dolichol and dolichylphosphate, i.e., dephosphorylation catalyzed by dolichylphosphate phosphatase. Initially, studies were carried out using cell-free preparations of embryos at various stages of development to measure *in vitro* the dolichylphosphate phosphatase activity. Unlike the kinase, this enzyme activity is low early in development and increases in the later stages (Fig. 14). Consistent with these *in vitro* findings on changes in the activity of dolichyl-

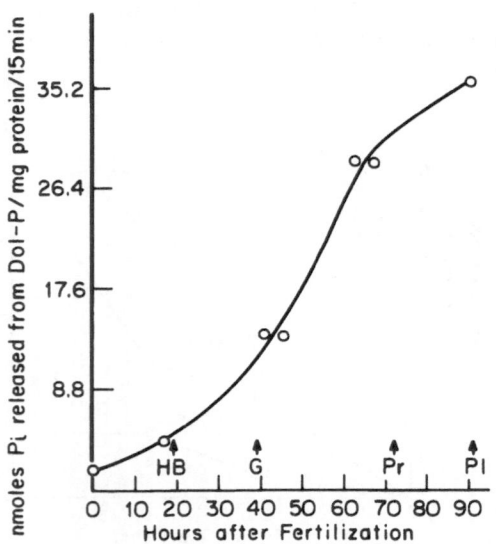

Figure 14. Dolichylphosphate phosphatase activity over the course of development.

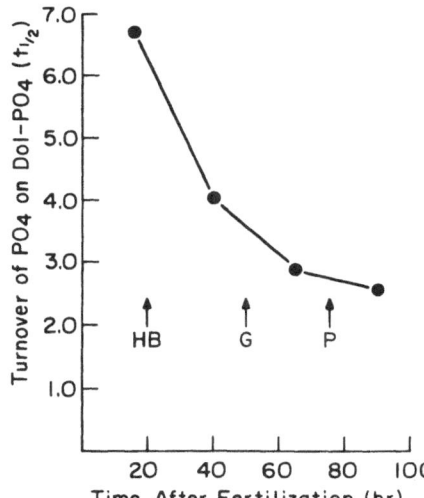

Figure 15. Turnover of the phosphate residue of dolichylphosphate over the course of development.

phosphate phosphatase, we have found that the half-life of .the phosphate group on dolichylphosphate changes markedly over the course of development. The phosphate group of dolichylphosphate was labeled *in vivo* by cultivating embryos in the presence of $^{32}P_i$ and then switching them to unlabeled medium. The results shown in Figure 15 indicate that during the subsequent chase period the half-life of the phosphate residue moiety decreases markedly as the embryo advances from hatching blastula to pluteus stage. Thus, it appears that these two enzymes, a kinase and a phosphatase, are differentially regulated over the course of development. As summarized in Figure 16, the kinase activity increases markedly before gastrulation, and then

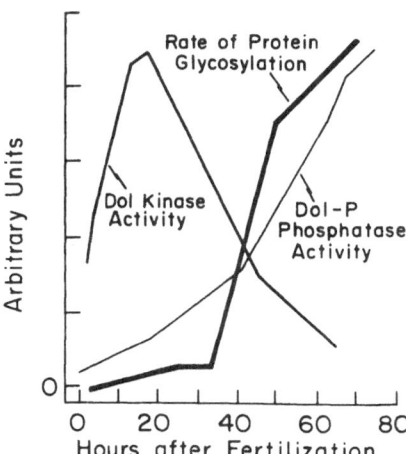

Figure 16. Relationship between dolichol kinase and dolichylphosphate phosphatase activities and glycoprotein synthesis over the course of development.

decreases as the phosphatase activity increases. All these findings point to the control of the level of dolichol and dolichylphosphate in developing embryos by two factors—the level of *de novo* synthesis and the relative activities of the two enzymes that interconvert the two compounds.

5. Regulation of the Synthesis of Glycosylatable Proteins

It has been known for some time that the asparagine residues in polypeptides that become glycosylated are always part of the sequence -Asn-X-Ser/Thr-, in which X is any amino acid except Pro (Struck and Lennarz, 1980). This is not to say that all such tripeptide sequences are glycosylated in proteins, but it is a structural requirement for their glycosylation. It became of interest to determine whether the onset of glycoprotein synthesis at gastrulation is regulated by the availability of new messages for glycosylatable proteins containing such tripeptide sequences. To investigate this, we used an *in vitro* translation/glycosylation system. We developed assays to detect the products encoded in the mRNAs for the glycoproteins, as shown schematically in Figure 17. After translation, Con A-agarose was used to separate the newly translated and glycosylated polypeptides from the large number of unglycosylated translation products. The separated glycoproteins were then analyzed by sodium dodecyl sulfate polyacrylamide electrophoresis (SDS-PAGE). A series of controls established that, using the procedure outlined, only glycoproteins were selected by Con A-agarose, and their synthesis was dependent on the presence of dog pancreas membranes and sea urchin mRNA.

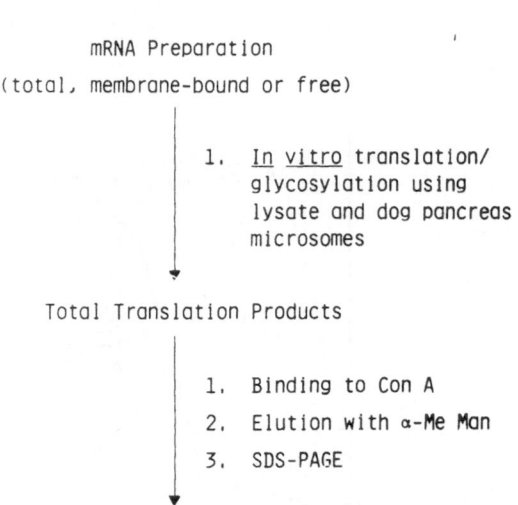

mRNA Preparation
(total, membrane-bound or free)

1. *In vitro* translation/
    glycosylation using
    lysate and dog pancreas
    microsomes

Total Translation Products

1. Binding to Con A
2. Elution with α-Me Man
3. SDS-PAGE

Glycosylated Translation Products

Figure 17. Protocol for detecting *in vitro* translation and glycosylation products encoded by sea urchin mRNA.

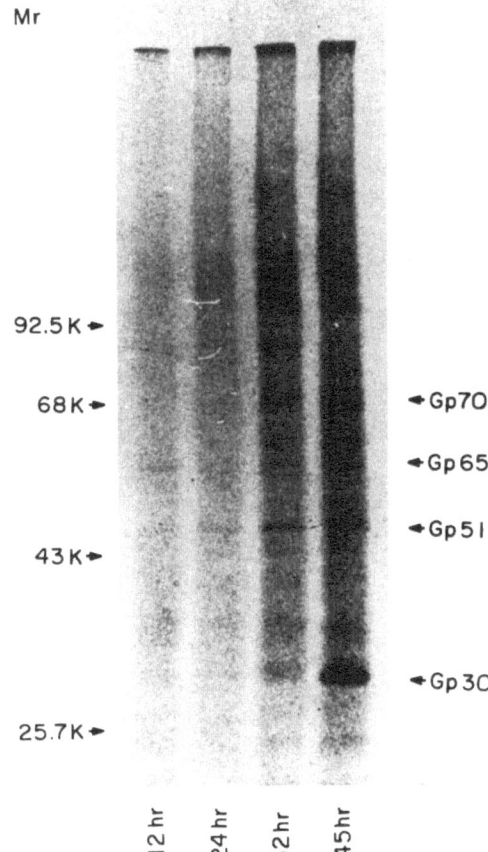

Figure 18. Sodium dodecyl sulfate polyacrylamide gel electrophoresis (SDS-PAGE) analysis of the glycoproteins synthesized from membrane-bound message over the course of development. (From J. Lau, M. Kelly, and W. J. Lennarz, 1983, unpublished observations.)

Using this system, we first asked whether the messages for glycoproteins were found on membrane-bound polysomes, since in all other systems studied this seems to be the case. When the distribution of messages for glycoproteins was studied in free and membrane-bound polysomes isolated from gastrula-stage embryos, only the membrane-bound polysomes were found to contain messages for glycoproteins (J. Lau, M. Kelly, and W. J. Lennarz, 1983, unpublished observations). Having established that membrane-bound polysomes do indeed contain messages for glycoproteins, we examined the distribution of these messages on membranes over the course of development. As shown in Figure 18, messages for four glycoproteins (Gp70, 65, 51, and 30) first appear in membrane-bound form at the mesenchyme blastula stage. At earlier stages of development (including the egg, which is not shown), there is little or no detectable membrane-bound message. We then asked whether these messages were present at all at early stages of development. We used two-dimensional gel electrophoresis and compared the products synthesized from membrane-bound message at the gastrula stage with those formed using total message in the egg. The four glycoprotein products encoded in the membrane-associated messages from gastrula stage embryos are readily detectable (Fig. 19B). However, when we examined the glycoprotein products of total mRNA from eggs, striking differences were seen (Fig. 19A). Although messages for two of the glycoproteins, Gp51 and Gp65, are present, messages for Gp30 and Gp70 are absent. Thus, it appears that in this developmental system there are two types of regulation. One is the apparent activation of so-called silent messages that are made during oogenesis; this appears to be the case with the messages for Gp51 and Gp65. The other type is new transcription or conversion of messages from a nontranslatable form to one that serves as a functional message; this mechanism seems to be operative for the messages for Gp30 and Gp70.

We have recently begun to investigate whether the new glycoprotein messages that are first expressed at gastrulation are localized to specific cell types of the sea urchin embryo (Lau *et al.*, 1983). To do this we adopted the procedure reported by Harkey and Whiteley (1980) for separating cell types as outlined in Figure 20. This procedure, which was first used for blastula-stage embryos, allows one to separate by centrifugation the released ectodermal cells from the mesodermal cells (in this case, primary mesenchyme cells) by virtue of their entrapment inside the sac formed by the basement membrane. Using gastrula-stage embryos, this same treatment yields similar separations, but here the presumptive ectodermal cells released from the basement membrane are separated from a mixture of the mesodermal cells and the endodermal cells that are part of the developing gut tube (Fig. 20). The results shown in Figure 21 were obtained when the glycoproteins encoded by total RNA isolated from the blastula-stage presumptive ectodermal cells (Fig. 21A) were compared with the

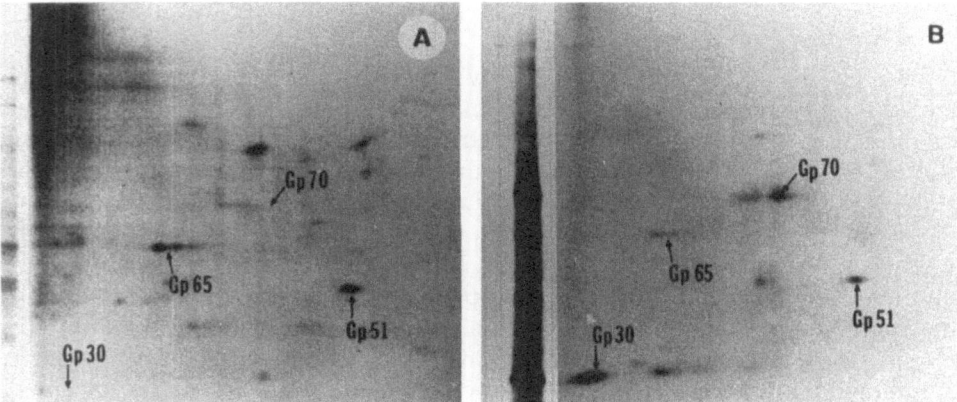

Figure 19. Two-dimensional gel analysis of the glycoproteins synthesized from membrane-bound message isolated from gastrula- stage embryos (B) and from total message isolated from eggs (A). (From J. Lau, M. Kelly, and W. J. Lennarz, 1983, unpublished observations.)

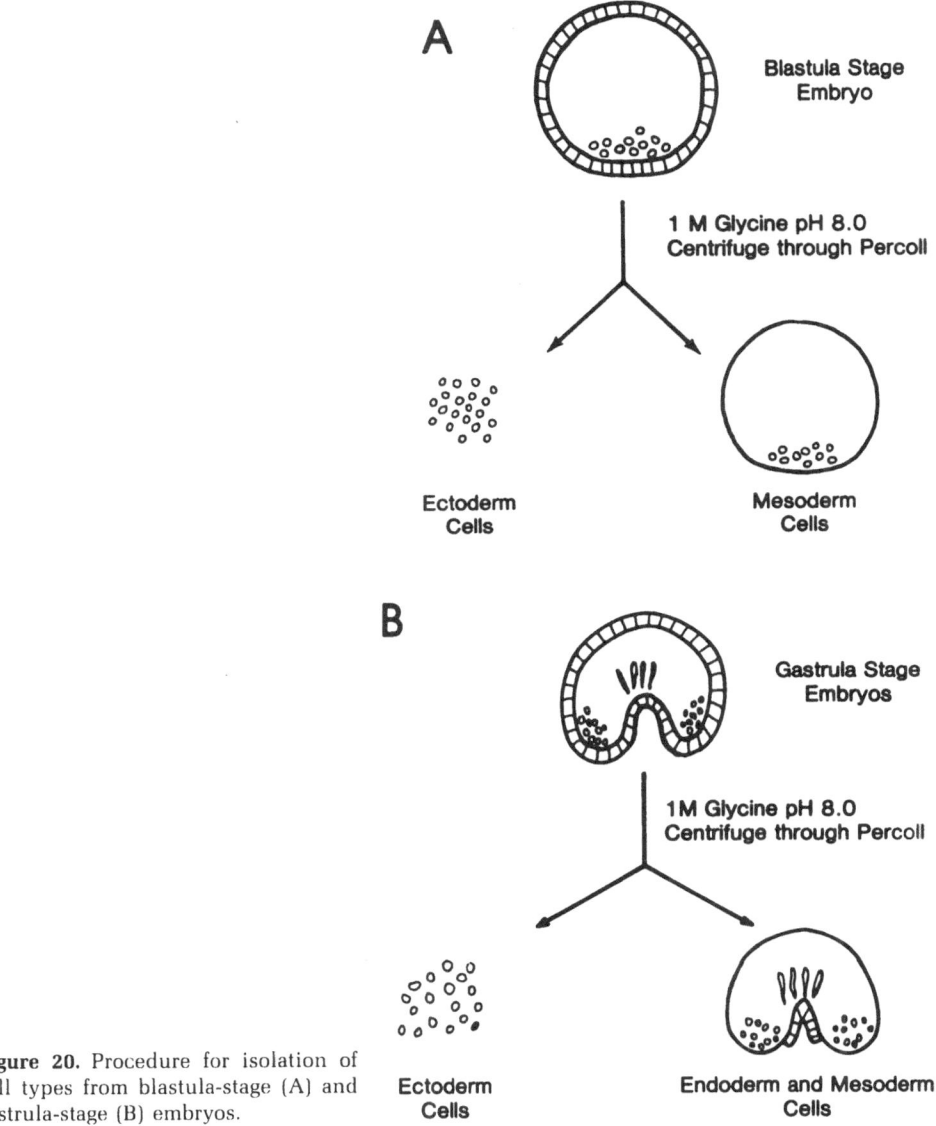

Figure 20. Procedure for isolation of cell types from blastula-stage (A) and gastrula-stage (B) embryos.

glycoproteins encoded by mRNA from blastula stage mesodermal cells (Fig. 21B). It is apparent that there are only a few differences between the two cell types in terms of messages for glycoproteins. There may be slight enrichment in the mesodermal cell preparation of the messages for a group of glycoproteins designated GpA, but the differences are rather minimal.

The situation was very different when we carried out similar experiments with cell types separated at the gastrula stage. There is differential enrichment of glycoprotein messages in ectodermal cells (Fig. 22A), as compared with the

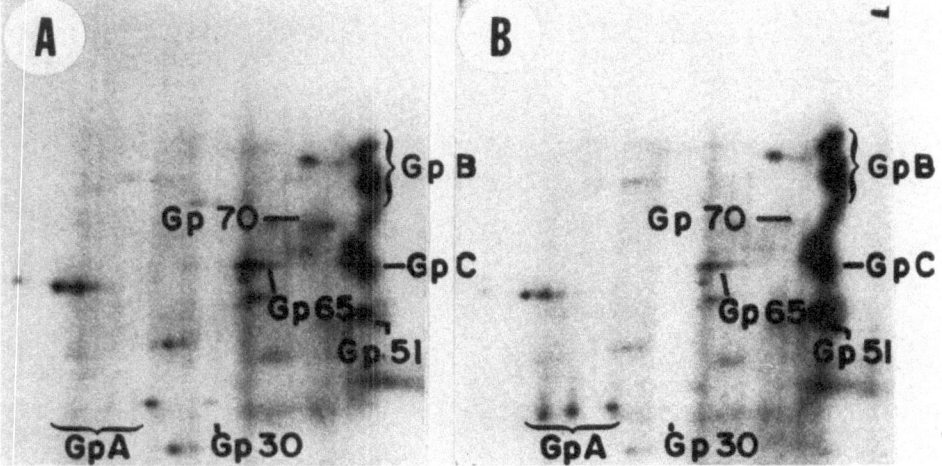

Figure 21. Two-dimensional gel analysis of the glycoproteins synthesized from total mRNA isolated from presumptive ectodermal cells (A) and mesodermal cells (B) isolated from blastula stage embryos.

messages from the mixture of mesodermal and endodermal cells (Fig. 22B). GpA is virtually absent from the ectodermal cell preparation, and only low levels of Gp30 and Gp51 are detected. By contrast, the message for Gp51 is highly enriched in the mesodermal/endodermal cell preparation.

It is clear, then, that the differentiation that occurs at gastrulation results in expression of new messages, and some of these are localized to specific cell

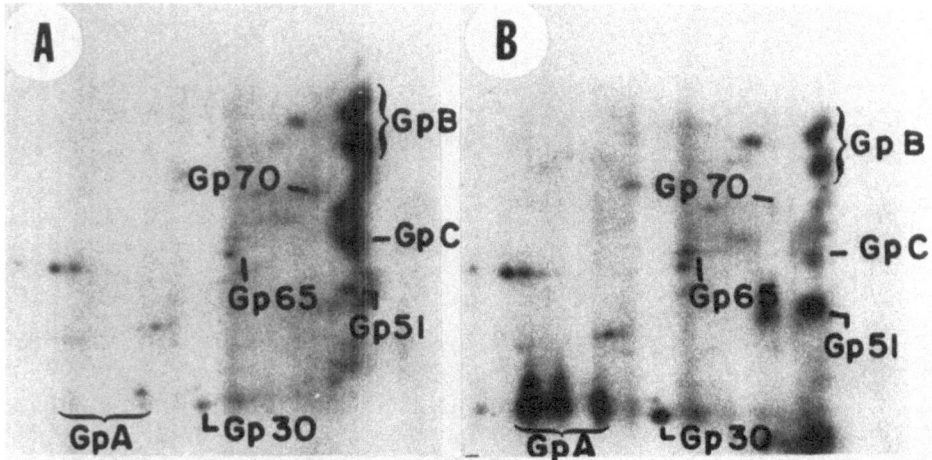

Figure 22. Two-dimensional gel analysis of the glycoproteins synthesized from total mRNA isolated from presumptive ectodermal cells (A) and mesodermal cells (B) isolated from gastrula-stage embryos.

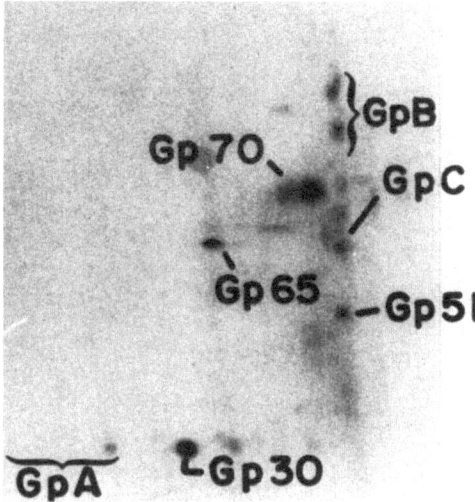

Figure 23. Two-dimensional gel analysis of the glycoprotein products encoded in membrane-bound mRNA isolated from gastrula-stage embryos.

types. Finally, it should be pointed out that in these two experiments we examined the products synthesized using total mRNA. When one examines, instead, the glycoprotein synthesized by gastrula-stage membrane-bound message, the results shown in Figure 23 are obtained. Comparison of these with the results of analysis of the glycoprotein products of total message (see Fig. 21) reveals that the messages for GpA, for another group of proteins known as GpB, and for GpC are almost absent. Thus, these glycoprotein messages are not in membrane-bound form at the gastrula stage and are presumably stored messages to be used at some later stage of development.

6. The Role of Glycoproteins in the Gastrulation Process

Finally, it is of interest to consider the question of the function of N-linked glycoproteins at gastrulation. Studies from a number of laboratories have shown that secondary mesenchyme cells at the tip of the advancing gut tube send out long pseudopodia that interact with the blastocoel wall and the endodermal cells of the developing gut tube (Akasaka *et al.*, 1980). These interactions are in a dynamic state, and it seems possible that they serve to direct or actually pull the gut tube to the inner wall of the blastocoel. Our hypothesis is that N-linked glycoproteins are somehow involved in this cell–cell recognition process, perhaps by interacting with a complementary binding protein. We plan to test this hypothesis by preparing antibodies to the N-linked glycoproteins first made at gastrulation. In addition, we will prepare glycopeptides derived from these glycoproteins by Pronase digestion. As outlined in Figure

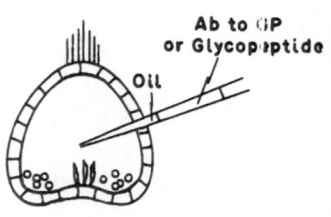

**Ab to GP
or Glycopeptide**

Oil

Figure 24. Schematic diagram postulating a role for glycoproteins (GP) and binding proteins (BP) in cell-recognition events in gastrulation. See text for further details.

24, the expectation would be that if the gastrulation process involves binding protein–glycoprotein interactions, microinjection of antibodies to the glycoproteins would block gastrulation. A similar effect would be expected upon microinjection of glycopeptides, which should act as ligands for the putative binding proteins. If our hypothesis proves correct, we should then be in a position to localize more precisely the glycoproteins to the cell surface of specific cell types by the use of tagged antibodies. In addition, it should be possible to use the glycoproteins as tools to isolate the complementary binding protein. Antibodies to these binding proteins should enable us to determine their possible localization to specific cell types.

7. Conclusions

The developing sea urchin embryo has proved an excellent model for studying the regulation of glycoprotein synthesis. The results obtained thus far indicate that the turn-on of glycoprotein synthesis at gastrulation is preceded by a number of other enzyme-mediated processes.

1. Synthesis of dolichol and dolichylphosphate is initiated at fertilization. This biosynthetic process continuously increases in rate until just before gastrulation, when the rate becomes constant and then declines.
2. Immediately after fertilization, the activity of dolichol kinase, which is present in the egg, increases at least twofold, peaks at mesenchyme blastula, and then progressively declines over the course of further development.
3. By contrast, dolichylphosphate phosphatase is not present in the egg but increases gradually over the course of development.
4. The messages for glycoprotein are first found in membrane-bound, translatable form just before gastrulation. However, regulation of these messages appears to be quite different, depending on the particular

glycoprotein, since some are present in the egg in silent form, while others are newly transcribed at the time of gastrulation.

5. These messages appear to be uniformly distributed in the two separable cell types obtained from blastula-stage embryos. However, by the gastrula stage, when the messages are first expressed, segregation of some of them either to the ectoderm or to the mesoderm/endoderm is observed.

Despite the good progress that has been made in understanding the regulation of glycoprotein synthesis in the developing sea urchin embryo, we still have little understanding of glycoprotein function in gastrulation. It is hoped that future studies will provide an answer to this interesting and important question.

ACKNOWLEDGMENTS. The author gratefully acknowledges the essential contribution of the following coworkers to these studies: Dr. E. Gayle Schneider, Dr. A. Heifetz, Dr. D. Carson, Dr. D. Rossignol, and Dr. J. Lau. This work was supported by a grant from the National Institutes of Health (HD 18600). Dr. William J. Lennarz, who is a Robert A. Welch Professor of Chemistry, gratefully acknowledges the Robert A. Welch Foundation. The assistance of Ms. Diana Welch with editorial work and manuscript preparation is gratefully acknowledged.

References

Akasaka, K., Amemiya, S., and Terayama, H., 1980, Scanning electron microscopical study of the inside of sea urchin embryos (Pseudocentrotus depressus), Exp. Cell Res. 129:1–13.

Carson, D. D., and Lennarz, W. J., 1979, Inhibition of polyisoprenoid and glycoprotein biosynthesis causes abnormal embryonic development, Proc. Natl. Acad. Sci. USA 76:5709–5713.

Carson, D. D., and Lennarz, W. J., 1981, Relationship of dolichol synthesis to glycoprotein synthesis during embryonic development, J. Biol. Chem. 256:4679–4686.

Hanover, J. A., and Lennarz, W. J., 1981, Transmembrane assembly of membrane and secretory glycoproteins, Arch. Biochem. Biophysics 211(1):1–19.

Harkey, M. A., and Whiteley, A. H., 1980, Isolation, culture, and differentiation of echinoid primary mesenchyme cells, Wilhelm Roux Arch. Dev. Biol. 189:111–122.

Heifetz, A., and Lennarz, W. J., 1979, Biosynthesis of N-glycosidically linked glycoproteins during gastrulation of sea urchin embryos, J. Biol. Chem. 254:6119–6127.

Lennarz, W. J., 1983, Glycoprotein synthesis and embryonic development, in: Critical Reviews of Biochemistry, Vol. 14, No. 4 (G. Fasman, ed.), pp. 257–272, CRC Press, Boca Raton, Florida.

Rossignol, D. P., Lennarz, W. J., and Waechter, C. J., 1981, Induction of phosphorylation of dolichol during embryonic development of the sea urchin, J. Biol. Chem. 256:10538–10542.

Rossignol, D. P., Scher, M., Waechter, C. J., and Lennarz, W. J., 1983, Metabolic interconversion of dolichol and dolichylphosphate during development of the sea urchin, J. Biol. Chem. 258:9122–9127.

Schneider, E. G., Nguyen, H. T., and Lennarz, W. J., 1978, The effect of tunicamycin, an inhibitor of protein glycosylation, on embryonic development in the sea urchin, J. Biol. Chem. 253:2348–2355.

Struck, D. K., and Lennarz, W. J., 1980, The role of saccharide lipids in glycoprotein synthesis, in: The Biochemistry of Glycoproteins and Proteoglycans (W. J. Lennarz, ed.), pp. 35–83, Plenum Press, New York.

Chapter 5

The Morphological Oncogenic Signature

Reorganization of Epithelial Cytoarchitecture and Metabolic Regulation by Tumor Promoters and by Transformation

EDWARD G. FEY and SHELDON PENMAN

1. Introduction

An often underappreciated function of metazoan cells is the construction of complex tissue with its attendant elaborate architecture. We propose that the information for architectural expression is mediated by the skeletal networks of each cell. These networks establish cell morphology and phenotype and may also serve to transduce morphological signals regulating gene expression. In this view, the transduction of cell shape and surface contact signals is central to the establishment of differentiated cell organization and function. We further postulate that alterations in cytoskeletal organization may result in the breakdown of controlled gene expression in differentiated cells and tissues. In this study we present evidence suggesting that the tumor promoter 12-0-tetradecanoyl phorbol 13-acetate (TPA) induces a specific alteration in the morphology and cytoskeletal organization of differentiated epithelial cell colonies. This morphological "signature" closely resembles the phenotype expressed by permanently transformed cells. These results suggest that such an alteration of cytoarchitecture may be an obligatory step in the establishment of neoplastic growth. Certainly gross changes in cellular organization are well-established diagnostic markers for neoplasia (Weil, 1978) and serve as the basis of pathological examination.

EDWARD G. FEY and SHELDON PENMAN • Department of Biology, Massachusetts Institute of Technology, Cambridge, Massachusetts 02139.

Study of the function of the cellular skeleton networks has been severely hampered by the inability of conventional electron microscopy and conventional fractionation to visualize and characterize internal cell structures. These impediments have largely been surmounted by the advent of embedment-free techniques for electron microscopy that, when coupled to appropriate sequential fractionation, yield cell structures that are clearly characterized both biochemically and morphologically (Capco *et al.*, 1984).

Our own interest in developing and employing these new techniques has been directed toward understanding the genesis of multicellular organization—the *sine qua non* of metazoan cell biology. Our study of carcinogenesis stemmed from the use of neoplastic transformation to perturb normal cell organization and function. Rather unexpectedly, we have been led to conclude that there exists a morphological and biochemical phenotype characteristic of oncogenic phenomena. The widespread observation of this phenotype (within the bounds outlined here) has led us to propose that this morphological signature is common to carcinogen response and permanent transformation.

The two-stage model of carcinogenesis involves separate initiation and promotion steps. Initiation is thought to be a permanent genetic alteration, while promotion is a reversible event that must occur repeatedly before tumorigenesis occurs (Boutwell, 1977; Weinstein, 1981). One of the most potent tumor promoters on mouse skin is TPA (Hecker, 1978). Although TPA alters a number of biochemical and morphological characteristics of cells in culture (Weinstein *et al.*, 1977; Colburn *et al.*, 1979; Fischer *et al.*, 1979; Laskin *et al.*, 1981), it has been difficult to establish which of these effects is relevant to tumor promotion *in vivo*.

In almost every study in which cell shape was examined, TPA has been shown to alter cell morphology significantly. Fibroblasts display a marked morphological response to TPA (Sivak and Van Duuren, 1967; Driedger and Blumberg, 1980; Rifkin *et al.*, 1979). This response seems to be similar to the shape change induced by viral transformation (Driedger and Blumberg, 1980). TPA also enhances the morphological changes induced by viral transformation (Bissell *et al.*, 1979). In these reports, fibroblast cells treated with TPA exhibit a rapid loss of orientation characterized by extensive overlapping and the development of many long, thin processes. These morphological changes reflect alterations of the cytoskeleton (Rifkin *et al.*, 1979).

Nonfibroblastic cells, particularly those with extensive intermediate filament networks, display a more pronounced shape change in response to TPA. When myotubes are treated with TPA, there is a prompt breakdown of myofibrils, paralleled by the appearance of dense regions of 10 nm-filament bundles (Croop *et al.*, 1980). Bundles of 10-nm filaments are also the predominant feature in macrophages treated with TPA (Phaire-Washington *et al.*, 1980).

Differentiated epithelia are particularly sensitive to morphological alterations induced by TPA. Epidermal keratinocytes display a breakdown of epithelial organization and develop a fibroblastic morphology with many filamentous processes (Parkinson and Emmerson, 1982; Hennings *et al.*, 1982). Gross morphological changes induced by TPA are particularly apparent in the well differentiated Madin Darby canine kidney (MDCK) distal kidney tubule cell line (Ohuchi and Levine, 1980; Ojakian, 1981). These studies have shown low concentrations of TPA to induce extensive changes in MDCK morphology concomitant with a breakdown of tight junctional complexes and loss of the differentiated property of vectorial ion transport characteristic of this cell type.

Aberrations in cell morphology are also characteristic of most tumor cell types (Robbins and Angell, 1971). Work from this laboratory has suggested that alterations in cell shape and the underlying cytoskeletal elements correspond and may be causally related to the degree of loss of anchorage dependence characteristic of tumorigenic cells (Wittlesberger et al., 1981). More recently we have shown that the alterations in cytoarchitecture induced by TPA are also induced by all tumor promoters studied so far (Fey and Penman, 1984).

In this study we examine alterations of the cytoskeletal framework seen in whole mounts of detergent-extracted epithelial cells and some concomitant biochemical changes in regulating mechanisms induced by nanomolar levels of the phorbol ester TPA. The MDCK cell line was chosen for these studies because of its high degree of differentiation, closely resembling "normal" kidney distal tubule cells (Taub et al., 1981), its remarkably stable karyotype, and its ability to form differentiated epithelial colonies that are nontumorigenic when injected into nude mice (McRoberts et al., 1981).

Utilizing the highly selective sequential fractionation procedure described by Fey et al. (1984), we examine the progression of morphological changes caused by TPA on both the cytoskeletal framework and on the further purified nuclear matrix-intermediate filament (NM-IF) core structure. The response of protein synthesis in these cells to an enforced change in shape, such as induced by suspension culture, suggests that TPA causes a partial and temporary release from anchorage dependence. Alterations in cytoskeletal architecture similar to those induced by TPA and other tumor promoters are manifested in MDCK cells that have been permanently transformed by chemical carcinogens. These observations lead naturally to a consideration of the possible role of alterations in cell organization in tumorigenic progression and to postulating a morphological signature associated with oncogenic phenomena.

2. Promoter-Induced Alterations of Epithelial Morphology: Scanning Electron Microscopy

Fundamental to understanding cytoskeletal organization is the use of embedment-free electron microscopy, in this case detergent-extracted cell whole mounts. In order to observe the architectural elements, soluble proteins and lipids are removed with Triton X-100 in a physiological buffer. The extraction with Triton X-100 does not itself effect changes in gross cell morphology. Figure 1 shows scanning electron micrographs (SEM) of the apical surfaces of MDCK cell colonies 2 hr after treatment with 5 ng/ml TPA (Fig. 1b). Control MDCK colonies with normal apical morphology are shown in Figure 1a. The retention of many characteristics of apical morphology in control MDCK colo-

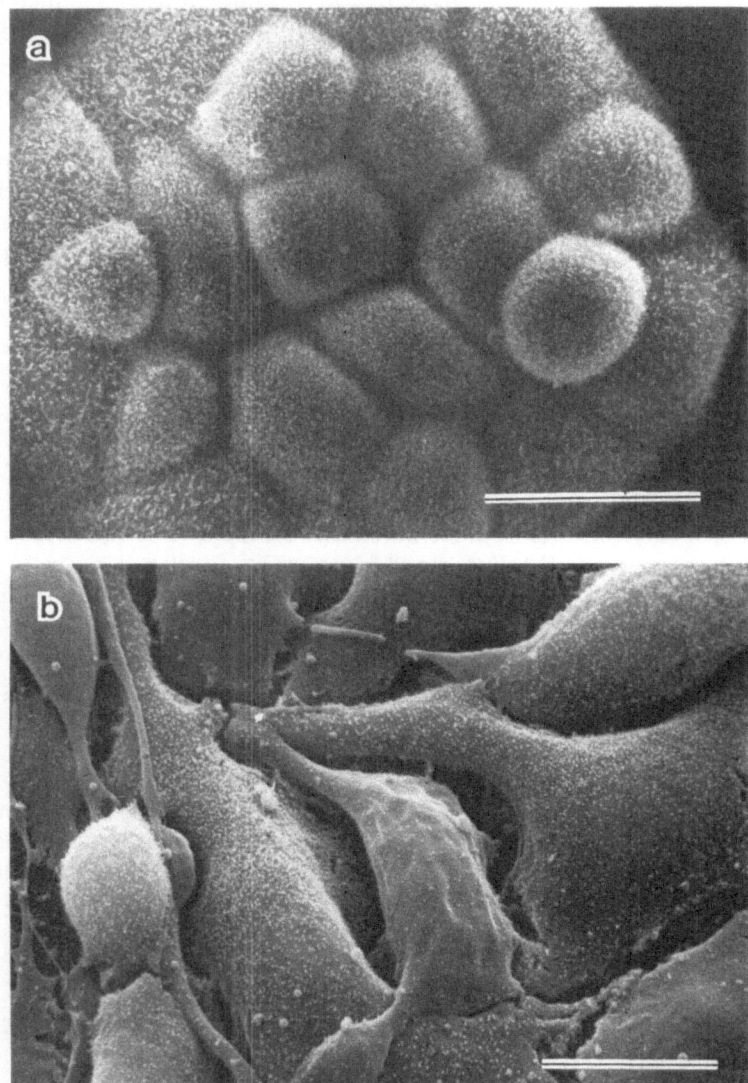

Figure 1. Scanning electron micrographs of control and TPA-treated MDCK colonies. Intact MDCK colonies were incubated with 5 ng/ml TPA for 2 hr and prepared as described in experimental procedures. Control (a) and TPA-treated whole cells (b) are compared, showing the rapid break-down of epithelial colony organization induced by TPA, as observed by Ojakian (1981). Higher magnification of the apical surfaces of whole cells (c) or of cells after extraction with 0.5% Triton X-100 in a physiological buffer (Fey et al., 1984) (d) 2 hr after incubation in 5 ng/ml TPA shows retention of overall cell shape by the lipid-depleted skeletal framework. Scale bar: (a,b) 10 μm; (c,d) 1.0 μm.

nies after extraction with Triton X-100 has been previously shown (Fey *et al.*, 1984). A comparison of detergent-extracted and unextracted MDCK colonies (Fig. 1c,d) shows that the removal of both phospholipid and soluble proteins has little effect on the aberrant morphology induced in these cells by TPA. The highly structured and polygonally organized normal MDCK cells (Fig. 1a) have become flattened and fibroblastic in appearance (Fig. 1b). The formerly con-

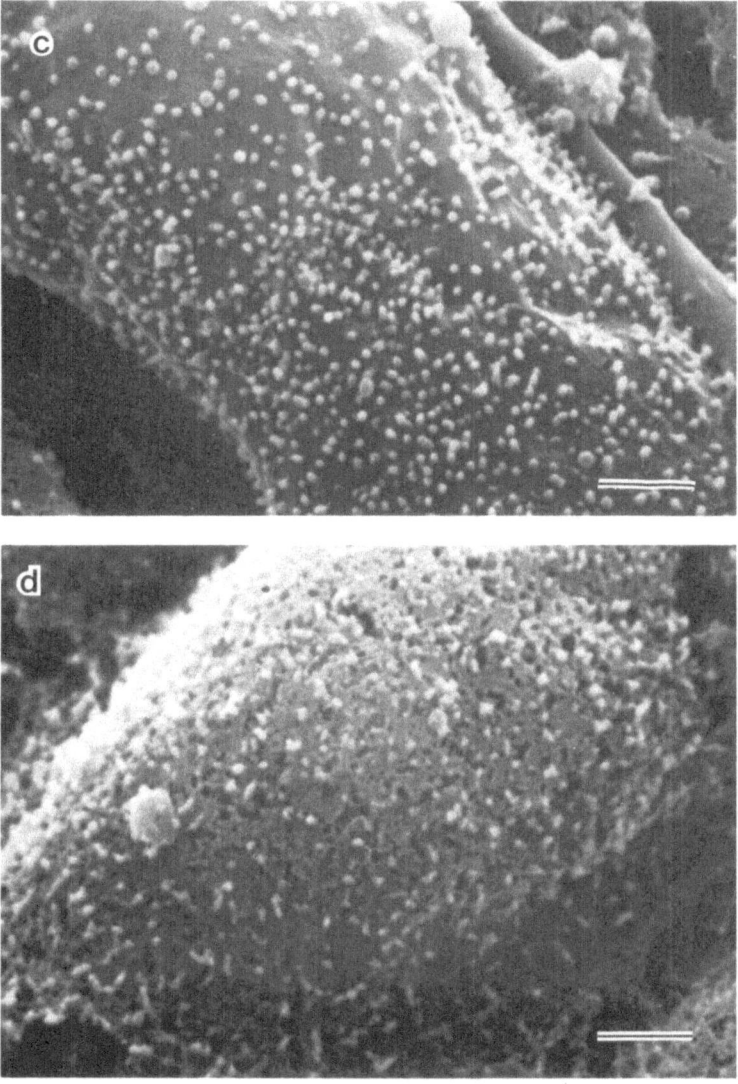

Figure 1. (*Continued*)

tiguous cells are now well separated and connected only by long processes. When the apical surface is viewed at a higher magnification (Fig. 1c,d), the lamina is seen to be a flat surface studded with relatively few microvilli that are well separated, and considerably shortened. In contrast to normal cells, the plasma membrane morphology of the TPA-treated cells is heterogeneous, displaying widely varying densities of microvilli and a diversity of surface curvatures. These characteristics of the plasma membrane surface morphology are retained in the plasma lamina that remains after extraction of membrane phospholipid and soluble protein with Triton X-100. It is possible that many of the surface-related phenomena resulting from exposure to TPA, such as release of plasminogen activator (Wigler and Weinstein, 1976) or the reduction of EGF binding (Lee and Weinstein, 1979) may actually stem from the primary reorganization of the plasma lamina.

3. Cytoskeletal Reorganization in Response to TPA

The cytoskeletal elements are visualized by transmission electron microscopic (TEM) evaluation of detergent-extracted cell whole mounts of MDCK colonies. The cytoskeletal alterations underlying the gross morphological changes induced by TPA are clearly seen by this technique, as compared with their obscure, at best, visualization in the conventional embedded thin section. After 2 hr in TPA, the MDCK cytoarchitecture, shown in Figure 2, exhibits the characteristic flattening and enlongation of cells concomitant with the formation of filamentous bundles. The intracellular associations, characteristic of control MDCK colonies (Fig. 2a), are broken down (Fig. 2b,c). The formerly uniform density of cytoskeletal fibers has been significantly altered, leaving large regions of the epithelium depleted of fibers, while other adjacent regions contain numerous densely packed fibers arranged in parallel bundles (Fig. 3b,c).

4. Morphological Alterations Retained in the NM-IF Scaffold

The profound effects of TPA on the nuclear matrix-intermediate filament scaffold can be clearly seen in the chromatin-depleted preparations of MDCK colonies. It has been previously shown that the high salt/nuclease treatment

Figure 2. Transmission electron micrographs of MDCK skeletal frameworks 2 hr after exposure to 5 ng/ml TPA. MDCK cells were grown on gold grids, treated with TPA, extracted with Triton X-100, and prepared as described by Fey et al. (1984). Control skeletal frameworks (a) show a uniform distribution of skeletal elements throughout the cytoplasm with well-defined cell borders (arrow). After 2 hr in 5 ng/ml TPA (b) many regions of the cytoplasm are depleted of filaments and dense, curved regions of filamentous bundles appear, often in association with nuclei. This is the characteristic morphological signature of tumor-promoting agents. A higher magnification of filament distribution in TPA-treated cells (c) details the asymmetrical distribution of cytoplasmic filaments. Scale bars: 1.0 μm.

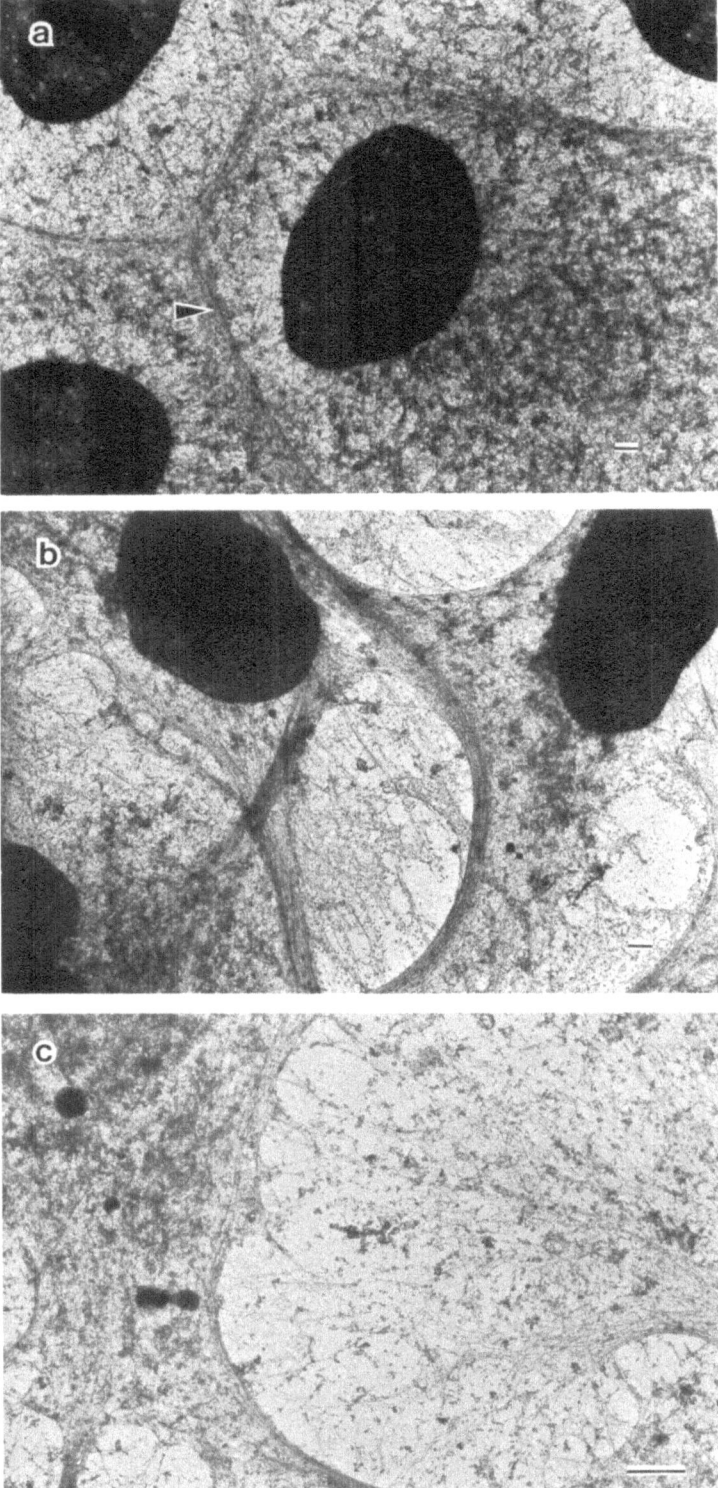

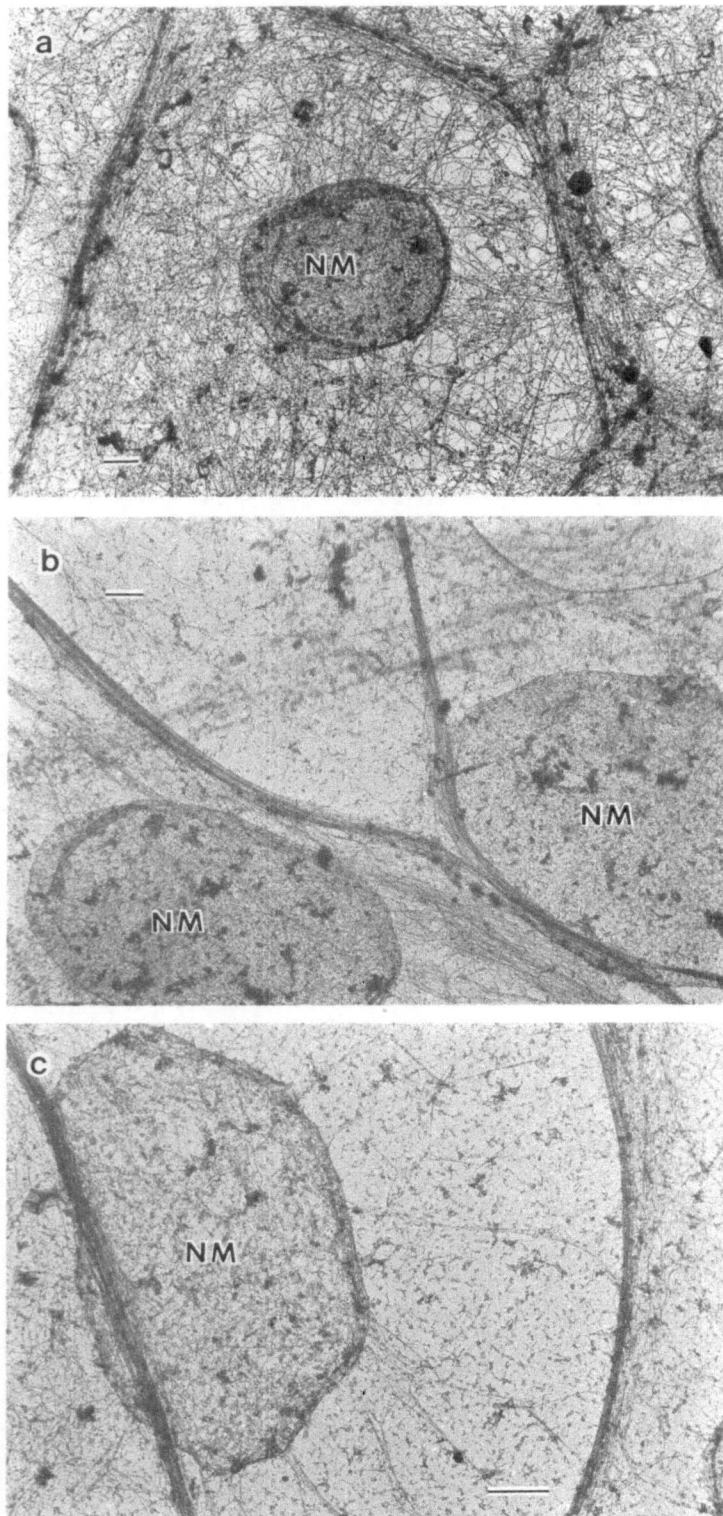

reveals the NM-IF structure, which retains many aspects of epithelial morphology, including residual cell junctions containing desmosomal proteins (Fey et al., 1984). In NM-IF preparations of control MDCK colonies, filaments are observed in a generally radial distribution with many filaments forming direct connections between the nuclear matrix and the intercellular junctions (Fig. 3a). When the nuclear matrix is prepared from cells previously exposed to TPA for 2 hr, the structural association between the cytoplasmic filaments and the matrix network is highly distorted (Fig. 3b,c). The filamentous bundles, seen in the complete cytoskeleton, are retained in this cytokeratin-rich fraction, and are observed to form direct connections with the nuclear matrix. Thus, they appear to be derived from a collapse or rearrangement of the normal cytokeratin network. The distortion of nuclear morphology observed in Figure 2 is faithfully retained by the nuclear matrix preparation of the same cells (Fig. 3b,c). In these cells the nuclear matrix appears stretched by large bundles of filaments which are oriented along the axis of nuclear elongation. As in normal cells, the protein of these structures makes up only 5% of the total cellular protein.

5. Immunofluorescent Localization of Cytokeratin and Desmosomal Proteins in NM-IF Scaffolds after TPA

The localization of specific proteins in the NM-IF scaffold of MDCK cell colonies and their rearrangement after exposure to promoters was determined using immunofluorescence microscopy. Photomicrographic images of untreated MDCK colonies and of colonies from identical cultures after exposure for 2 hr to 5 ng/ml TPA are shown in Figures 4a,c,e, and 4b,d,f, respectively. The distribution of total protein in these preparations was determined using fluorescein isothiocyanate (FITC) as a reagent for a general protein stain (Fig. 4a,b). The NM-IF scaffolds from a control MDCK colony show the characteristic epithelial polygonal organization of cell borders and nuclear matrices localized in the center of each cell. Exposure to TPA (Fig. 4b) produces rapid and profound changes in colony organization. The geometric cell organization is lost, and dense cytoplasmic processes are formed. These processes are often observed in close association with the nuclear matrix. The immunofluorescent images of control and TPA-treated NM-IF scaffolds stained with antibody to cytokeratin show these processes to be rich in cytokeratins (Fig. 4c,d). By contrast, the cytokeratin distribution in control MDCK colonies is somewhat diffuse and is formed of intricate networks of filaments concentrated near the cell borders (Fig. 4c). After exposure to TPA, the keratin staining shows pat-

Figure 3. Transmission electron micrograph of MDCK cores after exposure to 5 ng/ml TPA. Chromatin-depleted NM-IF preparations were made of control (a) and TPA-treated MDCK cells as described by Fey et al. (1984). The control nuclear matrix (a) (NM) is connected to a radial pattern of fibers that terminates in well-defined cell boundaries. After a 2-hr incubation in 5 ng/ml TPA (b), the appearance of filament bundles and some nuclear distortion are evident. These characteristics are intensified after 24-hr incubation in TPA (c). Scale bars: 1.0 μm.

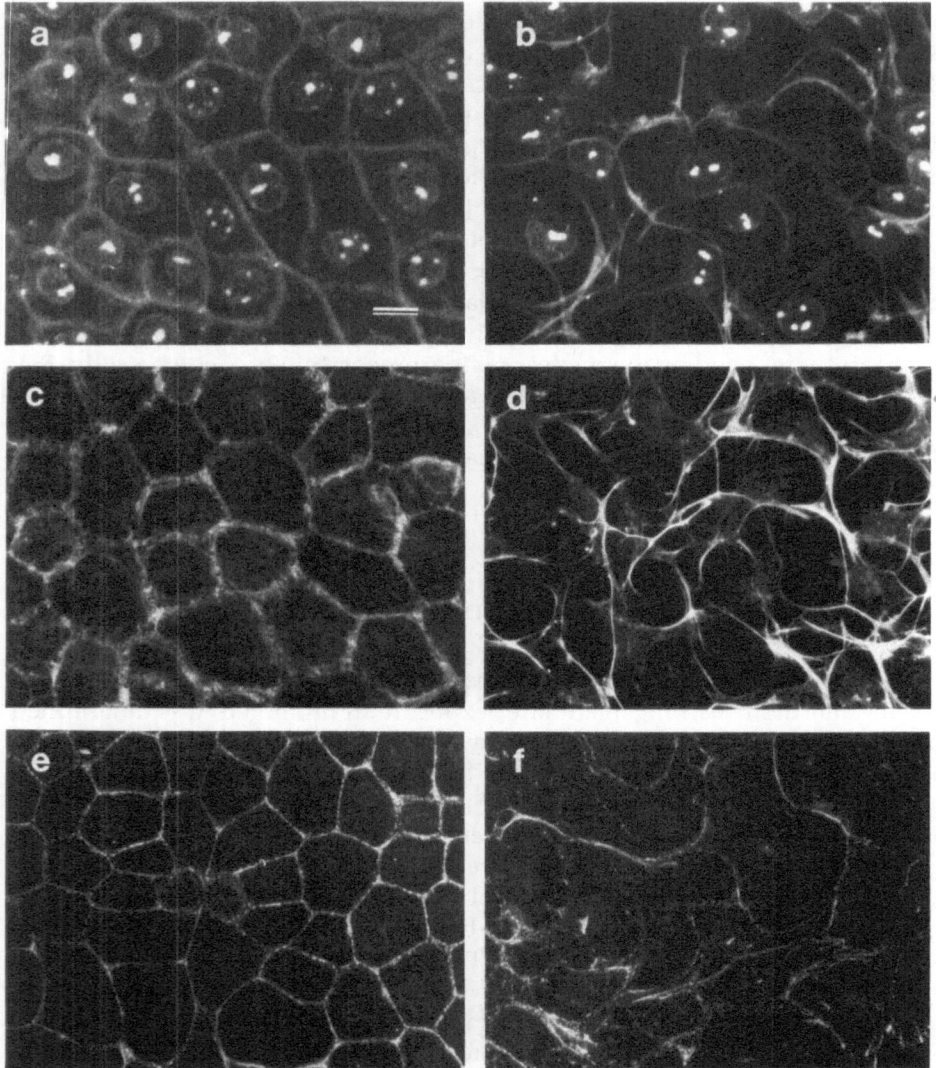

Figure 4. Immunofluorescent localization of cytokeratins and desmosomal proteins in the nuclear matrix-intermediate filament (NM-IF) core of MDCK cells treated with TPA (5 ng/ml) for 2 hr. MDCK cell colonies were incubated with 5 ng/ml TPA for 2 hr and then fractionated. Control colonies (a, c, e) and MDCK colonies exposed to TPA (b, d, f) are compared. The NM-IF structures were fixed and stained using a double antibody technique (Fey et al., 1984). Total protein (a, c) was stained using fluorescein isothiocyanate. In the control MDCK colony (a), the nuclear matrix, cytoplasmic filaments, and cellular junctions form regular polygonal cells. After exposure to TPA (b), the NM-IF scaffolds are elongated with filament bundles associated with the distorted nuclei. Cytokeratin fluorescence of control NM-IF preparations (c) has a regular pattern. Filaments radiate from the nuclear matrix and terminate at the cell–cell borders. The desmosomal fluorescence pattern of a control NM-IF colony (e) is localized as a punctate pattern at the cell borders. Both the cytokeratin (d) and desmosomal (f) fluorescence is altered after treatment with TPA. The cytokeratins form extensive bundles and the desmosomal fluorescence is observed along these cytokeratin-rich structures. Scale bar: 20 μm.

terns of dense cables in processes that correspond to those observed in Figures 3b,c and 4b. The desmosomal localization in control MDCK colonies was determined by staining with an antibody to total desmosomal proteins (Fig. 4e). The regular, punctate staining pattern corresponds to the polygonal cell borders observed previously. The localization of desmosomal proteins in MDCK cells after exposure to TPA (Fig. 4f) is strikingly altered. The desmosomal proteins are observed in long punctate arrays that are colinear with the cytokeratin cables.

6. Protein Synthesis and Distribution after Exposure to TPA

The distribution within the different cell structural fractions of proteins synthesized both before and during exposure to TPA was examined by measuring the incorporation of [^{35}S]methionine into specific subcellular protein fractions. As shown in Table I, the quantitative distribution of protein labeled for 24 hr before exposure to TPA is virtually identical to the distribution of protein label in control cultures. Approximately 55% of the labeled protein is soluble in Triton X-100 (SOL), 26% is in the high salt-labile cytoskeleton (CSK), and 13.5% of the protein is released by nuclease digestion and extraction with 0.25 M ammonium sulfate (CHROM); the structure remaining—the NM-IF scaffold—is composed of 5.5% of the total cellular protein. This protein distribution does not change significantly after exposure to 5 ng/ml TPA. Essentially the same result is obtained for proteins labeled after prior exposure to TPA. These results can be interpreted as showing that despite the radical morphological reorganization, the cell structural elements retain their normal connectivity.

The protein fractions obtained after incubation in 5 ng/ml TPA for 2 hr were analyzed by electrophoresis on 10% polyacrylamide gels and compared

Table I. Protein Distribution (%) in Control and TPA-Treated Monolayers[a]

Fraction	24-hr Prelabel[b]		1-hr Postlabel[c]	
	Control[d]	TPA[d]	Control	TPA
SOL	55.0	54.8	56.9	56.6
CSK	26.0	26.1	26.0	26.1
CHROM	13.5	13.6	13.5	13.6
NM-IF	5.5	5.5	3.6	3.7

[a]See text discussion in Section 6 for details.
[b]MDCK colonies labeled with 15 μCi/ml [^{35}S]methionine (10 mCi/ml; 1107 Ci/mmol; New England Nuclear) for 24 hr prior to incubation with TPA.
[c]MDCK colonies were incubated with TPA for 2 hr then pulse-labeled with 15 μCi/ml [^{35}S]methionine for 1 hr prior to fractionation.
[d]Control colonies were incubated with 0.1% DMSO for 2 hr. TPA colonies incubated with 5 ng/ml TPA in 0.1% DMSO for 2 hr.

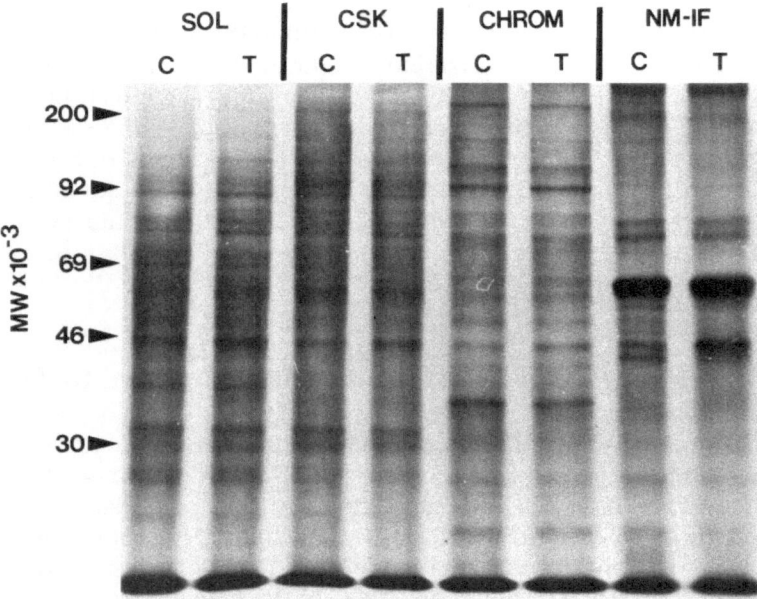

Figure 5. PAGE analysis of protein fractions labeled after exposure to 5 ng/ml TPA for 2 hr. MDCK colonies were exposed to 0.1% DMSO (C) or 5 ng/ml TPA (T) for 2 hr before a 60-min pulse label with 15 μCi/ml [35S]methionine (10 mCi/ml; 1107 Ci/mmol; New England Nuclear). Cells were fractionated into soluble (SOL), cytoskeleton (CSK), chromatin (CHROM) and nuclear matrix-intermediate filament (NM-IF) fractions as described previously (Fey *et al.*, 1984). These fractions were compared by electrophoresis on a 10% polyacrylamide gel. The proteins obtained in each fraction are virtually unchanged by treatment with TPA.

with those synthesized in an untreated culture. Figure 5 shows a fluorogram of the [35S]methionine-labeled proteins from each fraction. In each fraction, the pattern of proteins obtained from cells exposed to TPA is virtually identical to that of proteins from untreated cells.

7. Alteration of Protein Synthesis in Suspension Culture

The work of Folkman and Moscona (1978) showed a close correspondence between cell shape and the rate of cell growth. Folkman proposed that a breakdown in cell shape-related growth control underlies many, and perhaps most, neoplastic phenomena. Wittelsberger *et al.* (1981) further suggested that as cells become increasingly transformed, they progressively lose the ability to modulate specific metabolic processes in response to changes in cell shape. These shape signals are most likely transduced, at least in part, by cytoskeletal elements. TPA reorganizes these structural elements, and thus apparently alters their ability to effect metabolic control. In the experiment described in Figure 6, we examined one aspect of shape responsiveness; the decline of

protein synthesis following cell suspension after prior exposure to TPA (Fig.
6b) and in untreated MDCK cells (Fig. 6a).

The effect of TPA on anchorage-dependent protein synthesis was studied
by suspending MDCK colonies, as single cells, in methylcellulose for the periods of time indicated in Figure 6. The cells were then pelleted through the
methylcellulose, resuspended in culture medium, and pulse-labeled with
[35S]methionine for 1 hr before fractionation as described in Table I. These

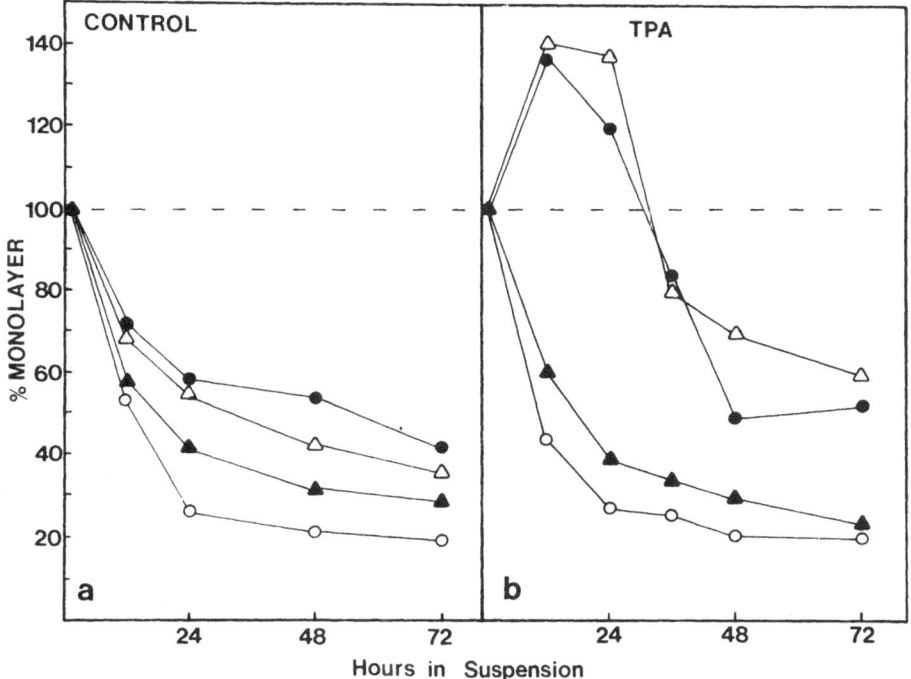

Figure 6. Fraction-specific stimulation of protein synthesis in MDCK cells exposed to 5 ng/ml TPA
24 hr before suspension in methylcellulose. Cells were suspended in methylcellulose and protein
synthesis was assayed at each time point by a 1-hr pulse lable of 15 μCi/ml [35S]methionine as
described in Table I. The cells were then fractionated into soluble (SOL)(●————●), cytoskeleton
(CSK)(○————○), chromatin (CHROM)(▲————▲), and nuclear matrix-intermediate filament
(NM-IF)(△————△) fractions, and total [35S]methionine incorporation was determined after pre-
cipitation in cold 7% trichloroacetic acid (TCA). The shutdown of protein synthesis in each fraction was expressed as a percentage of the rate of protein synthesis exhibited by that fraction in a
monolayer control plate labeled with [35S]methionine. (a) The individual protein fractions each
shut down protein synthesis at a different rate. (b) MDCK monolayers were exposed to 5 ng/ml TPA
24 hr before suspension in methylcellulose and analyzed for the shutdown of protein synthesis as
in (a). Both the soluble (SOL) and nuclear matrix-intermediate filament (NM-IF) fractions displayed
marked stimulation of anchorage-independent protein synthesis that was approximately 140% the
monolayer value after 12 hr in suspension. All data points represent cpm/cell and are expressed as
percentages of the monolayer value for the same fraction. By contrast, the rates of protein synthesis
shutdown in the cytoskeleton (CS) and chromatin (CHROM) fractions of TPA-treated cells (b) are
virtually identical to those of control CSK and CHROM fractions (a).

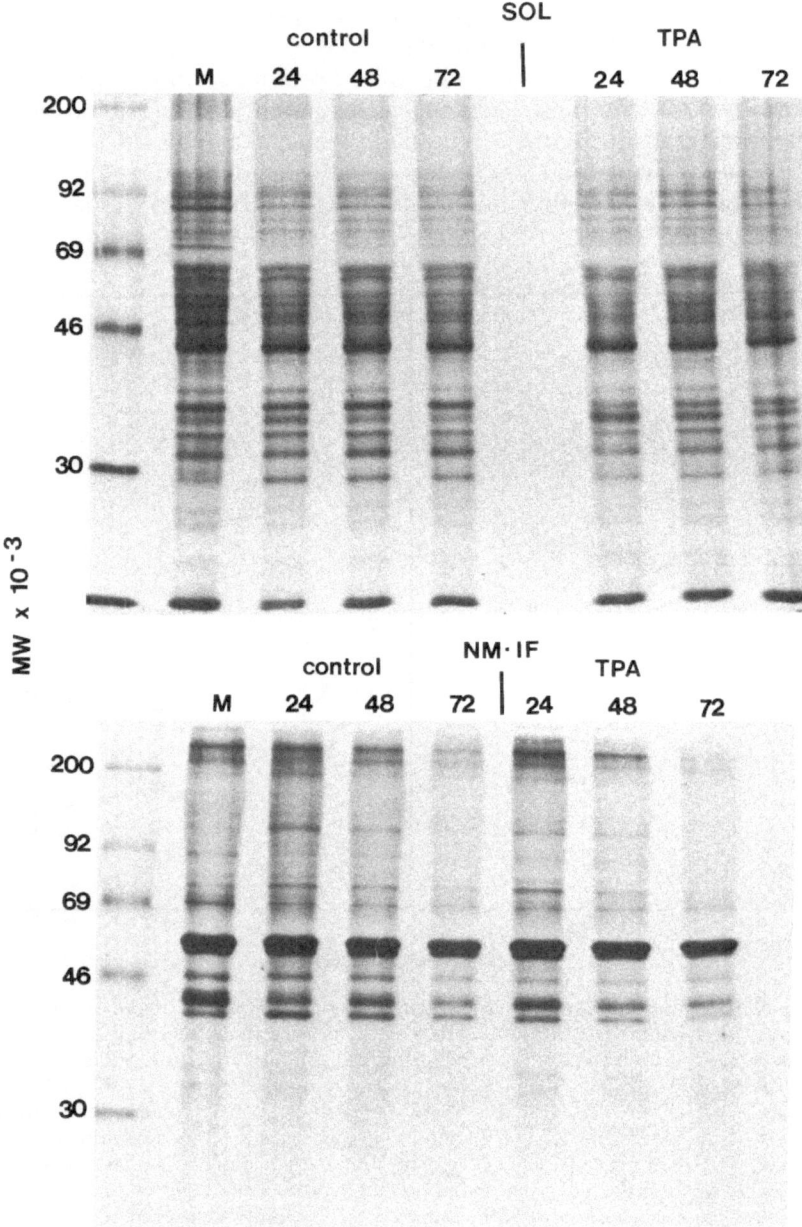

Figure 7. PAGE analysis of [³⁵S]methionine-labeled proteins synthesized during suspension in methylcellulose. Protein samples labeled with [³⁵S]methionine were obtained from each fraction and time point in the experiment described in Figure 6 and analyzed by electrophoresis on 10% polyacrylamide gels. The pattern of labeled proteins observed in the monolayer sample (M) is characteristic for each protein subfraction. This sample is compared with labeled proteins from an equivalent number of cells after suspension in methylcellulose for the time indicated and pulse labeling with [³⁵S]methionine as described in Figure 6. Proteins from the soluble (SOL), cytoskeleton (CSK), chromatin (CHROM), and nuclear matrix-intermediate filament (NM-IF) frac-

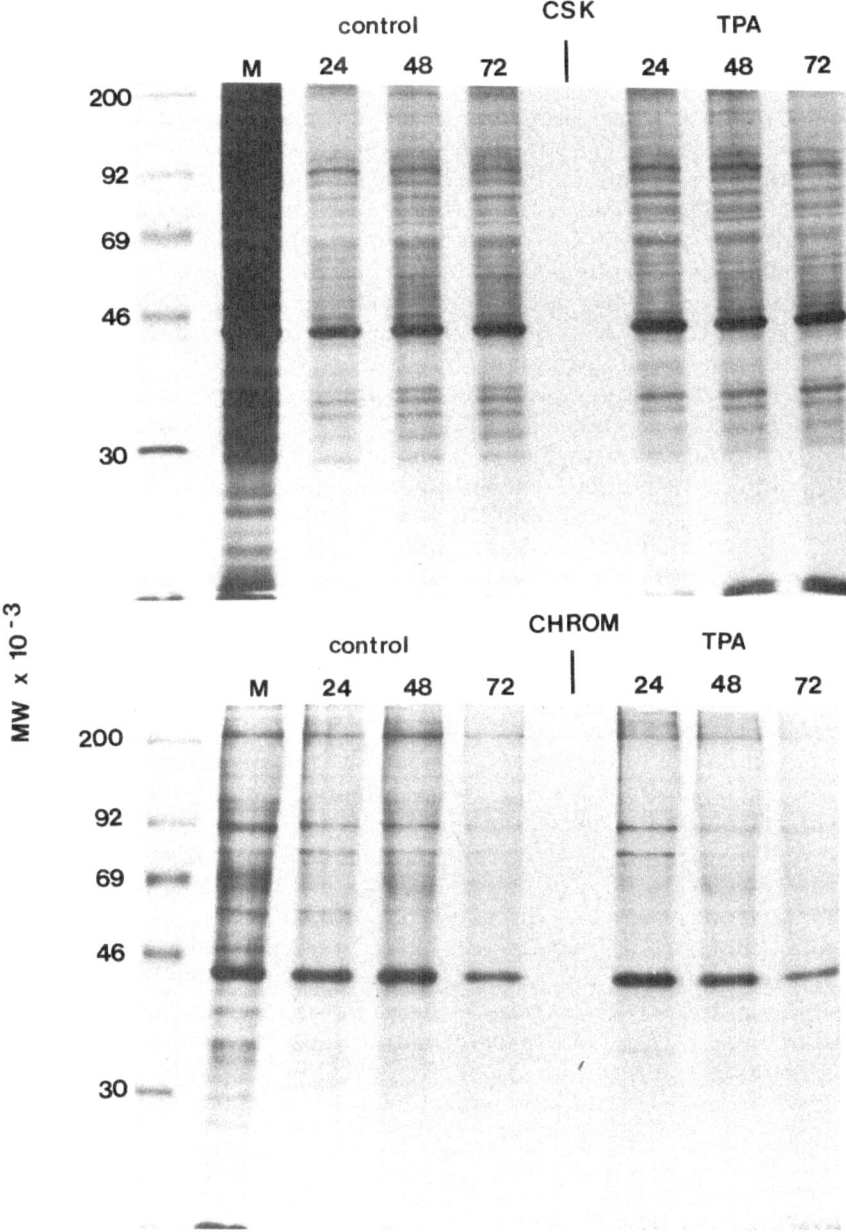

tions are compared. The M protein pattern is largely similar to the patterns observed in MDCK cells suspended in methylcellulose for up to 72 hr. Incubation with 5 ng/ml TPA before and during suspension has little effect on the pattern of proteins synthesized in each fraction. The intensity of the protein bands corresponds to the rate of protein synthesis shutdown observed in Figure 6. The results of this experiment suggest that the four fractions obtained using the differentiated fractionation procedure are retained during suspension in methylcellulose. These fractions appear to be regulated as intact and distinct metabolic units. The soluble and NM-IF fractions appear to be separately stimulated by treatment with TPA.

anchorage-dependent cells shut down protein synthesis in a complex manner. Normal MDCK cells execute a complex, fraction-specific, shutdown of protein synthesis during suspension (Fig. 6). The cytoskeleton (CSK) fraction shuts down most rapidly, while synthesis of the soluble (SOL) proteins declines slowly to a rate equivalent to 50% of the monolayer value at 24 hr after suspension. The shutdown of both chromatin (CHROM) and protein synthesis of NM-IF fractions is intermediate between the soluble and cytoskeletal fractions. Exposure of MDCK colonies to TPA 24 hr before suspension in methylcellulose drastically changes the normal shutdown of protein synthesis induced by suspension culture. As shown in Figure 6b, both the soluble (SOL) and NM-IF proteins display a temporary but marked release from anchorage-dependent regulation and are actually stimulated by suspension. However, under the same conditions, the cytoskeleton and chromatin fractions shut down protein synthesis as in control suspended MDCK cells.

The data represented in Figure 6 show that TPA specifically alters the regulation of protein synthetic rates in some fractions, while other fractions in the same cells remain unaffected. Thus, the synthesis of proteins for these fractions is differentially affected by tumor-promoting agents, at the same time altering the cell architecture. While overall protein synthesis for each structural fraction is regulated independently, the proteins within each fraction appear to be coordinately controlled. The proteins synthesized after varying times in suspension were analyzed by electrophoresis using 10% polyacrylamide gels.

The fluorograms exhibited in Figure 7 show the protein profiles for each of the four cellular subfractions after 24, 48, and 72 hr in suspension. This analysis was carried out for untreated MDCK cells and for MDCK cells exposed to 5 ng/ml TPA. Each gel slot was loaded with protein obtained from an equivalent number of cells. In this analysis, the shutdown of protein synthesis was examined for individual proteins within each fraction. The profiles in Figure 7 reflect the suspension data presented in Figure 6. The soluble and NM-IF fractions are stimulated at 24 hr in the TPA-treated cells. Here again, this stimulation is not observed for any particular protein band, but rather proteins throughout the entire fraction are regulated in concert. Thus, each fraction is regulated in a unified way, i.e., most proteins in each fraction are expressed or inhibited in unison. This coordinate regulation supports our earlier contention that the fractionation procedure, originally developed to preserve morphological integrity, may also produce cellular subfractions that reflect significant biochemical and metabolic divisions within the cell.

8. Induction of Morphological Signature in Permanently Transformed MDCK Cells

Despite their near-normal phenotype, MDCK cells are easily transformed to malignant behavior with complete carcinogens. Cell cultures were incubated for 12 hr with (+)-r-7,t-8-dihydroxy-t-9,10-epoxy-7,8,9,10-tetrahydrobenzo(a)-pyrene at 1 ng/ml, selected for growth in soft agar, subcloned and selected two additional times. Approximately 30 clones were obtained, all of which displayed phenotypes markedly different from normal and strikingly similar to the phenotype induced by TPA (Fig. 8) and other tumor promoters. None of the clones that display anchorage independence grows as polygonal epithelia. These results support the hypothesis that morphologically observed cytoskeletal alterations may be obligatory steps in establishing the neoplastic state.

9. Summary

The dramatic changes in morphology induced by nanomolar doses of tumor-promoting agents, especially in epithelial cells, have been noted previously (Driedger and Blumberg, 1980; Rifkin et al., 1979; Croop et al., 1980; Phaire-Washington et al., 1980; Ohuchi and Levine, 1980; Ojakian, 1981; Fey and Penman, 1984). This chapter shows the effect of the tumor promoter TPA on the underlying skeletal framework, which is involved in the maintenance of both cell and epithelial tissue morphology. It should be emphasized, however, that similar results are obtained for all the tumor promoters as well as for the complete, ultimate carcinogens examined so far. The organization of the cytoskeletal elements involved in these morphological changes is faithfully retained during the fractionation procedure employed here, as is evident from SEM and TEM analysis of Triton-extracted cells (Figs. 1–4).

A number of promoting agents have been compared, and the degree of disorganization viewed in these whole mounts appears to parallel the potency of the promoting agents as measured by other assays (Fey and Penman, 1984). Also, the inactive analogues of phorbol ester have no effect on cell structure (Rifkin et al., 1979; Ojakian, 1981; Fey and Penman, 1984).

We suggest that the effect of TPA on the cytoskeleton occurs early as compared with many of the commonly studied biochemical responses and may indeed underlie many of the previously described cellular responses to promoting agents, such as mitogenic stimulation. TPA-induced alterations in NM-IF scaffold occur in the absence of both protein and RNA synthesis (Fey and Penman, 1984). By contrast, plasminogen activator, stimulated by TPA (Wigler and Weinstein, 1976), is completely blocked by pretreatment with both cycloheximide and actinomycin D (Weinstein et al., 1977; Ojakian, 1981). Ornithine decarboxylase, another enzyme that is rapidly induced by tumor promoters, is inhibited by both cycloheximide and actinomycin D in the presence of TPA (O'Brien, 1976). Thus two of the early biochemical markers for tumor-promoter activity are separable from the induction of cytoskeletal alterations by TPA.

One of the most striking features of the response to promoting agents is the adoption of the transformed phenotype, in which cells lose growth control and cease being organized into meaningful tissue structure. The alteration of desmosomal and junctional associations and the concomitant change in cytokeratin organization are clearly related to the breakdown of epithelial organization (Fig. 4). The phenotype is completely reversible, although it takes about 3 days

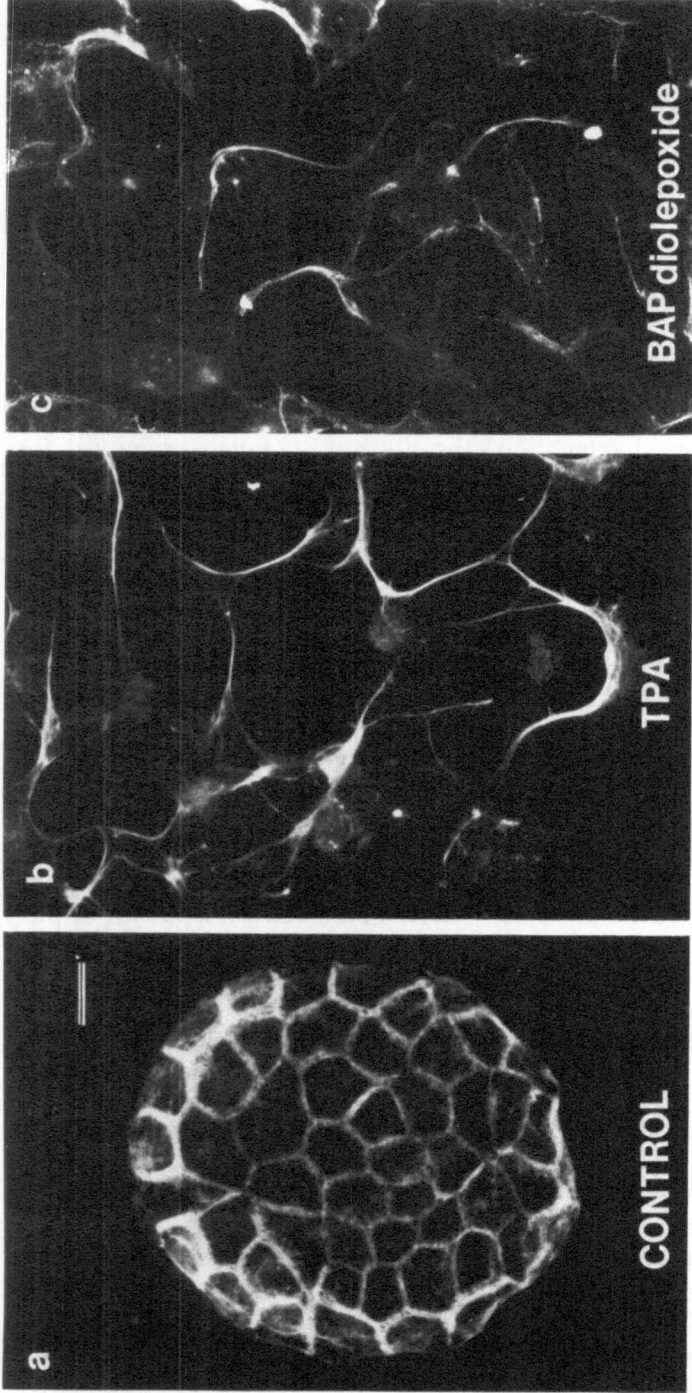

Figure 8. Immunofluorescent localization of cytokeratins in the NM-IF scaffolds of MDCK colonies after exposure to TPA and after permanent transformation by exposure to benzo(a)pyrene diolepoxide. MDCK colonies were treated with 0.1% DMSO (a) or 5 ng/ml TPA (b) for 2 hr before being extracted, fixed and stained with antibodies to cytokeratin as described in Figure 4. The morphological alterations in the NM-IF observed here are comparable to those observed above. (c) MDCK colonies were permanently transformed by 24-hr exposure to the ultimate carcinogen (+)-r-7-t-8-dihydroxy-t-9, 10-epoxy-7,8,9,10-tetrahydrobenzo(a)-pyrene (1 ng/ml). Cells were then passaged and allowed to grow to confluence. These plates were trypsinized and seeded onto soft agar. Transformed foci were isolated and subcloned. Fully transformed clones were identified by their ability to grow in methylcellulose. The morphology of these established, transformed cells (c) is remarkably similar to that promptly induced by short-term exposure to tumor promoters (b). Scale bar: 50 μm.

for the MDCK line to reestablish normal morphology (data not shown). By contrast, cells that have undergone permanent transformation and are anchorage-independent retain the morphological signature characteristic of tumor promoters in an irreversible manner.

It is possible that the partial loss of protein synthesis control in suspension (Fig. 6) and the disorganization of the cytoskeleton are intimately related and that after exposure to TPA, cells can no longer receive or interpret the architectural signals that normally control both their growth and tissue interaction.

One of the more surprising results is that whereas both the cytoskeletal framework and the NM-IF scaffold are rapidly reorganized by TPA or similar agents, the distribution of proteins into specific cellular subfractions remains unchanged. Structural fractions such as the cytoskeleton and NM-IF appear to have essentially the same proteins as they did before administration of the promoting agent. Considering the drastic effects on morphology, the biochemical measurements suggest that large changes in organization are possible, while the basic composition and connectivity of the structural elements remain unchanged.

ACKNOWLEDGMENTS. This work was supported by grant funds from the National Science Foundation and National Institutes of Health to Sheldon Penman. We are grateful to Gabriella Krochmalnic and David Cummings for assistance in electron microscopy. We wish to thank Patricia Turner for preparing the manuscript. We also wish to thank Dr. J. Rheinwald and Dr. M. Steinberg for generously providing respectively anti-bovine keratin and anti-desmosomal antibodies.

References

Bissell, M. J., Hatie, C., and Calvin, M. 1979, Is the product of the Src gene a promoter?, *Proc. Natl. Acad. Sci. USA* **76**: 348–352.

Boutwell, R. K., 1977, The role of the induction of ornithine decarboxylase in tumor promotion, in: *Origins of Human Cancer*, Book B (H. H. Hiatt, J. D. Watson, and J. A. Winsten, eds.), pp. 773–783, Cold Spring Harbor Laboratory, New York.

Capco, D. G., Krochmalnic, G., and Penman, S., 1984, A new method of preparing embedment-free sections for transmission electron microscopy: Applications to the cytoskeletal framework and other three-dimensional networks, *J. Cell Biol.* **98**:1878–1885.

Colburn, N. H., Farmer, B. F., Nelson, K. A., and Yuspa, S. H., 1979, Tumor promoter induces anchorage independence irreversibly, *Nature (Lond.)* **281**:589–591.

Croop, J., Toyama, Y., Dlugosz, A. A., and Holtzer, H., 1980, Selective effects of phorbol 12-myristate, 13-acetate on myofibrils and 10nm filaments, *Proc. Natl. Acad. Sci. USA* **77**:5273–5277.

Driedger, P. E., and Blumberg, P. M., 1980, Specific binding of phorbol ester tumor promoters, *Proc. Natl. Acad. Sci. USA* **77**:567–571.

Fey, E. G., and Penman, S., 1984, Tumor promoters induce a specific morphological signature in the nuclear matrix-intermediate filament scaffold of Madin Darby canine kidney (MDCK) cell colonies, *Proc. Natl. Acad. Sci. USA* **81**:4409–4413.

Fey, E. G., Wan, K. M., and Penman, S., 1984, The epithelial framework and nuclear matrix–intermediate filament scaffold: Three-dimensional organization and protein composition, *J. Cell Biol.* **98**:1973–1984.

Fischer, P. B., Dorsch-Hasler, K., Weinstein, I. B., and Ginsberg, H. S., 1979, Tumor promoters enhance anchorage-independent growth of adenovirus-transformed cells without altering the integration pattern of viral sequences, *Nature (Lond.)* **281**:591–594.

Folkman, J., and Moscona, A., 1978, The role of cell shape in growth control, *Nature (Lond.)* **273**:345–349.

Hecker, E., 1978, Structure–activity relationships in diterpene esters irritant and cocarcinogenic to mouse skin, in: *Carcinogenesis*, Vol. 2: *Mechanisms of Tumor Promotion and Cocarcinogenesis* (T. J. Slaga, A. Sivak, and R. K. Boutwell, eds.), pp. 11–48, Raven Press, New York.

Hennings, H., Lichti, U., Holbrook, K., and Yuspa, S. H., 1982, Role of differentiation in determining responses of epidermal cells to phorbol esters, in: *Carcinogenesis*, Vol. 7 (E. Hecker, ed.), pp. 319–323, Raven Press, New York.

Laskin, J. D., Mufson, R. A., Piccinini, L., Engelhardt, D. L., and Weinstein, I. B., 1981, Effects of the tumor promoter 12-0-tetradecanoyl-phorbol-13-acetate on newly synthesized proteins in mouse epidermis, *Cell* **25**:441–449.

Lee, L. S., and Weinstein, I. B., 1979, Mechanism of tumor promoter inhibition of cellular binding of epidermal growth factor, *Proc. Natl. Acad. Sci. USA* **76**:5168–5172.

McRoberts, J. A., Taub, M., and Saier, M. H., 1981, The Madin Darby canine kidney (MDCK) cell line, in: *Functionally Differentiated Cell Lines* (G. Sato, ed.), pp. 117–139, Alan R. Liss, New York.

O'Brien, T. G., 1976, The induction of ornithine decarboxylase as an early, possibly obligatory, event in mouse skin carcinogenesis, *Cancer Res.* **36**:2644–2653.

Ohuchi, K., and Levine, L., 1980, Alpha-tocopherol inhibits 12-0-tetradecanoyl-phorbol-13-acetate stimulated deacylation of cellular lipids, prostaglandin production, and changes in cell morphology of Madin-Darby canine kidney cells, *Biochim. Biophys. Acta* **619**:11–19.

Ojakian, G. K., 1981, Tumor promoter-induced changes in the permeability of epithelial cell tight junctions, *Cell* **23**:95–103.

Parkinson, E. K., and Emmerson, A., 1982, The effects of tumor promoters on the multiplication and morphology of cultured human epidermal keratinocytes, *Carcinogenesis* **3**:525–531.

Phaire-Washington, L., Silverstein, S. C., and Wang, E., 1980, Phorbol myristate acetate stimulates microtubule and 10-nm filament extension and lysosome redistribution in mouse macrophages, *J. Cell Biol.* **86**:641–655.

Rifkin, D. B., Crowe, R. M., and Pollack, R., 1979, Tumor promoters induce changes in the chick embryo fibroblast cytoskeleton, *Cell* **18**:361–368.

Robbins, S. L., and Angell, M., 1971, Neoplasia, in: *Basic Pathology*, pp. 63–109, W. B. Saunders, Philadelphia.

Sivak, A., and Van Duuren, B. L., 1967, Phenotypic expression of transformation: Induction in cell culture by a phorbol ester, *Science* **157**:1443–1444.

Taub, M., U. B., Chuman, L., Rindler, M. J., Saier, M. H., and Sato, G., 1981, Alterations of growth requirements of kidney epithelial cells in defined medium associated with malignant transformation, *J. Supramol. Stuct. Cell Biochem.* **15**:63–72.

Weil, R., 1978, Viral "tumor antigens"? A novel type of mammalian regulatory protein, *Biochim. Biophys. Acta* **516**:301–308.

Weinstein, I. B., 1981, Current concepts and controversies in chemical carcinogenesis, *J. Supramol. Struct. Cell Biochem.* **17**:99–120.

Weinstein, I. B., Wigler, M., and Pietropaolo, C., 1977, The action of tumor-promoting agents in cell culture, in: *Origins of Human Cancer*, Book B (N. H. Hiatt, J. D. Watson, and J. A. Winsten, eds.), pp. 751–772, Cold Spring Harbor Laboratory, New York.

Wigler, M., and Weinstein, I. B., 1976, Tumor promoter induces plasminogen activator, *Nature (Lond.)* **259**:232–233.

Wittelsberger, S. C., Kleene, K., and Penman, S., 1981, Progressive loss of shape-responsive metabolic controls in cells with increasingly transformed phenotype, *Cell* **24**:859–866.

Chapter 6

Growth Control in Capillary Endothelium

JUDAH FOLKMAN

1. Introduction

A network of capillary blood vessels supplies the nutritional needs of every part of the body. The endothelial cells lining these capillaries comprise a tissue of approximately 2 kg in a 70-kg human. These endothelial cells perform a variety of metabolic, immunological, and hemostatic functions (Ryan and Ryan, 1981; Pober et al., 1983; Loskutoff and Levin, 1984), but they rarely proliferate. Their rate of replication is usually so low that turnover times are measured in thousands of days (Folkman and Cotran, 1976; Denekamp and Hobson, 1984). By contrast, the turnover of intestinal mucosa is approximately 3 days and that of bone marrow about 5 days. However, with an appropriate stimulus, capillary endothelial cells can begin rapid proliferation and achieve turnover times almost as fast as bone marrow cells. Capillary endothelial cells proliferate most actively during angiogenesis, i.e., when new capillaries are being formed.

Physiological angiogenesis occurs for brief periods (days) during ovulation and during repair of the uterus after menstruation. It also occurs during the development of the placenta. Pathological angiogenesis occurs during wound healing, in chronic inflammation such as arthritis, and in a variety of other disease states such as diabetic retinopathy, neovascular glaucoma, and psoriasis. The most intense and persistent angiogenesis is that associated with tumor growth (Folkman, 1985).

Angiogenesis has become a subject of increasing interest and investigation during the past decade, largely because it is an important component of tumor growth. A series of experiments demonstrating that most solid tumors may be angiogenesis-dependent (Folkman, 1972; Gimbrone et al., 1972) prompted investigators to enter this field. The phenomenon of angiogenesis remained largely inaccessible to study, however, until the development of new methods during the 1970s.

JUDAH FOLKMAN • Department of Surgery, Children's Hospital, and Department of Surgery, Department of Anatomy and Cellular Biology, Harvard Medical School, Boston, Massachusetts 02115.

2. Methods for the Study of Angiogenesis

Before 1970 the principal method for studying angiogenesis was to observe blood vessel growth in a wound in a rabbit ear or hamster cheek pouch into which a transparent chamber had been inserted. Tumors were also implanted in these chambers (for review, see Peterson, 1979). However, it was difficult to use these chambers as a bioassay for testing the angiogenic activity of diffusible tumor products. Four new methods were developed in our laboratory:

1. The corneal micropocket technique permits linear measurement of growing capillaries. Tumors of about 1 mm³ are implanted in a pocket made in the cornea of an anesthetized rabbit at a distance of 1–2 mm from the normal vascular bed at the edge of the cornea (Gimbrone et al., 1974). New capillaries grow at right angles from the edge of the cornea toward the tumor at approximately 0.2 mm/day. They can be studied with a slit-lamp stereoscope. Subsequently this technique has been adapted by others to the mouse and rat corneas (Muthukkaruppan and Auerbach, 1979; Fournier et al., 1981).

2. A method was developed to provide sustained release of large molecules such as polypeptides and proteins from implantable polymers. This technique permitted the substitution of soluble tumor extracts for tumor implants in the corneal micropocket (Langer and Folkman, 1976). Macromolecules from these implants can be released at constant rates of nanograms to micrograms per day for periods of weeks to months (Murray et al., 1983).

3. The chorioallantoic membrane of chick embryos in a shell (Klagsbrun et al., 1976) or without a shell (Taylor and Folkman, 1982) was developed as a rapid and inexpensive method of screening for angiogenic factors and for angiogenic inhibitors (Crum et al., 1985). A test sample is dissolved in 10 μl 0.45% methylcellulose that is air dried on a 2-mm Teflon mold and then placed on the chorioallantoic membrane. The development of neovascularization or of an avascular zone can usually be detected 48 hr later.

4. Capillary endothelial cells were cloned and passaged in long-term culture (Folkman et al., 1979). They are growth-factor dependent, which makes them useful for detecting endothelial mitogens (Shing et al., 1984). These cells also form capillarylike tubes and branches in vitro (Folkman and Haudenschild, 1980). These results suggest that the formation of hollow capillary networks in vivo may be governed by a program encoded in the endothelial cell. The program is composed of a series of sequential steps.

3. Events of Capillary Growth

When capillary tube formation in vitro was analyzed together with intracorneal capillary growth in vivo, an orderly series of events was observed regardless of the type of angiogenic stimulus. New capillaries were generally found to originate from small venules or other capillaries lacking smooth muscle (Ausprunk and Folkman, 1977). Endothelial cells within these venules extrude from the vessel by local degradation of the basement membrane. Endothelial cells exposed to an angiogenic factor in vitro secrete specific collagenases, which degrade vascular basement membrane (Kalebic et al., 1983; Madri and Williams, 1983; Moscatelli et al., 1981). Endothelial cells continue to escape through this opening in the wall of the parent vessel and move in tandem toward the angiogenic stimulus (e.g., a tumor implant). A hollow lumen is formed either by a process of endothelial cell curvature (Ausprunk and Folkman, 1977; Folkman and Haudenschild, 1980) or by intraendothelial vacuole formation (Bär and Wolff, 1972; Folkman and Haudenschild, 1980). The hollow sprout thus formed continues to elongate by proliferation of endothelial cells in its mid-section and migration of endothelial cells at its tip. Two sprouts join to form a loop. New sprouts may arise from a loop. Blood flow begins within the loop. Pericytes appose themselves along the new capillaries.

There are still many loose ends to our understanding of this process. We do not know where pericytes come from. We do not have a picture of how a new capillary loop, which initially carries venous blood, eventually becomes arterialized. The mechanism of lumen formation is unknown. We do not understand why newly formed capillaries regress when an angiogenic stimulus is removed.

4. The Role of Mast Cells and Heparin in Angiogenesis

The program for capillary formation, which is expressed by endothelial cells when they are challenged by an angiogenic stimulus, can be modulated by either mast cells or heparin. Mast cells appear to act as helper cells for endothelium, and heparin can potentiate angiogenesis, i.e., increase the number of new capillaries and their rate of growth for a given angiogenic stimulus.

This new function of heparin was found to be independent of its anticoagulant activity, and its discovery originated from studies of the attraction of mast cells to tumors. Mast cells increased about 40-fold in the neighborhood of a new tumor implant in the chick embryo (Kessler et al., 1976). They accumulated about 24 hr in advance of new capillaries converging on the tumor. Isolated mast cells could not initiate angiogenesis by themselves. However, mast cells or mast cell-conditioned medium (Azizkhan et al., 1980) significantly increased the locomotion (Zetter, 1980) of capillary endothelial cells in vitro. Heparin was the only mast cell product responsible for stimulating the locomotion of endothelial cells (Azizkhan et al., 1980), and heparin potentiated angiogenesis in vivo (Taylor and Folkman, 1982).

These experiments gave the first indication that some part of the heparin molecule was acting as a positive regulator of capillary blood vessel growth—but they did not reveal a possible mechanism.

5. Angiogenic Factors

The first isolation of an angiogenic factor was reported in 1971 (Folkman *et al.*, 1971). A crude angiogenic preparation was extracted from the cytoplasm and subsequently from the nuclei (Tuan *et al.*, 1973) of a rat tumor. Other reports of tumor-derived angiogenic factors followed (Fenselau and Mello, 1976; Phillips *et al.*, 1976). Purification of these factors was hampered at the time, however, because the components of capillary growth had not yet been elucidated and bioassays were inadequate.

The availability of cultured endothelial cells and the realization that endothelial locomotion and proliferation were major components of angiogenesis led to efforts to purify endothelial growth factors. The first such mitogen was fibroblast growth factor (FGF) isolated from brain (Gospodarowicz *et al.*, 1978). Subsequently, endothelial cell growth factor (ECGF) from hypothalamus (Maciag *et al.*, 1979), retina-derived growth factor (D'Amore *et al.*, 1981), eye-derived growth factor (Barritault *et al.*, 1981), and a cartilage-derived factor (Klagsbrun and Smith, 1980) were reported but only in a partially purified state.

5.1. Heparin-Binding Angiogenic Factors

The discovery in 1983 that these polypeptide endothelial growth factors had a high affinity for heparin led to a breakthrough in their purification (Shing *et al.*, 1983, 1984). The first endothelial mitogen to be completely purified was derived from a rat chondrosarcoma by a rapid method in which the essential step was heparin-affinity chromatography (Shing *et al.*, 1983, 1984). This endothelial growth factor was also angiogenic in the chick embryo and rat cornea bioassays (Shing *et al.*, 1985). Since its introduction, the method of heparin-affinity chromatography has been used to purify most of the known angiogenic endothelial growth factors. Two of these growth factors have been completely sequenced: basic FGF (Esch *et al.*, 1985) and acidic FGF (Esch *et al.*, 1984; Thomas *et al.*, 1985); they share 53% sequence homology. Thus, there is a class of heparin-binding angiogenic factors distinguished by the following criteria: (1) Heparin binding is their only common biochemical property, (2) heparin-affinity chromatography is an essential purification step, and (3) heparin interacts with some of these factors to enhance their mitogenic activity *in vitro* and their angiogenic activity *in vivo*. Acidic FGF (Schreiber *et al.*, 1985) and basic FGF (Lobb *et al.*, 1986) represent two prototypes of this class of heparin-binding growth factors.

Those angiogenic endothelial mitogens that are related to acidic FGF have been found mainly in brain (Lobb and Fett, 1984; Maciag *et al* 1984; Thomas *et al.*, 1984), hypothalamus (Klagsbrun and Shing, 1985), retina (D'Amore and Klagsbrun, 1984), and eye (Courty *et al.*, 1985). They elute from heparin-Sepharose with about 1.0 M NaCl and have a pI value of about 5. Those angiogenic endothelial mitogens related to basic FGF elute from heparin–Sepharose with about 1.5 M NaCl and have a pI value of 8–10. They are more widespread in normal tissues and have also been found in tumors. They have been isolated from brain (Bohlen et al., 1984; Gospodarowicz et al., 1984), hypothalamus (Klagsbrun and Shing, 1985), retina (D'Amore and Klagsbrun, 1984), cartilage (Sullivan and Klagsbrun, 1985), bone (Hauschka et al., 1985), corpus luteum (Gospodarowicz et al., 1985), macrophages (Baird et al., 1985), and several human and animal tumors (Shing et al., 1984; Davidson et al., 1985).

5.2. Angiogenin

Recently, another angiogenic factor named angiogenin was purified from a human tumor cell line carried in serumfree culture (Fett et al., 1985). A basic polypeptide of 14,400 M_r, it is angiogenic in the chick embryo and the rabbit cornea. Its amino acid sequence and cDNA sequence have also been reported (Kurachi et al., 1985; Strydom et al., 1985). It differs from the heparin-binding angiogenic factors because (1) it does not bind to heparin; (2) it is apparently not an endothelial mitogen in vitro; (3) it is unrelated to FGF, and (4) it has 35% sequence homology to ribonuclease, yet is not an enzyme.

5.3. Other Angiogenic Factors

Other angiogenic factors that are not endothelial mitogens have been partially or completely purified. These include a factor secreted by adipocytes (Castellot et al., 1982), a factor from wound fluid macrophages (Banda et al., 1982), and prostaglandins PGE_1 and PGE_2 (BenEzra, 1978; Ziche and Gullino, 1982; Form and Auerbach, 1983). Other angiogenic factors of interest (Keegan et al., 1982; Kumar et al., 1983) await final purification and characterization before it can be determined whether they represent a third class of angiogenic factors different from the heparin-binding factors and from angiogenin.

6. Angiogenesis Inhibitors

The concept that solid tumors are angiogenesis dependent and that anti-angiogenesis could be a potential form of anticancer therapy was put forward in 1972 (Folkman, 1972). Although this idea rested on several experimental studies, it was difficult to test because there were no known angiogenesis inhibitors at the time. On the basis of clinical observations, the first angiogenesis inhibitor was found in cartilage (Eisenstein et al., 1973; Brem and Folkman, 1975). A soluble factor was isolated from cartilage that inhibited angiogenesis when the factor was administered locally (Langer et al., 1976), regionally (Langer et al., 1980), or systemically (Kaminski et al., 1978). However, this factor has not yet been completely purified. Protamine was also found to be an angiogenesis inhibitor when administered locally in the chick embryo or the rabbit cornea

(Taylor and Folkman, 1982). However, it was too toxic for prolonged systemic use as an antitumor agent. Platelet factor 4 was also found to inhibit angiogenesis in the chick embryo (Folkman *et al.*, 1983b; Taylor and Folkman, 1982), but its availability was found only in minute quantities, and its immunogenicity precluded long-term studies in tumor-bearing mice.

Whereas heparin alone acts generally as a positive regulator of angiogenesis, heparin in the presence of cortisone or hydrocortisone inhibited angiogenesis in the chick embryo and in the rabbit cornea (Folkman *et al.*, 1983a). The antiangiogenic effect of heparin in the presence of corticosteroids was independent of its anticoagulant activity. However, heparin from different manufacturers differed significantly in antiangiogenic activity. When used locally in the chick embryo or rabbit cornea, the less active heparins could be made effective by increasing their dose. But, when administered systemically in tumor-bearing mice, heparin had to be given orally to eliminate the anticoagulant effect. Tumor regression was observed in some types of tumors (melanoma, Lewis lung, and reticulum cell sarcoma) when the most active heparin was used (Abbott Panheprin). For other heparins, however, raising the oral dose eventually led to resumption of tumor growth. This curious effect may be the result of the accumulation of different heparin fragments in the plasma after oral administration (Larsen *et al.*, 1986). Certain heparin fragments may promote angiogenesis, whereas others can inhibit it (in the presence of corticosteroids).

A new class of corticosteroids has recently been discovered that inhibits angiogenesis in the presence of heparin but lacks any of the other corticoid functions, such as glucocorticoid or mineralocorticoid activity (Crum *et al.*, 1985). These are now known as angiostatic steroids. They occur naturally (e.g., as tetrahydrocortisol in plasma) and are also made synthetically.

7. Conclusions

During the past 10 years there has been increased interest in trying to understand the phenomenon of angiogenesis. This new field was initiated because of the possibility that tumor growth could be angiogenesis dependent. Subsequent studies from our laboratory and from many others have led to new methods for studying angiogenesis, the elucidation of the events of capillary growth, the purification of angiogenic factors, and the discovery of angiogenesis inhibitors. We are now in a position to ask more fundamental questions: What governs secretion of angiogenic factors from normal tissues and tumor tissues? What prevents the endothelial mitogens found in normal tissues from stimulating endothelial proliferation under resting conditions? How does the vascular system develop in the embryo, and how do veins and arteries arise from capillary tubes? What is the mechanism of capillary regression? How is the normal endothelial cell population maintained in capillaries? What is the physiological role of the natural angiostatic steroids that circulate in plasma?

These and many other questions suggest exciting avenues for future research.

ACKNOWLEDGMENTS. The research summarized here was supported in part by NIH Grant 5R01-CA37395 and by a grant to Harvard University from the Monsanto Company.

References

Ausprunk, D. H., and Folkman, J., 1977, Migration and proliferation of endothelial cells in preformed and newly formed blood vessels during tumor angiogenesis, *Microvasc. Res.* **14**:53–65.

Azizkhan, R. G., Azizkhan, J. C., Zetter, B. R., and Folkman, J., 1980, Mast cell heparin stimulates migration of capillary endothelial cells in vitro, *J. Exp. Med.* **152**:931–944.

Baird, A., Mormede, P., and Bohlen, P., 1985, Immunoreactive fibroblast growth factor in cells of peritoneal exudate suggests its identity with macrophage-derived growth factor, *Biochem. Biophys. Res. Commun.* **126**:358–364.

Banda, M. J., Knighton, D. R., Hunt, T. K., and Werb, Z., 1982, Isolation of a nonmitogenic angiogenic factor from wound fluid, *Proc. Natl. Acad. Sci. USA* **79**:7773–7777.

Bär, T., and Wolff, J. R., 1972, The formation of capillary basement membranes during internal vascularization of the rat's cerebral cortex, *Z. Zellforsch.* **133**:231–248.

Barritault, D., Arruti, C., and Courtois, Y., 1981, Is there a ubiquitous growth factor in the eye? Proliferation induced in different cell types by eye-derived growth factors, *Differentiation* **18**:29–42.

BenEzra, D., 1978, Neovasculogenic ability of prostaglandins, growth factors, and synthetic chemoattractants, *Am. J. Ophthalmol.* **86**:455–461.

Bohlen, P., Baird, A., Esch, F., Ling, N., and Gospodarowicz, D., 1984, Isolation and partial molecular characterization of pituitary fibroblast growth factor, *Proc. Natl. Acad. Sci. USA* **81**:5364–5368.

Brem, H., and Folkman, J., 1975, Inhibition of tumor angiogenesis mediated by cartilage, *J. Exp. Med.* **141**:427–439.

Castellot, J. J., Jr., Karnovsky, M. J., and Spiegelman, B. M., 1982, Differentiation dependent stimulation of neovascularization and endothelial cell chemotaxis by 3T3 adipocytes, *Proc. Natl. Acad. Sci. USA* **79**:5597–5601.

Courty, J., Loret, C., Moenner, M., Chevallier, B., Lagente, O., Courtois, Y., and Barritault, D., 1985, Bovine retina contains three growth factor activities with different affinities for heparin: Eye-derived growth factors, I, II, III, *Biochimie* **67**:265–269.

Crum, R., Szabo, S., and Folkman, J., 1985, A new class of steroids inhibits angiogenesis in the presence of heparin or a heparin fragment, *Science* **230**:1375–1378.

D'Amore, P. A., and Klagsbrun, M., 1984, Endothelial cell mitogens derived from retina and hypothalamus: Biochemical and biological similarities, *J. Cell Biol.* **99**:1545–1549.

D'Amore, P. A., Glaser, B. M., Brunson, S. K., and Fenselau, A. H., 1981, Angiogenic activity from bovine retina: Partial purification and characterization, *Proc. Natl. Acad. Sci. USA* **78**:3068–3072.

Davidson, J. M., Klagsbrun, M., Hill, K. E., Buckley, A., Sullivan, R., Brewer, P. S., and Woodward, S. C., 1985, Accelerated wound repair, cell proliferation, and collagen accumulation are produced by a cartilage-derived growth factor, *J. Cell Biol.* **100**:1219–1227.

Denekemp, J., and Hobson, B., 1984, Endothelial proliferation in normal tumour blood vessels, *Microvasc. Res.* **27(3)**:388.

Eisenstein, R., Sorgente, N., Soble, L. W., Miller, A., and Kuettner, K. E., 1973, The resistance of certain tissues to invasion: Penetrability of explanted tissues by vascularized mesenchyme, *Am. J. Pathol.* **73**:765–774.

Esch, F., Ling, N., Bohlen, P., Baird, A., Benoit, R., and Guillemin, R., 1984, Isolation and characterization of the bovine hypothalamic corticotropin-releasing factor, *Biochem. Biophys. Res. Commun.* **122**:899–905.

Esch, F., Baird, A., Ling, N., Ueno, N., Hill, F., Denoroy, L., Klepper, R., Gospodarowicz, D., Bohlen, P., and Guillemin, R., 1985, Primary structure of bovine pituitary basic fibroblast growth factor

(FGF) and comparison with the amino-terminal sequence of bovine brain acidic FGF, *Proc. Natl. Acad. Sci. USA* **82**:6507–6511.

Fenselau, A., and Mello, R. J., 1976, Growth stimulation of cultured endothelial cells by tumor cell homogenates, *Cancer Res.* **36**:3269–3273.

Fett, J. W., Strydom, D. S., Lobb, R. R., Alderman, E. M., Bethune, J. L., Riordan, J. F., and Vallee, B. L., 1985, Lysozyme: A major secretory product of a human colon carcinoma cell line, *Biochemistry* **24**:965–975.

Folkman, J., 1972, Anti-angiogenesis: New concept for therapy of solid tumors, *Ann. Surg.* **175**:409–416.

Folkman, J., 1985, Tumor angiogenesis, *Adv. Cancer Res.* **43**:175–203.

Folkman, J., and Cotran, R., 1976, Relation of vascular proliferation to tumor growth, *Int. Rev. Exp. Pathol.* **16**:207–248.

Folkman, J., and Haudenschild, C., 1980, Angiogenesis in vitro, *Nature (Lond.)* **288**:551–556.

Folkman, J., Merler, E., Abernathy, C., and Williams, G., 1971, Isolation of a tumor factor responsible for angiogenesis, *J. Exp. Med.* **133**:275–288.

Folkman, J., Haudenschild, C. C., and Zetter, B. R., 1979, Long-term culture of capillary endothelial cells, *Proc. Natl. Acad. Sci. USA* **76(10)**:5217–5221.

Folkman, J., Langer, R., Linhardt, R., Haudenschild, C., and Taylor, S., 1983a, Angiogenesis inhibition and tumor regression caused by heparin or a heparin fragment in the presence of cortisone, *Science* **221**:719–725.

Folkman, J., Taylor, S., and Spillberg, C., 1983b, The role of heparin in angiogenesis, in: *Development of the Vascular System*, Ciba Foundation Symposium 100, pp. 132–149, Pitman, London.

Form, D. M., and Auerbach, R., 1983, PGE$_2$ and angiogenesis (41548), *Proc. Soc. Exptl. Biol. Med.* **172**:214–218.

Fournier, G. A., Lutty, G. A., Watt, S., Fenselau, A., and Patz, A., 1981, A corneal micropocket assay for angiogenesis in the rat eye, *Invest. Ophthalmol. Vis. Sci.* **21**:351–354.

Gimbrone, M. A., Jr., Leapman, S. B., Cotran, R. S., and Folkman, J., 1972, Tumor dormancy *in vivo* by prevention of neovascularization, *J. Exp. Med.* **136**:261–276.

Gimbrone, M. A., Jr., Cotran, R. S., Leapman, S. B., and Folkman, J., 1974, Tumor growth and neovascularization: An experimental model using the rabbit cornea, *J. Natl. Cancer Inst.* **52**:413–427.

Gospodarowicz, D., Bialecki, H., and Greenburg, G., 1978, Purification of the fibroblast growth factor activity from bovine brain, *J. Biol. Chem.* **253(10)**:3736–3743.

Gospodarowicz, D., Cheng, J., Lui, G. M., Baird, A., and Bohlen, P., 1984, Isolation of brain fibroblast growth factor by heparin–Sepharose affinity chromatography: Identity with pituitary fibroblast growth factor, *Proc. Natl. Acad. Sci. USA* **81**:6963–6967.

Gospodarowicz, D., Cheng, J., Lui, G. M., Baird, A., Esch, F., and Bohlen, P., 1985, Corpus luteum angiogenic factor is related to fibroblast growth factor, *Endocrinology* **117**:2283–2391.

Hauschka, P. V., Iafrati, S. E., Doleman, S. E., Sullivan, R. C., and Klagsbrun, M., 1985, Resolution of distinct classes of bone matrix-derived cell growth factors by heparin affinity chromatography, *Calcif. Tissue Int.* **37**:7a.

Kalebic, T., Garbisa, S., Glaser, B., and Liotta, L. A., 1983, Basement membrane collagen degradation by migrating endothelial cells, *Science* **221**:281–283.

Kaminski, M., Kaminska, G., and Majewski, S., 1978, Inhibition of new blood vessel formation in mice by systemic administration of human rib cartilage extract, *Experientia* **34**:490–491.

Keegan, A., Hill, C., Kumar, S., Phillips, P., Schor, A., and Weiss, J., 1982, Purified tumour angiogenesis factor enhances proliferation of capillary, but not aortic, endothelial cells in vitro, *J. Cell Sci.* **55**:261–276.

Kessler, D. A., Langer, R. S., Pless, N. A., and Folkman, J., 1976, Mast cells and tumor angiogenesis, *Int. J. Cancer* **18**:703–709.

Klagsbrun, M., and Shing, Y., 1985, Heparin affinity of anionic and cationic capillary endothelial cell growth factors: Analysis of hypothalamus-derived growth factors (HDGF) and fibroblast growth factors (FGF), *Proc. Natl. Acad. Sci. USA* **82**:805–809.

Klagsbrun, M., and Smith, S., 1980, The purification of a growth factor from bovine cartilage, *J. Biol. Chem.* **255**:10859–10866.

Klagsbrun, M., Knighton, D., and Folkman, J., 1976, Tumor angiogenesis activity in cells grown in tissue culture, *Cancer Res.* **36**:110–114.

Kumar, S., West, D., Daniel, M., Hancock, A., and Carr, T., 1983, Human lung tumour cell line adapted to grow in serum-free medium secretes angiogenesis factor, *Int. J. Cancer* **32**:461–464.

Kurachi, K., Davie, E. W., Strydom, D. J., Riordan, J. F., and Vallee, B. L., 1985, Sequence of the cDNA and gene for angiogenin, a human angiogenesis factor, *Biochemistry* **24**:5493–5499.

Langer, R., and Folkman, J., 1976, Polymers for the sustained release of proteins and other macromolecules, *Nature (Lond.)* **263**:797–800.

Langer, R., Brem, H., Falterman, K., Klein, M., and Folkman, J., 1976, Isolation of a cartilage factor which inhibits tumor neovascularization, *Science* **193**:70–72.

Langer, R., Conn, H., Vacanti, J., Haudenschild, C., and Folkman, J., 1980, Control of tumor growth in animals by infusion of an angiogenesis factor, *Proc. Natl. Acad. Sci. USA* **77**:4331–4335.

Larsen, A. K., Lund, D. P., Langer, R., and Folkman, J., 1986, Oral heparin results in the appearance of heparin fragments in the plasma of rats, *Proc. Natl. Acad. Sci. USA* **83**:2964–2968.

Lobb, R. R., and Fett, J. W., 1984, Purification of 2 distinct growth factors from bovine neural tissue by heparin affinity chromatography, *Biochemistry* **23(26)**:6295–6298.

Lobb, R., Sasse, J., Sullivan, R., Shing, Y., D'Amore, P., Jacobs, J., and Klagsbrun, M., 1986, Purification and characterization of heparin-binding endothelial cell growth factors, *J. Biol. Chem.* **261(4)**:1924–1928.

Loskutoff, D. J., and Levin, E., 1984, Properties of plasminogen activators produced by endothelial cells, in: *Biology of Endothelial Cells* (E. A. Jaffe, ed.), pp. 200–208, Martinus Nijhoff, Boston.

Maciag, T., Cerundolo, Ilsley, S., Kelley, P. R., and Forand, R., 1979, An endothelial cell growth factor from bovine hypothalamus: Identification and partial characterization, *Proc. Natl. Acad. Sci. USA* **76**:5674–5678.

Maciag, T., Mehlman, T., Friesel, R., and Schreiber, A., 1984, Heparin binds endothelial cell growth factor, the principal endothelial cell mitogen in bovine brain, *Science* **225**:932–935.

Madri, J. A., and Williams, S. K., 1983, Capillary endothelial cell cultures: Phenotypic modulation by matrix components, *J. Cell Biol.* **97**:153–165.

Moscatelli, D., Gross, J. L., and Rifkin, D. B., 1981, Angiogenic factors stimulate plasminogen activator and collagenase production by capillary endothelial cells, *J. Cell Biol.* **91**:201a.

Murray, J. B., Brown, L., Langer, R., and Klagsbrun, M., 1983, A micro sustained release system for epidermal growth factor, *In Vitro* **19(10)**:743–748.

Muthukkaruppan, V., and Auerbach, R., 1979, Angiogenesis in the mouse cornea, *Science* **205**:1416–1417.

Peterson, H.-I., 1979, Tumor blood flow compared with normal tissue blood flow, in: *Tumor Blood Circulation: Angiogenesis, Vascular Morphology and Blood Flow of Experimental Human Tumors*, pp. 103–135, CRC Press, Boca Raton, Florida.

Phillips, P., Steward, J. K., and Kumar, S., 1976, Tumour angiogenesis factor (TAF) in human and animal tumours, *Int. J. Cancer* **17**:549–558.

Pober, J., Collins, T., Gimbrone, M. A., Jr., Cotran, R., Gitlin, J. D., Fiers, W., Clayberger, C., Krensky, A., Burakoff, S., and Reiss, C. S., 1983, Lymphocytes recognize human vascular endothelial and dermal fibroblast Ia antigens induced by recombinant immune interferon, *Nature (Lond.)* **305**:726–729.

Ryan, J. W., and Ryan, U., 1981, Endothelial metabolism, in: *Microcirculation, Current Physiologic, Medical and Surgical Concepts* (R. M. Effros, H. Schmid-Schonbein, and J. Ditzel, eds.) p. 147, Academic Press, New York.

Schreiber, A. B., Kenney, J., Kowalski, J., Thomas, K. A., Gimenez-Gallego, G., Rios-Candelore, M., DiSalvo, J., Barritault, D., Courty, J., Courtois, Y., Moenner, M., Loret, C., Burgess, W. H., Mehlman, T., Friesel, R., Johnson, T., and Maciag, T., 1985, A unique family of endothelial cell polypeptide mitogens: The antigenic and receptor cross-reactivity of bovine endothelial cell growth factor, brain-derived acidic fibroblast growth factor, and eye-derived growth factor-II, *J. Cell Biol.* **101**:1623–1626.

Shing, S., Folkman, J., Murray, J., and Klagsbrun, M., 1983, Purification by affinity chromatography on heparin–Sepharose of a growth factor that stimulates capillary endothelial cell proliferation, *J. Cell Biol.* **97**:395a.

Shing, Y., Folkman, J., Sullivan, R., Butterfield, C., Murray, J., and Klagsbrun, M., 1984, Heparin affinity: Purification of a tumor-derived capillary endothelial cell growth factor, *Science* **223**:1296–1298.

Shing, Y., Folkman, J., Haudenschild, C., Lund, D., Crum, R., and Klagsbrun, M., 1985, Angiogenesis is stimulated by a tumor-derived endothelial cell growth factor, *J. Cell. Biochem.* **29**:275–287.

Strydom, D. J., Fett, J. W., Lobb, R. R., Alderman, E. M., Bethune, J. L., Riordan, J. F., and Vallee, B. L., 1985, Amino acid sequence of human tumor-derived angiogenin, 1985, *Biochemistry* **24**:5486–5493.

Sullivan, R., and Klagsbrun, M., 1985, Purification of cartilage-derived growth factor by heparin affinity chromatography, *J. Biol. Chem.* **260**:2399–2403.

Taylor, S., and Folkman, J., 1982, Protamine is an inhibitor of angiogenesis, *Nature (Lond.)* **297**:307–312.

Thomas, K. A., Rios-Candelore, M., and Fitzpatrick, S., 1984, Purification and characterization of acidic fibroblast growth factor from bovine brain, *Proc. Natl. Acad. Sci. USA* **81**:357–361.

Thomas, K. A., Rios-Candelore, M., Gimenez-Gallego, G., DiSalvo, J., Bennett, C., Rodkey, J., and Fitzpatrick, S., 1985, Pure brain-derived acidic fibroblast growth factor is a potent angiogenic vascular endothelial cell mitogen with sequence homology to interleukin 1, *Proc. Natl. Acad. Sci. USA* **82**:6409–6413.

Tuan, D., Smith, S., Folkman, J., and Merler, E., 1973, Isolation of the non-histone proteins of rat Walker carcinoma 256: Their association with tumor angiogenesis, *Biochemistry* **12**:3159–3165.

Zetter, B. R., 1980, Migration of capillary endothelial cells is stimulated by tumour-derived factors, *Nature (Lond.)* **285**:41–43.

Ziche, M., and Gullino, P. M., 1982, Angiogenesis and neoplastic progression in vitro, *J. Natl. Cancer Inst.* **69(2)**:483–487.

Chapter 7

The Growth Factor–Platelet–Transformation Connection

DANIEL F. BOWEN-POPE

1. Introduction

Until recently, the growth of diploid animal cells in monolayer culture required that the defined nutrient medium be supplemented with serum prepared from clotted whole blood (Temin *et al.*, 1972; Holley, 1975). The concentration of serum added to the nutrient medium was found to be a major determinant of the growth rate and final population density of these cultures (Holley, 1974). Recently, the purification of growth-promoting factors from serum and the development of chemically defined media able to support the growth of some cell types without the presence of serum (reviewed by Barnes and Sato, 1980) have supported the view that serum provides growth-stimulatory factors as well as nutrients or vitamins, or both.

A second system for studying the regulation of growth of cultured animal cells has been the alteration of growth control through transformation by oncogenic viruses. Recent evidence suggests that, in some cases, transformation may involve production of, and response to, a class of growth factor-like molecules, "transforming growth factors," that confer on test cells the ability to grow while suspended in soft agar. The presence in platelets of platelet-derived growth factor, epidermal growth factor, and transforming growth factor, supports the hypothesis that platelets are involved in the proliferative response to injury (Ross and Glomset, 1976). In addition, the presence of transforming growth factors in normal platelets suggests that the TGFs may be involved in normal processes of tissue repair, possibly by stimulating proliferation under circumstances in which other mitogens are not sufficient.

This chapter discusses the observations that platelets are the major blood repository of both TGF and of the mitogens platelet-derived growth factor and epidermal growth factor, reviews what is known about their properties and speculates on their possible functions *in vivo*.

DANIEL F. BOWEN-POPE • Department of Pathology, University of Washington, Seattle, Washington 98195.

2. Why Does Serum Stimulate the Growth of Cultured Animal Cells?

Serum was used to culture animal cells for several decades before the components contributing to its activity began to be isolated and characterized. The existence in platelets of the growth factor we now call platelet-derived growth factor (PDGF) was deduced from the observation that serum prepared from plasma (PDS) could not substitute for whole blood serum (WBS) in supporting rapid proliferation of cultured connective tissue cells (Balk, 1971; Kohler and Lipton, 1974; Ross et al., 1974; Westermark and Wasteson, 1976). Among the several differences (Fig. 1) between PDS and WBS, that which accounts for the difference in growth-promoting activity was found to be the release of platelet components into serum during clotting (Kohler and Lipton, 1974; Ross et al., 1974; Westermark and Wasteson, 1976). The recognition that growth-promoting activity was present in platelets led to the identification of PDGF as one of the serum mitogens.

The ability of PDS to maintain cultured cells in a state of healthy quiescence (Vogel et al., 1978) was significant in its own right. Animal cells cultured in vitro using far lower than "physiological" (i.e., 100%) concentrations of serum multiplied much more rapidly than did comparable cells in vivo. Differences in the relationships between adjacent cells and between cell and substratum undoubtedly play a role in this difference. However, it was now clear that the interstitial fluid bathing cells in vivo is very different from the serum

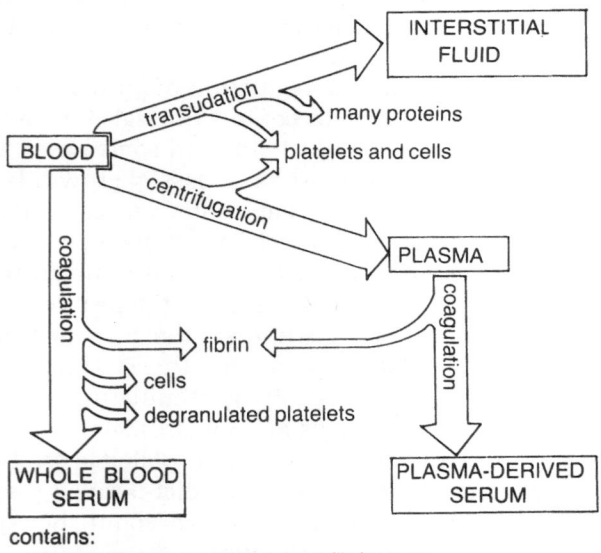

Figure 1. Derivation of cell-free fluids from whole blood.

used to culture cells *in vitro* and that some of the differences in growth state could be attributed to the presence of platelet mitogens in WBS. Whole blood serum is a pathological fluid, seen *in vivo* only in areas of vascular damage. Even plasma-derived serum, although devoid of platelet proteins, differs substantially from the interstitial fluid bathing connective tissue cells *in vivo*; it contains activated clotting cascade factors and many of the proteins filtered out during the process of transudation. Among the activated coagulation factors, thrombin has been shown to be active as a mitogen for certain cultured cells (Chen and Buchanan, 1975; Carney *et al.*, 1978) and to be an activator of release of PDGF from vascular endothelial cells (Harlan *et al.*, 1985).

3. Platelet Derived Growth Factor

3.1. Purification and Characterization of PDGF

Human PDGF has now been purified from washed platelets (Antoniades *et al.*, 1979; Heldin *et al.*, 1979a) and from platelet-rich plasma (Deuel *et al.*, 1981; Raines and Ross, 1982). Highly purified PDGF has been resolved by sodium dodecyl sulfate polyacrylamide gel electrophoresis (SDS-PAGE) into four species with apparent molecular weights of 27,000; 28,500; 29,000; and 31,000 (Raines and Ross, 1982). All species seem to be active in stimulating DNA synthesis and in binding to responsive cell types and show extensive peptide homology (Raines and Ross, 1982). Where individual species have been purified and compared, they have been found to have comparable mitogenic activity (Antoniades, 1981; Deuel *et al.*, 1981) and amino acid composition (Deuel *et al.*, 1981) but to differ in carbohydrate content (Deuel *et al.*, 1981).

Incubating PDGF under denaturing and reducing conditions eliminates its biological activity and converts the 27,000–31,000-M_r proteins to at least three smaller peptides with molecular weights of 14,400–17,500 (Heldin *et al.*, 1979a; Antoniades, 1981; Johnsson *et al.*, 1982; Raines and Ross, 1982). It therefore seems likely that the native PDGF molecule consists of two 14,000–17,000-M_r polypeptide chains covalently joined by disulfide bonds. Amino acid sequence data (Antoniades and Hunkapiller, 1983; Waterfield *et al.*, 1983) suggest that human PDGF consists of two different, but homologous subunits. Whether each molecule of PDGF consists of heterodimers or homodimers or both is unknown.

At neutral pH, PDGF is highly positively charged (pH-9.5–10.4, Heldin *et al.*, 1977; Deuel *et al.*, 1981) and is also quite hydrophobic (Heldin *et al.*, 1979b). This combination probably accounts for the ability of PDGF to adsorb nonspecifically to many surfaces, including those routinely used in handling biological materials, e.g., untreated glass and tissue culture plastic (Bowen-Pope and Ross, 1982; Raines and Ross, 1982; Smith *et al.*, 1982). PDGF is chemically very stable but is physiologically labile. Purified PDGF dissolved in 10 mM acetic acid with 0.25% BSA shows little or no loss of activity during storage at 4° C for 1 month. The activity of PDGF in sterile WBS *in vitro*, as determined by radioreceptor assay, is quite stable. *In vivo*, however, PDGF injected intravenously into baboons is rapidly removed from the blood compartment (Bowen-Pope *et al.*, 1984). Even during the brief period PDGF remains in the plasma its biological activity is substantially reduced by complex formation with binding proteins (Huang *et al.*, 1983; Raines *et al.*, 1984; Bowen-Pope *et al.*, 1984).

3.2. Biological Activity

The existence of PDGF was first deduced from its ability to stimulate cultured connective tissue cells to proliferate. This property was used to name the factor and to guide purification. Responsiveness to the mitogenic effects of PDGF seems to be characteristic of cultured connective tissue cells (reviewed in Bowen-Pope *et al.*, 1985), including vascular smooth muscle cells (Ross *et al.*, 1974; Rutherford and Ross, 1976; Witte *et al.*, 1978; Bowen-Pope and Ross, 1982) fibroblasts (Balk, 1971; Rutherford and Ross, 1976; Slayback and Cheung, 1977; Bowen-Pope and Ross, 1982), a glial cell line (Westermark and Wasteson, 1976; Heldin *et al.*, 1979*a*), chondrocytes (Bowen-Pope *et al.*, 1985), and the mouse fibroblastoid 3T3 cell lines often used to study growth control in cultured cells (Kohler and Lipton, 1974; Antoniades *et al.*, 1975; Vogel *et al.*, 1980; Bowen-Pope and Ross, 1982). Until PDGF was completely purified it was not possible to determine which of the other processes known to be stimulated by serum could be stimulated by PDGF itself and which might be stimulated by other serum components. It has now been shown that many cell processes are stimulated by pure PDGF—many of them soon after exposure of the cell to PDGF and long before initiation of DNA synthesis or cell division. Table I lists these processes in the order in which the effects of PDGF have been first detected. The order is only an approximation of the actual order in which these processes are initiated, since some assays have very poor temporal resolution. For example, the chemotaxis assay itself takes several hours—how soon within this period the process is initiated is not resolved.

3.3. The PDGF Receptor

Although the causal interrelationships, if any, between these biological effects of PDGF are unknown, it is unlikely that they are individually and directly regulated by PDGF. The basis for this belief is that PDGF seems to interact with cells through a single cell surface receptor. [125I]-PDGF shows a saturable binding to all PDGF-responsive cell types. The apparent affinity of the interaction is very high ($K_d = 1–3 \times 10^{-11}$ M, Bowen-Pope and Ross, 1982). The PDGF receptor is an approximately 170,000-M_r single subunit glycoprotein possessing PDGF-stimulatable tyrosine protein kinase activity (reviewed in Bowen-Pope *et al.*, 1985). The identification of the PDGF-binding site with the PDGF "receptor," i.e., with a cell component whose binding of PDGF is necessary for a physiological response to the PDGF, is strengthened by

Table I. Time Course of PDGF Effects[a,b]

Time first observed	Effect
1–2 min	Increased phosphorylation of the putative PDGF receptor in membrane preparations *in vitro*[1]
	Increased phosphorylation of cytoplasmic proteins in intact cells[2]
	Increased ion fluxes[3]
2–5 min	Decreased EGF binding[4]
	Increased amino acid transport[5]
	Increased release of arachidonate and PGE_2[6]
	Increased cAMP levels[7]
10–20 min	$\frac{1}{2}$ for internalization of surface-bound [^{125}I]-PDGF[8]
20 min	Increased cell surface ruffles and microvilli[9]
40 min	Increased protein synthesis[10]
$1\frac{1}{2}$ hr	First degraded PDGF appears in the culture medium[11]
1–2 hr	Increased RNA synthesis[12]
2–6 hr	Increased pinocytosis[13]
	Increased cholesterol metabolism and biosynthesis[14]
	Chemotactic response[15]
$\geq$12 hr	Cells elongated with filamentous protrusions[16]
	Increased rates of DNA synthesis and cell proliferation[17]
	Cell death in absence of PDGF or other mitogen[18]

[a]The effects of PDGF have been listed in order of their earliest detection after exposure of the cells to PDGF. Studies using different cell types appear to produce comparable results, so data from all cell types studied have been combined. Only results obtained using purified or partially purified PDGF are reported.

[b]*References:*

[1]Ek and Heldin (1982)
Ek et al. (1982)
Nishimura et al. (1982)
Pike et al. (1983)

[2]Nishimura and Deuel (1981)
Cooper et al. (1982)

[3]Rozengurt and Heppel (1975)
Mendoza et al. (1980)
Cassel et al. (1983)

[4]Heldin et al. (1982)
Bowen-Pope et al. (1983)

[5]Owen et al. (1982)

[6]Habenicht et al. (1981)
Shier and Durkin (1982)
Rozengurt et al. (1983)

[7]Rozengurt et al. (1983)

[8]Nilsson et al. (1983)
Williams et al. (1982)
Rosenfeld et al. (1984)

[9]Schmidt et al. (1982)

[10]Pledger et al. (1981)
Owen et al. (1982)
Olashaw and Pledger (1983)

[11]Bowen-Pope and Ross (1982)
Heldin et al. (1982)
Huang et al. (1982)
Rosenfeld et al. (1984)

[12]Abelson et al. (1979)
Cochran et al. (1983)
Linzer and Nathans (1983)

[13]Davies and Ross (1978)
Habenicht et al. (1980a)

[14]Chait et al. (1980)
Habenicht et al. (1980a,b)
Leslie et al. (1982)
Witte et al. (1982)

[15]Bernstein et al. (1982)
Deuel et al. (1982)
Grotendorst et al. (1982)
Seppa et al. (1982)
Senior et al. (1983)

[16]Kohler and Lipton (1974)
Ross et al. (1974)
Antoniades et al. (1975)
Rutherford and Ross (1976)
Westermark and Wasteson (1976)
Pledger et al. (1977)
Slayback and Cheung (1977)

[17]Slayback and Cheung (1977)
Witte et al. (1978)
Heldin et al. (1979a)
Vogel et al. (1980)

[18]Scher et al. (1982)

the close correlation between the concentration dependence of [^{125}I]-PDGF binding and stimulation of [^{3}H]thymidine incorporation, by the absence of PDGF binding sites on PDGF-nonresponsive cell types, and by the great specificity of the interaction (Heldin *et al.*, 1981; Bowen-Pope and Ross, 1982; Huang *et al.*, 1982; Williams *et al.*, 1982). Human [^{125}I]-PDGF binds with comparable affinity to cells from species as evolutionarily distant as chicken and

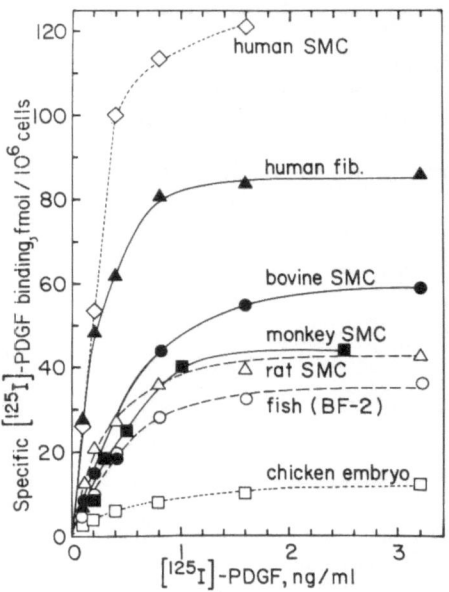

Figure 2. Specific binding of [^{125}I]-PDGF to cultured cells from different species. Monolayer cultures of each cell type were prepared in 24-well culture trays. When the cultures were confluent, the medium was replaced with culture medium containing 2% plasma-derived serum or 2% calf whole blood serum from which the PDGF had been removed by cation-exchange chromatography. After 2 days, the cultures were rinsed once with ice cold saline and then incubated for 4 hr at 4°C with gentle oscillation in 1 ml of medium containing 0.25% BSA and the concentration of [^{125}I]-PDGF shown on the abcissa; or with [^{125}I]-PDGF plus a 100-fold excess of unlabeled PDGF to determine nonspecific binding (always less than 15% of total binding); or with medium alone to determine cell number by electronic particle counting. The cell types examined were as follows: ($\diamond$) Adult human aortic smooth muscle; ($\blacktriangle$) adult human foreskin fibroblasts; ($\bullet$) bovine aortic smooth muscle; ($\blacksquare$) adult monkey (*Macaca nemestrina*) aortic smooth muscle; ($\triangle$) adult rat aortic smooth muscle; ($\bigcirc$) bluegill fry fish cell line BF-2 (American Type Culture Collection); ($\square$) secondary cultures of chicken embryo cells.

fish (Fig. 2) and sera from species ranging from humans to lampreys compete with human [^{125}I]-PDGF for binding to the PDGF receptors on mouse 3T3 cells (Singh *et al.*, 1982). It has even been reported that pure PDGF, at concentrations comparable to those which are mitogenic and chemotactic for mammalian cells, is chemotactic for *Tetrahymena*, a free-living unicellular protozoan (Andersen *et al.*, 1984). It thus appears that the PDGF receptor system was developed early in evolution and served a function important enough to ensure its survival in almost unmodified form.

4. Epidermal Growth Factor in Platelets

PDGF is not the only growth factor in platelets. In the assay for mitogenic activity used to guide its purification, PDGF accounts for about one-half the mitogenic activity present in human platelets. The fact that all the groups working to purify a mitogen from human platelets ended up purifying the same mitogen probably reflects, in part, the partitioning of the remaining mitogenic activity among several different mitogens, none prominent enough to be pursued. Our own identification of one of these other mitogens was a result of the observation that PDGF reduces [^{125}I]-EGF binding to human fibroblasts and to mouse 3T3 cells. The major significance of this effect, shown to result from a rapid, temperature-independent decrease in the affinity of [^{125}I]-EGF binding (Bowen-Pope *et al.*, 1983; Collins *et al.*, 1983), is the suggestion that these two growth-factor systems, each with its own receptor, can interact relatively directly at a primary level. Whether interaction at this level is involved in normal physiological control mechanisms is unknown. However, the observed effects of PDGF on EGF binding also suggested a possible explanation for the poor agreement between different reports of the EGF concentration in blood. Assay methods based on [^{125}I]-EGF binding to cell types expressing both EGF and PDGF receptors would overestimate the EGF content of a sample if PDGF were present, e.g., if WBS were being assayed. We found that this artifact could indeed reproduce some of the very high estimates of EGF levels in blood (Bowen-Pope and Ross, 1983). EGF levels in plasma were lower than our detection limit, and the EGF apparently detected in WBS could be greatly reduced by treatment with an antibody to PDGF, or by using for the assay a cell type with EGF receptors that did not express PDGF receptors. In addition we found that WBS did seem to contain some "true" EGF activity and that this seemed to be present largely in the platelets. At the same time, Oka and Orth (1983) reported a detailed study of EGF activity in human platelets in which, using a radioimmunoassay confirmed by radioreceptor and mitogenicity assays, they described the platelet as the major repository of EGF activity in humans.

The demonstration of EGF activity in platelets is significant when considering the possible physiological sources and roles of EGF in vivo. EGF was first shown to be present in astonishing concentration (0.06% of tissue wet weight) in the submaxillary glands of male mice (Savage and Cohen, 1972). As a consequence of its facile purification from this source, and of the brilliant subsequent investigative work on its mechanism of action, EGF has become, from a biochemical or cell biological point of view, one of the paradigmatic polypeptide mitogens. Nevertheless, considerable uncertainty exists as to its physiological source and role in vivo. Most species do not have high concentrations of EGF in their submaxillary glands. Even in the male mouse, surgical ablation of the salivary glands did not reduce levels of EGF found in the blood (Byyny et al., 1974). The human form of EGF, called urogastrone before its homology to mouse EGF was recognized (Gregory, 1975) and now often cautiously referred to as urogastrone/EGF, was discovered on the basis of its ability to inhibit gastric acid secretion by stomach mucosal cells, an activity of EGF that may account for its presence in the salivary gland in some species. In any case, current evidence regarding the presence of EGF in human platelets suggests that the platelet should be considered as one of the repositories of EGF in vivo, and possibly as a more likely source of mitogenically acting EGF than is the salivary gland. The significance of the presence of growth factors in platelets is discussed in Section 7.

5. Transforming Growth Factors

Most normal nonhematopoietic cells do not multiply in vitro unless they are provided with a "solid" substratum upon which to attach. The ability to multiply while suspended in soft agar is the in vitro property of cells that best distinguishes neoplastic cells from their normal counterparts (Macpherson and Montagnier, 1968). The operational definition of a "transforming growth factor" (TGF) has thus been taken to be the ability to permit cell growth in the absence of attachment (Sporn et al., 1981). The first TGFs to be completely characterized were discovered as a consequence of their ability to bind to EGF receptors (Todaro et al., 1976). Purification and sequencing of TGF-α from transformed cells of mouse, rat, and human origin indicated that the sequence of these TGFs is highly conserved (the three sequences are essentially identical) and is distantly related to that of EGF (Marquardt et al., 1983). Since TGF-α is not encoded by viral sequences, its synthesis by transformed cells seems to result from viral induction of a cellular gene. Use of a cDNA probe for human TGF-α mRNA has demonstrated expression of the gene in several human tumor cell lines, with the exception of cell lines of hematopoietic origin (Derynck et al., 1985b). Considerably lower levels of TGF-α mRNA have also been detected in certain adult human tissues (Lee et al., 1985). The use of very highly purified, or recombinant TGF-α has shown that EGF and TGF-α have identical biological activities in most systems, for example in stimulating precocious eyelid opening in newborn mice (Smith et al., 1985) and in stimulating mitogenesis in cell culture (Carpenter et al., 1983). The initial reports that TGF-α (then called sarcoma growth factor) had unique transforming activity probably reflected the presence of both TGF-α and TGF-β (see below) in the preparations (Anzano et al., 1983).

Other factors produced by transformed cells have now been discovered that do not bind to the EGF receptor (Roberts et al., 1981, 1982). Their activity in inducing growth in soft agar is stimulated 100-fold or more by added EGF or TGF-α, making them virtually EGF dependent. This EGF dependent class of transforming growth factor (TFG-β) has been purified, sequenced, and cloned. Human TGF-β is a dimer of identical 12,500 M_r subunits. The amino acid sequence of TGF-β seems to be very highly conserved, for example mouse and human TGF-βs differ in a single amino acid (Derynck et al., 1985b and personal communication). By comparison, mouse and human EGFs are only 69% homologous (Carpenter and Cohen, 1979). This suggests that TGF-β probably plays a role in some key cell function. A 2.4 Kb mRNA for TGF-β has been detected in all cell lines examined (Derynck et al., 1985a) and TFG-β activity has been detected in both embryonic and adult tissues (reviewed in Roberts et al., 1983; Moses et al., 1984). TGF-β has been detected in normal tissues at low levels (Roberts et al., 1981). Secretion by certain normal cell lines is increased about 40-fold by transformation with Maloney or Harvey Sarcoma Viruses (Anzano et al., 1985a). TGF-β has also been found in normal tissues at low levels (Roberts et al., 1981). The tissue richest in TGF-β was recently reported to be the platelet (Childs et al., 1982; Assoian et al., 1983). Given the presence of EGF activity in platelets, this TGF-β is presumably fully competent. Indeed, high concentrations of serum were known to stimulate growth in soft agar; the concentration of serum in soft agar growth assays is chosen to optimize detection of exogenous TGFs (presumably because of its content of supporting factors, like EGF) without contributing too high a background of its own.

6. PDGF-like Molecules Produced by Transformed Cells

The TGF-β purified from platelets is distinct from PDGF. Its molecular weight is too low (10,000 vs. 30,000 for PDGF), and it is not mitogenic for attached cells (Assoian *et al.*, 1983). Nevertheless, there are reports that purified PDGF itself is active in inducing growth in soft agar (Dicker *et al.*, 1981; P. L. Kaplan and Ozanne, 1983; McClure, 1983). In view of conflicting reports that PDGF is not active in inducing growth in soft agar (Childs *et al.*, 1982; Roberts *et al.*, 1982), it appears that the activity of PDGF may depend on the precise conditions of assay, including the cell type used and the presence of synergistic or antagonistic factors (Assoian *et al.*, 1984).

Nevertheless, several other observations link PDGF to many forms of oncogenic transformation. PDGF was the first growth factor (or hormone of any sort) that has been shown to be homologous to a viral oncogene product. Viral oncogenes are those viral genetic elements that are necessary for the acute transforming activity of the virus. In almost all instances, these viral oncogenes have normal cellular homologues (reviewed by Duesberg, 1983; Bishop, 1983), but in no previous case had the normal function of the cellular homologue been established. When the partial amino acid sequence of PDGF was determined (Antoniades and Hunkapiller, 1983; Waterfield *et al.*, 1983), it was quickly observed that the sequence of one PDGF peptide was virtually identical to that of the deduced transforming protein of Simian sarcoma virus (SSV) (Doolittle *et al.*, 1983; Waterfield *et al.*, 1983). This sequence homology immediately suggests a mechanism for transformation by SSV—that synthesis of the virally encoded PDGF analogue provides a continuous growth stimulus for infected PDGF-responsive connective tissue cells and renders the cells independent of exogenous mitogens.

The second observation linking PDGF with oncogenic transformation is that PDGF-like molecules are produced by cells transformed by a wide variety of oncogenic viruses, the oncogenes of which are not homologous to each other or to PDGF (Bowen-Pope *et al.*, 1984). For this study, we obtained matched sets of "normal" PDGF-responsive cells and parallel lines transformed by known viruses or agents. All cultures were examined for (1) expression of PDGF receptors, and (2) production of PDGF-like molecules able to bind to the PDGF receptor of test diploid human fibroblasts. We found that in all cases tested, the transformed cells expressed far fewer receptors than did their counterparts. Since medium conditioned by the transformed cells contained sufficient PDGF-like activity to reduce the number of available PDGF receptors on the untransformed cells to a phenocopy of the transformed cells, it is reasonable to speculate that the decreased expression of available receptors is a consequence of production of PDGF-like molecules. It is also reasonable to speculate that the consequence for the transformed cell of binding its own PDGF is the same as that for normal cells, i.e., a mitogenic signal is transmitted. If so, such "autocrine" production of, and response to, PDGF could account for the relative mitogen independence of these transformed cells—they do not become quiescent when incubated in low concentrations of plasma-derived serum. Attempts have been made to test the possibility that PDGF-producing transformed cells are responding in an autocrine fashion to this PDGF. Betsholtz et al. (1984) found that anti-PDGF antibodies prevented stimulation of normal cells by PDGF-producing osteosarcoma cells in coculture, but that the antibody did not inhibit the proliferation of the osteosarcoma cells themselves. Huang et al. (1984) reported partial inhibition of the proliferation of simian sarcoma virus transformed mouse 3T3 cells and rat NRK cells but saw no effect on transformed monkey cells. Garrett et al. (1984) reported that the decreased expression of PDGF receptors on SSV-transformed NRK cells, which in the autocrine hypothesis is explained as a consequence of downregulation by bound PDGF, could indeed be reversed by incubation with suramin, a substance which prevents PDGF from binding to its receptor. These observations suggest that secreted PDGF probably does function as an autocrine factor in some cases but that in other cases the PDGF must be either unimportant or must function in a cell compartment inaccessible to externally added antibodies.

Although production of PDGF-like molecules is a common concomitant of transformation and seems sufficient to explain some aspects (decreased available PDGF receptors and mitogen independence) of transformation of PDGF-responsive cells, it is not clear whether (or where) it is necessary or sufficient. In one case, that of bladder epithelial cells expressing an altered ras oncogene, it is unlikely that the observed PDGF production has a role in mitogenic stimulation in a simple autocrine fashion—nontransformed epithelial cells do not bind, or respond to, PDGF. In one of the other examples we have investigated—3T3 cells transformed by Kirsten or Moloney murine sarcoma viruses—the cells are known to also produce TGF-α (DeLarco and Todaro, 1978) and TGF-β (Roberts et al., 1982). Whether these different factors perform essential and different functions in transformation remains to be determined. The potential for complexity is almost unlimited; unlike the case of SSV, in which the limited coding potential of the viral oncogene is devoted to making a PDGF analogue, the other viral oncogenes that result in production of PDGF must be inducing expression of a cellular PDGF gene. There is reason to suspect that other growth-factor genes are induced as well. It might be characteristic of the acutely transforming, relatively wide host-range oncogenes that they activate a large number of host cell genes and benefit from them in a cumulative opportunistic fashion.

7. The Platelet as a Circulating Paracrine Cell—A Role for Transforming Growth Factors in Normal Animals

Human blood platelets are small (2–5 μm diameter), non-nucleated cells produced by controlled fragmentation of multinucleated megakaryocytes resident in the bone marrow (Weiss, 1975). Their role in hemostasis has been recognized for 100 years (see Baumgartner and Muggli, 1976); by aggregating together, with or without fibrin, they can form a barrier to blood flow. In addition, the coagulation cascade that ultimately results in the formation of fibrin from fibrinogen stimulates, and is also stimulated by, platelet activation.

In addition to its role in hemostasis, the platelet probably functions to deliver growth factors to sites of tissue damage. As discussed above, platelets contain PDGF, EGF, TGF(s), and probably other growth factors as well. Within the platelet, the growth-promoting activity has been localized to the α-granules by studying platelets from patients with α-granule defects (Gerrard et al., 1980), by fractionating subcellular organelles (D. R. Kaplan et al., 1979), and by studying the differential release of granule components by different platelet-activating agents (Witte et al., 1978). Essentially all the PDGF present in human blood circulates sequestered within the α-granules (Bowen-Pope et al., 1984) unless the α-granules are activated to release their contents, e.g., during clotting in vitro to prepare WBS, or as a result of platelet contact with an area of exposed connective tissue in vivo at a site of vascular damage. This release of growth factors at sites of tissue damage provides a mechanism for localizing initiation of tissue repair, including synthesis of extracellular matrix proteins and proliferation of connective tissue cells (Ross and Glomset, 1976).

The observations that platelets are a major tissue repository of TGFs in adults (Childs et al., 1982; Assoian et al., 1983) and that platelets contain the EGF necessary to support maximal response to EGF-dependent TGFs (Bowen-Pope and Ross, 1983; Oka and Orth, 1983) suggest that TGFs may be involved in extending the circumstances under which platelet factors can stimulate cell proliferation. The original proposal (Ross and Glomset, 1976) that platelet-derived mitogens might be involved in stimulating cell proliferation at sites of vascular damage was based on the observation that platelets contain a mitogen(s) for cultured connective tissue cells. These studies used monolayer cultures and led to purification of PDGF as the major such mitogen in human platelets. At the same time, ability to grow suspended in soft agar was recognized as the property of cultured cells that most closely correlated with tumorigenicity of cells in vivo (Macpherson and Montagnier, 1968); stimulation of growth in soft agar was used to purify substances (TGFs) produced by transformed cells and postulated to be proximal effectors of the transformed state (Sporn et al., 1981). Although TGFs have been studied to date largely because of their association with the transformed phenotype, it is likely that they have a normal physiological role as well. TGFs produced by transformed cells are encoded by cellular genes and are present in normal platelets. Sporn et al. (1983) have shown that TGF-β stimulates matrix deposition and cell proliferation in wound chambers in vivo (i.e., promotes some of the activities necessary for wound repair). The ability of TGFs to stimulate growth in soft agar, under conditions in which mitogens such as PDGF may not be active, may permit them to stimulate the growth of cells in vivo whose environment is more comparable to suspension in soft agar than to adhesion and spreading on tissue culture plastic. The relationship of cells in vivo to each other, and to the more or less "solid" substrata that surround or border them, is varied and variable (Fig. 3). Do fibroblasts in the dermis or vascular smooth muscle cells in the multilayered tunica media behave in response to growth factors more like cells

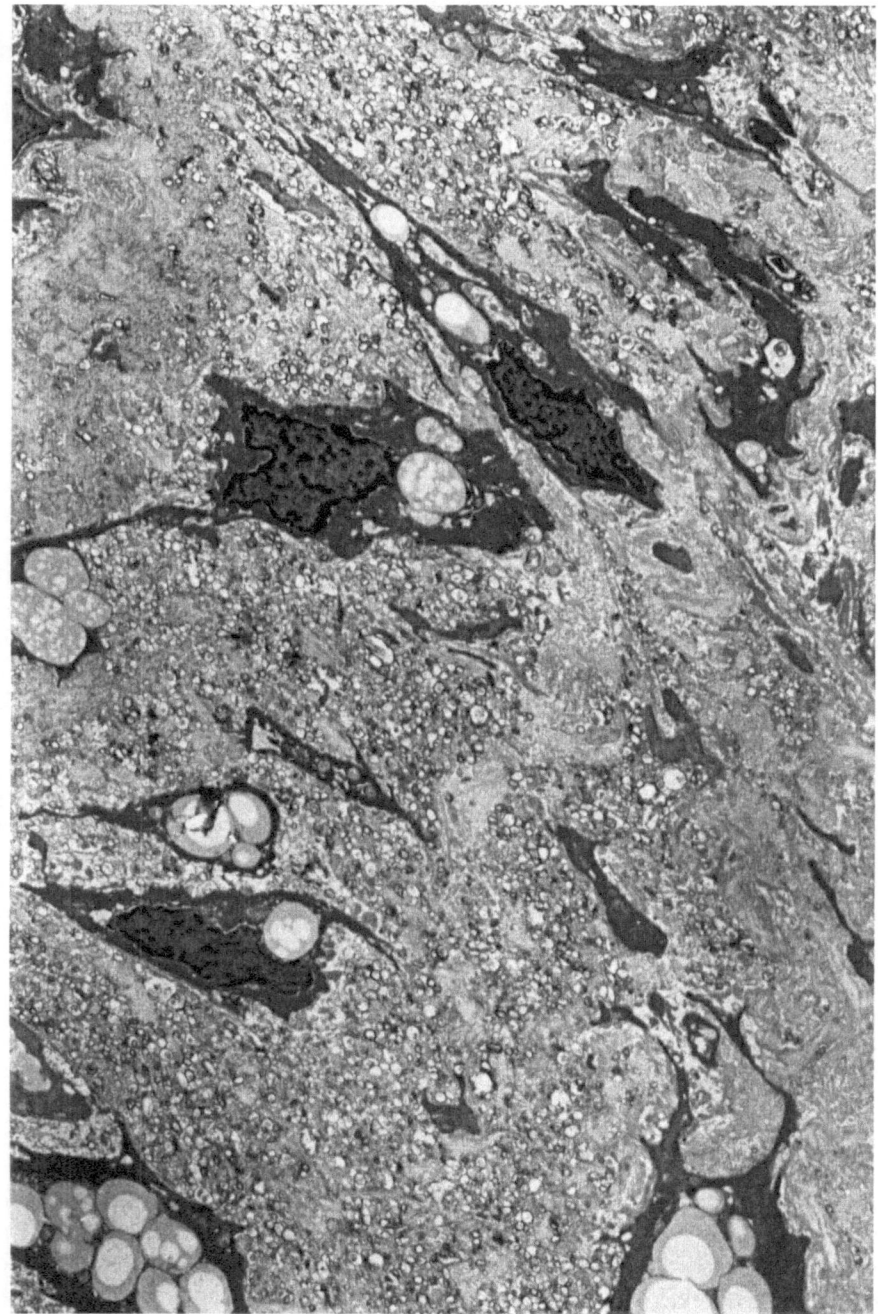

Figure 3. Smooth muscle cells surrounded by extracellular matrix. Electron micrograph of part of an atherosclerotic lesion induced by balloon catheter de-endothelization of an iliac artery from a monkey maintained on a hyperlipidemic diet (Ross and Harker, 1976). Many cells contain large lipid inclusions. The extracellular matrix contains many small lipid deposits. (Micrograph generously provided by R. Ross.)

attached to plastic or suspended in methylcellulose? Is the pattern of responsiveness altered when a tissue is "disorganized" by being damaged? Is it possible that many of the cells *in vivo* whose biosynthetic and proliferative rates "should" be stimulated by factors released by platelets are not in an environment that enables them to respond to "mitogens" alone, and that TGFs released from the platelet permit response *in vivo* just as it does *in vitro*?

References

Abelson, H. T., Antoniades, H. N., and Scher, C. D., 1979, Uncoupling of RNA and DNA synthesis after plasma stimulation of G_0-arrested BALB/c 3T3 cells, *Biochim. Biophys. Acta* **561**:269–275.

Andersen, A. A., Flodgaard, H., Klenow, H., and Leick, V., 1984, Platelet-derived growth factor stimulates chemotaxis and nucleic and synthesis in the protozoan *Tetrahymena*, *Biochim. Biophys. Acta* **782**:437–440.

Antoniades, H. N., 1981, Human platelet-derived growth factor (PDGF): Purification of PDGF-I and PDGF-II and separation of their reduced subunits, *Proc. Natl. Acad. Sci. USA* **78**:7314–7317.

Antoniades, H. N., and Hunkapiller, M. W., 1983, Human platelet-derived growth factor (PDGF): Amino-terminal amino acid sequence, *Science* **220**:963–965.

Antoniades, H. N., and Scher, C. D., 1977, Radioimmunoassay of a human serum growth factor for Balb/c 3T3 cells: Derivation from platelets, *Proc. Natl. Acad. Sci. USA* **74**:1973–1977.

Antoniades, H. N., Stathakos, D., and Scher, C. D., 1975, Isolation of a catonic polypeptide from human serum that stimulates proliferation of 3T3 cells, *Proc. Natl. Acad. Sci. USA* **72**:2635–2639.

Antoniades, H. N., Scher, C. D., and Stiles, C. D., 1979, Purification of human platelet-derived growth factor, *Proc. Natl. Acad. Sci. USA* **76**:1809–1813.

Anzano, M. A., Roberts, A. B., Smith, J. M., Sporn, M. B., and De Larco, J. E., 1983, Sarcoma growth factor from conditioned medium is composed of both type α and type β transforming growth factors, *Proc. Natl. Acad. Sci. USA* **80**:6264–6268.

Anzano, M. A., Roberts, A. B., De Larco, J. E., Wakefield, L. M., Assoian, R. K., Roche, N. S., Smith, J. E., Lazarus, J. E., and Sporn, M. B., 1985, Increased secretion of type β transforming growth factor accompanies viral transformation of cells, *Molec. Cell. Biol.* **5**:242–247.

Assoian, R. K., Komoriya, A., Meyers, C. A., Miller, D. M., and Sporn, M. B., 1983, Transforming growth factor-β in human platelets. Identification of a major storage site, purification, and characterization, *J. Biol. Chem.* **258**:7155–7160.

Assoian, R. K., Gortendorst, G. R., Miller, D. M., Sporn, M. B., 1984, Cellular transformation by coordinated action of three peptide growth factors from human platelets, *Nature* **309**:804–806.

Balk, S. D., 1971, Calcium as a regulator of the proliferation of normal, but not transformed, chicken fibroblasts in a plasma-containing medium. *Proc. Natl. Acad. Sci. USA* **68**:271–275.

Barnes, D., and Sato, G., 1980, Serum-free cell culture: A unifying approach, *Cell* **22**:649–655.

Baumgartner, H. R., and Muggli, R., 1976, Adhesion and aggregation: Morphological demonstration and quantitation in vivo and in vitro, in: *Platelets in Biology and Pathology* (J. I. Gordon, ed.), pp. 23–60, Amsterdam, Elsevier/North Holland Biomedical Press.

Bernstein, L. R., Antoniades, H. N., and Zetter, B. R., 1982, Migration of cultured vascular cells in response to plasma and platelet-derived factors, *J. Cell. Sci.* **56**:71–82.

Betsholtz, C., Westermark, B., Ek, B., and Heldin, C-H., 1984, Coexpression of a PDGF-like factor and PDGF receptor in a human osteosarcoma cell line: Implications for autocrine receptor activation, *Cell* **39**:447–457.

Bishop, M. J., 1983, Cellular oncogenes and retroviruses, *Annu. Rev. Biochem.* **52**:301–354.

Bowen-Pope, D. F., and Ross, R., 1982, Platelet-derived growth factor. II: Specific binding to cultured cells, *J. Biol. Chem.* **257**:5161–5171.

Bowen-Pope, D. F., and Ross, R., 1983, Is epidermal growth factor present in human blood? Alteration of EGF binding specificity in the radioreceptor assay, *Biochem. Biophys. Res. Commun.* **114:**1036–1041.

Bowen-Pope, D. F., DiCorleto, P. E., and Ross, R., 1983, Interactions between receptors for platelet-derived growth factor and epidermal growth factor, *J. Cell Biol.* **96:**679–683.

Bowen-Pope, D. F., Malpass, T. W., Foster, D. M., and Ross, R. 1984, Platelet-derived growth factor *in vivo:* Levels, activity and rate of clearance, *Blood* **64:**458–469.

Bowen-Pope, D. F., Seifert, R. A., Ross. R., 1985, The platelet-derived growth factor receptor, in: *Cell Proliferation: Recent Advances* (A. L. Boynton and H. L. Leffert, eds.), pp. 287–312, New York, Academic Press.

Burke, J. M., 1982, Cultured retinal glial cells are insensitive to platelet-derived growth factor, *Exp. Eye Res.* **35:**663–669.

Carney, D. H., Glenn, K. C., and Cunningham, D. D., 1978, Conditions which affect initiation of animal cell division by trypsin and thrombin, *J. Cell. Physiol.* **95:**13.

Carpenter, G., and Cohen, S., 1979, Epidermal growth factor, *Ann. Rev. Biochem.* **48:**193–216.

Carpenter, G., Stoscheck, C. M., Preston, Y. A., and De Larco, J. E., 1983, Antibodies to the epidermal growth factor receptor block the biological activities of sarcoma growth factor, *Proc. Natl. Acad. Sci. USA* **80:**5627–5630.

Cassell, D., Rothenberg, P., Zhuang, Y.-X., Deuel, T.-F., and Glaser, L., 1983, Platelet-derived growth factor stimulates Na$^+$/H$^+$ exchange and induces cytoplasmic alkalinization in NR6 cells, *Proc. Natl. Acad. Sci. USA* **80:**6224–6228.

Chait, A., Ross, R., Albers, J. J., and Bierman, E. L., 1980, Platelet-derived growth factor stimulates activity of low density lipoprotein receptors, *Proc. Natl. Acad. Sci. USA* **77:**4084–4088.

Chen, L. B., and Buchanan, J. M., 1975, Mitogenic activity of blood components. I. Thrombin and prothrombin, *Proc. Natl. Acad. Sci. USA* **72:**131.

Childs, C. B., Proper, J. A., Tucker, R. F., and Moses, H. L., 1982, Serum contains a platelet-derived transforming growth factor, *Proc. Natl. Acad. Sci. USA* **79:**5312–5316.

Cochran, B. H., Reffel, A. C., and Stiles, C. D., 1983, Molecular cloning of gene sequences regulated by platelet-derived growth factor, *Cell* **33:**939–947.

Collins, M. K. L., Sinnett-Smith, J. W., and Rozengurt, E., 1983, Platelet-derived growth factor treatment decreases the affinity of the epidermal growth factor receptors of Swiss 3T3 cell, *J. Biol. Chem.* **258:**11689–11693.

Cooper, J. A., Bowen-Pope, D. F., Raines, E., Ross, R., and Hunter, T., 1982, Similar effects of platelet-derived growth factor and epidermal growth factor on the phosphorylation of tyrosine in cellular proteins, *Cell* **31:**263–273.

Davies, P. F., and Ross, R., 1978, Mediation of pinocytosis in cultured arterial smooth muscle and endothelial cells by platelet-derived growth factor, *J. Cell Biol.* **79:**663–671.

DeLarco, J. E., and Todaro, G. J., 1978, Growth factors from murine sarcoma virus-transformed cells, *Proc. Natl. Acad. Sci. USA* **78:**4001–4005.

Derynck, R., Roberts, A. B., Eaton, D. H., Winkler, M. E., and Goeddel, D. V., 1985a, Human transforming growth factor-α: Precursor sequence, gene structure, and heterologous expression, in: *Cancer Cells 3* (J. Feramisco, B. Ozanne, and C. Stiles, eds.), pp. 79–85, Cold Spring Harbor Laboratory, Cold Spring Harbor, New York.

Derynck, R., Jarrett, J. A., Chen, E. Y., Eaton, D. H., Bell, J. R., Assoian, R. K., Roberts, A. B., Sporn, M. B., and Goeddel, D. V., 1985b, Human transforming growth factor-β: cDNA sequence and expression in tumor cell lines, *Nature* (in press).

Deuel, T. F., Huang, J. S., Proffitt, R. T., Baenziger, J. U., Chang, D., and Kennedy, B. B., 1981, Human platelet-derived growth factor—Purification and resolution into two active protein fractions, *J. Biol. Chem.* **256:**8896–8899.

Deuel, T. F., Senior, R. M., Huang, J. S., and Griffin, G. L., 1982, Chemotaxis of monocytes and neutrophils to platelet-derived growth factor, *J. Clin. Invest.* **69:**1046–1049.

Dicker, P., Pohjanpelto, P., Pettican, P., and Rozengurt, E., 1981, Similarities between fibroblast-derived growth factor and platelet-derived growth factor, *Exp. Cell Res.* **135:**221–227.

Doolittle, R. F., Hunakpiller, M. W., Hood, L. E., Devare, S. G., Robbins, K. C., Aaronson, S. A., and Antoniades, H. N., 1983, Simian sarcoma virus *onc* gene, v-*sis*, is derived from the gene (or genes) encoding a platelet-derived growth factor, *Science* **221:**275–277.

Duesberg, P. H., 1983, Retroviral transforming genes in normal cells?, *Nature (Lond.)* **304:**219–226.

Ek, B., and Heldin, C. A., 1982, Characterization of tyrosine-specific kinase activity in human fibroblast membranes stimulated by platelet-derived growth factor, *J. Biol. Chem.* **257**:10486–10492.

Ek, B., Westermark, B., Wasteson, Å., and Heldin, C. H., 1982, Stimulation of tyrosine-specific phosphorylation by platelet-derived growth factor, *Nature (Lond.)* **295**:419–420.

Garrett, J. S., Caughlin, S. R., Nimian, H. L., Tremble, P. M., Giels, G. M., and Williams, L. T., 1984, Blockade of autocrine stimulation in simian sarcoma virus-transformed cells reverses down-regulation of platelet-derived growth factor receptors, *Proc. Natl. Acad. Sci. USA* **81**:7466–7470.

Gerrard, J. M., Phillips, D. R., Rao, G. H. R., Plow, E. F., Walz, D. A., Ross, R., Harker, L. A., and White, J. G., 1980, Biochemical studies of two patients with the gray platelet syndrome, *J. Clin. Invest.* **66**:102–109.

Gregory, H., 1975, Isolation and structure of urogastrone and its relationship to epidermal growth factor, *Nature (Lond.)* **275**:325–327.

Grotendorst, G. R., Chang, T., Seppa, H. E. J., Kleinman, H. K., and Martin, G. R., 1982, Platelet-derived growth factor is a chemoattractant for vascular smooth muscle cells, *J. Cell. Phys.* **113**:261–266.

Habenicht, A. J. R., Glomset, J. A., Ross, R., and Gronwald, R., 1980a, Increased fluid pinocytosis, induced by action of platelet-derived growth factor on quiescent arterial smooth muscle cells, does not require increased cholesterol biosynthesis, *BBA Libr.* **631**:495–498.

Habenicht, A. J. R., Glomset, J. A., and Ross, R., 1980b, Relation of cholesterol and mevalonic acid to the cell cycle in smooth muscle and Swiss 3T3 cells stimulated to divide by platelet-derived growth factor, *J. Biol. Chem.* **255**:5134–5140.

Habenicht, A. J. R., Glomset, J. A., King, W. C., Nist, C., Mitchell, C. D., and Ross, R., 1981, Early changes in phosphatidylinositol and arachidonic acid metabolism in quiescent Swiss 3T3 cells stimulated to divide by platelet-derived growth factor, *J. Biol. Chem.* **256**:12329–12335.

Harlan, J. M., Bowen-Pope, D. F., Thompson, P. J., and Ross, R., 1985, Alpha thrombin induces secretion of PDGF-like molecules from cultured human endothelial cells, *Clin. Res.* **33**:342a.

Heldin, C.-H., Wasteson, Å., and Westermark, B., 1977, Partial purification and characterization of platelet factors stimulating the multiplication of normal human glial cells, *Exp. Cell Res.* **109**:429–437.

Heldin, C.-H., Westermark, B., and Wasteson, Å., 1979a, Platelet-derived growth factor: Purification and partial characterization, *Proc. Natl. Acad. Sci. USA* **76**:3722–3726.

Heldin, C.-H., Westermark, B., and Wasteson, Å., 1979b, Purification and characterization of human growth factors, in: *Hormones and Cell Culture* (G. Sato and R. Ross, eds.), pp. 17–31, *Cold Spring Harbor Conference of Cell Proliferation*, Vol. 6, Cold Spring Harbor Laboratory, Cold Spring Harbor, New York.

Heldin, C.-H., Westermark, B., and Wasteson, Å., 1981, Specific receptors for platelet-derived growth factor on cells derived from connective tissue and glia, *Proc. Natl. Acad. Sci. USA* **78**:3664–3668.

Heldin, C.-H., Wasteson, Å., and Westermark, B., 1982, Interaction of platelet-derived growth factor with its fibroblast receptor: Demonstration of ligand degradation and receptor modulation, *J. Biol. Chem.* **257**:4216–4221.

Holley, R. W., 1974, Serum factors and growth control, in: *Control of Proliferation in Animal Cells* (B. Clarkson and R. Baserga, eds.), pp. 13–18, *Cold Spring Harbor Symposium*, Cold Spring Harbor Laboratory, Cold Spring Harbor, New York.

Holley, R. W., 1975, Control of growth of mammalian cells in cell culture, *Nature (Lond.)* **258**:487–490.

Huang, J. S., Huang, S. S., Kennedy, B., Deuel, T. F., 1982, Platelet-derived growth factor: Specific binding to target cells, *J. Biol. Chem.* **257**:8130–8136.

Huang, J. S., Huang, S. S., and Deuel, T. F., 1983, Human platelet-derived growth factor: radioimmunoassay and discovery of a specific plasma-binding protein, *J. Cell Biol.* **97**:383–388.

Huang, J. S., Huang, S. S., and Deuel, T. F., 1984, Transforming protein of simian sarcoma virus stimulates autocrine growth of SSV-transformed cells through PDGF cell-surface receptors, *Cell* **39**:79–87.

Johnsson, A., Heldin, C.-H., Wasteson, Å., and Westermark, B., 1982, Platelet-derived growth factor: Identification of constituent polypeptide chains, *Biochem. Biophys. Res. Commun.* **104**:66–74.

Kaplan, D. R., Chao, F. C., Stiles, C. D., Antoniades, H. N., and Scher, C. D., 1979, Platelet alpha granules contain a growth factor for fibroblasts, *Blood* **53**:1043–1052.

Kaplan, P. L., and Ozanne, B., 1983, Cellular responsiveness to growth factors correlates with a cell's ability to express the transformed phenotype, *Cell* **33**:931–938.

Kohler, N., and Lipton, A., 1974, Platelets as a source of fibroblast growth-promoting activity, *Exp. Cell Res.* **87**:297–301.

Lee, D. C., Rose, T. M., Webb, N. R., and Todaro, G. J., 1985, Cloning and sequence analysis of a cDNA for rat transforming growth factor-alpha, *Nature* **313**:489–491.

Leslie, C. C., Antoniades, H. N., and Geyer, R. P., 1982, Stimulation of phopholipid and cholesterol ester synthesis by platelet-derived growth factor in normal and homozygous familial hypercholesterolemia human skin fibroblasts, *Biochem. Biophys. Acta* **711**:290–304.

Linzer, D. I. H., and Nathans, D., 1983, Growth-related changes in specific mRNAs of cultured mouse cells, *Proc. Natl. Acad. Sci. USA* **80**:4271–4275.

Macpherson, I., and Montagnier, L., 1968, Agar suspension culture for the selective assay of cells transformed by polyomavirus, *Virology* **23**:291–294.

Marquardt, H., Hunkapiller, M. W., Hood, L. E., Twardzik, D. R., DeLarco, J. E., Stephenson, J. R., and Todaro, G. J., 1983, Transforming growth factors produced by retrovirus-transformed rodent fibroblasts and human melanoma cells: Amino acid sequence homology with epidermal growth factor, *Proc. Natl. Acad. Sci. USA* **80**:4684–4688.

McClure, D. B., 1983, Anchorage-independent colony formation of SV40 transformed BALB/c 3T3 cells in serum-free medium: Role of cell- and serum-derived factors, *Cell* **32**:999–1006.

Mendoza, S. A., Wigglesworth, N. M., Pohjanpelto, P., and Rozengurt, E., 1980, Na entry and Na–K pump activity in murine, hamster, and human cells—Effect of monensin, serum, platelet extract, and viral transformation, *J. Cell. Phys.* **103**:17–27.

Moses, H. L., Childs, C. B., Halper, J., Shipley, G. D., and Tucker, R. F., 1984, Role of transforming growth factors in neoplastic transformation. in: *Control of Cell Growth and Proliferation* (C. M. Veneziale, ed.), pp. 147–167, Van Nostrand Reinhold Co., New York.

Nilsson, J., Thyberg, J., Heldin, C.-H., Wasteson, A., and Westermark, B., 1983, Surface binding and internalization of platelet-derived growth factor in human fibroblasts, *Proc. Natl. Acad. Sci. USA* **80**:5592–5596.

Nishimura, J., and Deuel, T. F., 1981, Stimulation of protein phosphorylation in Swiss mouse 3T3 cells by human platelet-derived growth factor, *Biochem. Biophys. Res. Commun.* **103**:355–361.

Nishimura, J., Huang, J. S., and Deuel, T. F., 1982, Platelet-derived growth factor stimulates tyrosine-specific protein kinase activity in Swiss mouse 3T3 cell membranes, *Proc. Natl. Acad. Sci. USA* **79**:4303–4307.

Oka, Y., and Orth, D. N., 1983, Human plasma epidermal growth factor/beta urogastrone is associated with blood platelets, *J. Clin. Invest.* **72**:249–259.

Olashaw, N. E., and Pledger, W. J., 1983, Association of platelet-derived growth factor-induced protein with nuclear material, *Nature (Lond.)* **306**:272–274.

Owen, A. J., Geyer, R. P., and Antoniades, H. N., 1982, Human platelet-derived growth factor stimulates amino acid transport and protein synthesis by human diploid fibroblasts in plasma-free medium, *Proc. Natl. Acad. Sci. USA* **79**:3203–3207.

Pike, L. J., Bowen-Pope, D. F., Ross, R., and Krebs, E. G., 1983, Characterization of platelet-derived growth factor-stimulated phosphorylation in cell membranes, *J. Biol. Chem.* **258**:9383–9390.

Pledger, W. J., Stiles, C. D., Antoniades, H. N., and Scher, C. D., 1977, Induction of DNA synthesis in BALB/c 3T3 cells by serum components: Re-evaluation of the commitment process, *Proc. Natl. Acad. Sci. USA* **74**:4481–4485.

Pledger, W. J., Hart, C. A., Locatell, K. L., and Scher, C. D., 1981, Platelet-derived growth factor-modulated proteins: Constitutive synthesis by a transformed cell line, *Proc. Natl. Acad. Sci. USA* **78**:4358–4362.

Raines, E., and Ross, R., 1982, Platelet-derived growth factor. I. High yield purification and evidence for multiple forms, *J. Biol. Chem.* **257**:5154–5160.

Raines, E. W., Bowen-Pope, D. F., and Ross, R., 1984, Plasma binding proteins for platelet-derived growth factor that inhibit its binding to cell-surface receptors, Proc. Natl. Acad. Sci. USA **87**:3424–3428.

Roberts, A. B., Anzano, M. A., Lamb, L. C., Smith, J. M., and Sporn, M. D., 1981, New class of transforming growth factors potentiated by epidermal growth factor: Isolation from non-neoplastic tissues, Proc. Natl. Acad. Sci. USA **78**:5339–5343.

Roberts, A. B., Anzano, M. A., Lamb, L. C., Smith, J. M., Frolik, C. A., Marquardt, H., Todaro, G. J., and Sporn, M. B., 1982, Isolation from murine sarcoma cells of novel transforming growth factors potentiated by EGF, Nature (Lond.) **295**:417–419.

Roberts, A. B., Frolik, C. A., Anzano, M. A, and Sporn, M. B., 1983, Transforming growth factors from neoplastic and nonneoplastic tissues, Fed. Proc. **42**:2621–2626.

Rosenfeld, M. E., Bowen-Pope, D. F., and Ross, R., 1984, Platelet-derived growth factor: Morphological and biochemical studies of binding, internalization, and degradation, J. Cell Physiol. **121**:263–274.

Ross, R., and Glomset, J. A., 1976, The pathogenesis of atherosclerosis, N. Engl. J. Med. **295**:369–377, 420–425.

Ross, R., and Harker, L., 1976, Hyperlipidemia and atherosclerosis, Science **193**:1094–1100.

Ross, R., and Glomset, J. A., Kariya, B., and Harker, L., 1974, Platelet-dependent serum factor that stimulates the proliferation of arterial smooth muscle cells in vitro, Proc. Natl. Acad. Sci. USA **71**:1207–1210.

Rozengurt, E., and Heppel, L. A., 1975, Serum rapidly stimulates ouabain-sensitive [86]Rb[+] influx in quiescent 3T3 cells, Proc. Natl. Acad. Sci. USA **72**:4492–4495.

Rozengurt, E., Stroobant, P., Waterfield, M. D., Deuel, T. F., and Keehan, M., 1983, Platelet-derived growth factor elicits cyclic AMP accumulation in Swiss 3T3 cells: Role of prostaglandin production, Cell **34**:265–272.

Rutherford, R. B., and Ross, R., 1976, Platelet factors stimulate fibroblasts and smooth muscle cells quiescent in plasma serum to proliferate, J. Cell Biol. **69**:196–203.

Savage, C. R., and Cohen, S. J., 1972, Epidermal growth factor and a new derivative. Rapid isolation procedures and biological and chemical characterization, J. Biol. Chem. **247**:7609–7611.

Scher, C. D., Young, S. A., and Locatell, K. L., 1982, Control of cytolysis of BALB/c 3T3 cells by platelet-derived growth factor: A model system for analyzing cell death, J. Cell. Phys. **113**:211–218.

Schmidt, R. A., Glomset, J. A., Wight, T. N., Habenicht, A. J. R., and Ross, R., 1982, A study of the influence of mevalonic acid and its metabolites on the morphology of Swiss 3T3 cells, J. Cell Biol. **95**:144–153.

Senior, R. M., Griffin, G. L., Huang, J. S., Walz, D. A., and Deuel, T. F., 1983, Chemotactic activity of platelet alpha granule proteins for fibroblasts, J. Cell Biol. **96**:382–385.

Seppa, H., Grotendorst, G., Seppa, S., Schiffmann, E., and Martin, G. R., 1982, Platelet-derived growth factor is chemotactic for fibroblasts, J. Cell Biol. **92**:584–588.

Shier, W. T., and Durkin, J. P., 1982, Role of stimulation of arachidonic acid release in the proliferative response of 3T3 mouse fibroblasts to platelet-derived growth factor, J. Cell. Phys. **112**:171–181.

Shupnik, M. A., Antoniades, H. N., and Tashjian, A. H., 1982, Platelet-derived growth factor increases prostaglandin production and decreases in epidermal growth factor receptors in human osteosarcoma cells, Life Sci. **30**:347–353.

Singh, J. P., Chaikin, M. A., and Stiles, C. D., 1982, Phylogenetic analysis of platelet-derived growth factor by radioreceptor assay, J. Cell Biol. **95**:667–671.

Slayback, J. R. B., and Cheung, L. W. Y., 1977, Comparative effects of human platelet growth factor on the growth and morphology of human fetal and adult diploid fibroblasts, Exp. Cell Res. **110**:462–466.

Smith, J. C., Singh, J. P., Lillquist, J. S., Goon, D. S., and Stiles, C. D., 1982, Growth factors adherent to cell substrate are mitogenically active in situ, Nature (Lond.) **296**:154–156.

Smith, J. M., Sporn, M. B., Roberts, A. B., Derynck, R., Winkler, M. E., and Gregory, H., 1985, Human transforming growth factor-α causes precocious eyelid opening in newborn mice, Nature **315**:515–516.

Sporn, M. B., Newton, D. L., Roberts, A. B., DeLarco, J. E., and Todaro, G. J., 1981, Retinoids and suppression of the effects of polypeptide transforming factors—A new molecular approach to chemoprevention of cancer, in: *Molecular Actions and Targets for Cancer Chemotherapeutic Agent* (A. C. Sartorelli, J. S. Lazo, and J. R. Bertino, eds.), pp. 541–544, Academic Press, New York.

Sporn, M. B., Roberts, A. B., Skull, J. H., Smith, J. M., and Ward, J. M., 1983, Polypeptide transforming growth factors isolated from bovine sources and used for wound healing in vivo, *Science* **219:**1329–1331.

Temin, H. M., Pierson, R. W., and Dulak, N. C., 1972, The role of serum in the control of multiplication of avian and mammalian cells in culture, in: *Growth, Nutrition and Metabolism of Cells in Culture*, Vol. 5 (G. H. Rothblat and V. J. Cristofalo, eds.), pp. 50–81, Academic Press, New York.

Todaro, G. J., DeLarco, J. E., and Cohen, S., 1976, Transformation by murine and feline sarcoma viruses specifically blocks binding of EGF to cells, *Nature (Lond.)* **264:**26–31.

Vogel, A., Raines, E., Kariya, B., Rivest, M. J., and Ross, R., 1978, Coordinate control of 3T3 cell proliferation by platelet-derived growth factor and plasma components, *Proc. Natl. Acad. Sci. USA* **75:**2810–2814.

Vogel, A., Ross, R., and Raines, E., 1980, Role of serum components in density-dependent inhibition of growth of cells in culture. Platelet-derived growth factor is the major serum determinant of saturation density, *J. Cell Biol.* **85:**377–385.

Waterfield, M. D., Scrace, G. T., Whittle, N., Stroobant, P., Johnsson, A., Wasteson, Z., Westermark, B., Heldin, C.-H., Huang, J. S., and Deuel, T. F., 1983, Platelet derived growth factor is structurally related to the putative transforming protein p28^{sis} of simian sarcoma virus, *Nature (Lond.)* **304:**35–39.

Weiss, H. J., 1975, Platelet physiology and abnormalities of platelet function, *N. Engl. J. Med.* **293:**531–541.

Westermark, B., and Wasteson, Z., 1976, A platelet factor stimulating human normal glial cells, *Exp. Cell Res.* **98:**170–174.

Wharton, W., Leoff, E., Pledger, W. J., and O'Keefe, E. J., 1982, Modulation of the epidermal growth factor receptor by platelet-derived growth factor and by choleragen: Effects on mitogenesis, *Proc. Natl. Acad. Sci. USA* **79:**5567–5571.

Williams, L. T., Tremble, P., and Antoniades, H., 1982, Platelet-derived growth factor binds specifically to receptors on vascular smooth muscle cells and the binding becomes nondissociable, *Proc. Natl. Acad. Sci. USA* **79:**5867–5870.

Witte, L. D., Kaplan, K. L., Nossel, H. L., Lages, B. A., Weiss, H. G., and Goodman, D. S., 1978, Studies of the release from human platelets of the growth factor for cultured human arterial smooth muscle cells, *Circ. Res.* **42:**402–409.

Witte, L. D., Cornicelli, J. A., Miller, R. W., and Goodman, D. S., 1982, Effects of platelet-derived and endothelial cell-derived growth factors on the low density lipoprotein receptor pathway in cultured human fibroblasts, *J. Biol. Chem.* **257:**5392–5401.

Chapter 8

Intercellular Recognition and Adhesion in Desmosomes

GARY GORBSKY

1. Introduction

The regulation of intercellular adhesion has been long believed to have important roles in embryonic morphogenesis, in the maintenance of normal adult tissues, and in cancer. Intercellular adhesion is mediated by components of the plasma membrane and the extracellular matrix. In certain regions on the apposed plasma membranes of two cells, surface molecules coalesce to form ultrastructurally recognizable intercellular junctions. These are presumed to be sites of enhanced intercellular adhesion. The purpose of this chapter is to describe the current state of knowledge and methodology in understanding the biogenesis and function of the major adhesive intercellular junctions of vertebrates.

2. Cell Junctions and Intercellular Adhesion

By strictly morphological criteria, most vertebrate junctions can be grouped into one of four categories: (1) the tight junction, also called the *zonula occludens*; (2) the gap junction; (3) the *adhaerens*-type junction, called in various tissues the *zonula adhaerens*, belt desmosome, *fascia adhaerens*, or intermediate junction; and (4) the spot desmosome, also called the *macula adhaerens* or simply desmosome (for reviews of junctional morphology and classification, see McNutt and Weinstein, 1973; Staehelin, 1974; Weinstein *et al.*, 1976). The synapse may also be considered a specialized intercellular junction. For the purposes of this discussion, adhaerens junction is used as a general term for junctions of the zonula adhaerens–fascia adhaerens–intermediate junction class, and the term *desmosome* is reserved for the spot desmosome or macula adhaerens.

GARY GORBSKY • High Voltage Electron Microscopy Laboratory and Department of Molecular Biology, University of Wisconsin—Madison, Madison, Wisconsin 53706.

Presumably all cell junctions as well as ultrastructurally unspecialized cell surfaces contribute to some extent in intercellular adhesion. However, because of their clear structural association with cytoskeletal elements, the adhaerens junction and the desmosome likely have the most significant roles in maintaining tissue integrity and in providing traction for cell motility. Until recently, studies of the biochemistry of cellular adhesion have rarely been directed toward understanding the molecular nature of cellular junctions, perhaps because junctions were considered relatively slow-forming, static structures. However, adhaerens junctions are formed within seconds of contact by fibroblasts in culture (Heaysman and Pegrum, 1973) and ultrastructurally recognizable, incipient desmosomes are observed minutes after recombining dissociated epithelial cells (Dembitzer et al., 1980).

In addition, because junctions of various types are present on a wide variety of tissues, they have sometimes been thought to be too nonspecific, i.e., found in too many different kinds of tissues to be considered important in regulating morphogenetic events. For some time, studies of the reaggregative potential of dissociated tissue cells have led to generalizations that cells exhibit considerable selectivity in their adhesive preferences (for reviews, see Marchase et al., 1976; Maslow, 1976; Greig and Jones, 1977; Denburg, 1978; Frazier and Glaser, 1979; Jones, 1980; Garrod and Nicol, 1981). Observations of cellular behavior and biochemical efforts on the purification of vertebrate cell adhesion molecules have, taken together, provided evidence that the adhesive preferences of cells from different tissues are in some cases mediated through qualitatively different cell surface adhesion molecules (Takeichi, 1977; Rutishauser et al., 1978; Hyafil et al., 1980; Knudsen et al., 1981; Grunwald et al., 1982; Yoshida and Takeichi, 1982; Damsky et al., 1983; Gallin et al., 1983; Ocklind et al., 1983; Ogou et al., 1983). Clearly, the molecular interactions of cell junctions exhibit a very high level of qualitative specificity. A half-desmosome is never found directly apposed to a half-gap junction, although both types of junction may be found near one another along a cell border. Moreover, the existence of qualitatively different junctional adhesion molecules in different tissues is ultrastructurally evident; desmosomes are abundant in epithelia and cardiac tissue but are absent in mesenchymal and neuronal tissues. Perhaps differences in the kinds or relative amounts of junctional components on cells from different tissues contribute to tissue specificity in regulating cellular adhesive preferences.

The adhaerens junction and the desmosome are of similar morphology but were distinguished early in ultrastructural studies of cell junctions (Farquhar and Palade, 1963). The most reliable structural distinction is the fact that adhaerens junctions are attached to the 7-nm actin filaments while desmosomes are linked to the 10-nm intermediate filaments. While a number of the protein components of the cytoplasmic region of adhaerens junctions have been identified (Geiger et al., 1981; Tokuyasu et al., 1981; Chen and Singer, 1982; Maher and Singer, 1983; reviewed in Geiger, 1983), only preliminary information about their intercellular adhesive components is available (Colaco and Evans, 1981; Maher and Singer, 1983). The adhaerens junction is therefore considered only briefly in this discussion. By contrast, the major structural proteins of the desmosome have been characterized (Skerrow and Matoltsy, 1974a,b; Gorbsky and Steinberg, 1981; Franke et al., 1982, 1983c; Mueller and Franke, 1983) and the intercellular adhesive elements isolated and analyzed (Gorbsky and Steinberg, 1981; Cohen et al., 1983). These are considered in detail (Sections 4.2 and 4.3).

3. The Adhaerens Junction

Adhaerens junctions are widely distributed and are pleiomorphic in size and shape among different tissues. Encircling the apical regions of columnar epithelia, the adhaerens junction is referred to as a zonula adhaerens or belt desmosome. Between cardiac muscle cells it occurs in sheets and is hence termed the fascia adhaerens. In many cell types in vivo and in vitro these junctions are found as smaller, focal attachments between cells. On the basis of morphological and biochemical criteria, adhaerens junctions are also thought to be related to the focal contacts or adhesion plaques by which cells in culture attach to plastic or glass.

With conventional electron microscopy, cross-sectioned adhaerens junctions are characterized by a 15–25-nm intercellular space often occupied by some roughly filamentous, electron-dense material (Fig. 1). Cell–cell crossbridges in these junctions have been clearly visualized by the rapid freeze-deep etch technique (Hirokawa and Heuser, 1981). Underlying these intercellular components, actin filaments insert into a fuzzy plaque on the inner surface of the plasma membrane. Thus, this junction probably contributes to the coordination of contractile activity among groups of cells.

Fractions enriched in adhaerens junctions have been prepared from rodent heart tissue and analyzed by electrophoresis (Kensler and Goodenough, 1980; Colaco and Evans, 1981; Maher and Singer, 1983). Very little characterization of putative intercellular adhesive components is yet available. Cell surface components have been identified within the adhesive contacts of tissue culture cells with their substrata (Culp et al., 1979; Lark and Culp, 1982; Oesch and Birchmeier, 1982; Neyfakh and Svitkina, 1983). Possibly these same macromolecules may also contribute to intercellular adhesion in adhaerens junctions between cells.

In contrast to the dearth of information about the adhesive elements of adhaerens junctions, a number of proteins associated with the cytoplasmic region of the junction have been identified (reviewed in Geiger, 1983). These include α-actinin (Craig and Pardo, 1979; Geiger et al., 1981) vinculin (Geiger et al., 1981), and a recently described 200,000-M_r protein (Maher and Singer, 1983). These proteins have also been localized at the termini of microfilament bundles in the focal contacts of tissue culture cells (Lazarides and Burridge, 1975; Schollmeyer et al., 1976; Geiger, 1979; Burridge and Feramisco, 1980; Chen and Singer, 1982; Maher and Singer, 1983).

Immunoelectron microscopy on ultrathin frozen sections has been used to

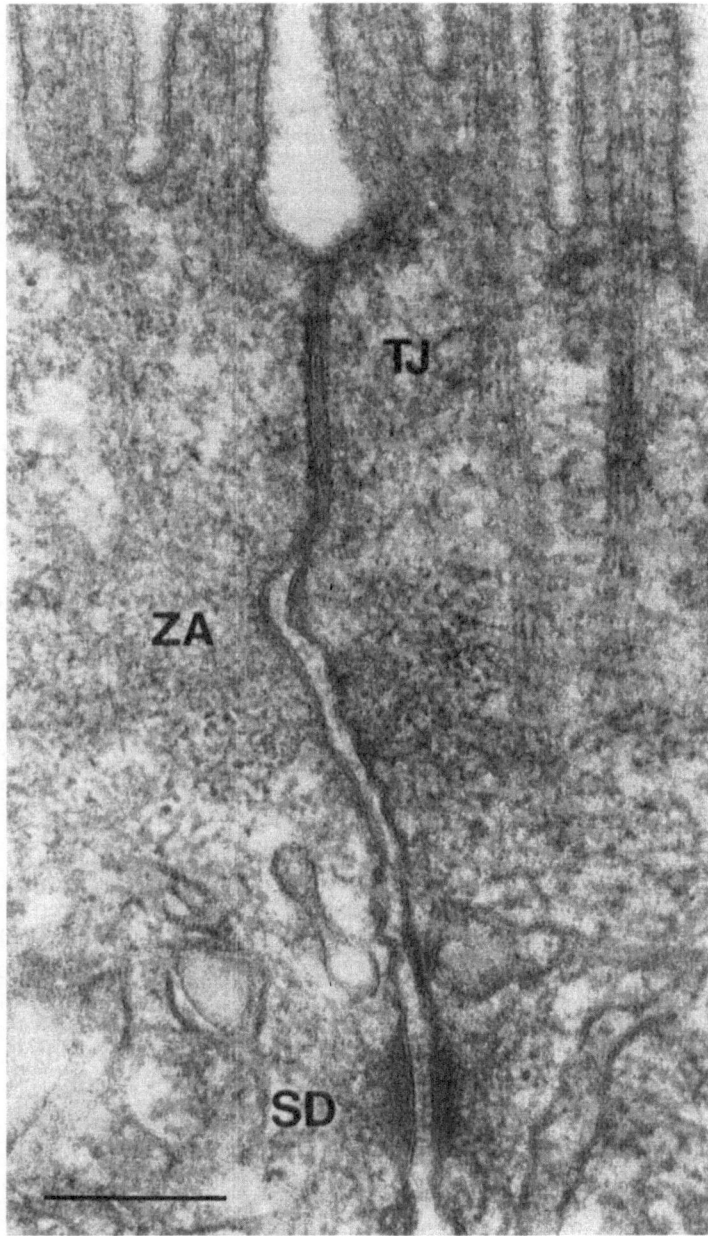

Figure 1. Electron micrograph of section through the border between two intestinal epithelial cells of a *Xenopus laevis* tadpole. This area at the apical region of the cells shows a junctional complex containing a tight junction (TJ), a zonula adhaerens (ZA) and a spot desmosome (SD). Scale bar: 0.2 μm. (Reproduced from Hull and Staehelin, 1979, by copyright permission of Rockefeller University Press.)

compare the localization of α-actinin and vinculin in adhaerens junctions of intestinal epithelium (Geiger *et al.*, 1981) and cardiac tissue (Tokuyasu *et al.*, 1981). Vinculin was found in the junctional plaques very near the cell surface, while α-actinin occupied a region somewhat farther from the plasma membrane toward the cell interior. The precise role of these macromolecules in the attachment of actin filaments to the plasma membrane is not yet understood.

4. Molecular Architecture of the Desmosome

4.1. Structure of the Desmosome

The desmosome is the most ultrastructurally elaborate of the cellular junctions (for reviews, see Overton, 1975; Skerrow, 1978; Skerrow and Skerrow, 1980; Arnn and Staehelin, 1981). It occupies a round or oval region of the cell surface, about 0.1–1.0 μm in diameter. In cross section, desmosomes are characterized by a 25–35-nm intercellular space, usually bisected by an electron-dense central stratum (Fig. 2). Some studies have revealed regularly spaced crossbridges linking this central stratum to the cell membranes (Rayns *et al.*, 1969; Hull and Staehelin, 1979; Franke *et al.*, 1983a). Freeze-fracture replicas show the presence of irregularly arranged intramembrane particles 8–15 nm in diameter (Breathnach *et al.*, 1976; Kelly and Shienvold, 1976; Shimono and Clementi, 1976; Caputo and Peluchetti, 1977; Leloup *et al.*, 1979; Kelly and Kuda, 1981; Kitajima *et al.*, 1983). The hemidesmosome is a related structure that attaches epithelial cells to the basal lamina. In cross section it resembles half a normal desmosome, although certain fine structural features are distinct (Kelly, 1966; Shienvold and Kelly, 1976; Shimono and Clementi, 1976; Caputo and Peluchetti, 1977).

In the desmosome, on the cytoplasmic face of the plasma membrane is a dense plaque sometimes revealed as a mat of closely packed fine filaments (McNutt and Weinstein, 1973). This plaque is the region of attachment of the intermediate filaments. These filaments are observed to loop through the plaque (Kelly, 1966; Kelly and Kuda, 1981) or to pass parallel to it (see Overton, 1962; Lentz and Trinkaus, 1971; Hull and Staehelin, 1979; Kartenbeck *et al.*, 1983), suggesting that the filaments and plaque associate laterally. Evidence based on electron microscopic images of epidermal cells *in vivo* (Leloup *et al.*, 1979) and *in vitro* (Jones *et al.*, 1982) and of fractions from disrupted cells containing desmosome intermediate filament complexes (Franke *et al.*, 1981b) suggests that the filaments may also terminate in the desmosomal plaque.

In favorably oriented and stained sections of desmosomes, it is clear that different strata exist within the desmosomal plaque (Odland, 1958; Snell, 1965; Raknerud, 1975; Shida *et al.*, 1982). Two dense regions, one directly adjacent to the plasma membrane, are separated by an electron-lucent area. The dense region most interior to the cell may represent the closest approach of the intermediate filaments to the plasma membrane. Freeze-fracture methods have permitted observation of linear elements called "traversing filaments" that appear

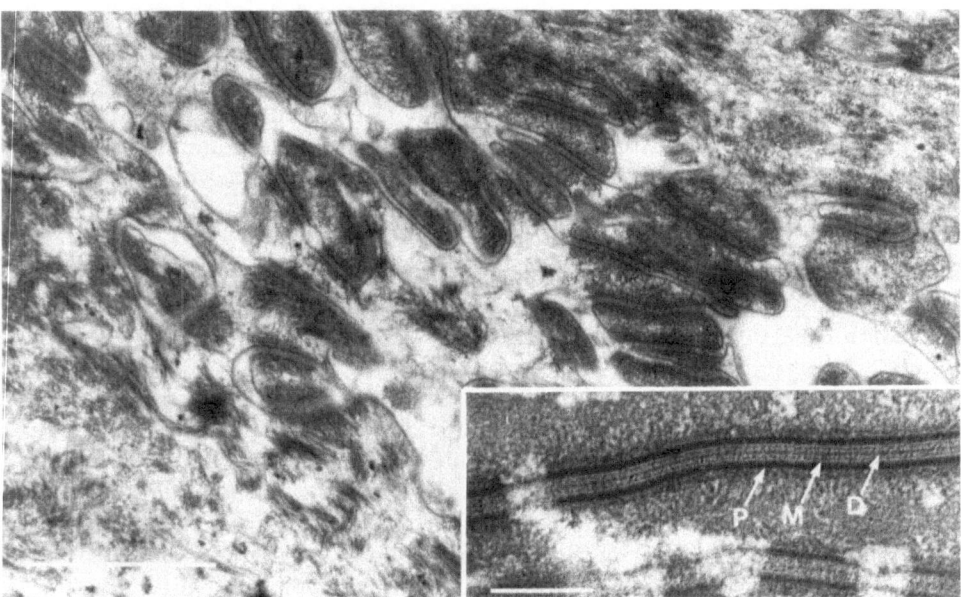

Figure 2. Electron micrograph through the border between two epidermal cells of bovine muzzle. Note the highly interdigitated intercellular boundary showing numerous desmosomes. Scale bar: 1.0 μm. Inset: Desmosome at high magnification showing intracellular plaques (P), plasma membranes (M), and intercellular components or desmoglea (D). Scale bar: 0.2 μm (Reproduced from Gorbsky and Steinberg, 1981, by copyright permission of Rockefeller University Press.)

to join the tonofilaments to the membrane or to the intercellular region of the desmosome (Kelly and Shienvold, 1976; Leloup *et al.*, 1979; Kelly and Kuda, 1981). Some investigators have proposed that tonofilaments are unraveled into *protofilaments* accompanying their attachment to the desmosomal plaque (Leloup *et al.*, 1979; Arnn and Staehelin, 1983).

4.2. Biochemical and Structural Mapping of the Glycosylated Desmosomal Proteins

Early ideas on the biochemistry of the intercellular adhesive(s) of desmosomes were drawn from histochemical studies and from tests on the susceptibility of desmosomes to chemical cleavage. These early studies showed the presence of carbohydrates in the intercellular region and implicated calcium and protein as mediators of intercellular attachment.

Staining with ruthenium red (Kelly, 1966), phosphotungstic acid (Hopwood *et al.*, 1977), and periodate-silver methenamine (Rambourg and Leblond, 1967; Hopwood *et al.*, 1977) suggested that the intercellular region of the desmosome is rich in carbohydrate. Sedar and Forte (1964) reported that desmosomes in gastric mucosa were cleaved when divalent cations were removed by EDTA but reformed when calcium was re-added. Muir (1967) observed the splitting of desmosomes when heart muscle was perfused with calcium-free media. Early ultrastructural studies showed that trypsin-treatment cleaved desmosomes in chick blastoderm (Overton, 1962, 1968) and in chick corneal epithelium (Overton, 1973). Borysenko and Revel (1973) examined desmosomes in various tissues and reported differences in their susceptibility to cleavage. In general, desmosomes from columnar epithelia were cleaved by EDTA and unaffected by trypsin, while those of stratified squamous epithelia and glandular epithelia showed the reverse sensitivity.

Thorough understanding of the chemistry of desmosomal adhesion requires analysis of isolated desmosomes and their components. Bovine snout epidermis has provided a rich and readily available source of material. Skerrow and Matoltsy (1974a) extracted slices of bovine epidermis with low pH to solubilize the keratin proteins. After sucrose-gradient centrifugation of the insoluble residue, a desmosome fraction was obtained. Analysis by sodium dodecyl sulfate-polyacrylamide gel electrophoresis (SDS-PAGE) revealed some 24 protein bands, two of which were found positive by periodic acid-Schiff (PAS) reaction (Skerrow and Matoltsy, 1974b). Drochmans et al. (1978) described an alternative procedure using high pH, but electrophoretic analysis demonstrated that desmosome fractions obtained by this method are composed largely of keratin polypeptides.

Gorbsky and Steinberg (1981) modified the procedure used by Skerrow and Matoltsy to obtain large quantities of purified whole desmosomes (Fig. 3A). By means of metrizamide gradient centrifugation, these workers removed the cytoplasmic plaque elements of the desmosome and purified the intercellular-plasma membrane regions. These were called desmosomal cores (Fig. 3B). Characterization of these fractions by SDS-PAGE revealed that desmosomes contain a limited number of polypeptide components (Fig. 4). Blotting analyses with monoclonal antibodies and lectins and mapping of peptides by partial proteolysis indicated that the desmosomal cores are composed of three distinct glycoproteins, two of which form multiple bands when analyzed by SDS-PAGE (Cohen et al., 1983). These desmosomal intercellular proteins have been named desmoglein I, II, and III (Gorbsky et al., 1985). Desmoglein I (DGI) is a closely spaced triplet with a molecular weight of about 150,000. Desmoglein II (DGII) is composed of two broad bands of 118,000 and 97,000 M_r. Desmoglein III (DGIII) is a single band of 22,000 M_r. Some or all of these molecules must make up the structural adhesive elements of desmosomes.

In an effort to elucidate the structural interactions of desmosomal components, Shida et al. (1983) have used high-resolution immunoelectron microscopy to map the locations of the DGI and DGII polypeptides within desmosomes in intact tissue. The carbohydrate moieties of DGI and DGII were also localized through the use of a ferritin conjugate of the lectin concanavalin A (Con A) (Shida et al., 1982). Polyclonal antibodies to DGI and DGII bound in the intercellular region and, surprisingly, in the desmosomal plaque as well. The lectin conjugates labeled only along the intercellular midline. These results indicate that DGI and DGII are transmembrane elements extending from the

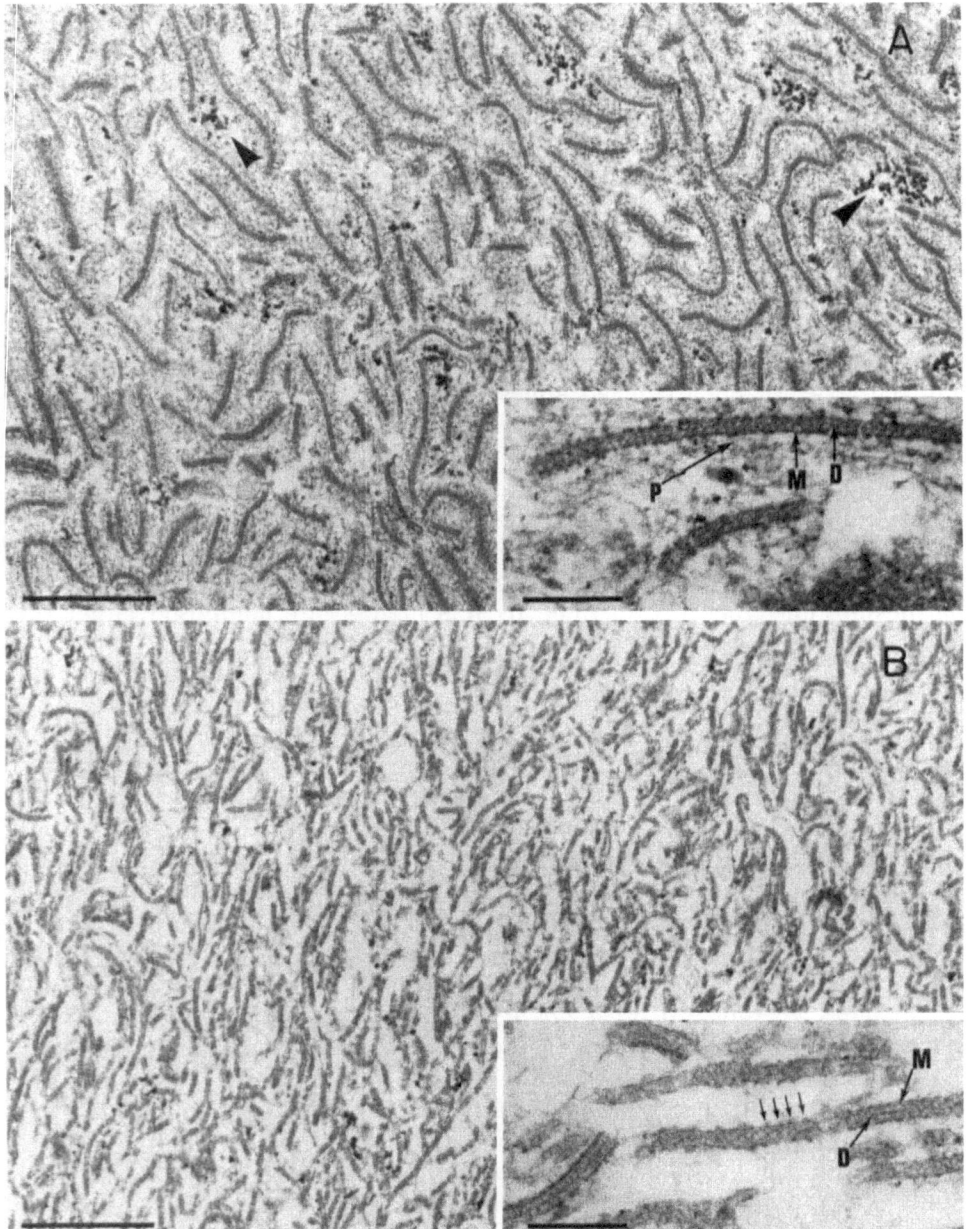

Figure 3. (A) Purified whole desmosomes. Loose filamentous material presumably derived from the desmosomal plaques is apparent on the former cytoplasmic side of the plasma membrane. Some electron- dense contaminants are present (arrowheads). Scale bar: 1.0 μm. Inset: Whole desmosomes at high magnification. Filamentous material presumably derived from the desmosomal plaque (P), plasma membranes (M) and intercellular components or desmoglea (D) are present. Scale bar: 0.2 μm. (B) Desmosome cores. Filamentous plaque material is largely absent. Scale bar: 1.0 μm. Inset:

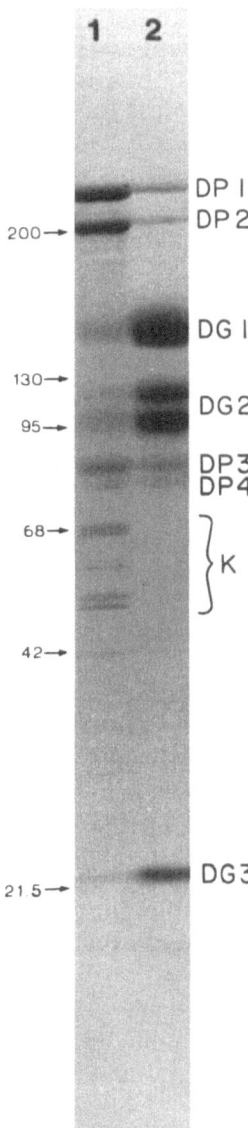

Figure 4. Coomassie blue-stained SDS-PAGE gels of whole desmosomes (lane 1) and desmosome cores (lane 2). Note the enrichment of the desmogleins (DG) and depletion of the desmoplakins (DP) and keratins (K) in the desmosome core fraction. Mobilities of some molecular-weight standards are indicated at left. (Reproduced from Gorbsky and Steinberg, 1981, by copyright permission of Rockefeller University Press.)

Desmosome cores at high magnification. Plaque materials are almost entirely absent, leaving only small insertions (arrows), sometimes spaced regularly on the cytoplasmic side of the plasma membranes. Scale bar: 0.2 μm. (Reproduced from Gorbsky and Steinberg, 1981, by copyright permission of Rockefeller University Press.)

region of the plaque, where the tonofilaments insert, to the intercellular zone, where the carbohydrate components are concentrated. Similar localization studies of DGIII, the 22,000-M_r component, will complete the ultrastructural mapping of the major intercellular proteins of the desmosome. Because of the transmembrane nature of DGI and DGII, it is possible that these desmosomal glycoproteins both mediate intercellular adhesion and participate in attachment of the membrane surface to the cytoskeletal intermediate filaments. Precisely how they function in these roles remains to be determined.

4.3. Biochemical and Structural Mapping of the Nonglycosylated Desmosomal Proteins

In electrophoretic gels of purified whole desmosomes, two bands of 240,000 and 210,000 M_r are most prominent. These polypeptides are not glycosylated, suggesting that they are intracellular (Skerrow and Matoltsy, 1974b; Gorbsky and Steinberg, 1981; Cohen et al., 1983). Skerrow and Matoltsy (1974b) first proposed that these polypeptides are components of the desmosomal plaque. Gorbsky and Steinberg (1981) found that metrizamide gradient centrifugation of purified whole desmosomes stripped away most of the plaque material, leaving desmosomal cores, with the intercellular components sandwiched between plasma membranes. With the loss of most of the ultrastructurally recognizable plaque material, the 240,000- and 210,000-M_r polypeptides and two other nonglycosylated proteins, of 81,000 and 77,000 M_r, were selectively depleted. From this evidence Gorbsky and Steinberg (1981) concluded that all four polypeptides originated in the desmosomal plaque.

The biochemical relationships of the nonglycosylated desmosomal polypeptides, termed desmoplakins (Mueller and Franke, 1983), have been studied in some detail. Mapping by partial proteolysis has revealed that the 240,000-M_r desmoplakin (DPI) and the 210,000-M_r desmoplakin (DPII) show similarity in sequence (Mueller and Franke, 1983; Gorbsky et al., 1984), a result supported by immunoblotting studies using polyclonal antibodies (Cowin and Garrod, 1983; Mueller and Franke, 1983; Gorbsky et al., 1984). The 81,000-M_r desmoplakin (DPIII) and the 77,000-M_r protein, presumed to be another desmoplakin, are distinct from each other and from DPI and DPII (Franke et al., 1983c; Gorbsky et al., 1984).

Using immunoperoxidase and immunogold staining for electron microscopic evaluation, Franke et al. (1982, 1983a) reported that polyclonal antibodies to DPI–DPII bound to the intracellular plaque region in desmosomes. This same group reported that these antibodies also labeled the plaque region of hemidesmosomes (Mueller and Franke, 1983). Gorbsky et al. (1985) used polyclonal antibodies for high-resolution mapping of the ultrastructural distribution of DPIII on thin sections of intact bovine epidermal tissue. These workers showed that DPIII was also located within the desmosomal plaque, extending from the plasma membrane to the region of insertion of the intermediate filaments.

In addition to the major components of the desmosomal plaque some minor elements have been identified by immunolocalization. These include D1, a bovine-specific component or components defined by an antibody that reacts with several high-molecular-weight protein bands (Franke *et al.*, 1981*a*) and plectin, a 300,000-M_r polypeptide associated with the cytoplasmic region of desmosomes and other cell junctions (Wiche *et al.*, 1983).

5. Desmosomes in Intercellular Recognition

5.1. Tissue Heterogeneity of Desmosomal Proteins

Immunofluorescence studies with polyclonal antibodies have demonstrated that the desmogleins and desmoplakins are antigenically similar among a variety of epithelia and cardiac tissue in a wide range of species (Franke *et al.*, 1982, 1983*b*; Cowin and Garrod, 1983). This finding suggests that desmosomes in different tissues may use similar molecular mechanisms for intercellular adhesion and cytoskeletal attachment. However, ultrastructural studies suggest that there may be a limited capacity for cells from widely different tissues to form desmosomes cooperatively (Overton, 1973, 1974). Desmosomes were formed between cells from chick corneal epithelium and mouse epidermis, two related tissues from different species, whereas none were seen between cells from chick corneal epithelium and chick heart (Overton and Kapmarski, 1974). Thus the antigenic similarity shown in desmosomes from different tissues through labeling studies with polyclonal antibodies may or may not imply functional similarity in the recognition properties of desmosomal proteins.

Other evidence also suggests that considerable molecular heterogeneity may exist among the desmosomal components of different tissues. As noted previously, desmosomes from various epithelia showed differences in their susceptibility to chemical cleavage (Borysenko and Revel, 1973). Studying three related bovine epithelia—epidermis, cornea, and esophagus—Giudice *et al.* (1984) found that the desmoglein proteins exhibit certain differences in electrophoretic mobility and reactivity to a group of monoclonal antibodies. Franke *et al.* (1982) reported that polyclonal antibodies to epidermal DPI–DPII, the high-molecular-weight desmoplakin doublet, labeled only a single high-molecular-weight band in extracts of heart tissue. Possibly this difference between tissues is related to the fact that the desmosome-associated intermediate filaments are composed of keratin in the epidermis and of desmin in the heart (Lazarides, 1980).

Because of their abundance, desmosomes of stratified squamous epithelia, especially epidermis, have been the chosen materials for most biochemical studies. Isolations of desmosome-enriched fractions from another type of epithelial tissue, the liver, have been described (Hubbard and Ma, 1983; Hixson *et al.*, 1983), but extensive characterizations of desmosome-specific components have not yet been reported.

Tissues that lack ultrastructurally recognizable desmosomes (e.g., mesenchyme and neuronal tissue) were unlabeled by polyclonal antibodies to desmosomal components (Cowin and Garrod, 1983; Mueller and Franke, 1983). Geiger et al. (1983) performed double immunofluorescent labeling comparing the distribution of vinculin and the D1 set of bovine desmosomal plaque polypeptides. Almost no overlap in distribution was detected, suggesting that adhaerens junctions and desmosomes are not biochemically related.

5.2. Biochemical Mechanisms of Desmosomal Recognition

Although desmosomal proteins have been separated and isolated, it has heretofore not been possible to reassemble a desmosome from its components. Thus very little is known about the biochemistry of the interaction of desmosomal constituents, either laterally within one cell surface or across the intercellular space. As a further complication, unlike cytoplasmic organelles, the synthesis of desmosomes and other intercellular junctions requires cooperative assembly by two adjacent cells. Much of what is understood about the mechanisms of intermolecular recognition in desmosomes has been inferred from pharmacological, ultrastructural, and cellular behavior studies.

In many cellular systems a recurrent theme in studies of cell adhesion biochemistry is the importance of calcium (Ueda and Takeichi, 1976; Brackenbury et al., 1977; Magnani et al., 1981; Grunwald et al., 1980, 1982; Hyafil et al., 1981; Thomas and Steinberg, 1981; Yoshida and Takeichi, 1982; see also Chapter 9, this volume). Hennings et al. (1980, 1983) showed that neonatal mouse keratinocytes cultured in media containing low levels of calcium (0.05–0.09 mM) did not form desmosomes and grew as monolayers. Increasing the calcium concentration to physiological levels (1.2 mM) caused the cells to assemble desmosomes and to become stratified into layers approximating their normal in vivo differentiation pathway. Desmosome assembly was observed minutes after the increase in calcium concentration and did not require protein synthesis (Hennings and Holbrook, 1983). This finding suggests that in low calcium, preformed desmosomal components are present on the cell surface or within the cytoplasm. An increase in the calcium level may then permit interaction of the desmosomal precursors, leading to junction assembly.

Overton (1982) provided evidence that the carbohydrate portions of the desmogleins are not essential for desmosome assembly in cells from chick corneal epithelium. Tunicamycin, a drug that inhibits the maturation of N-linked oligosaccharides, did not affect the frequency of desmosome formation, provided that a protease inhibitor was also present in the cell culture medium. Without the protease inhibitor, desmosomes were not formed, indicating that in these cells, mature desmosomal carbohydrates function to stabilize desmosomes against proteolytic attack but are not strictly required for desmosome assembly. King and Tabiowo (1981) also saw no change in desmosomal morphology in epidermal cells cultured 4 days in tunicamycin.

The formation of desmosomes is restricted to cardiac and epithelial tissues. In culture, desmosomes do not form between cells that lack them in *vivo* and cells that normally contain desmosomes (Overton, 1974). A similar phenomenon occurs *in vivo* in the epidermis. Among the keratinocytes, the epithelial cells that compose the bulk of the epidermis, are numerous dendritic mesodermally derived cells called Langerhans cells. These cells are thought to perform immunoreactive functions (Stingl *et al.*, 1980; Thorbecke *et al.*, 1980). Numerous desmosomes are formed on cell surfaces between keratinocytes. However, none occur on cell borders between keratinocytes and Langerhans cells (Snell, 1965). Thus, both *in vitro* and *in vivo*, mutual competence and participation by both neighbor cells are prerequisites to desmosome assembly. (The hemidesmosome is a special case discussed below.)

In certain tissues, cells from older embryos make more desmosomes than do cells from younger embryos (Hay and Revel, 1969; Overton, 1973; Overton and Kapmarski, 1974). Overton (1977) found that confrontation of tissue cells with different inherent desmosomal frequencies could be used to modulate the frequency of junctions made at heterotypic cell borders. When cells that formed many desmosomes (older cells) were apposed to cells that formed few (younger cells), an intermediate number of desmosomes were assembled along the heterotypic cell borders. Importantly, young cells were thus induced to form more desmosomes than they would normally produce. Moreover, the frequency of desmosomes was altered only at the heterotypic borders; on the surfaces of the cells where they were apposed to cells of like type, the desmosome frequency was unchanged. These experiments suggest that the induction signals for desmosome assembly may occur locally, involving molecules in or near the region of the cell surface that will give rise to the desmosome.

5.3. Ultrastructural Events in Desmosome Assembly

In contrast to the dearth of information on the biochemistry of the molecular interactions in desmosome assembly, the ultrastructural events have been described for a number of systems, including the reaggregation of dissociated cells (Overton, 1962, 1973; Dembitzer *et al.*, 1980), keratinocytes in monolayer culture (Hennings *et al.*, 1980; Hino *et al.*, 1982; Hennings and Holbrook, 1983), the maturation of the normal epidermis (Leloup *et al.*, 1979), the morphogenesis of epithelia in embryos (Overton, 1962; Hay and Revel, 1969; Lentz and Trinkaus, 1971), and wound healing (Krawczyk and Wilgram, 1973). Some such studies suggest that the earliest events of desmosome assembly are the formation of intercellular connecting strands (Krawczyk and Wilgram, 1973; Leloup *et al.*, 1979). As the intercellular elements become more distinct, the cytoplasmic plaque then becomes visible, and association of the intermediate filaments occurs. One possible scheme is that close apposition of cell surfaces results in the interconnection of adhesion molecules from adjacent cells. Being initially mobile within the plasma membranes, a co-patching of these molecules serves to concentrate them into a small region of the cell border. This concentration of cell surface elements then signals initiation of assembly of the cytoplasmic desmosomal elements and the attachment of the intermediate filaments. A form of this model for the assembly of intercellular junctions was initially proposed by Campbell and Campbell (1971).

In apparent contradiction to the above model, certain ultrastructural studies of developing chick corneal epithelium emphasize the frequent appearance of unpaired cytoplasmic densities that resemble desmosomal plaques (Hay and Revel, 1969; Overton, 1973). From this evidence, a different mechanism of desmosomal assembly can be described. Cytoplasmic and/or surface components are assembled unilaterally by a cell. Only later, through matching of plaques or induction of a complementary structure in an adjacent cell, are symmetrical desmosomes obtained. Thus, in this model, assembly of the cytoplasmic plaque may precede the interconnection of the intercellular adhesive elements. Overton and DeSalle (1980) have shown that after trypsin-mediated cell dissociation, cleaved desmosomal remnants, if maintained on the cell surface by inhibition of phagocytosis (Overton and Culver, 1973), seem to become incorporated into newly formed whole desmosomes. From this evidence, it appears that "half-desmosomes" maintained on a cell surface may pair up or induce complementary structures in adjacent cells.

The mechanisms of desmosome assembly used in normal morphogenesis are as yet unresolved. Possibly different mechanisms are used in different tissues or at different times during development. Studies of assembly have heretofore been limited to inferred sequences, on the basis of interpretations of static electron microscopic images. Immature desmosomes are difficult to distinguish from other sources of electron density, including other junction types. The availability of specific antibodies to the intercellular and cytoplasmic components of desmosomes should greatly facilitate progress in elucidating the sequence and mechanism of desmosomal assembly.

5.4. Breakdown and Turnover of Desmosomes

In order for cells to move past their neighbors, they must possess mechanisms of disengaging their intercellular connections. Very little is known about how cells *in vivo* sever their desmosomal attachments. In addition, it is unclear to what degree normal turnover of desmosomes or their components takes place in migrating or in stationary cells. Whole symmetrical desmosomes have been reported to be taken up in phagocytic vesicles in epidermis (Allen and Potten, 1975) and in tumor cells (Hayashi and Ishimaru, 1981). Whether this phenomenon occurs with sufficient frequency to account for a significant portion of desmosomal turnover is uncertain.

In vitro, "half-desmosomes" produced by the cleavage of desmosomes with proteases or calcium chelators are generally internalized by phagocytosis (Overton, 1968; Berry and Friend, 1969; Kartenbeck *et al.*, 1982). Except in certain disease conditions (Hashimoto and Lever, 1970; Wolff and Schreiner, 1971), cleavage and phagocytosis of desmosome halves have not been described *in vivo*. Little structural evidence is available for other mechanisms of desmosomal turnover or remodeling. However, other possibilities exist. These include lateral dispersion of desmosomal components in the plasma membrane, release and exchange of intercellular desmosomal connections, "sliding" of whole desmosomes along the cell surface in the membrane, and sloughing of desmosomes into the intercellular space. The availability of antibodies to enhance the detectability of desmosomal components should greatly facilitate studies of normal turnover of desmosomes. The application of antibody probes to study desmosomal phagocytosis induced by calcium chelators has been described for a bovine kidney cell line (Kartenbeck et al., 1982). Interestingly, these investigators found distinctly different modes of internalization for the cytoplasmic plaque proteins of desmosomes and adhaerens junctions.

Another programmed desmosomal breakdown occurs during the normal maturation and shedding of epithelial cells in the gut and on body surfaces. Accompanying terminal differentiation in epidermal cells, desmosomes undergo a series of structural alterations (Snell, 1965; Allen and Potten, 1975). Presumably, these changes contribute to the protective function of the terminally differentiated epidermis or are required to loosen intercellular adhesions in preparation for the eventual sloughing of the dead cells from the skin surface.

6. The Hemidesmosome

Although resembling in structure an unpaired desmosome half, hemidesmosomes are quite different in function, serving to join epithelial cells to the basal lamina. They are characteristically found situated over dense specializations in the basal lamina (Kelly, 1966; Flickinger, 1970; Shienvold and Kelly, 1976) and are often associated with dermally derived fibers called anchoring fibrils (Susi et al., 1967; Palade and Farquhar, 1965; Briggaman et al., 1971). No macromolecules restricted exclusively to the space between hemidesmosomes and the basal lamina have yet been identified. Type IV collagen, laminin, and bullous pemphigoid antigen commonly occur along the epithelial–basal lamina boundary (Yaoita et al., 1978; Stanley et al., 1981; Laurie et al., 1982), but these proteins are found both under and adjacent to hemidesmosomes. In an ultrastructural study of the time course of basal lamina formation by epithelia, Briggaman et al. (1971) reported that basal lamina was initially organized under the hemidesmosomes and later extended to become continuous along the epidermal–dermal interface.

Differences in the sensitivity of desmosomes and hemidesmosomes to cleavage by certain proteases and chelators of divalent cations suggest differences in their adhesive chemistry. The protease dispase can be used to separate intact epithelial sheets from their basal laminae, thus cleaving hemidesmosomes without disturbing desmosomal connections (Green et al., 1979; Gipson et al., 1983). Hemidesmosomes are apparently also more susceptible to cold trypsin (Rawles, 1963) and EDTA (Dodson and Hay, 1971).

Gipson et al. (1983) found that epithelial sheets obtained by dispase treatment of rabbit corneas had fully formed desmosomes but lacked hemidesmosomes. If the epithelium was recombined with corneal basal lamina in culture, hemidesmosomes reformed within a few hours in their proper position over the anchoring fibrils. By contrast, no hemidesmosomes formed even after many hours when corneal epithelium was cultured on foreign extracellular matrices, such as Descemet's membrane and the lens capsule, that lacked anchoring fibrils. The precise biochemical nature of anchoring fibrils is unknown. The presence of anchoring fibrils is not an essential prerequisite, since hemidesmosome formation in culture has been observed in epithelial cells grown on gelatin (Billig et al., 1982) and plastic substrata (Christophers and Wolff, 1975; Jepsen et al., 1980).

Trinkaus-Randall and Gipson (1984) found physiological calcium levels and calmodulin activity essential for hemidesmosome formation. In media containing normal calcium, calmodulin antagonists prevented the appearance of hemidesmosomes in epithelial sheets cultured on appropriate basal laminae. The inhibition was eliminated when the antagonist was removed and exogenous calmodulin was added to the culture.

7. Desmosomes and Hemidesmosomes in Morphogenetic Processes

7.1. Desmosomes in Early Development

Desmosomes are observed in epithelia during early embryogenesis in fish (Lentz and Trinkaus, 1971), amphibians (Baker and Schroeder, 1967; Perry, 1975), birds (Overton, 1962), and mammals (Potts, 1966; Jackson et al., 1980). Here they presumably function in initiating and maintaining the cohesiveness of tissue sheets during the spreading and bending events of epithelial morphogenesis. Moreover, the developmentally regulated destruction of desmosomal attachments must take place during certain epithelial morphogenetic processes such as the separation of the neural tube and the segregation of sence organ primordia. Unfortunately, little more is known about the precise role of desmosomes in the morphogenetic events of early embryogenesis.

Evidence for directed cellular migration driven or guided by differences in desmosome frequency is available from in vitro studies of Overton (1977) and Wiseman and Strickler (1981). These workers mixed cells from tissues with different inherent frequencies of desmosomes and compared their cell-sorting behavior. If desmosomes dominate the adhesive interactions of these cells, then, according to Steinberg's differential adhesion hypothesis (Steinberg, 1963, 1970), cells that form more desmosomes should sort out as an internal mass surrounded by the cells that form fewer desmosomes. This behavior was observed as predicted, suggesting that desmosomal interactions may indeed contribute to morphogenetic events guided by differences in cellular adhesion.

7.2. Desmosomes in Cellular Renewal—The Epidermis

In adult vertebrates, certain tissues such as the intestinal epithelium and the epidermis continue to undergo large-scale population renewal and cellular migration. In epidermis and other stratified squamous epithelia, cells continually produced by mitoses in the basal layers migrate through the upper layers during their differentiation, finally to be sloughed off into the environment. Throughout this migration, the cells are interconnected by numerous desmosomes.

In certain regions of mammalian epidermis, cells are stacked in highly ordered arrays (MacKenzie, 1969, 1972) termed epidermal proliferative units (Potten, 1974) (Fig. 5). Structural studies suggest that only basal cells near the edge of the epidermal proliferative unit are able to enter the upper layers (Christophers, 1971; Allen and Potten, 1974). These cells leave the basal lamina, losing their hemidesmosomal attachments, and intercalate between the flattened upper layer cells and the 10 or so rounded cells that make up the basal dividing layer of the epidermal proliferative unit. This migration and change in cell shape clearly involve the regulated dissolution and formation of many desmosomal attachments. It should be noted that some limited cellular rearrangement may be possible without cleavage of desmosomes. Desmosomes and other junctions may be somewhat mobile in the plasma membrane and thus may be capable of "sliding" laterally over the cell surface. For example, in cultured PtK_2 cells, microinjection of anti-keratin antibodies caused lateral aggregation of all keratin-intermediate filaments together with the accumulation of all desmosomes into a single patch on the cell surface (Klymkowsky et al., 1983). However, this lateral sliding phenomenon cannot account for cellular migrations involving neighbor exchanges such as those that appear to occur in the epidermis.

The number of desmosomes on cell surfaces increases significantly as epidermal cells leave the basal layer and move into the upper living cell layers (Klein-Szanto, 1977). Increased synthesis of desmosomal adhesive elements combined with the cessation of the production of hemidesmosomes might in principle create an adhesive gradient drawing cells from the basal layer into the upper layers (Skerrow, 1978). Watt (1984) recently found that cultured epidermal cells destined to differentiate terminally show altered adhesive behavior with respect to undifferentiated, proliferative cells.

7.3. Wound Closure

One of the final events of wound healing in skin is the migration of the epidermis across the wound surface, a phenomenon termed wound closure. During this process, the migrating epidermal cells must break and re-form many desmosomes and hemidesmosomes (Odland and Ross, 1968; Croft and Tarin, 1970; Martinez, 1971; Krawczyk and Wilgram, 1973). In the normal epidermis, the matrix proteins commonly underlying the basal layer cells are type IV collagen, laminin, and bullous pemphigoid antigen (Yaoita et al., 1978; Hintner et al., 1980; Stanley et al., 1981; Laurie et al., 1982). In wounds, however, the migrating epithelium moves over a substrate rich in other extracellular matrix components such as fibronectin and fibrin (Grinnell et al., 1981;

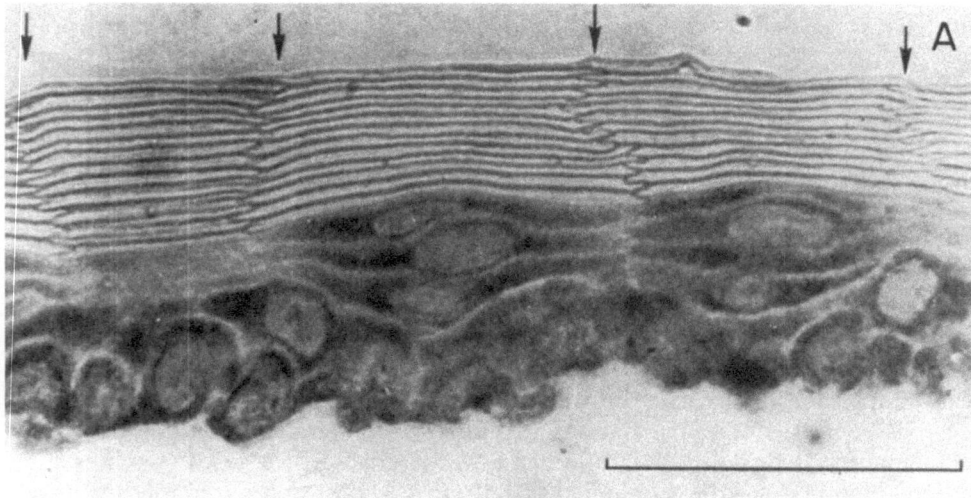

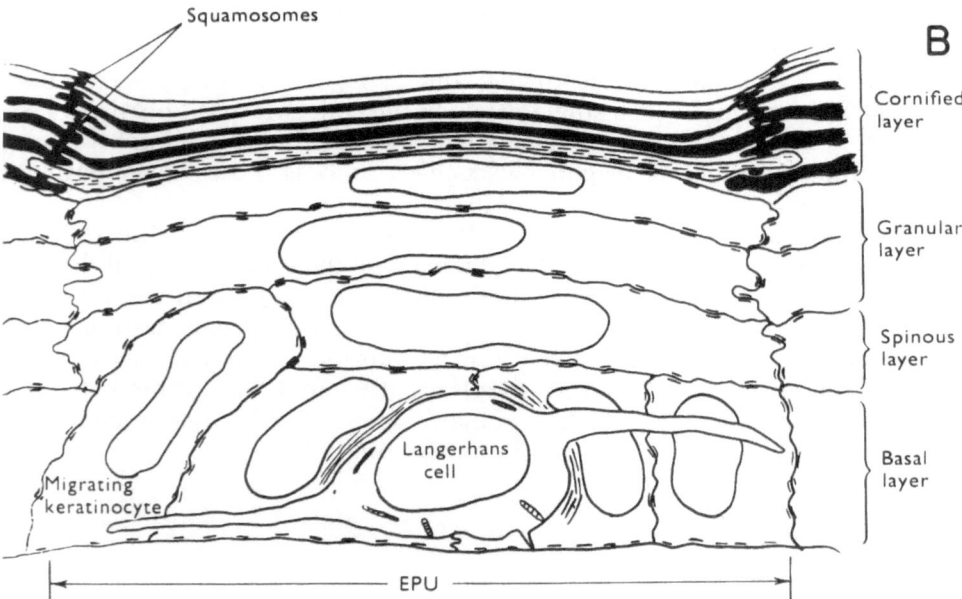

Figure 5. (A) Light micrograph of section through hamster ear epidermis after expansion in alkali. Arrows denote boundary between epidermal proliferative units. Note the highly ordered stacks of cornified outer cell layers and intermediate living cell layers. (Reproduced with permission from MacKenzie, © 1972 by Year Book Medical Publishers Inc., Chicago.) (B) Scheme of the epidermal proliferative unit. A migrating keratinocyte from the basal layer is shown entering the upper living cell layers, the spinous and granular layers. The keratinocytes, but not the mesodermally derived Langerhans cell, contain numerous desmosomes. Cells of the dead cornified layers are joined together by modified desmosomes and at their edges by thickened ridges called squamosomes. (Reproduced from Allen and Potten, 1974, by copyright permission of The Company of Biologists Ltd.)

Clark *et al.,* 1982, 1983; Donaldson and Mahan, 1983; Fujikawa *et al.,* 1981, 1984; Repesh *et al.,* 1982). These proteins are presumably deposited by blood plasma during the formation of the eschar or by invading fibroblasts. Fibronectin also occurs in high concentration at the epidermal–dermal border in fetal skin, at times when significant morphogenetic events such as hair follicle formation are taking place (Gibson *et al.,* 1983).

Donaldson and Mahan (1983) found that coverslips coated with fibronectin, fibrinogen, or collagen supported rapid migration of epithelial sheets when inserted into wounds in newt epidermis. However, if fibronectin or fibrinogen coverslips were pretreated with specific antibodies, or if bovine serum albumin (BSA)-coated coverslips were used, little migration was observed. This finding suggests that specific interactions of epithelial cell surface receptors with extracellular matrix components may be necessary for epithelial migration. Thus, wound closure may provide an accessible and physiologically important system to test whether desmosomal or hemidesmosomal cell surface components interact with extracellular matrix elements during epithelial morphogenesis.

8. Desmosomes and Cancer

The altered behavior of cancer cells, their invasive and metastatic properties, have long implicated changes in cellular adhesion as a primary consequence of malignant transformation (Coman, 1944; Fidler, 1975; Nicolson *et al.,* 1976; Hayashi and Ishimaru, 1981). Since most human cancers are of epithelial origin and since desmosomes are abundant in epithelia, alterations in desmosomal function may contribute significantly to malignant behavior. Many early studies attempted to correlate changes in the frequency of cellular junctions with the malignant phenotype (Shingleton *et al.,* 1968; Alroy *et al.,* 1981; for review, see Weinstein *et al.,* 1976, and references cited therein). In general, it was found that various tumors may show increased, decreased, or unchanged frequencies of desmosomes and other junctions, rendering rather difficult any conclusions regarding the role of intercellular junctions in malignant behavior.

While simple quantitation by electron microscopic sampling of tumors has not proved very enlightening, it appears likely that chemical alterations in junctional components, even if not ultrastructurally evident, may have significant roles in malignant behavior. For example, transformation of cultured cells with Rous sarcoma virus perturbs the cytoskeleton, causing cells to become more rounded and decreasing their adherence to the substratum (Wang and Goldberg, 1976; Vasiliev and Gelfand, 1977; Pastan and Willingham, 1978; David-Pfeuty and Singer, 1980). Accompanying this transformation, the phosphorylation level of tyrosines in vinculin, a component of adhaerens junctions and focal contacts, is increased greatly through the action of the viral protein kinase (Rohrschneider *et al.,* 1982; Sefton *et al.,* 1982). Mueller and Franke (1983) reported that the high-molecular-weight desmoplakins of desmosomes are phosphoproteins. Whether they are altered upon transformation is as yet unknown. In a brief report, Hixson *et al.* (1983), using monoclonal antibodies to rat liver desmosomes, found that certain transplantable rat hepatocarcinomas showed weakened or altered immunofluorescent staining patterns. The further use of antibodies to study and isolate the junctional components of malignant cells may help reveal whether biochemical changes in intercellular junctions are involved in the aberrant behavior of cancer cells.

9. Concluding Remarks: Future Problems

The advent of concerted study of the biochemistry of adhesive cellular junctions, in particular the adhaerens junctions and desmosomes, is very recent. Components of the cytoplasmic portion of adhaerens junctions have been identified, but little is known of the biochemistry of the intercellular adhesive elements. By contrast, it is likely that all the major intercellular and cytoplasmic components of desmosomes have been identified. Nevertheless, many questions of desmosomal structure and function remain unanswered: How are the various desmosomal components structurally interconnected? How are they integrated with the cytoskeleton? What is the chemical nature of the intercellular adhesive?

Other problems amenable to solution with the tools and methods currently being developed concern the dynamics of junction assembly and turnover and their regulation during organogenesis. What are the inductive signals for junction assembly? How do neighbor cells coordinate the assembly processes? How are the size and frequency of cellular junctions controlled? Is calcium involved in regulating junction assembly *in vivo*? One puzzling observation, possibly relevant to this last point, is the fairly frequent observation of structural attachment of mitochondria to desmosomal plaques (Deane and Wurzelmann, 1965; Asmussen, 1980; Rassat *et al.*, 1981). Might these interactions reflect examples of calcium regulation or energy input during desmosome synthesis?

Finally, future studies should explore possible functions, in addition to cellular attachment, performed by adhesive junctions *in vivo*. Findings that intermediate filaments are associated with nuclear pores (Jones *et al.*, 1982) naturally invite speculation about the role of these cytoskeletal elements in information transfer to and from the nucleus. If so, might desmosomes serve as mediators of information from the cell surface, possibly transferring mechanical signals from cell to cell within a tissue? The availability of large quantities of purified junctional material, the production of specific antibody probes, and the use of systems for regulating and studying junction assembly and function *in vitro* and *in vivo* should improve our understanding of the role of adhaerens junctions and desmosomes in morphogenesis and disease.

ACKNOWLEDGMENTS. I gratefully acknowledge the supervision, assistance, and encouragement of Malcolm S. Steinberg. I thank my other colleagues Stephen Cohen, Hisato Shida, Mariko Shida, and George Giudice for their help and advice. Special thanks go to Doris White and Edward Kennedy for their invaluable technical assistance. I also thank Carmen Huston for her help in preparing this manuscript. The author's research referred to in this chapter was supported by research grant 5RO1 GM26047 from the National Institute of General Medical Sciences awarded to M. S. Steinberg and by training grant CA09167 from the National Cancer Institute. G. G. was the recipient of a predoctoral fellowship from the National Science Foundation.

References

Allen, T. D., and Potten, C. S., 1974, Fine-structural identification and organization of the epidermal proliferative unit, *J. Cell Sci.* **15**:291–319.

Allen, T. D., and Potten, C. S., 1975, Desmosomal form, fate, and function in mammalian epidermis, *J. Ultrastruct. Res.* **51**:94–105.

Alroy, J., Pauli, B. U., and Weinstein, R. S., 1981, Correlation between numbers of desmosomes and the aggressiveness of transitional cell carcinoma in human urinary bladder, *Cancer* **47**:104–112.

Arnn, J., and Staehelin, L. A., 1981, The structure and function of spot desmosomes, *Int. J. Dermatol.* **20**:330–339.

Arnn, J., and Staehelin, L. A., 1983, Ultrastructural and immunofluorescence investigations of spot desmosomes, *J. Cell Biol.* **97**:85a.

Asmussen, I., 1980, Mitochondrial–desmosomal junctions in liver of man, *J. Submicrosc. Cytol.* **12**:149–152.

Baker, P. C., and Schroeder, T. E., 1967, Cytoplasmic filaments and morphogenetic movement in the amphibian neural tube, *Dev. Biol.* **15**:432–450.

Berry, M. N., and Friend, D. S., 1969, High yield preparation of isolated rat liver parenchymal cells. A biochemical and fine structural study, *J. Cell Biol.* **43**:506–520.

Billig, D., Nicol, A., McGinty, R., Cowin, P., Morgan, J., and Garrod, D., 1982, The cytoskeleton and substratum adhesion in chick embryonic corneal epithelial cells, *J. Cell Sci.* **57**:51–71.

Borysenko, J. Z., and Revel, J. P., 1973, Experimental manipulation of desmosome structure, *Am. J. Anat.* **137**:403–422.

Brackenbury, R., Thiery, J. P., Rutishauser, U., and Edelman, G. M., 1977, Adhesion among neural cells of the chick embryo. I. An immunological assay for molecules involved in cell–cell binding, *J. Biol. Chem.* **252**:6835–6840.

Breathnach, A. S., Gross, M., Martin, B., and Stolinski, C., 1976, A comparison of membrane fracture faces of fixed and unfixed glycerinated tissue, *J. Cell Sci.* **21**:437–448.

Briggaman, R. A., Dalldorf, F. G., and Wheeler, C. E., 1971, Formation and origin of basal lamina and anchoring fibrils in adult human skin, *J. Cell Biol.* **51**:384–395.

Burridge, K., and Feramisco, J. R., 1980, Microinjection and localization of a 130K protein in living fibroblasts: A relationship of actin and fibronectin, *Cell* **19**:587–595.

Campbell, R. D., and Campbell, J. H., 1971, Origin and continuity of desmosomes, in: *Origin and Continuity of Cell Organelles* (J. Reinert and H. Ursprug, eds.), pp. 261–298, Springer-Verlag, Berlin, Heidelberg.

Caputo, R., and Peluchetti, D., 1977, The junctions of normal human epidermis, *J. Ultrastruct. Res.* **61**:44–61.

Chen, W.-T., and Singer, S. J., 1982, Immunoelectron microscopic studies of the sites of cell–substratum and cell–cell contacts in cultured fibroblasts, *J. Cell Biol.* **95**:205–222.

Christophers, E., 1971, The columnar structure of the epidermis: Possible mechanism of differentiation, *Z. Zellforsch. Mikrosk. Anat.* **114**:441–447.

Christophers, E., and Wolff, H. H., 1975, Differential formation of desmosomes and hemidesmosomes in epidermal cell cultures treated with retinoic acid, *Nature (Lond.)* **256**:209–210.

Clark, R. A. F., Lanigan, J. M., DellaPelle, P., Manseau, E., Dvorak, H. F., and Colvin, R. B., 1982, Fibronectin and fibrin provide a provisional matrix for epidermal cell migration during wound reepithelialization, *J. Invest. Dermatol.* **79:**264–269.

Clark, R. A. F., Winn, H. S., Dvorak, H. F., and Colvin, R. B., 1983, Fibronectin beneath re-epithelializing epidermis *in vivo:* Sources and significance, *J. Invest. Dermatol.* **80:**26s–30s.

Cohen, S. M., Gorbsky, G., and Steinberg, M. S., 1983, Immunochemical characterization of related families of glycoproteins in desmosomes, *J. Biol. Chem.* **258:**2621–2627.

Colaco, C. A. L. S., and Evans, W. H., 1981, A biochemical dissection of the cardiac intercalated disk: Isolation of subcellular fractions containing *fascia adherentes* and gap junctions, *J. Cell Sci.* **52:**313–325.

Coman, D. R., 1944, Decreased mutual adhesiveness, a property of cells from squamous cell carcinomas, *Cancer Res.* **4:**625–629.

Cowin, P., and Garrod, D. R., 1983, Antibodies to epithelial desmosomes show wide tissue and species cross-reactivity, *Nature (Lond.)* **302:**148–150.

Craig, S. W., and Pardo, J. V., 1979, Alpha-actinin localization in the junctional complex of intestinal epithelial cells, *J. Cell Biol.* **80:**203–210.

Croft, C. B., and Tarin, D., 1970, Ultrastructural studies on wound healing in mouse skin. I. Epithelial behavior, *J. Anat.* **106:**63–77.

Culp, L. A., Murray, B. A., and Rollins, B. J., 1979, Fibronectin and proteoglycans as determinants of cell–substratum adhesion, *J. Supramol. Struct.* **11:**401–427.

Damsky, C. H., Richa, J., Solter, D., Knudsen, K., and Buck, C. A., 1983, Identification and purification of a cell surface glycoprotein mediating intercellular adhesion in embryonic and adult tissue, *Cell* **34:**455–466.

David-Pfeuty, T., and Singer, S. J., 1980, Altered distributions of the cytoskeletal proteins vinculin and α-actinin in cultured fibroblasts transformed by Rous sarcoma virus, *Proc. Natl. Acad. Sci. USA* **77:**6687–6691.

Deane, H. W., and Wurzelmann, S., 1965, Electron microscopic observations on the postnatal differentiation of the seminal vesicle epithelium of the laboratory mouse, *Am. J. Anat.* **117:**91–134.

Dembitzer, H. M., Herz, F., Schermer, A., Wolley, R. C., and Koss, L. G., 1980, Desmosome development in an *in vitro* model, *J. Cell Biol.* **85:**695–702.

Denburg, J. L., 1978, The biochemistry of intercellular recognition, *Adv. Comp. Physiol. Biochem.* **7:**105–226.

Dodson, J. W., and Hay, E. D., 1971, Secretion of collagenous stroma by isolated epithelium grown *in vitro*, *Exp. Cell Res.* **65:**215–220.

Donaldson, D., and Mahan, J. T., 1983, Fibrinogen and fibronectin as substrates for epidermal cell migration during wound closure, *J. Cell Sci.* **62:**117–128.

Drochmans, P., Freudenstein, C., Wanson, J.-C., Laurent, L., Keenan, T. W., Stadler, J., Leloup, R., and Franke, W. W., 1978, Structure and biochemical composition of desmosomes and tonofilaments isolated from calf muzzle epidermis, *J. Cell Biol.* **79:**427–443.

Farquhar, M. G., and Palade, G. E., 1963, Junctional complexes in various epithelia, *J. Cell Biol.* **17:**375–412.

Fidler, I. J., 1975, Mechanism of cancer invasion and metastasis, in: *Cancer*, Vol. 4 (F. F. Becker, ed.), pp. 101–131, Plenum Press, New York.

Flickinger, C. J., 1970, Extracellular specializations associated with hemidesmosomes in the fetal rat urogenital sinus, *Anat. Rec.* **168:**195–202.

Franke, W. W., Schmid, E., Grund, C., Mueller, H., Engelbrecht, I., Moll, R., Stadler, J., and Jarasch, E.-D., 1981a, Antibodies to high molecular weight polypeptides of desmosomes: Specific localization of a class of junctional proteins in cells and tissues, *Differentiation* **20:**217–241.

Franke, W. W., Winter, S., Grund, C., Schmid, E., Schiller, D. L., and Jarasch, E.-D., 1981b, Isolation and characterization of desmosome-associated tonofilaments from rat intestinal brush border, *J. Cell Biol.* **90:**116–127.

Franke, W. W., Moll, R., Schiller, D. L., Schmid, E., Kartenbeck, J., and Mueller, H., 1982, Desmoplakins of epithelial and myocardial desmosomes are immunologically and biochemically related, *Differentiation* **23:**115–127.

Franke, W. W., Kapprell, H.-P., and Mueller, H., 1983a, Isolation and symmetrical splitting of desmosomal structures in 9 M urea, *Eur. J. Cell Biol.* **32:**117–130.

Franke, W. W., Moll, R., Mueller, H., Schmid, E., Kuhn, C., Krepler, R., Artlieb, U., and Denk, H., 1983b, Immunocytochemical identification of epithelium-derived human tumors with antibodies to desmosomal plaque proteins, *Proc. Natl. Acad. Sci. USA* **80:**543–547.

Franke, W. W., Mueller, H., Mittnacht, S., Kapprell, H.-P., and Jorcano, J. L., 1983c, Significance of two desmosome plaque-associated polypeptides of molecular weights 75 000 and 83 000, *EMBO J.* **2:**2211–2215.

Frazier, W., and Glaser, L., 1979, Surface components and cell recognition, *Ann. Rev. Biochem.* **48:**491–523.

Fujikawa, L. S., Foster, C. S., Harris, T. J., Lanigan, J. M., and Colvin, R. B., 1981, Fibronectin in healing rabbit corneal wounds, *Lab. Invest.* **45:**120–129.

Fujikawa, L. S., Foster, C. S., Gipson, I. K., and Colvin, R. B., 1984, Basement membrane components in healing rabbit corneal epithelial wounds: Immunofluorescence and ultrastructural studies, *J. Cell Biol.* **98:**128–138.

Gallin, W. J., Edelman, G. M., and Cunningham, B. A., 1983, Characterization of L-CAM, a major cell adhesion molecule from embryonic liver cells, *Proc. Natl. Acad. Sci. USA* **80:**1038–1042.

Garrod, D. R., and Nicol, A., 1981, Cell behavior and molecular mechanisms of cell–cell adhesion, *Biol. Rev.* **56:**199–242.

Geiger, B., 1979, A 130K protein from chicken gizzard: Its localization at the termini of microfilament bundles in cultured chicken cells, *Cell* **18:**193–205.

Geiger, B., 1983, Membrane–cytoskeleton interaction, *Biochem. Biophys. Acta* **737:**305–341.

Geiger, B., Dutton, A. H., Tokuyasu, K. T., and Singer, S. J., 1981, Immunoelectron microscope studies of membrane–microfilament interactions: Distributions of α-actinin, tropomyosin, and vinculin of intestinal epithelial brush border and chicken gizzard smooth muscle cells, *J. Cell Biol.* **91:**614–628.

Geiger, B., Schmid, E., and Franke, W. W., 1983, Spatial distribution of proteins specific for desmosomes and *adhaerens* junctions in epithelial cells demonstrated by double immunofluorescence microscopy, *Differentiation* **23:**189–205.

Gibson, W. T., Couchman, J. R., and Weaver, A. C., 1983, Fibronectin distribution during the development of fetal rat skin, *J. Invest. Dermatol.* **81:**480–485.

Gipson, I. K., Grill, S. M., Spurr, S. J., and Brennan, S. J., 1983, Hemidesmosome formation *in vitro*, *J. Cell Biol.* **97:**849–857.

Giudice, G. J., Cohen, S. M., Patel, N., and Steinberg, M. S., 1984, Immunological comparison of desmosomal components from several bovine tissues, *J. Cell Biochem.* **26:**35–45.

Gorbsky, G., and Steinberg, M. S., 1981, Isolation of the intercellular glycoproteins of desmosomes, *J. Cell Biol.* **90:**243–248.

Gorbsky, G., Cohen, S. M., Shida, H., Giudice, G. J., Steinberg, M. S., 1985, Isolation of the non-glycosylated proteins of desmosomes and immunolocalization of a third plaque protein: Desmoplakin. III. *Proc. Natl. Acad. Sci. USA* **82:**810–814.

Green, H., Kehinde, O., and Thomas, J., 1979, Growth of cultured human epidermal cells into multiple epithelia suitable for grafting, *Proc. Natl. Acad. Sci. USA* **76:**5665–5668.

Greig, R. G., and Jones, M. N., 1977, Mechanisms of intercellular adhesion, *BioSystems* **9:**43–55.

Grinnell, F., Billingham, R. E., and Burgess, L., 1981, Distribution of fibronectin during wound healing *in vivo*, *J. Invest. Dermatol.* **76:**181–189.

Grunwald, G. B., Geller, R. L., and Lilien, J., 1980, Enzymatic dissection of embryonic cell adhesive mechanisms, *J. Cell Biol.* **85:**766–776.

Grunwald, G. B., Pratt, R. S., and Lilien, J., 1982, Enzymatic dissection of embryonic cell adhesive mechanisms. III. Immunological identification of a component of the calcium-dependent adhesive system of embryonic chick neural retina cells, *J. Cell Sci.* **55:**69–83.

Hashimoto, K., and Lever, W. F., 1970, An ultrastructural study of cell junctions in pemphigus vulgaris, *Arch. Dermatol.* **101:**287–298.

Hay, E. D., and Revel, J.-P., 1969, Fine Structure of the Developing Avian Cornea, in: *Monographs in Developmental Biology*, Vol. 1 (A. Wolsky and P. S. Chen, eds.), pp. 121–122, Basel, Switzerland.

Hayashi, H., and Ishimaru, Y., 1981, Mophological and biochemical aspects of adhesiveness and dissociation of cancer cells, *Int. Rev. Cytol.* **70**:139−215.

Heaysman, J. E. M., and Pegrum, S. M., 1973, Early contacts between fibroblasts, *Exp. Cell Res.* **78**:71−78.

Hennings, H., and Holbrook, K. A., 1983, Calcium regulation of cell−cell contact and differentiation of epidermal cells in culture, *Exp. Cell Res.* **143**:127−142.

Hennings, H., Michael, D., Cheng, C., Steinert, P., Holbrook, K., and Yuspa, S. H., 1980, Calcium regulation of growth and differentiation of mouse epidermal cells in culture, *Cell* **19**:245−254.

Hennings, H., Holbrook, K. A., and Yuspa, S. H., 1983, Potassium mediation of calcium induced terminal differentiation of epidermal cells in culture, *J. Invest. Dermatol.* **81**:505−555.

Hino, H., Kobayasi, T., and Asboe-Hansen, G., 1982, Desmosome formation in normal human epidermal cell culture, *Acta Dermatol.* **62**:185−191.

Hintner, H., Fritsch, P. O., Foidart, J.-M., Stingl, G., Schuler, G., and Katz, S. I., 1980, Expression of basement membrane zone antigens at the dermo-epidermal junction in organ cultures of human skin, *J. Invest. Dermatol.* **74**:200−204.

Hirokawa, N., and Heuser, J. E., 1981, Quick-freeze, deep-etch visualization of the cytoskeleton beneath surface differentiations of intestinal epithelial cells, *J. Cell Biol.* **91**:399−409.

Hixson, D. C., McEntire, K. D., Donahue, K. L., and Chesner, J. E., 1983, Detection of alterations in adhesive interactions during hepatocarcinogenesis using monoclonal antibodies against rat liver desmosomes, *J. Cell Biol.* **97**:85a.

Hopwood, D., Logan, K. R., and Milne, G., 1977, Mucosubstances in the normal human esophageal epithelium, *Histochemistry* **54**:67−74.

Hubbard, A. L., and Ma, A., 1983, Isolation of rat hepatocyte plasma membranes. II. Identification of membrane-associated cytoskeletal proteins, *J. Cell Biol.* **96**:230−239.

Hull, B. E., and Staehelin, L. A., 1979, The terminal web. A reevaluation of its structure and function, *J. Cell Biol.* **81**:67−82.

Hyafil, F., Babinet, C., and Jacob, F., 1981, Cell−cell interactions in early embryogenesis: A molecular approach to the role of calcium, *Cell* **26**:447−454.

Jepsen, A., MacCallum, D. K., and Lillie, J. H., 1980, Fine structure of subcultivated stratified squamous epithelium, *Exp. Cell Res.* **125**:141−152.

Jackson, B., Grund, C., Schmid, E., Burki, K., Franke, W. W., and Illmensee, K., 1980, Formation of cytoskeletal elements during mouse embryogenesis, *Differentiation* **17**:161−179.

Jones, B. M., 1980, Regulation of the contact behavior of cells, *Biol. Rev.* **55**:207−235.

Jones, J. C. R., Goldman, A. E., Steinert, P. M., Yuspa, S., and Goldman, R. D., 1982, Dynamic aspects of the supramolecular organization of intermediate filament networks in cultured epidermal cells, *Cell Motil.* **2**:197−213.

Kartenbeck, J., Schmid, E., Franke, W. W., and Geiger, B., 1982, Different modes of internalization of proteins associated with *adhaerens* junctions and desmosomes: Experimental separation of lateral contacts induces endocytosis of desmosomal plaque material, *EMBO J.* **1**:725−732.

Kartenbeck, J., Franke, W. W., Moser, J. G., and Stoffels, U., 1983, Specific attachment of desmin filaments to desmosomal plaques in cardiac myocytes, *EMBO J.* **2**:735−742.

Kelly, D. E., 1966, Fine structure of desmosomes, hemidesmosomes, and an adepidermal globular layer in developing newt epidermis, *J. Cell Biol.* **28**:51−72.

Kelly, D. E., and Kuda, A. M., 1981, Traversing filaments in desmosomal and hemidesmosomal attachments: Freeze-fracture approaches toward their characterization, *Anat. Rec.* **199**:1−14.

Kelly, D. E., and Shienvold, F. L., 1976, The desmosome: Fine structural studies with freeze-fracture replication and tannic acid staining of sectioned epidermis, *Cell Tissue Res.* **172**:309−323.

Kensler, R. W., and Goodenough, D. A., 1980, Isolation of mouse myocardial gap junctions, *J. Cell Biol.* **86**:755−764.

King, I. A., and Tabiowo, A., 1981, Effect of tunicamycin on epidermal glycoprotein and glycosaminoglycan synthesis *in vitro*, *Biochem. J.* **198**:331−338.

Kitajima, Y., Eguchi, K., Mori, S., and Yaoita, H., 1983, Membrane characterization of cultured human keratinocytes by freeze-fracture electron microscopy, *Cell Tissue Res.* **234**:561−572.

Klein-Szanto, A. J. P., 1977, Serologic baseline data of normal human epidermis, *J. Invest. Dermatol.* **68**:73−78.

Klymkowsky, M. W., Miller, R. H., and Lane, E. B., 1983, Morphology, behavior, and interaction of cultured epithelial cells after the antibody-induced disruption of keratin filament organization, *J. Cell Biol.* **96**:494–509.

Knudsen, K. A., Rao, P. E., Damsky, C. H., and Buck, C. A., 1981, Membrane glycoproteins involved in cell–substratum adhesion, *Proc. Natl. Acad. Sci. USA* **78**:6071–6075.

Krawczyk, W. S., and Wilgram, G. F., 1973, Hemidesmosome and desmosome morphogenesis during epidermal wound healing, *J. Ultrastruct. Res.* **45**:93–101.

Lark, M. W., and Culp, L. A., 1982, Selective solubilization of hyaluronic acid from fibroblast substratum adhesion sites, *J. Biol. Chem.* **257**:14073–14080.

Laurie, G. W., Leblond, C. P., and Martin, G. R., 1982, Localization of type IV collagen, laminin, heparin sulfate, proteoglycan and fibronection to the basal lamina of basement membranes, *J. Cell Biol.* **95**:340–344.

Lazarides, E., 1980, Intermediate filaments as mechanical integrators of cellular space, *Nature (Lond.)* **283**:249–256.

Lazarides, E., and Burridge, K., 1975, α-Actinin: Immunofluorescent localization of a muscle structural protein in nonmuscle cells, *Cell* **6**:289–298.

Leloup, R., Laurent, L., Ronveaux, M.-F., Drochmans, P., and Wanson, J.-C., 1979, Desmosomes and desmogenesis in the epidermis of calf muzzle, *Biol. Cell.* **34**:137–152.

Lentz, T. L., and Trinkaus, J. P., 1971, Differentiation of the junctional complex of surface cells in the developing *Fundulus* blastoderm, *J. Cell Biol.* **48**:455–472.

MacKenzie, I. C., 1969, Ordered structure of the stratum corneum in mammalian skin, *Nature (Lond.)* **222**:881–882.

MacKenzie, I. C., 1972, The ordered structure of mammalian epidermis, in: *Epidermal Wound Healing* (H. I. Maibach and D. T. Rouee, eds.), Year Book Medical Publishers, Chicago.

Magnani, J. L., Thomas, W. A., and Steinberg, M. S., 1981, Two distinct adhesion mechanisms in embryonic chick neural retina cells. I. A kinetic analysis, *Dev. Biol.* **81**:96–105.

Maher, P., and Singer, S. J., 1983, A 200-kd protein isolated from the *fascia adherens* membrane domains of chicken cardiac muscle cells is detected immunologically in fibroblast focal adhesions, *Cell Motil.* **3**:419–429.

Marchase, R. B., Vosbeck, K., and Roth, S., 1976, Intercellular adhesive specificity, *Biochim. Biophys. Acta* **457**:385–416.

Martinez, I. R., 1971, Fine structural studies of migrating epithelial cells following incision wounds, in: *Epidermal Wound Healing* (H. I. Maiabach and D. T. Rouee, eds.), pp. 323–342, Year Book Medical Publisher, Chicago.

Maslow, D. E., 1976, *In vitro* analysis of surface specificity in embryonic cells, in: *The Cell Surface in Animal Embryogenesis and Development* (G. Poste and G. Nicolson, eds.), pp. 697–739, North-Holland, New York.

McNutt, N. S., and Weinstein, R. S., 1973, Membrane ultrastructure at mammalian intercellular junctions, *Proc. Biophys. Mol. Biol.* **26**:47–101.

Mueller, H., and Franke, W. W., 1983, Biochemical and immunological characterization of desmoplakins I and II, the major polypeptides of the desmosomal plaque, *J. Mol. Biol.* **163**:647–671.

Muir, A. R., 1967, The effects of divalent cations on the ultrastructure of the perfused rat heart, *J. Anat.* **101**:239–261.

Neyfakh, A. A., and Svitkina, T. M., 1983, Isolation of focal contact membrane using saponin, *Exp. Cell Res.* **149**:582–586.

Nicolson, G. L., Birdwell, C. R., Brunson, K. W., and Robbins, J. C., 1976, Cellular interactions in the metastatic process, in: *Membranes and Neoplasia: New Approaches and Strategies* (V. T. Marchesi, ed.), pp. 237–244, Alan R. Liss, New York.

Ocklind, C., Forsum, U., and Öbrink, B., 1983, Cell surface localization and tissue distribution of a hepatocyte cell–cell adhesion glycoprotein (Cell-CAM 105), *J. Cell Biol.* **96**:1168–1171.

Odland, G. F., 1958, The fine structure of the interrelationship of cells in the human epidermis, *J. Biophys. Biochem. Cytol.* **4**:529–538.

Odland, G., and Ross, R., 1968, Human wound repair: Epidermal regeneration, *J. Cell Biol.* **39**:135–151.

Oesch, B., and Birchmeier, W., 1982, New surface component of fibroblast's focal contacts identified by a monoclonal antibody, *Cell* **31:**671–679.

Ogou, S.-I., Yoshida-Noro, C., and Takeichi, M., 1983, Calcium-dependent cell–cell adhesion molecules common to hepatocytes and teratocarcinoma stem cells, *J. Cell Biol.* **97:**944–948.

Overton, J., 1962, Desmosome development in normal and reassociating cells in the early chick blastoderm, *Dev. Biol.* **4:**532–548.

Overton, J., 1968, The fate of desmosomes in trypsinized tissue, *J. Exp. Zool.* **168:**203–214.

Overton, J., 1973, Experimental manipulation of desmosome formation, *J. Cell Biol.* **56:**636–646.

Overton, J., 1974, Selective formation of desmosomes in chick cell reaggregates, *Dev. Biol.* **39:**210–225.

Overton, J., 1975, Experiments with junctions of the *adhaerens* type, *Curr. Topics Dev. Biol.* **10:**1–34.

Overton, J., 1977, Formation of junctions and cell sorting in aggregates of chick and mouse cells, *Dev. Biol.* **55:**103–116.

Overton, J., 1982, Inhibition of desmosome formation with tunicamycin and with lectin in corneal cell aggregates, *Dev. Biol.* **92:**66–72.

Overton, J., and Culver, N., 1973, Desmosomes and their components after cell dissociation and reaggregation in the presence of cytochalasin B, *J. Exp. Zool.* **185:**341–356.

Overton, J., and DeSalle, R., 1980, Control of desmosome formation in aggregating embryonic chick cells, *Dev. Biol.* **75:**168–176.

Overton, J., and Kapmarski, R., 1974, Hybrid desmosomes in aggregated chick and mouse cells, *J. Exp. Zool.* **192:**33–42.

Palade, G. E., and Farquhar, M. G., 1965, A special fibril of the dermis, *J. Cell Biol.* **27:**215–224.

Pastan, I., and Willingham, M., 1978, Cellular transformation and the "morphologic phenotype" of transformed cells, *Nature (Lond.)* **274:**645–650.

Perry, M. M., 1975, Microfilaments in the external surface layer of the early amphibian embryo, *J. Embryol. Exp. Morphol.* **33:**127–146.

Potten, C. S., 1974, The epidermal proliferative unit (EPU). The possible role of the central basal cell, *Cell Tissue Kinet.* **7:**77–80.

Potts, M., 1966, The attachment phase of ovoimplantation, *Am. J. Obstet. Gynecol.* **96:**1122–1128.

Raknerud, N., 1975, The ultrastructure of the interfollicular epidermis of the hairless (hr/hr) mouse, *J. Ultrastruct. Res.* **52:**32–51.

Rambourg, A., and Leblond, C. P., 1967, Electron microscope observations on the carbohydrate-rich cell coat present at the surface of cells in the rat, *J. Cell Biol.* **32:**27–53.

Rassat, J., Robenek, H., and Themann, H., 1981, Structural relationship between desmosomes and mitochondria in human livers exhibiting a wide range of diseases, *Am. J. Pathol.* **105:**207–211.

Rawles, M. E., 1963, Tissue interactions in scale and feather development as studied in dermal epidermal recombinations, *J. Embryol. Exp. Morphol.* **11:**765–789.

Rayns, D. G., Simpson, F. O., and Ledingham, J. M., 1969, Ultrastructure of desmosomes in mammalian intercalated disc: Appearances after lanthanum treatment, *J. Cell Biol.* **47:**322–326.

Repesh, L. A., Fitzgerald, T. J., Furcht, L. T., 1982, Fibronectin involvement in granulation tissue and wound healing in rabbits, *J. Histochem. Cytochem.* **30:**351–358.

Rohrschneider, L., Rosok, M., and Shriver, K., 1982, Mechanism of transformation by Rous sarcoma virus: Events within adhesion plaques, *Cold Spring Harbor Symp. Quant. Biol.* **46:**953–965.

Rutishauser, U., Thiery, J.-P., Brackenbury, R., and Edelman, G. M., 1978, Adhesion among neural cells of the chick embryo, *J. Cell Biol.* **79:**371–381.

Schollmeyer, J. E., Furcht, L. T., Goll, D. E., Robson, R. M., and Stromer, M. M., 1976, Localization of contractile proteins in smooth muscle cells and in normal and transformed fibroblasts, in: *Cell Motility* (R. Goldman, T. Pollard, and J. Rosenbaum, eds.), pp. 361–388, Cold Spring Harbor Laboratory, Cold Spring Harbor, New York.

Sedar, A. W., and Forte, J. G., 1964, Effects of calcium depletion on the junctional complex between oxyintic cells of gastric glands, *J. Cell Biol.* **22:**173–188.

Sefton, B. M., Hunter, T., Nigg, E. A., Singer, S. J., and Walter, G., 1982, Cytoskeletal targets for viral transforming proteins with tyrosine protein kinase activity, *Cold Spring Harb. Symp. Quant. Biol.* **46:**939–951.

Shida, H., Gorbsky, G., Shida, M., and Steinberg, M. S., 1982, Ultrastructural and biochemical identification of Con A receptors in the desmosome, *J. Cell. Biochem.* **20:**113–126.

Shida, H., Cohen, S. M., Guidice, G. J., and Steinberg, M. S., 1983, Quantitative electronmicroscopic immunocytochemistry of desmosomal antigens, *J. Cell Biol.* **97:**85a.

Shienvold, F. L., and Kelly, D. E., 1976, The hemidesmosome: New fine structural features revealed by freeze-fracture techniques, *Cell Tissue Res.* **172:**289–307.

Shimono, M., and Clementi, F., 1976, Intercellular junctions of oral epithelium. I. Studies with freeze-fracture and tracing methods of normal rat keratinized oral epithelium, *J. Ultrastruct. Res.* **56:**121–136.

Shingleton, H. M., Richart, R. M., Wiener, J., and Spiro, D., 1968, Human cervical intraepithelial neoplasia: Fine structure of dysplasia and carcinoma in situ, *Cancer Res.* **28:**695–706.

Skerrow, C. J., 1978, Intercellular adhesion and its role in epidermal differentiation, *Invest. Cell Pathol.* **1:**23–37.

Skerrow, C. J., and Matoltsy, A. G., 1974a, Isolation of epidermal desmosomes, *J. Cell Biol.* **63:**515–523.

Skerrow, C. J., and Matoltsy, A. G., 1974b, Chemical characterization of isolated epidermal desmosomes, **63:**524–530.

Skerrow, C. J., and Skerrow, D., 1980, Desmosomes and filaments in mammalian epidermis, in: *Cell Adhesion and Motility* (A. S. G. Curtis and J. D. Pitts, eds.), pp. 445–464, Cambridge University Press, Cambridge.

Snell, R. S., 1965, The fate of epidermal desmosomes in mammalian skin, *Z. Zellforschg.* **66:**471–487.

Staehelin, L. A., 1974, Structure and function of intercellular junctions, *Int. Rev. Cytol.* **39:**191–283.

Stanley, J. R., Hawley-Nelson, S. H., Yuspa, S. H., Shevach, E. M., and Katz, S. I., 1981, Characterization of bullous pemphigoid antigen: A unique basement membrane protein of stratified squamous epithelium, *Cell* **24:**897–903.

Steinberg, M. S., 1963, Reconstruction of tissues by dissociated cells, *Science* **141:**401–408.

Steinberg, M. S., 1970, Does differential adhesion govern self-assembly processes in histogenesis? Equilibrium configurations and the emergence of a hierarchy among populations of embryonic cells, *J. Exp. Zool.* **173:**395–434.

Stingl, G., Katz, S. I., Green, I., and Shevach, E. M., 1980, The functional role of Langerhans cells, *J. Invest. Dermatol.* **74:**315–318.

Susi, F. R., Belt, W. D., and Kelly, J. W., 1967, Fine structure of fibrillar complexes associated with the basement membrane in human oral mucosa, *J. Cell Biol.* **34:**686–690.

Takeichi, M., 1977, Functional correlation between cell adhesive properties and some cell surface proteins, *J. Cell Biol.* **75:**464–474.

Thomas, W. A., and Steinberg, M. S., 1981, Two distinct adhesion mechanisms in embryonic neural retina cells. II. An immunological analysis, *Dev. Biol.* **81:**106–114.

Thorbecke, G. J., Silberberg-Sinakin, I., and Flotte, T. J., 1980, Langerhans cells as macrophages in skin and lymphoid organs, *J. Invest. Dermatol.* **75:**32–43.

Tokuyasu, K. T., Dulton, A. H., Geiger, B., and Singer, S. J., 1981, Ultrastructure of chicken cardiac muscle as studied by double immunolabeling in electron microscopy, *Proc. Natl. Acad. Sci. USA* **78:**7619–7623.

Trinkaus-Randall, V., and Gipson, I. K., 1984, The role of calcium and calmodulin in hemidesmosome formation in vitro, *J. Cell Biol.* **98:**1565–1571.

Ueda, M. J., and Takeichi, M., 1976, Two mechanisms in cell adhesion revealed by effects of divalent cations, *Cell Struct. Funct.* **1:**377–388.

Vasiliev, J. M., and Gelfand, I. M., 1977, Mechanisms of morphogenesis in cell cultures, *Int. Rev. Cytol.* **50:**159–274.

Wang, E., and Goldberg, A. R., 1976, Changes in microfilament organization and surface topography upon transformation of chick embryo fibroblasts with Rous sarcoma virus, *Proc. Natl. Acad. Sci. USA* **73:**4065–4069.

Watt, F. M., 1984, Selective migration of terminally differentiating cells from the basal layer of cultured human epidermis, *J. Cell Biol.* **98:**16–21.

Weinstein, R. S., Merk, F. B., and Alroy, J., 1976, The structure and function of intercellular junctions in cancer, *Adv. Cancer Res.* **23**:23–89.

Wiche, G., Krepler, R., Artlieb, U., Pytela, R., and Denk, H., 1983, Occurrence and immunolocalization of plectin in tissues, *J. Cell Biol.* **97**:887–901.

Wiseman, L. L., and Strickler, J., 1981, Desmosome frequency: Experimental alteration may correlate with differential cell adhesion, *J. Cell Sci.* **49**:217–223.

Wolff, K., and Schreiner, E., 1971, Ultrastructural localization of pemphigus autoantibodies within the epidermis, *Nature (Lond.)* **229**:59–61.

Yaoita, H., Foidart, J. M., and Katz, S. I., 1978, Localization of the collagenous component in skin basement membrane, *J. Invest. Dermatol.* **70**:191–193.

Yoshida, C., and Takeichi, M., 1982, Teratocarcinoma cell adhesion: Identification of a cell-surface protein involved in calcium-dependent cell aggregation, *Cell* **28**:217–224.

Chapter 9

Dual Adhesive Recognition Systems in Chick Embryonic Cells

WILLIAM A. THOMAS

1. Introduction

One of the fundamental processes of animal development is morphogenesis, the sculpting of embryonic pattern and form from ever-changing pools of multi-plying, differentiating cells. That it is a process of staggering complexity is compellingly obvious, and even the most casual observer might be drawn to ask how a series of events so involved can possibly be directed and coordinated. In many cases, the inherent complexity of developmental interactions defies direct analysis within the natural context—the growing organism. To understand morphogenesis in its larger sense, then, one is constrained to start small, to ask answerable questions about the individual aspects of morphogenesis that together generate the cascade of events we observe. It is within this context that the molecular analysis of cellular adhesive interactions can provide a beginning for a more comprehensive understanding of developmental processes.

A major component of morphogenesis is the translocation of cells, singly or in groups, from one locus in the embryo to another. Both from the paths followed by migratory cell populations and from the locations and configurations of their eventual target structures, it is apparent that the observed migrations are anything but random. The relative constancy of a given morphogenetic scenario from individual to individual argues convincingly for the existence of underlying directional themes guiding cells in their movements and structural associations.

Pioneering work first by Wilson (1907) using marine sponges, and later by Holtfreter using amphibian embryos (Holtfreter, 1939; Townes and Holtfreter, 1955), demonstrated that much of the information required for orderly association of cells is intrinsic to the cells themselves. Holtfreter attributed the potential for self-organization to *tissue affinities*, a term for which he himself offered no mechanistic explanation, but that was later interpreted by Weiss (1941) to

WILLIAM A. THOMAS • Department of Biology, Wake Forest University, Winston-Salem, North Carolina 27109.

describe selective adhesive properties of the cell surface. This concept, simple though it may seem, has been of great value in structuring our analysis of cell and tissue arrangements both *in vitro* and *in vivo*.

If considered from the perspective of a single migratory cell, the translocations introduced above may be viewed as a sequential exchange of neighbors, a continual leaving of one local environment to enter another. Each step in the sequence requires the restructuring of adhesive associations with the cell's changing local environment. Existing bonds to the "rear" must be exchanged for newly forming ones "ahead." Since the direction of migration can actually be described in terms of adhesions made and lost, it seems intuitively obvious that a knowledge of the adhesive forces at work can greatly aid our understanding of cell rearrangements.

Owing largely to the extensive thermodynamic analyses of *in vitro* cell sorting carried out by Steinberg and co-workers (Steinberg, 1962a,b,c, 1963, 1964, 1970; Phillips, 1969; Phillips *et al.*, 1977; Phillips and Davis, 1978), the "tissue affinities" or associative "preferences" of cells described above can indeed be convincingly ascribed to differences in the relative intensities of adhesions among them. It was seen that adherent cells rearrange or "sort out" to approach an equilibrium configuration in which the adhesive free energy, or potential to form new adhesive bonds, is minimized. This analytical approach proved powerful in correlating cellular spatial organization with measurable adhesive properties of constituent cell populations. By virtue of its biophysical nature, however, the approach makes no attempt to define the molecular basis of the adhesive forces with which it treats. It remains, then, to analyze in detail the adhesion mechanisms themselves, focusing especially on the means by which their activity is regulated in time and space, to understand more completely the role of adhesive interactions in morphogenesis.

It is from the foregoing perspective that the subject of adhesive recognition in chick embryonic tissues will be viewed. Data leading to our current knowledge of dual adhesion mechanisms in the chick are presented, the focus initially being on studies on the 7-day neural retina. Results from other chick tissues, neural and non-neural, are then summarized to demonstrate the more general applicability of the retinal data. Finally, the data are discussed in relationship to the concept of *recognition* in developing systems.

2. Dual Systems of Embryonic Adhesion

2.1. Neural Retina

2.1.1. Background

The chemical basis for chick embryonic retinal cell adhesion has been the object of many studies spanning more than two decades. This can be attributed in large part to several basic characteristics that make the neural retina nearly ideal for such investigations. An essentially pure preparation of retinas can be obtained with ease by simple dissection; a variety of dissociation procedures offer excellent yields of living, single cells; and the tissue, although relatively uncomplicated at early embryonic stages, undergoes significant developmental changes both in its own structure and in its associations with other tissues as the embryo matures.

Despite the relatively widespread early use of the retina for adhesion studies, no real consensus concerning its adhesive characteristics had developed for some time. There persisted instead a surprising disparity in results obtained by different laboratories. An array of retinal cell components accumulated, some of which inhibited (Merrell et al., 1975; Jakoi and Marchase, 1979), while others stimulated (Lilien and Moscona, 1967; Lilien, 1968; Hausman and Moscona, 1975, 1976; Lilien and Rutz, 1977) retinal cell adhesion. In one case, the stimulation depended on specific sequential interaction of the cell surface with two distinct external components (Rutz and Lilien, 1979). A third class of components was shown to possess the antigenicity of a putative surface adhesion molecule (Rutishauser et al., 1976; Thiery et al., 1977). An extensive review of these various early results can be found in Moscona and Hausman (1977).

More recently, an increasingly coherent body of results has begun to emerge from various laboratories working in this field. This consensus has arisen in large part from the adoption of an experimental approach initiated in the work of Urushihara et al. (1977) on baby hamster kidney (BHK) cell adhesion and subsequently employed in one manifestation or another by several other groups. Our own choice to employ this approach grew from the conviction that the inconsistencies in earlier findings reflected an unappreciated complexity in the adhesive behavior of retinal cells. It therefore seemed essential to characterize the phenomenology of the system before undertaking an analysis of underlying adhesion mechanisms at the molecular level.

2.1.2. Kinetic Analysis

Using the above approach, we developed five standard dissociation procedures, each employing a unique combination of Ca^{2+} and trypsin and yielding a preparation of dissociated retinal cells displaying unique aggregation kinetics (Magnani and Steinberg, 1978; Magnani, 1979; Thomas, 1979; Magnani et al., 1981) (see Table I). Very similar work had been completed by Takeichi and co-workers (Takeichi, 1977; Urushihara et al., 1977, 1979; Takeichi et al., 1979) in Japan; the results from the two laboratories clearly demonstrated the existence of two distinct embryonic retinal adhesion systems—one Ca^{2+}-independent (CI) and the other Ca^{2+}-dependent (CD). Although assay conditions used by the two groups were not identical (Takeichi, 1977; Thomas and Steinberg, 1980), both did in essence follow the disappearance of single cells from a cell suspension over the course of a 60-min assay. Three other groups also presented data that, despite more significant differences in tissue preparation and/or assay conditions, were basically in accord with those cited above. Brackenbury et al. (1981) and McClay and Marchase (1982) both used cell-to-monolayer attachment assays to test the adhesion of cells dissociated using

Table I. Summary of Preparative Procedures for and Adhesion Mechanisms Present on the Various Retinal Cell Preparations[a,b]

Cell preparation	Dissociation medium			Adhesion mechanism	
	Trypsin per retina (NFU)	Ca^{2+} (mM)	EGTA (mM)	CI	CD
E	—	—	1.3	+	(+)
LTC	68	2.5	—	+	+
LTE	68	—	1.3	+	−
TC	3000	2.5	—	−	+
TE	3000	—	1.3	−	−

[a]Dissociation in EGTA, E; dissociation in trypsin plus EGTA, TE; dissociation in trypsin plus Ca^{2+}, TC; dissociation in low trypsin plus EGTA, LTE; dissociation in low trypsin plus Ca^{2+}, LTC; Ca^{2+}-independent mechanism, CI; Ca^{2+}-dependent mechanism, CD: present and active, +; present and undetected in the self-aggregation assay, (+); absent. −.
[b]From Magnani et al. (1981).

procedures very similar to those described by Urushihara et al. (1977) and Magnani et al. (1981). Grunwald et al. (1980) measured the rate of cell aggregation by monitoring the appearance of 100-cell aggregates. Moreover, they used chymotrypsin rather than trypsin to prepare cells displaying only CI aggregation. Each of the three groups, however, demonstrated clear CI and CD adhesion.

To underscore the import of this commonality of results, it is worth noting that the different assays employed focus on quite different temporal phases of cell aggregation. Three of the assays (Takeichi, 1977; Thomas and Steinberg, 1980; Brackenbury et al., 1981) monitor cell adhesions formed under constant shear during the first 30–60 min of aggregation. By contrast, the assay used by Grunwald et al. (1980) is designed to test for potentially slower-forming associations dependent on protein synthesis, and it yields its first data point after 1 hr. Finally, the centrifugation/monolayer assay used by McClay and Marchase (1982) provides for precise regulation of the contact time between interacting cells and has been used to study the very early events of cell-to-cell adhesion. Despite these differences in the bias of each assay, each of the five laboratories demonstrated ostensibly the same two distinct modes of retinal cell adhesion. These findings strongly indicate major roles for the observed CI and CD systems in retinal cell interactions.

A synthesis of information deriving from both the dissociation of intact retinas and the reaggregation of resulting single cells permits one to categorize not only the two classes of retinal adhesion, but the molecular mechanisms underlying them as well. Within that context, it is appropriate to indicate that the five standard dissociation procedures do not selectively dissociate different retinal cell subpopulations but rather yield in each case a cell suspension containing more than 90% of the cells in the intact retina. Moreover, some of the cell preparations are interconvertible with nearly a 100% yield. This indicates that nearly all cells from the intact 7–8-day retina possess both adhesion mechanisms and that expression of only a single mechanism is typically effected (*in vitro* at least) through inactivation or removal of the other. With that introduction, let us turn briefly to a description of the CI and CD adhesion characteristics.

2.1.2a. Ca^{2+}-Independent Aggregation. In our hands, the reaggregation kinetics of mechanically dissociated retinal cells are essentially unaffected by the presence or absence of Ca^{2+} in the aggregation medium and are thus defined as Ca^{2+}-independent, or CI (Fig. 1). An intact retina does dissociate more readily if pretreated with a Ca^{2+} chelator (1.3 mM EGTA), but such treatment has no effect on subsequent reaggregation. The adhesion mechanism responsible for CI aggregation is fully functional without proteolytic activation and is moreover deactivated by trypsin (with or without Ca^{2+}) at a rate proportional to both the trypsin concentration and the duration of exposure. For example, the CI mechanism is completely removed after a 15-min incubation in 0.044% trypsin (Fig. 2) but is not detectably altered by exposure to 0.001% trypsin for the same period of time. Because of this last observation, the CI mechanism is considered relatively resistant to tryptic cleavage. After complete removal of the CI mechanism by proteolysis, retinal cells typically undergo a 30-min lag period before exhibiting the first detectable, renewed CI adhesion (Fig. 2). Full recovery of CI adhesion after its proteolytic removal requires several hours. Finally, CI aggregation proceeds nearly as well at 4°C as at 37°C.

2.1.2b. Ca^{2+}-Dependent Aggregation. Ca^{2+}-dependent, or CD, adhesion is distinctly different from CI adhesion. The retinal CD mechanism is extremely sensitive to proteolysis, being completely removed by a brief exposure (less

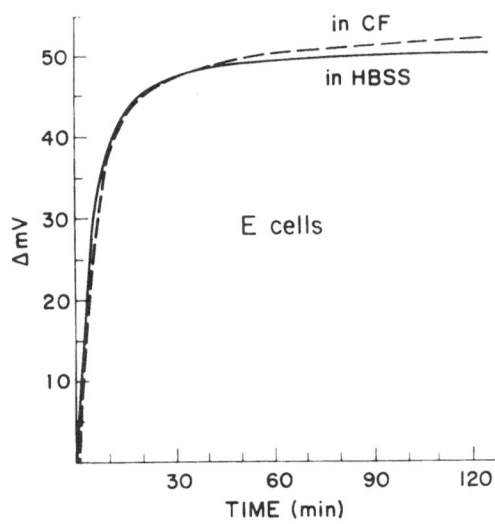

Figure 1. Ca^{2+}-independent aggregation of retinal E cells prepared by mechanical dissociation after incubation in EGTA (see Table I). Aggregation in the presence (HBSS:———) and absence (CF:––––) of Ca^{2+}. HBSS, Hanks balanced salt solution; CF, Ca^{2+}-free HBSS. (From Magnani et al., 1981.)

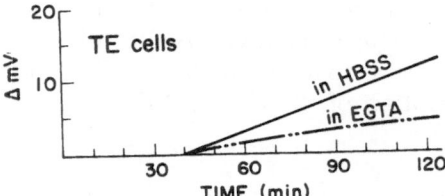

Figure 2. Postlag, Ca^{2+}-dependent, and Ca^{2+}-independent aggregation of retinal TE cells prepared by dissociation after incubation in trypsin plus EGTA (see Table I). Aggregation in the presence (HBSS·———) and absence (EGTA: $- \cdot \cdot - \cdot \cdot -$) of Ca^{2+}. HBSS, Hanks balanced salt solution; EGTA, Ca^{2+}-free HBSS with 1.3 mM EGTA. (From Magnani *et al.*, 1981.)

than 15 min) to 0.001% trypsin in the absence of Ca^{2+} (1.3 mM EGTA) (e.g., see Fig. 2). In continued contrast to the CI adhesion mechanism, not only is the CD mechanism protected by Ca^{2+} (1.3 mM) from removal by a relatively high concentration (0.044%) of trypsin (Fig. 3; cf. Fig. 2)—it actually requires proteolysis in the presence of Ca^{2+} for "normal" activity. For example, although the CD adhesion mechanism is present on virtually all cells in the intact retina, mechanically dissociated cells display no detectable CD aggregation in the aggregometer assay (see Fig. 1) until treated secondarily with trypsin plus Ca^{2+} (Fig. 3). Moreover, activity of the CD mechanism then remains completely dependent on the continued presence of Ca^{2+}. Again to illustrate, CD cells aggregate normally even in 0.044% trypsin as long as Ca^{2+} is present, but the activated CD mechanism is quickly lost upon removal of Ca^{2+}, even in the absence of further proteolysis (Fig. 3). After proteolytic removal of the CD mechanism, a 30-min lag precedes the onset of renewed CD adhesive activity (Fig. 2), and several hours are required for complete restoration of CD adhesion. Unlike that of the CI mechanism, activity of the CD mechanism is nearly abolished at 4°C but is completely restored by return of the cells to 37°C.

Although the details of this summary of CI and CD aggregation come from work done by our group (Magnani *et al.*, 1981), they are, with minor exceptions, in accord with results from other laboratories. For example, Urushihara *et al.* (1979) report a limited but detectable CD component to the aggregation of mechanically dissociated cells in their assay. This finding is in fact consistent with our results. We know the CD mechanism to be present in a relatively inactive form on mechanically dissociated cells. The assay used by Takeichi *et al.* (1979) employs lower shear forces than does our aggregometer, and it is in

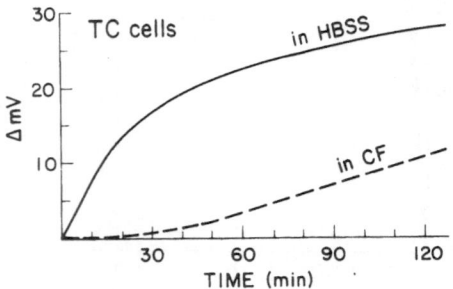

Figure 3. Immediate Ca^{2+}-dependent and postlag Ca^{2+}-independent aggregation of retinal TC cells prepared by dissociation after incubation in trypsin plus Ca^{2+} (see Table I). Aggregation in the presence (HBSS:———) and absence (CF: ————) of Ca^{+}. HBSS, Hanks balanced salt solution; CF, Ca^{2+}-free HBSS. (From Magnani *et al.*, 1981.)

related, lower shear assays that we also detect a limited CD activity. McClay and Marchase (1982) report a temperature-insensitive component to retinal CD adhesion in apparent contrast to statements in the summary presented herein. However, their assay separates in time the bringing together and subsequent forcible separation of interacting cells, providing for no-shear contact intervals of controlled duration. They have used this assay to focus primarily on the earliest stages in cell–cell interactions. By contrast, transient cell collisions and shear-dependent separations are simultaneously acting interdependent parameters in our aggregometer, and thus our assay tends to average the effects of such events occurring over longer periods of time. Although we observe limited low-temperature CD adhesion in the aggregometer, in our hands such adhesion is far more pronounced under other conditions that provided for uninterrupted cell contacts of more extended duration. Finally, Grunwald et al. (1980) reported that deactivation of the CD mechanism upon removal of Ca^{2+} requires the simultaneous action of trypsin, either that bound to the cells from prior treatment or that added exogenously for the purpose. After repeated efforts to explore this particular point, we were unable to resolve the discrepancy in our two sets of results.

In summary, despite obvious differences in assay conditions and relatively minor discrepancies in the outcome of certain experiments, the results outlined represent a concurrence of data from five different laboratories. When originally presented, the data demonstrated a previously unappreciated complexity in retinal adhesion phenomena and thus helped explain some of the confusion in the earlier data. CD and CI aggregation were demonstrated to be largely, if not completely, responsible for the observed in vitro retinal adhesion phenomena and, given the importance of dissociation and assay conditions in determining the relative activities of these two mechanisms, it was no longer surprising that different laboratories employing rather different procedures should have obtained such varied results. It seemed quite possible in fact that the various adhesion factors (both inhibitory and stimulatory) described earlier might be solubilized components of these two primary systems. No concerted attempt to test this possibility by systematically correlating the earlier and newer data has yet been reported, but the original isolation of a cell adhesion molecule (CAM) as a proteolytic fragment from culture supernatants (Rutishauser et al., 1976) is certainly consistent with such a hypothesis.

2.1.3. Molecular Analysis

The logical extension of the foregoing work is the characterization of molecules associated with the two adhesion mechanisms. Much of the success realized in this endeavor has resulted from the application of an immunological approach introduced by Beug et al. (1970) in their analysis of cellular adhesion in Dictyostelium discoideum. The ability of an antibody preparation to interfere with a particular adhesion event is interpreted to indicate a specific reaction with the molecule(s) responsible for that adhesion. The inhibition of adhesion can thus be used as an assay in the purification of that molecule.

Following this approach, we prepared anti-retinal antibodies whose Fab fragments selectively inhibited one or the other (CI or CD) mode of retinal cell adhesion (Fig. 4) (Thomas and Steinberg, 1981). Adsorption experiments combining these antibodies with variously prepared retinal cell populations (see Table I) succeeded in (1) substantiating the hypothesis that distinct and different molecules were associated with each mode of aggregation, and (2) verifying our conclusions regarding the presence or absence of those molecules on cells from each of the five standard cell preparations (Magnani *et al.*, 1981). Through the combined use of immunological and biochemical techniques, considerable progress has been made in the identification and detailed analysis of those molecules.

2.1.3.a. Ca²⁺-Independent Adhesion Mechanism. Rutishauser *et al.* (1976) had shown that Fab from antibodies against a particular cell surface antigen (CAM) would inhibit aggregation of dissociated retinal cells; a signifi-

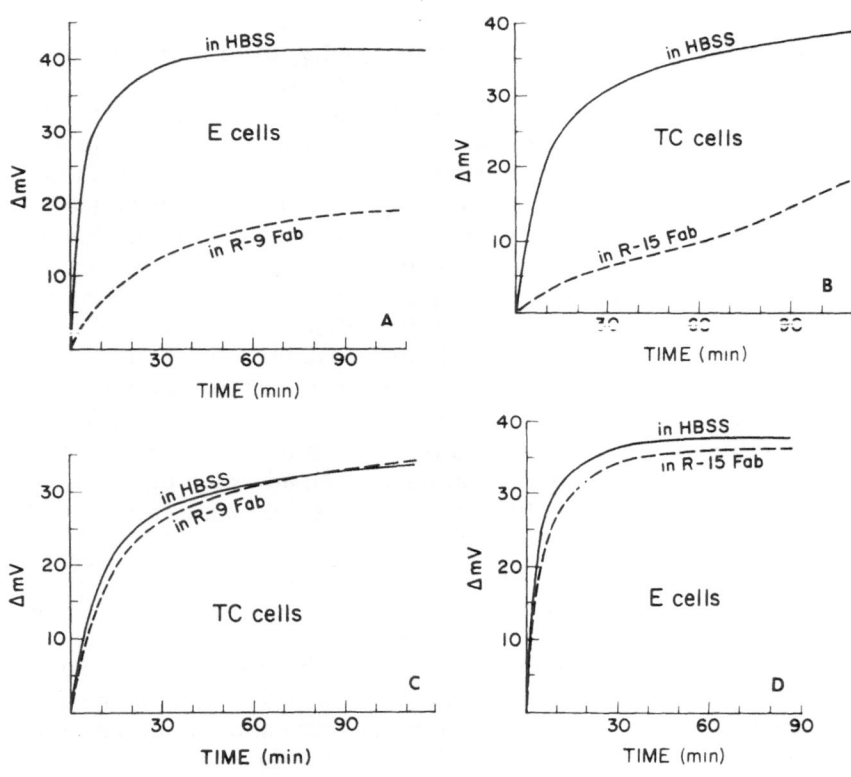

Figure 4. Selective inhibition of retinal Ca^{2+}- independent (E cell, A + D) or Ca^{2+}-dependent (TC cell, B + C) aggregation by anti-E cell (R-9, A + C) or anti-TC cell (R-15, B + D) Fab. R-9 and R-15 Fabs are used, respectively, at 2 mg/ml and 4 mg/ml in HBSS as indicated. HBSS, Hanks balanced salt solution. (From Thomas and Steinberg, 1981.)

cant body of literature gradually emerged from that group characterizing CAM and its contribution to cell interactions. Initially, no connection was established between CAM and either of the two retinal adhesion mechanisms described here. In subsequent work, however, Brackenbury et al. (1981) demonstrated a clear correspondence between CAM and the retinal CI adhesion mechanism.

Cell adhesion molecule (CAM) was originally identified in the 10-day chick neural retina and in that tissue had a molecular weight of 140,000 in sodium dodecyl sulfate (SDS) gels (Rutishauser et al., 1976; Thiery et al., 1977). Several forms of lower molecular weight, predominantly at 120,000 and 65,000, were also detected. More recent work has been carried out with CAM purified from 14-day (Hoffman et al., 1982; Cunningham et al., 1983; Edelman, 1983) chick brain, and from this source the molecule has been shown to be a glycoprotein with an apparent molecular weight ranging from 200,000 to 250,000 in SDS gels. [By the same criteria, work performed by Hoffman et al. (1982) on retinal CAM showed its weight to range from 140,000 to 200,000.] It is composed of a single polypeptide chain existing in at least two forms (140,000- and 170,000-M_r respectively), with the remainder of its mass attributable to its carbohydrate portion. This latter is rich (nearly 80%) in sialic acid, which, by virtue of its relative insensitivity to neuraminidase, appears to be attached through unusual linkages. In solution, CAM displays a limited proteolytic activity, which, under conditions of low salt and high temperature, cleaves the molecule autolytically to a polypeptide of 65,000. CAM self-associates in aqueous solution and binds Ca^{2+} independently to cells either as a molecule in solution or as a constituent of lipid vesicles (Rutishauser et al., 1982). Finally, CAM undergoes a conversion with developmental age from the embryonic form rich in sialic acid to several adult forms significantly poorer in sialic acid (Rothbard et al., 1982).

As a result of intensive studies from Edelman's laboratory, much is known about CAM and its role in cell interactions. It is preferentially located on neural cell processes, is not associated with a particular cell surface junctional complex, but is critical for tissue integrity and normal neurite fasciculation (Rutishauser et al., 1978a,b). Culturing retina in anti-CAM Fab' disrupts the orderly layering typical of normal retina development (Buskirk et al., 1980). While neural explants exhibit near-normal levels of neurite outgrowth in the presence of anti-CAM Fab', the extent of neurite–neurite association is markedly reduced over that seen in the control explants (Rutishauser et al., 1978b; Rutishauser and Edelman, 1980).

Finally, the distribution of CAM in the early chick embryo appears to correlate well with a number of developmental events (Thiery et al., 1982b). For example, CAM appears to be conspicuously absent during the migratory phase of neural crest cell translocations but reappears quite strikingly upon condensation of crest cells into cell clusters at the target locale (Duband et al., 1982; Thiery et al., 1982a).

2.1.3b. Ca^{2+}-Dependent Adhesion Mechanism. Similar though less extensive progress has been made in the analysis of molecules associated with the

CD adhesion system of the embryonic chick retina. Grunwald *et al.* (1980, 1981) demonstrated that retinal cells dissociated and subsequently maintained so as to lack the CI adhesion system nonetheless reassociate into cell aggregates with structural features very similar to those of intact, undissociated tissue of comparable embryonic age. Although not definitive, these results strongly suggest that the CD mechanism may have a significant role in directing or stabilizing such tissue organization. It should be noted in passing that the correlation in these studies between activity of the CD mechanism and preservation of junctional complexes (Grunwald *et al.*, 1981) hints at a possible role for components of the CD system in the organization of those specialized adhesion structures.

Work from the same laboratory has identified several cell surface molecules that appear to be intimately associated with CD adhesion activity (Cook and Lilien, 1982; Grunwald *et al.*, 1982). Prominent among these is a glycoprotein Ca^{2+}-Trypsin (CAT) with a molecular weight of 130,000 and an isoelectric point of 4.8. CAT is protected both from tryptic removal and from surface iodination by the presence of Ca^{2+}, and its removal and subsequent recovery at the cell surface correlate with CD activity (Geller and Lilien, 1981). Moreover, antibody yielding Fab' fragments that inhibit CD adhesion immunoprecipitates CAT from a supernate of detergent-solubilized retinal cells (Grunwald *et al.*, 1982).

Work from the laboratory of Marchase has identified another molecule, ligatin, which has also been implicated in the CD adhesion of neural retina cells (Jakoi and Marchase, 1979; Marchase *et al.*, 1982). Ligatin is present on the cell surface as 4.5-nm filaments that depolymerize upon treatment with EGTA to yield 10,000-M_r monomers. Ligatin added exogenously to retinal cells inhibits their reassociation (Marchase *et al.*, 1981), and removal of ligatin from the cell surface causes a decrease in the subsequent CD reaggregation (Marchase *et al.*, 1982). Although the precise role of ligatin in CD adhesion is unknown, its filamentous nature and close association with mannose-rich glycoproteins suggest that it may serve as a base plate for the attachment of peripheral glycoproteins to the external cell surface. Similar molecules have been identified in a variety of other tissues (Jakoi *et al.*, 1976, 1981; Gaston *et al.*, 1982).

One additional molecule, cognin, has also been implicated in retinal cell adhesion. This glycoprotein of approximately 50,000 M_r has an isoelectric point of approximately 4 (Hausman and Moscona, 1975). The presence of cognin on the cell surface correlates with adhesive activity of the cells (Hausman and Moscona, 1979), and exogenously added cognin increases the size of 16-hr cell aggregates (Hausman and Moscona, 1975). However, the association of cognin with either CI or CD adhesion has not been tested.

2.2. Dual Adhesion Mechanisms in Other Embryonic Chick Tissues

The considerable progress made to date in identifying and characterizing components of the CI and CD adhesion mechanisms in the chick neural retina has been of great importance in increasing our understanding of the molecular basis of cell interactions in that tissue. However, the findings gain added importance from the fact that very similar results have been obtained from studies using a variety of other tissues from diverse sources. Let us turn now to consider some of the additional examples from the chick embryo in an effort to explore the more general significance of the results described above in the retinal system.

2.2.1. Neural Tissues

Numerous earlier studies (Lilien and Moscona, 1967; Lilien, 1968; Hausman and Moscona, 1975, 1976; Merrell *et al.*, 1975; Lilien and Rutz, 1977; Jakoi and Marchase, 1979; Rutz and Lilien, 1979) described the isolation from embryonic chick neural tissue of aggregation factors that would enhance specifically the aggregation of the neural tissue from which the factor had been obtained. The discovery of two distinct adhesion mechanisms in the neural retina colored our interpretation of data from earlier studies on that tissue, and it seemed appropriate to reanalyze other chick neural tissues in a similar fashion. The intimate connections between the retina and optic tectum during development of the visual system made that latter tissue an obvious first choice for comparison to the retina.

2.2.1a. Optic Tectum. Analysis of the optic tectum using the techniques developed for study of the retina demonstrated clearly that behavior of cells from the optic tectum was indistinguishable in every respect from that of identically prepared retinal cells (Thomas *et al.*, 1981a,b). Aggregation kinetics of tectal cells dissociated using the five standard retinal dissociation procedures (Table I) were identical to those of comparably prepared retina cells (Fig. 5). Moreover, anti-retinal Fabs inhibited tectal cell aggregation with the same specificity and specific activity (Fig. 6). Thus, by every criterion employed, the optic tectum appears to possess dual adhesion mechanisms (CD and CI) physiologically and immunologically identical to those of the retina. We shall return to this observation presently.

2.2.1b. Other Neural Tissues. Other neural tissues studied in a similar fashion yielded results that differed somewhat from tissue to tissue (Thomas *et al.*, 1981a). Each tissue investigated (cerebrum, cerebellum, and spinal cord) clearly possessed two distinct adhesion mechanisms—one CD, the other CI— similar to those of the retina and tectum. However, the concentration of trypsin, Ca^{2+}, or EGTA required to optimize the activity of each mechanism varied considerably from tissue to tissue (W. A. Thomas, unpublished results). Moreover, while anti-retinal Fabs did inhibit aggregation of cells from each of the three tissues, that inhibition was variable from tissue to tissue, showing typically different concentration dependence and less specificity than that observed with both the retina and tectum (W. A. Thomas, unpublished results). Our comparative analysis of the mechanisms from the different neural tissues

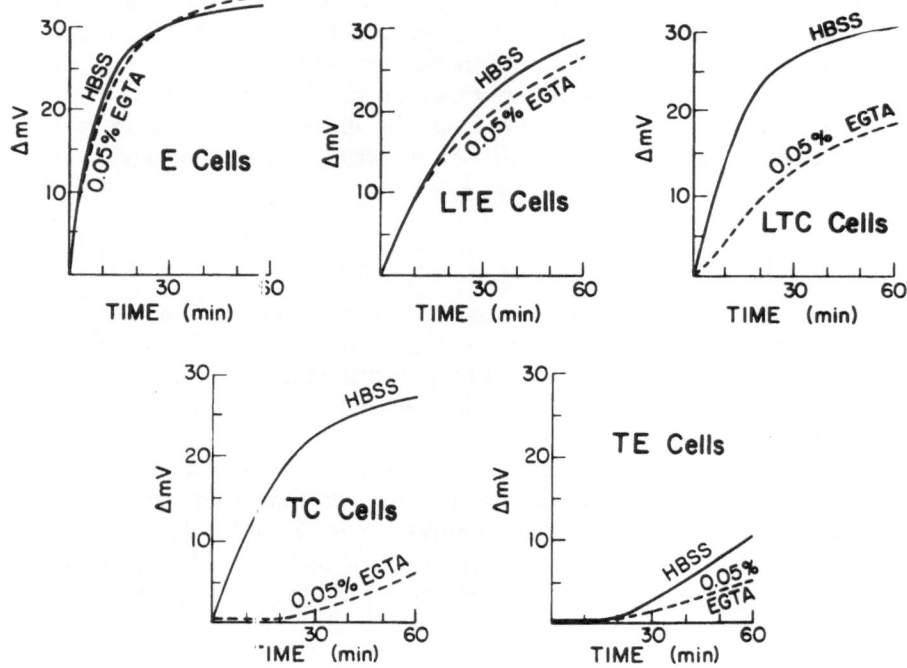

Figure 5. Aggregation kinetics of cells prepared from 7-day optic tectum using the five standard dissociation procedures developed for neural retina (see Table I). The aggregation kinetics of each tectal cell preparation are identical to those of comparably prepared retinal cells. Aggregation in the presence (HBSS:————) or absence (0.05% EGTA: - - - -) of Ca^{2+}. HBSS, Hanks balanced salt solution; 0.05% EGTA, Ca^{2+}-free HBSS containing 0.05% EGTA

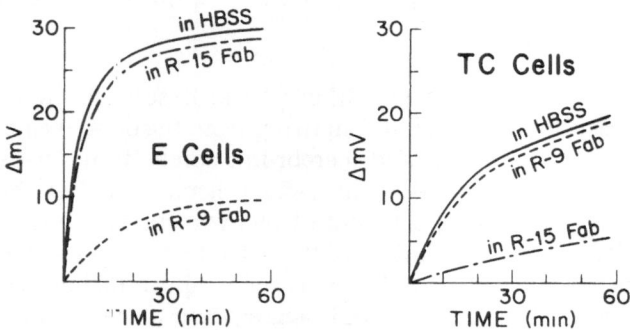

Figure 6. Inhibition of optic tectum cell aggregation by anti-retina Fabs at 3 mg/ml. (Left) Optic tectum E cells in HBSS (————), in anti-retina E cell (R-9) Fab (- - - -), in anti-retina TC cell (R-15) Fab (– · — · –). (Right) Optic tectum TC cells in HBSS (————), in anti-retina E cell (R-9) Fab (- - - -), in anti-retina TC cell (R-15) Fab (– · — · –). (From Thomas et al., 1981.)

has been limited by the heterogeneity of the immunological probes with which the work has been carried out; it must be explored more thoroughly with monoclonal antibodies currently under development. A clear trend is nonetheless evident in the work to date. The CD and CI mechanisms present in each of the five chick neural tissues investigated are related both physiologically and immunologically, but they seem (with the exception of those from the retina and tectum) not to be identical. These results are consistent with those of Edelman's group showing on the one hand a generalized distribution of CAM-related molecules in a number of embryonic chick neural tissues (Rutishauser et al., 1978a; Thiery et al., 1982b) and on the other hand subtle but distinct molecular differences in purified CAM from two such neural sources (Hoffman et al., 1982).

2.2.2. Nonneural Tissues

The results outlined in Section 2.1 suggested the possibility that such a pattern—that is, two general classes of adhesion mechanism varied in detail from tissue to tissue—might not represent an isolated occurrence in one or two neural tissues, but rather a basic organizational theme common to tissues throughout the chick embryo. Work from our laboratory on three nonneural tissues (7-day heart and liver, and 3¾-day limb bud) has demonstrated the existence of CI and CD adhesion mechanisms in each (Thomas et al., 1981b). However, the details in each case are markedly different from those derived from the neural tissues. For example, in the case of limb bud, which is most like the generalized neural case, we have been unable to prepare cells possessing only the CI adhesion mechanism (Fig. 7). This finding suggests that the sensitivities of the limb bud CI and CD mechanisms to trypsin and Ca^{2+} are quite different from those in the neural tissues. Moreover, while anti-retinal Fabs do inhibit limb bud cell aggregation, they do so only partially and with little or no selectivity for either mechanism (W. A. Thomas, unpublished results). The results for liver and heart differ even more radically from those of the neural cell case. Dissociation conditions required to evoke CD and CI adhesion in these two tissues are very different from those employed in the retina. Heart represents the extreme case in which CI aggregation is observed only in those cells dissociated with collagenase in the complete absence of other proteolytic activity. Moreover anti-retinal Fabs have no inhibitory effect at all on heart and liver cell aggregation (CI or CD) (Thomas et al., 1981a).

Embryonic heart and limb bud have not been used for other studies of precisely this kind, so no body of literature exists for comparison. On cells from embryonic chick skeletal muscle Grumet et al. (1982) detected molecules that cross-react with antibodies directed against retinal CAM and that also appear to mediate CI adhesion between retinal and muscle cells. The adhesive properties of embryonic chick liver cells have been the subject of several other studies and, although dual CI and CD mechanisms have not been reported, CD adhesion has been well documented (Grady and McGuire, 1976; McGuire and Burdick, 1976). In particular, Bertolotti et al. (1980) and Nielson et al. (1981)

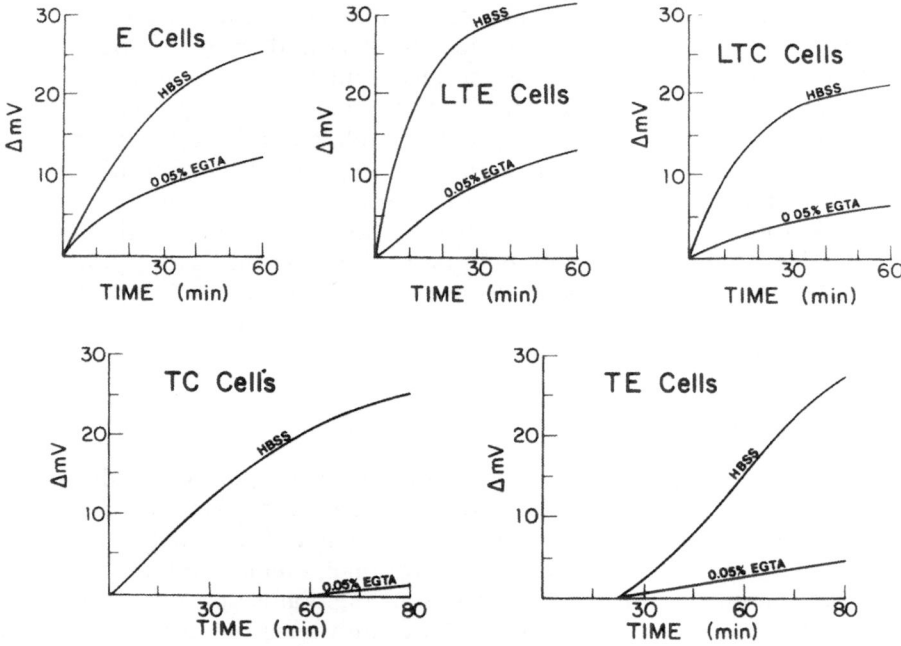

Figure 7. Aggregation kinetics of cells prepared from 3 3/4-day limb bud using modifications of the five standard dissociation procedures developed for neural retina (see Table I), (see also Thomas *et al.*, 1981). Cells exhibiting only Ca^{2+}-independent aggregation could not be prepared. (Compare E cells and LTE cells from limb bud, Fig. 7, and optic tectum, Fig. 5). Aggregation in the presence (HBSS) or absence (0.05% EGTA) of Ca^{2+}. HBSS, Hanks balanced salt solution; 0.05% EGTA, Ca^{2+}-free HBSS with 0.05% EGTA.

described a molecule (68,000 and 66,000 M_r, respectively) that is associated with the CD aggregation of hepatocytes from embryonic livers (10- and 9-day, respectively). Fab' from antibody against this molecule (L-CAM, Bertolotti *et al.*, 1980) inhibits the aggregation of those cells but has virtually no effect on the aggregation of retinal cells. Despite significant differences in both the age of the starting material and the conditions of tissue dissociation, it is probable that the molecule described in these studies is also associated with the liver CD adhesion described in our work.

Finally, to emphasize the broader applicability of these results, it should be noted that such findings are by no means restricted to the chick embryo. The first report of dual CI and CD adhesion mechanisms was made for BHK cells (Urushihara *et al.*, 1977) in a fibroblastic hamster kidney tissue culture cell line. Subsequent work from the same laboratory has made significant strides not only in demonstrating dual mechanisms in cells from a variety of sources, but in purifying and characterizing molecular components of the two adhesion systems in several of those cell lines (Takeichi, 1977, 1981; Atsumi and Uno, 1979; Takeichi *et al.*, 1979, 1981; Urushihara *et al.*, 1979; Atsumi and Takeichi, 1980;

Urushihara and Takeichi, 1980; Ogou *et al.*, 1982; Yoshida and Takeichi, 1982). The elements of these various systems have been shown to be glycoproteins that migrate in the 130,000–150,000-M_r region of SDS gels. Correlations between results from these studies and those in other laboratories (Hyafil *et al.*, 1980, 1981; Damsky *et al.*, 1981) raise intriguing questions concerning the extent of the commonality in adhesion mechanisms from very different sources. Answers to those questions will have to await not only more detailed analyses of the various molecules under study by the different groups but direct exchange among those groups as well.

3. Adhesive Specificity and Cellular "Recognition"

As one considers the development of a particular multicellular organism, one is quickly impressed with the reproducibility not only of the general body plan, but of the detailed sequence of morphogenetic events as well. Directed cell growth (as in an axonal tip) and cell and tissue movements are repeated with an accuracy that would seem to require a kind of road map of specific information made available at critical times (Goodman *et al.*, 1982; Ho and Goodman, 1982; Taghert *et al.*, 1982) (see Chapter 14). One is led to inquire into both the nature of this information and the means by which it might be presented and processed into the events we observe. As suggested earlier, molecular, and more precisely adhesive, interactions at the cell surfaces are very likely candidates to serve as the source of this basic information.

Results from early reaggregation studies suggested that *in vitro* reassociation of vertebrate cells proceeded essentially by tissue type (Townes and Holtfreter, 1955; Trinkaus and Groves, 1955; Moscona, 1956, 1957, 1961*a*) rather than by species (Moscona, 1957, 1961*a,b*, 1962; Garber and Moscona, 1964; Levak-Svajger and Moscona, 1964). That is, cells of two different tissues from a single species would sort out from one another in coaggregation experiments, whereas under the same conditions, cells of a given tissue type from two different species would intermix. An early but persistent interpretation of this observation grew from what was initially the lock-and-key model of molecular specificity proposed by Weiss (1947) and later developed into the chemospecificity model proposed by Sperry (1963). This interpretation postulated the existence of "tissue-specific" cell surface molecules that would "recognize" only their specific receptors. Thus only cells from a particular tissue would share a specific class of ligands and receptors and would therefore form adhesive bonds.

The above-stated hypothesis seemed to explain an array of "specific" cell interactions and drew support from the results of several types of experiments. Many investigators demonstrated measurable differences in the rates of adhesion between different tissues from avian and mammalian embryonic sources (Roth and Weston, 1967; Roth, 1968; Roth *et al.*, 1971; Barbera *et al.*, 1973; Walther *et al.*, 1973; Gottlieb and Glaser, 1975; Grady and McGuire, 1976; McGuire and Burdick, 1976). The data typically demonstrated a faster rate of adhesion of single cells to aggregates or monolayers from the same tissue than to aggregates or monolayers from another tissue. Through differential adsorption of anti-chick serum, Goldschneider and Moscona (1972) demonstrated distinct classes of embryonic chick cell surface antigens. Finally, a variety of factors were identified that demonstrated tissue-selective enhancement of embryonic cell aggregation (Lilien and Moscona, 1967; Lilien, 1968; Garber and Moscona, 1972; Hausman and Moscona, 1975).

Upon closer examination, the case for "tissue-specific recognition" appears less than convincing; the greater portion of the data mentioned above were in fact not truly tissue-specific. For example, cells from different tissues typically did not fail to cross-adhere in tests of specificity, but rather showed different rates of adhesion in the heterotypic and homotypic cases (Walther *et al.*, 1973; Cassiman and Bernfield, 1976; McGuire and Burdick, 1976). It was demonstrated that cells of a given tissue type from two different species would in fact sort out from one another (Burdick and Steinberg, 1969; Burdick, 1970). The antigens described by Goldschneider and Moscona were typically not restricted to a single tissue but were instead shared by groups of tissues with some common feature. Finally, there were basic difficulties in interpreting some of the data. The *in vitro* reorganization of dissociated cells into "pseudo-tissues" is driven by the minimization of surface free energy (Steinberg, 1962a,b,c), and the equilibrium conditions dictated by this consideration were shown to be unrelated to the rates with which cells associated in collecting assays (Moyer and Steinberg, 1976). In other words, the rate of adhesion is typically an inappropriate parameter for measuring the preferential associations of cells *in vitro*, and presumably, *in vivo* as well.

In essence, then, the concept of "tissue-specific recognition" has frequently been misconceived and ill used. Because cells of different kinds share certain adhesion systems, molecular specificity à la Weiss need not necessarily give rise to specificity at the cellular level, while demonstrable selectivity at the cellular level may arise not only from qualitative differences in adhesion systems but even from quantitative differences in the distribution of a single class of cell surface ligand (Steinberg, 1978). Thus the issue of cellular recognition is a complex one involving considerations at multiple levels of organization. Analyses of the dual adhesion mechanisms in the embryonic chick have generated some rather surprising results that offer a new insight into the means by which preferential cell associations may be regulated.

3.1. Cross-Adhesion Experiments

3.1.1. Homotypic Cases

Chick embryonic retinal cells possessing either the CI or CD adhesion mechanism can be prepared by manipulating the concentrations of trypsin and Ca^{2+} in the dissociation medium. This fortunate fact makes it possible to obtain information concerning the molecular specificity of cell adhesions by performing experiments at the cellular level. The adhesion behavior of cells possessing only one adhesion mechanism reflects the activity of that mechanism alone, and so coaggregation experiments performed with different combinations of cells from a given tissue prepared so as to possess one or the other adhesion mechanism can yield considerable insight into the functional relationship of those two mechanisms.

Labeling a retinal cell preparation with fluorescein isothiocyanate (FITC) distinguishes that population from another but has no effect on the adhesion characteristics of the labeled cells (Springer and Barondes, 1978; Takeichi *et al.*, 1979). Thus labeled and unlabeled CI cells aggregate together indiscriminately (as do labeled and unlabeled CD cells) and are completely intermixed in resulting aggregates. As expected, in experiments in which retinal CI and CD cell suspensions are mixed (one labeled, one not) and permitted to coaggregate (Fig. 8), cells from each preparation self-aggregate as before. Unexpectedly, however, virtually no cross-adhesion occurs between CI and CD cells (Fig.

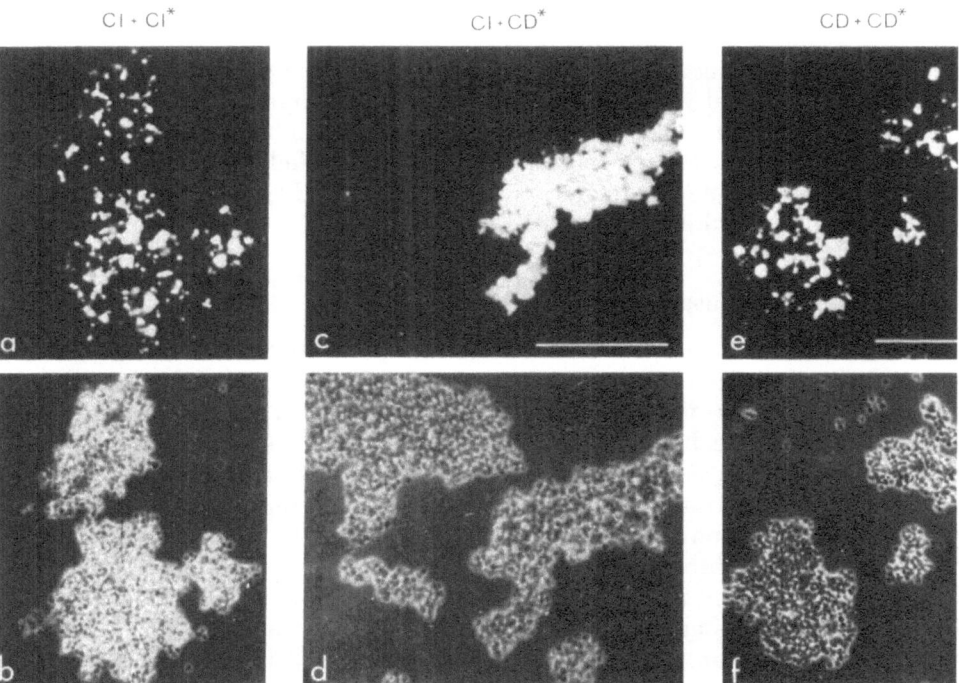

Figure 8. Retinal cell cross-adhesion. Paired fluorescence (a,c,e) and phase-contrast (b,d,f) photographs of aggregates formed by coaggregation of variously prepared FITC-labeled (*) and unlabeled retinal cells. (a,b) FITC-labeled and unlabeled LTE cells—(CI mechanism only). (c,d) FITC-labeled TC cells—CD mechanism only; and unlabeled LTE cells —CI mechanism only. (e,f) FITC-labeled and unlabeled TC cells—CD mechanism only. Each pair of photographs shows a single microscopic field that represents accurately the distribution of labeled and unlabeled cells in each of several hundred aggregates. Scale bar: 100 μm. (From Thomas *et al.*, 1981.)

8c,d). Aggregates are not mixed, but contain exclusively cells from one or the other preparation (Takeichi *et al.*, 1979; Thomas *et al.*, 1981b). Using a mono-layer assay, Brackenbury *et al.* (1981) obtained results generally in agreement with those outlined above. These investigators found in certain cases that the nature (CI or CD) of adhesive interactions between monolayers and adhering probe cells depended on the nature of the cell type forming the monolayer. This finding is in contrast to those cited above (Takeichi *et al.*, 1979; Thomas *et al.*, 1981b) and, although not investigated further by Brackenbury *et al.* (1981), is probably due at least in part to the differences in assay conditions and cell preparation procedures that distinguish the two approaches.

It should be stressed that the adhesive selectivity we observe is essentially absolute and is neither rate nor shear dependent. Retinal CI and CD cells fail to cross-adhere under a wide range of assay conditions as long as they truly lack one or the other adhesion mechanism. Moreover, the result is not simply an artifact of proteolytic removal of one or the other adhesion mechanism. The adhesive selectivity of retinal (LTC) cells prepared so as to possess both mecha-nisms (see Table I) in an active form can be reversibly manipulated through the addition or removal of Fab fragments selectively inhibitory of one or the other adhesion mechanism (Thomas *et al.*, 1981b).

The above findings were not restricted to the chick neural retina. Identical results were obtained with each of the chick neural (optic tectum, cerebrum, cerebellum, and spinal cord) and non-neural tissues (limb bud, heart and liver) tested (Thomas *et al.*, 1981a). In each tissue, FITC-labeled and unlabeled CI (or CD) cells self-aggregated normally, but in no tissue did CI and CD cells cross-adhere to form mixed aggregates.*

3.1.2. Heterotypic Cases

The above results demonstrated very clearly that the CD and CI adhesion mechanisms of any given tissue (homotypic cases) are independent and nonin-teracting. The experiments left untested the nature of the functional rela-tionship between mechanisms of the same class (CI or CD) from two different tissues (heterotypic cases). Results from combined kinetic and immunological studies had demonstrated clear tissue-dependent differences both in the sen-sitivity of each mechanism to trypsin and Ca^{2+} and in the ability of anti-retinal Fab fragments to inhibit cell aggregation. Despite these very significant dif-ferences, it was found in cross-adhesion experiments that cells, regardless of tissue of origin, cross-adhered if and only if they shared at least one of the two categories of adhesion mechanism in the active state (Thomas *et al.*, 1981a). These results are consistent with earlier findings by Takeichi *et al.* (1979) that demonstrated similar mechanism-dependent adhesive interactions between chick neural retina and Chinese hamster V-79 cells. Taken together, the results of the above coaggregation experiments were initially rather surprising and

---

*Because limb bud cells could not be prepared so as to possess only the CI mechanism, tests of limb bud CI activity were performed in 0.05% EGTA in order to inactivate unwanted CD activity.

resulted in a new perspective on the issues of tissue-specific adhesion and cell "recognition."

3.2. Discussion

3.2.1. Review

Several observations and conclusions are fundamental to the altered perspective mentioned above. The first, already stated earlier, is that there exists a clear specificity of interaction at the molecular level. The two major categories of embryonic adhesion mechanism are completely independent of one another in their activity, and neither cross-reacts with any component of the other to form hybrid adhesive bonds. The second observation is that within this context, adhesive selectivity at the cellular level follows directly from the distribution of CI and CD adhesion activity. All cells in each of the tissues examined appear to possess both categories of adhesion mechanism initially. It is the selective activation, inactivation, or removal of one or the other mechanism that confers the observed adhesive selectively. The third observation is that this adhesive selectivity does not follow the conventionally expected pattern; it is in no functional sense tissue-specific. There do seem to be distinct molecular differences in the CI or CD mechanism from tissue to tissue, but these differences apparently do not affect function as tested in our assays.

Before addressing the significance of these conclusions, it is important to reiterate that the results upon which they are based are not limited to tissues from the embryonic chick. Takeichi and co-workers have performed extensive studies comparable to those outlined above, not only on chick neural retina (Takeichi et al., 1979), but on several transformed and untransformed fibroblastic cell lines (Takeichi et al., 1979, 1981). Their basic results are quite consistent with those from our work.

From the foregoing results it would seem that the theme of dual—or more generally, multiple—adhesion mechanisms is one that may prove to be surprisingly common, not only in tissues from the embryonic chick, but in other cell and tissue types from a variety of sources. Should this be borne out, the most likely explanation will reside in the inherent potential of such a system for subtle regulation of cellular adhesive properties.

In the most general sense, it is easy to envision how such regulation might occur. A cell with only a single adhesion mechanism is constrained to alter its adhesiveness by regulating the activity, number, or distribution of adhesion molecules on its surface. With two or more distinct mechanisms, noninteractive and independently regulated, the possibilities for adhesive specificity and for the specification of cell position and orientation are immensely compounded. The two adhesion mechanisms described above have very different sensitivities to proteolysis and Ca^{2+}, both parameters well within the capabilities of the cell to regulate at its surface. We have already seen that the adhesive selectivity of embryonic chick cells can be varied reversibly and at will in vitro. Given this potential, one is left to ask whether regulation of the dual adhesion mechanisms is in fact employed in normal development in the chick embryo. Fortunately, in at least one existing system that question can be posed effectively.

3.2.2. Retinotectal Mapping

The phenomenon of patterned retinotectal innervation provides an excellent opportunity to address the issue of adhesion-mediated specificity of cell interaction in a developing system. Several hypotheses have been presented to explain the precise regional matching between the retina and the optic tectum (Sperry, 1963; Barondes, 1970; Marchase et al., 1975; Prestige and Willshaw, 1975; Chung and Cooke, 1978; Horder and Martin, 1978; Meyer, 1978; Schmidt, 1978; Willshaw and von der Malsburg, 1979; Fraser, 1980; Law and Constantine-Paton, 1980, 1981; Whitelaw and Cowan, 1981). Prominent among them is one postulating the existence of complementary molecular gradients specifying the precise positional information necessary to guide the retinotectal mapping (Sperry, 1963; Marchase et al., 1975). Gradients of adhesiveness have in fact been described both in the nasotemporal (Halfter et al., 1981) and dorsoventral (Barbera, 1975; Gottlieb et al., 1976; McClay et al., 1977; Cafferata et al., 1979) axes of the retina and optic tectum, and the correlation between these gradients and the observed patterns of innervation strongly suggests a possible role for the former in generating the latter. A model has been proposed that describes in detail how modulation of adhesive parameters might actually generate the necessary information (Fraser, 1980). The possibility that cellular adhesiveness might play such a role, coupled with the virtual identity between the CI and CD adhesion mechanisms of the chick retina and those of the tectum, strongly suggested an involvement of one or both of those mechanisms in retinotectal patterning.

Earlier studies (Barbera, 1975; Gottlieb et al., 1976; McClay et al., 1977; Cafferata et al., 1979) had demonstrated that retinal cells now known to possess both adhesion mechanisms (CD and CI) would display preferential binding along the dorsoventral axis of the retina and tectum. Labeled cells dissociated from the dorsal half of a chick neural retina were shown to bind in larger numbers to (heterologous) collecting surfaces consisting either of intact ventral retinal halves or of cell monolayers prepared from those halves than to comparable collecting surfaces from the (homologous) dorsal half of the retina. Similarly, probe cells from the ventral half of the retina bound preferentially to collecting surfaces prepared from the dorsal retina half.

Upon repeating such experiments using probe cells and monolayers prepared so as to possess only one (CI or CD) of the two adhesion mechanisms, we found that only activity of the CI adhesion system correlates with the dorsoventral gradient of adhesiveness in the retina (see Fig. 9 and Table II). No such correlation has been demonstrated for the CD system, but its potential contribution to the nasotemporal gradient described by Halfter et al. (1981) remains to be examined. Within the context of this presentation, the possibility that positional information critical for retinotectal mapping might derive from a carte-

D/D*   V/D*

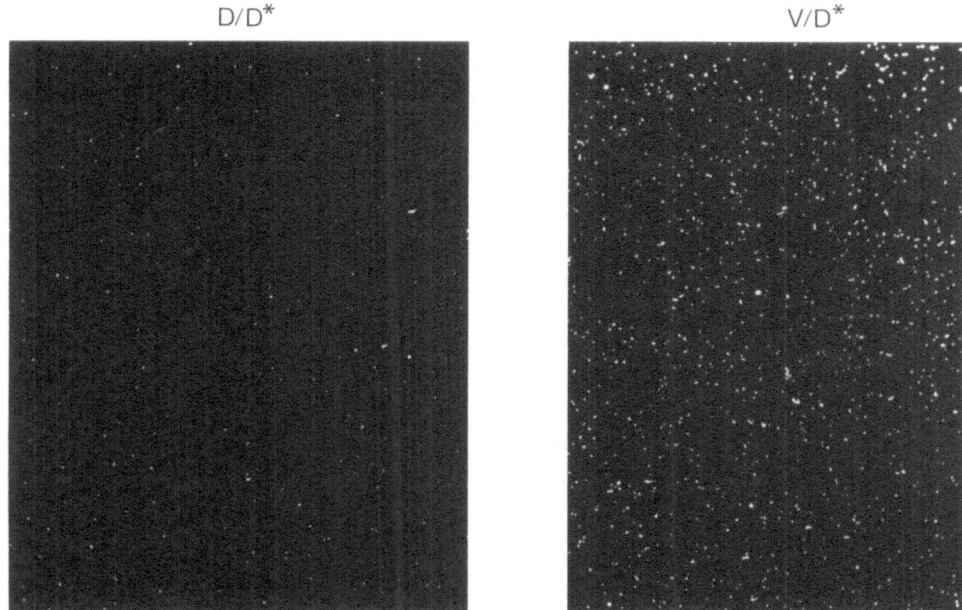

Figure 9. Differential Ca^{2+}-independent adhesions in the retinal dorsoventral axis. Fluorescence photographs of monolayer assays performed using LTE monolayers and FITC-labeled (*) LTE probe cells (Ca^{2+}-independent mechanism only) from dorsal and ventral extremes of 8-day neural retinas. 2.5×10^5 probe cells were centrifuged gently onto monolayers prepared on polylysine coated coverslips, allowed to incubate for 5 min at 37°C, and then placed on a gyratory shaker at 100 rpm for 10 min. Coverslips were washed and observed by fluorescence microscopy. Assays were carried out in Hanks balanced salt solution, and photographic fields were chosen at random.

Table II. Quantitation of Data Demonstrating Differential Ca^{2+}-Independent Adhesion in the Dorsoventral Axis of the Retina[a]

FITC-labeled LTE probe cells	Unlabeled LTE monolayer	Percentage probe cells bound
Dorsal	Ventral	52
Dorsal	Dorsal	11
Ventral	Dorsal	27
Ventral	Ventral	12

[a]For each test category (D*/V, D*/D, V*/V, V*/D), monolayer-bound FITC-labeled* probe cells from five photographic fields, each taken at random of a different monolayer (see Fig. 9), were counted, averaged, and calculated as a percentage of the total probe cell population (2.5×10^5 cells) initially applied to each monolayer.

sian coordinate system (Sperry, 1963) based on two distinct adhesion systems, one in each axis, is particularly appealing and thus deserving of careful consideration.

Interestingly, the afore-mentioned results underscored the likelihood that significant heterogeneity exists within the CI adhesion system (see Section 4). Although Rutishauser *et al.* (1978a) demonstrated that Fab' from anti-CAM disrupts the adhesion of retinal probe cells to retinal monolayers in dorsoventral binding experiments, their data do not show activity of CAM to be responsible for the observed dorsoventral gradient of adhesiveness in that tissue. Indeed, it is difficult to explain such a gradient in terms of any single adhesion molecule binding to itself. According to such a scheme, homologous binding (D/D or V/V) would be preferred over heterologous binding (D/V), since cells with the highest complement of adhesion molecules would bind most strongly to themselves. This is not observed. The actual results are more easily explained in terms of complementary gradients of a ligand and its receptor (Fig. 10) (Marchase *et al.*, 1975; Barbera, 1975), each having unique biochemical characteristics (Barbera, 1975; Marchase, 1977; Cafferata *et al.*, 1979).

The issue of molecular heterogeneity aside, it must be stated that the data from the work to date are not adequate to prove that the observed differential CI activity actually generates information capable of guiding retinotectal patterning. Direct involvement of the CI mechanism in this developmental process thus remains speculative. Furthermore, there exists a well-defined molecule whose known distribution across the retina qualifies it for serious consideration as another potential specifier of positional information across the retina.

Trisler *et al.* (1981) prepared a monoclonal antibody that exhibits a 35-fold difference in binding along the dorsoventral axis of 14-day chick retina, the greatest binding being localized at the dorsal extreme. The antigen (designated TOP for toponimic) recognized by this monoclonal antibody has an apparent molecular weight of 47,000 by SDS gel electrophoresis and is trypsin sensitive

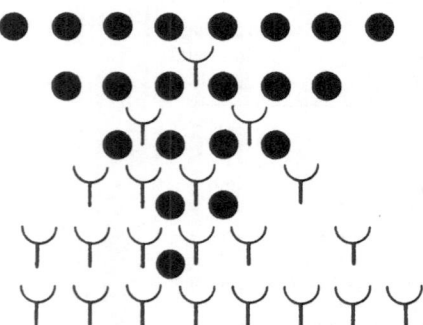

Figure 10. Simple model depicting complementary gradients of ligand (●) and receptor (Ⴘ) postulated to explain the observed dorsoventral gradient of adhesiveness in the retina. Dorsal cells have many ligand molecules but few receptors, whereas ventral cells have many receptors and few ligand molecules. The most stable connections would be formed between dorsal and ventral cells since that combination would generate the greatest number of ligand-receptor interactions (Ⴘ) (Adapted from Barbera, 1975.)

(Nirenberg et al., 1983). Cells from which TOP is removed by trypsinization reconstitute the antigen in culture to a level appropriate to the original position of those cells in the retina. Thus the positional information represented by the level of TOP is not dependent on the organizational integrity of the retina but is instead intrinsic to each cell.

No adhesive function has been demonstrated for TOP, and so the relationship between the dorsoventral gradient of adhesiveness and the comparable gradient in the distribution of TOP is unclear. Whether either (or both) of these systems of positional information actually contributes to the orderly mapping of the retina onto the tectum also remains to be tested. It is apparent from work already performed that several classes of molecules have the topographical distribution or the known adhesive characteristics to qualify them for a role in this developmental process. The real challenge lies ahead in determining not only what information is actually used, but how it is coordinated to assure the eventual outcome.

4. Conclusions

It has not been the intention in the foregoing analysis to imply that the dual adhesion mechanisms described therein are the sole mediators of adhesive interactions in the tissues discussed—that surely is not the case. It is apparent from the simple act of dissociating a neural retina that multiple mechanisms are likely to be at work. The enzymatic treatments that degrade the CI and CD adhesion mechanisms do not suffice to dissociate the tissue but must be complemented by trituration in order to disperse the cells of the still intact tissue. One is led to speculate that additional mechanisms relatively unaffected by the dissociation conditions may contribute more to maintaining tissue integrity than to bringing cells together.

Moreover, the assays employed in the analyses outlined above are for the most part short-term kinetic tests, the results of which may or may not reflect accurately adhesive forces relevant in longer-term development events. Although the dual mechanisms described herein appear to be the major contributors to short-term aggregation, many other adhesive interactions may contribute to longer-term adhesion-mediated events in the tissues examined.

Interactions between cell surfaces and extracellular molecules such as laminin, fibronectin, collagen, and glycosaminoglycans have been shown to play an important role in a growing number of morphogenetic events. Underhill and Dorfman (1978) demonstrated in a number of fibroblastic cell lines dual adhesion systems based in part on hyaluronic acid. Both fibronectin (Li and Sheffield, 1982) and fibronectin-binding proteins (Thomas and Giudice, 1983) have been detected in the embryonic chick neural retina. Coupled with the fact that fibronectin (and laminin) has been shown to be an effective substratum for support of retinal neurite outgrowth (Akers et al., 1981; Adler and Hewitt, 1983), this may suggest that matrix-mediated cues could contribute to in vivo guidance of retinal axonal growth. Finally, lectins (carbohydrate-binding proteins) have been identified in teratocarcinoma cells and in several embryonic tissues (Grabel *et al.*, 1979, 1982; Barondes *et al.*, 1981; Gaston *et al.*, 1982; Marchase *et al.*, 1982) and may participate in specifying cell "recognition."

In short, there are many other potential adhesion mechanisms whose contributions to embryonic cell interactions have not been explored in the approaches outlined above. It is quite likely that multiple mechanisms must be employed in a subtly regulated coordinate fashion to specify the precise "recognition" characteristic of a complex developmental system. However, although not alone in their function, the two classes of mechanisms (CI and CD) that have been identified and characterized clearly contribute significantly to the adhesive activity of the embryonic cells studied and thus represent a point of departure for analyzing the role of adhesive selectivity in morphogenetic movements.

Three major considerations have emerged from the results accumulated to date that are likely to be of continued, if not increased, importance in the future. The first of these is the growing indication that the adhesion mechanisms discussed here may be structural variations on a limited number of common themes. Dual CI and CD mechanisms are not restricted to the chick neural retina but are found in tissues throughout the embryo. The CI and CD molecules from different sources are clearly not identical, but they seem nonetheless to be functionally compatible. This general result is not restricted to embryonic chick cells, as similar dual mechanisms exist in transformed and normal fibroblastic tissue cells (Atsumi and Uno, 1979; Takeichi *et al.*, 1979; Urushihara *et al.*, 1977). In fact, mechanism-dependent adhesion between those fibroblastic cells tested (Chinese hamster V-79) and chick retina cells is apparently blind even to species differences (Takeichi *et al.*, 1979). The growing list of adhesion molecules purified from diverse sources shows many (although certainly not all) falling into the molecular weight range of 130,000–150,000, and future studies may show these to be related by more than size alone.

The second consideration fundamental to this analysis is that of molecular heterogeneity. Just as elements of commonality are shared by comparable adhesion mechanisms from different sources, there are clear indications that the molecules involved are not identical. Our work has demonstrated that mechanisms with the same class of activity (CI or CD) but from different tissue sources show not only very different sensitivities to trypsin and Ca^{2+}, but widely different reactivities with antibodies selective for one or the other retinal adhesion mechanism. Similar results have been obtained in other studies (Bertolotti *et al.*, 1980; Brackenbury *et al.*, 1981).

The issue of heterogeneity is in fact a relatively complex one that must be considered at more than a single level. At one level is the matter of bond formation. Are adhesive interactions homophilic, involving only a single class of molecule, or are they heterophilic, involving more than one? If the latter, are the bonds formed by a simple lock-and-key mechanism (bimolecular), or are multiple components required? The self-association properties of CAM suggest that homotypic bonding is a likely mode of interaction between CAM molecules on apposing cells (Rutishauser *et al.*, 1982). On the other hand, results from experiments exploring the CI adhesive gradient in the dorsoventral axis of the retina suggest a greater complexity to the CI adhesion mechanism, and as yet, too little is known about other embryonic adhesion molecules even to speculate on the possible modes of interaction.

A second level of heterogeneity closely linked to the first is the possible microheterogeneity of a single adhesion system (i.e., CD or CI) in a given tissue. Is each such system composed of a single molecular mechanism, or might there be a family of related molecules? Recent work with CAM has indicated that neither retinal nor brain CAM is monodisperse with regard to molecular weight. Both show considerable variation in the amount of bound sialic acid, and results from analysis of brain CAM indicate minor variations in the polypeptide chain as well (Hoffman *et al.*, 1982).

At a third level is the issue of heterogeneity from tissue to tissue within each class of mechanism. Work from several studies has demonstrated an apparent commonality of function within a given class of adhesion mechanism. That is, cells even from two quite different sources cross-adhere readily, but they do so if and only if they share at least one of two classes of adhesion mechanism in an active form. These results seem somewhat paradoxical from at least two points of view. First, they seem to argue convincingly against the existence of truly tissue-specific cell associations. Second, the results are obtained despite the significant physiological and immunological differences that can distinguish similar adhesion mechanisms from different sources.

It should be noted in passing that intriguing exceptions to the pattern of results generalized here do exist. For example, Takeichi *et al.* (1981) observed CD adhesion in several embryonal carcinoma and differentiated fibroblastic cell lines. However, the CD molecules shared by the carcinoma lines are functionally incompatible with and of a different molecular weight from the comparable molecules shared by the normal cell lines. Thus, the various carcinoma cells tested all coaggregate Ca^{2+}-dependently, as do the normal cells tested, but no CD adhesion occurs between any normal and carcinoma cells.

No such dramatic exceptions have been observed with embryonic chick tissues. Although no satisfactory explanation has been found for the observed results, several possible interpretations can help resolve the conceptual conflicts. The most facile one is that we are dealing with an artifact of our assay system. In truth, we lack strong evidence to show that molecules responsible for the CI or CD self-adhesion in each of two different cell types actually interact directly when cross-adhesions occur. Immunological tests indicate that retina cells use the same molecules for cross-adhesion to heart or liver that they use for self-adhesion to other retinal cells (Thomas *et al.*, 1981a), but we have no knowledge of the molecules on the non-neural cells with which the retinal molecules interact.

Despite such inadequacies in the available data, we do have a large and growing body of consistent results ostensibly demonstrating molecularly selective adhesive interactions, and the apparent structure–function paradox itself may hold an alternative explanation for the observed results. In describing the relatedness of various histocompatibility antigens, Hood *et al.* (1977) proposed an "area code hypothesis" to group related molecules into multigene families. Although existing data are simply not sufficient to warrant the serious proposal of such a complex hypothesis with regard to the dual CI and CD adhesion mechanisms, it is tempting to think of those mechanisms as families of related molecules sharing a common binding site linked to structural or regulatory segments that vary from tissue to tissue. Such a hypothesis does have the virtue of explaining the simultaneous commonality of function in the face of significant molecular heterogeneity. Although none of the data available specifically substantiates such a concept, the recent data describing the microheterogeneity of CAM (Hoffman *et al.*, 1982; Cunningham *et al.*, 1983; Edelman, 1983) cast such a speculation in new light. Further analysis will surely help clarify the functional significance of the observed molecular differences and either substantiate or invalidate the above speculation.

The third and final major consideration emerging from the body of results treated here is that of regulation of adhesiveness. One of the most appealing features of the dual adhesion mechanisms under consideration is the potential for exquisite control of cellular adhesiveness afforded by their differential sensitivities to proteolysis and free Ca^{2+}. In our experimental work on the various embryonic tissues, we took considerable pains to remove completely one or the other adhesion mechanism. Thus, a particular cell preparation possessed only a single adhesion mechanism, and experiments with that cell preparation were actually experiments on that mechanism alone. Only in this way could we obtain the clear, interpretable results that made possible the assigning of particular characteristics to one or the other mechanism. In doing so, however, we eliminated the inherent possibility for variable expression of either adhesion mechanism and dealt only with the two extreme cases—all or none—of what must surely be a complex matrix of combinations. The results discussed above, then, do not reflect the true spectrum of cell- or tissue-adhesive properties but simply demonstrate a limited set of adhesive properties imposed on the system by the experimental conditions. In principle, a cell *in vivo* would have limitless options in specifying its own adhesive characteristics. Any given combination of proteolytic activity and extracellular Ca^{2+} would dictate a unique set of adhesive properties reflecting the activities of the two (CI and CD) adhesion mechanisms combined in a particular ratio. Thus, despite the fact that the adhesion mechanisms themselves are not tissue-specific in their molecular recognition, they can be regulated individually to generate adhesive properties characteristic of any particular cell or tissue type.

Such variability in adhesive properties can in turn give rise to striking variations in the spatial arrangements of interacting cell populations (Steinberg, 1978). Simple modulation of a single adhesion mechanism is capable of generating many of the different patterns observed in *in vitro* sorting experiments, ranging from complete intermixing of two different cell populations to the sphere-within-a-sphere pattern observed when a strongly cohesive aggregated cell population is completely engulfed by a less cohesive one (Fig. 11, cases C/D and E/F). The presence of a second class of adhesion mechanism

Figure 11. Illustration of how the molecular specificity, the abundance, and the cellular distribution of adhesion receptor sites on cell surfaces can determine particular sets of intercellular adhesive intensities that in turn determine particular most stable cell associations or tissue configurations. Only lock-and-key binding is illustrated, but many other possibilities exist. Note that A has only "locks" and B has only "keys"; C and D share "locks" and "keys" in equal number; and E and F share the same "locks" and "keys"; but E has far more of both than does F. As a result, A and B have different adhesive specificities (i.e., each will bind to the other, but neither will bind to self); C and D share the same adhesive specificity and the same adhesive intensity; and E and F share the same adhesive specificity, but the adhesive intensity of E is greater than that of F. G cells and H cells have completely independent adhesion mechanisms, each with its own "locks and keys." G cells self-adhere, as do H cells, but G cells cannot adhere to H cells. (From Steinberg, 1978.)

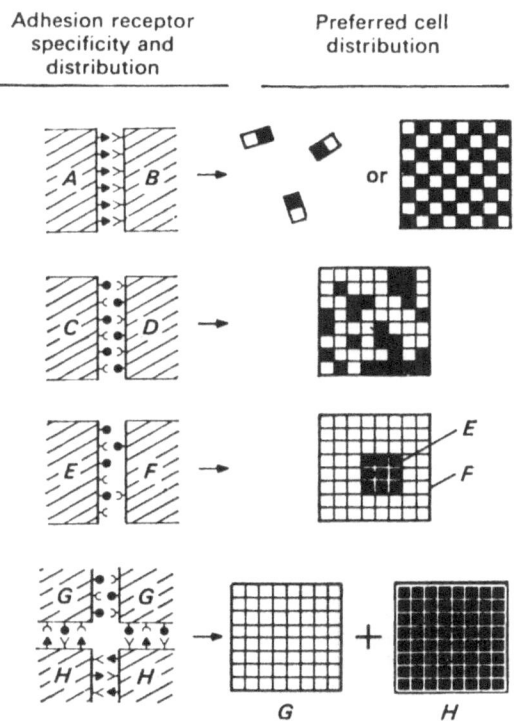

with its own unique molecular specificity allows for the generation of yet another spatial pattern observed *in vitro*—the independent aggregation of two different cell populations to form two classes of aggregates, each containing cells of only one kind (Fig. 11, case G/H). The reader will recall that cases C/D and G/H describe actual results obtained in the coaggregation experiments outlined in Section 3.1 (see Figure 8); case E/F is a generalized example of the spatial pattern resulting from adhesive properties intermediate to those of the two demonstrated extremes. It seems clear that these simple patterns of cell associations observed *in vitro* would have myriad manifestations *in vivo* reflecting the vastly more complex three-dimensional topography of that environment.

The making and breaking of adhesive bonds is fundamental not only to the relatively simple cell organizations described above, but also to morphogenetic processes of far greater complexity—the long-range translocations of migratory cells and the continuous restructuring of developing tissues and organs, to name only two. Because of their inherent potential to provide adhesive regulation through physiologically controlled variables (proteolysis and extracellular Ca^{2+}), the dual embryonic adhesion mechanisms in question seem ideally suited to have an important role in such morphogenetic events. We lack a clear demonstration that this potential is actually exploited in normal developmental processes. Fortunately, in recent years a combination of techniques and perspective has evolved that should enable us to address this issue directly. With the approaches made possible by such advances, we should expect to see the potential developmental roles of these dual adhesion mechanisms become the subject of future studies both fascinating and productive.

References

Adler, R., and Hewitt, A. T., 1983, Responses of cultured neural retinal cells to substratum-bound extracellular matrix molecules, *J. Cell Biol.* **97**:97a.

Akers, R. M., Mosher, D. F., and Lilien, J. E., 1981, Promotion of retinal neurite outgrowths by substratum-bound fibronectin, *Dev. Biol.* **86**:179–181.

Atsumi, T., and Takeichi, M., 1980, Cell association pattern in aggregates controlled by multiple cell–cell adhesion mechanisms, *Dev. Growth Diff.* **22**:133–142.

Atsumi, T., and Uno, K., 1979, Clonal teratocarcinoma stem cells have similar adhesion mechanisms to cells from differentiated tissues, *Cell Struct. Funct.* **4**:388.

Barbera, A. J., 1975, Adhesive recognition between developing retina cells and the optic tecta of the chick embryo, *Dev. Biol.* **46**:167–191.

Barbera, A. J., Marchase, R. B., and Roth, S., 1973, Adhesive recognition and retino-tectal specificity, *Proc. Natl. Acad. Sci. USA* **70**:2482–2486.

Barondes, S. H., 1970, Brain glycomacromolecules and interneuronal recognition, in: *The Neurosciences Second Study Program* (F. O. Schmidt, ed.), pp. 747–760, The Rockefeller University Press, New York.

Barondes, S. H., Beyer, E. C., Springer, W. R., and Cooper, D. N., 1981, Endogenous lectins in chickens and slime molds: Transfer from intracellular to extracellular sites, *J. Supramol. Struct. Cell. Biochem.* **16**:233–242.

Bertolotti, R., Rutishauser, U., and Edelman, G. M., 1980, A cell surface molecule involved in aggregation of embryonic liver cells, *Proc. Natl. Acad. Sci. USA* **77**:4831–4835.

Beug, H., Gerisch, G., Kempff, S., Riedel, V., and Cremer, G., 1970, Specific inhibition of cell contact formation in *Dictyostelium* by univalent antibodies, *Exp. Cell Res.* **63**:147–158.

Brackenbury, R., Rutishauser, U., and Edelman, G. M., 1981, Distinct calcium-independent and calcium-dependent adhesion systems of chicken embryo cells, *Proc. Natl. Acad. Sci. USA* **78**:387–391.

Burdick, M. L., 1970, Cell sorting out according to species in aggregates containing mouse and chick embryonic limb mesoblast cells, *J. Exp. Zool.* **175**:357–368.

Burdick, M. L., and Steinberg, M. S., 1969, Embryonic cell adhesiveness: Do species differences exist among warm-blooded vertebrates?, *Proc. Natl. Acad. Sci. USA* **63**:1169–1173.

Buskirk, D. R., Thiery, J.-P., Rutishauser, U., and Edelman, G. M., 1980, Antibodies to neural cell adhesion molecule disrupt histogenesis in cultured chick retinae, *Nature (Lond.)* **285**:488–489.

Cafferata, R., Panosian, J., and Bordley, G., 1979, Developmental and biochemical studies of adhesive specificity among embryonic retinal cells, *Dev. Biol.* **69**:108–117.

Cassiman, J. J., and Bernfield, M. R., 1976, Use of preformed cell aggregates and layers to measure tissue-specific differences in intercellular adhesion, *Dev. Biol.* **52**:231–245.

Chung, S. H., and Cooke, J., 1978, Observations on the formation of the brain and of nerve connections following embryonic manipulation of the amphibian neural tube, *Proc. Soc. Lond. [Biol.]* **201**:335–373.

Cook, J. H., and Lilien, J., 1982, The accessibility of certain proteins on embryonic chick neural retina cells to iodination and tryptic removal is altered by calcium, *J. Cell Sci.* **55**:85–104.

Cunningham, B. A., Hoffman, S., Rutishauser, U., Hemperly, J. J., and Edelman, G. M., 1983, Molecular topography of the neural cell adhesion molecule N-CAM: Surface orientation and location of sialic acid-rich and binding regions, *Proc. Natl. Acad. Sci. USA* **80**:3116–3120.

Damsky, C. H., Knudsen, K. A., Dorio, R. J., and Buck, C. A., 1981, Manipulation of cell–cell and cell–substratum interactions, *J. Cell Biol.* **89**:173–184.

Duband, J. L., Delouvee, A., Rosvasio, R. A., and Thiery, J.-P., 1982, Mechanisms of avian crest cell migration and homing, in: *Embryonic Development, Part B: Cellular Aspects, Progress in Clinical and Biological Research,* Vol. 85B (M. M. Burger and R. Weber, eds.), pp. 497–508, Alan R. Liss, New York.

Edelman, G. M., 1983, Cell adhesion molecules, *Science* **219**:450–457.

Fraser, S. E., 1980, A differential adhesion approach to the patterning of nerve connections, *Dev. Biol.* **79**:453–464.

Garber, B., and Moscona, A. A., 1964, Aggregation *in vitro* of dissociated cells. I. Recombination of skin in the chorioallantoic membrane from suspensions of embryonic chick and mouse skin cells. *J. Exp. Zool.* **155**:179–202.

Garber, B., and Moscona, A. A., 1972, Reconstruction of brain tissue from cell suspension. II. Specific enhancement of aggregation of embryonic cerebral cells by supernatant from homologous cells, *Dev. Biol.* **27**:235–243.

Gaston, S. M., Marchase, R. B., and Jakoi, E. R., 1982, Brain ligatin: A membrane lectin that binds acetylcholinesterase, *J. Cell Biochem.* **18**:447–459.

Geller, R. L., and Lilien, J., 1981, Repair of calcium-dependent neural cell adhesive system, *J. Cell Biol.* **91**:105a.

Goldschneider, I., and Moscona, A. A., 1972, Tissue-specific cell surface antigens in embryonic cells, *J. Cell Biol.* **53**:435–449.

Goodman, C. S., Raper, J. A., Ho. R. K., and Chang, S., 1982, Pathfinding by neuronal growth cones during grasshopper embryogenesis, in: *Fortieth Symposium of the Society for Developmental Biology* (S. Subtelny and P. Green, eds.), pp. 275–316, Alan R. Liss, New York.

Gottlieb, D. I., and Glaser, L., 1975, A novel assay of neuronal cell adhesion, *Biochem. Biophys. Res. Commun.* **63**:815–821.

Gottlieb, D. I., Rock, K., and Glaser, L., 1976, A gradient of adhesive specificity in developing avian retina, *Proc. Natl. Acad. Sci. USA* **73**:410–414.

Grabel, L. B., Rosen, S. D., and Martin, G. R., 1979, Teratocarcinoma stem cells have a cell surface carbohydrate binding component implicated in cell–cell adhesion, *Cell* **17**:477–484.

Grabel, L. B., Martin, G. R., and Rosen, S. D., 1982, Teratocarcinoma stem cell surface lectin: Characterization and proposed function, in: *Cellular Recognition, UCLA Symposia on Molecular and Cellular Biology,* Vol. 3 (W. A. Frazier, L. Glaser, and D. I. Gottieb, eds.), pp. 879–888, Alan R. Liss, New York.

Grady, S. R., and McGuire, E. J., 1976, Intercellular adhesive selectivity. III. Species selectivity of embryonic liver intercellular adhesion, *J. Cell Biol.* **71**:96–106.

Grumet, M., Rutishauser, U., and Edelman, G. M., 1982, Neural cell adhesion molecule is on embryonic muscle cells and mediates adhesion to nerve cells *in vitro, Nature (Lond.)* **195**:693–695.

Grunwald, G. B., Geller, R. L., and Lilien, J., 1980, Enzymatic dissection of embryonic cell adhesive mechanisms, *J. Cell Biol.* **85**:766–776.

Grunwald, G. B., Bromberg, R. E. M., Crowley, N. J., and Lilien, J., 1981, Enzymatic dissection of embryonic cell adhesive mechanisms. II. Developmental regulation of an endogenous adhesive system in the chick neural retina, *Dev. Biol.* **86**:327–338.

Grunwald, G. B., Pratt, R. S., and Lilien, J., 1982, Enzymatic dissection of embryonic cell adhesive mechanisms. III. Immunological identification of a component of the calcium-dependent adhesive system of embryonic chick neural retina cells, *J. Cell Sci.* **55**:69–83.

Halfter, W., Claviez, M., and Schwarz, V., 1981, Preferential adhesion of tectal membranes to anterior embryonic chick retina neurites, *Nature (Lond.)* **292**:67–78.

Hausman, R. E., and Moscona, A. A., 1975, Purification and characterization of the retina-specific cell-aggregating factor, *Proc. Natl. Acad. Sci. USA* **72**:916–920.

Hausman, R. E., and Moscona, A. A., 1976, Isolation of retina-specific cell-aggregating factor from membranes of embryonic neural retina tissue, *Proc. Natl. Acad. Sci. USA* **73**:3594–3598.

Hausman, R. E., and Moscona, A. A., 1979, Immunologic detection of retina cognin on the surface of embryonic cells, *Exp. Cell Res.* **119**:191–204.

Ho, R. K., and Goodman, C. S., 1982, Peripheral pathways are pioneered by an array of central and peripheral neurons, *Nature (Lond.)* **297**:404–406.

Hoffman, S., Sorkin, B. C., White, P. C., Brackenbury, R., Mailhammer, R., Rutishauser, U., Cunningham, B. A., and Edelman, G. M., 1982, Chemical characterization of a neural cell adhesion molecule purified from embryonic brain membranes, *J. Biol. Chem.* **257**:7720–7729.

Holtfreter, J., 1939, Gewebeaffinität: Ein Mittel der embryonalen Formbildung, *Arch. Exp. Zellforsch. Besonders Gewebezuecht.* **23**:169–209.

Hood, L., Huang, H. V., and Dreyer, W. J., 1977, The area-code hypothesis: The immune system provides clues to understanding the genetic and molecular basis of cell recognition during development, *J. Supramol. Struct.* **7**:531–559.

Horder, T. J., and Martin, K. A. C., 1978, Morphogenetics as an alternative to chemospecificity in the formation of nerve connections, in: *Cell–cell Recognition, Society for Experimental Biology Symposium* (A. S. G. Curtis, ed.), pp. 275–358, Cambridge University Press, Cambridge.

Hyafil, F., Morello, D., Babinet, C., and Jacob, F., 1980, A cell surface glycoprotein involved in the compaction of embryonal carcinoma cells and cleavage stage embryos, *Cell* **21**:927–934.

Hyafil, F., Babinet, C., and Jacob, F., 1981, Cell–cell interactions in early embryogenesis: A molecular approach to the role of calcium, *Cell* **26**:447–454.

Jakoi, E. R., and Marchase, R. B., 1979, Ligatin from embryonic chick neural retina, *J. Cell. Biol.* **80**:642–650.

Jakoi, E. R., Zampighi, G., and Robertson, J. D., 1976, Regular structures in unit membranes. II. Morphological and biochemical characterization of two water-soluble membrane proteins isolated from the suckling rat ileum, *J. Cell Biol.* **70**:97–111.

Jakoi, E. R., Kempe, K., and Gaston, S. M., 1981, Ligatin binds phosphohexose residues on acidic hydrolases, *J. Supramol. Struct. Cell Biol.* **16**:139–153.

Law, M. I., and Constantine-Paton, M., 1980, Right and left eye bands in frogs with unilateral tectal ablations, *Proc. Natl. Acad. Sci. USA* **77(4)**:2314–2318.

Law, M. I., and Constantine-Paton, M., 1981, Anatomy and physiology of experimentally produced striped tecta, *J. Neurosci.* **1**:741–759.

Levak-Svajger, B., and Moscona, A. A., 1964, Differentiation in grafts of aggregates of embryonic chick and mouse cells, *Exp. Cell Res.* **36**:692–695.

Li, H.-P., and Sheffield, J. B., 1982, Delayed appearance of fibronectin in stationary cultures of embryonic neural retina cells, *J. Cell Biol.* **95**:30a.

Lilien, J. E., 1968, Specific enhancement of cell aggregation *in vitro*, *Dev. Biol.* **17**:657–678.

Lilien, J., and Moscona, A. A., 1967, Cell aggregation: Its enhancement by a supernatant from cultures of homologous cells, *Science* **157**:70–72.

Lilien, J., and Rutz, R., 1977, A multicomponent model for specific cell adhesion, in: *Cell and Tissue Interactions* (J. W. Lash and M. M. Burger, eds.), pp. 187–195, Raven Press, New York.

Magnani, J. L., 1979, Mechanisms and regulation of cell adhesion among chick embryonic neural retinal cells, Ph.D. dissertation, Princeton University.

Magnani, J. L., and Steinberg, M. S., 1978, Mechanisms of intercellular adhesion among chick neural retina cells, *J. Cell Biol.* **79**:A56.

Magnani, J. L., Thomas, W. A., and Steinberg, M. S., 1981, Two distinct adhesion mechanisms in embryonic neural retina cells. I. A kinetic analysis, *Dev. Biol.* **81**:96–105.

Marchase, R. B., 1977, Biochemical investigations of retinotectal adhesive specificity, *J. Cell Biol.* **75**:237–257.

Marchase, R. B., Barbera, A. J., and Roth, S., 1975, A molecular approach to retino-tectal specificity, in: *Cell Patterning, Ciba Foundation, Sympium*, pp. 315–327, Elsevier, Amsterdam.

Marchase, R. B., Harges, P., and Jakoi, E. R., 1981, Ligatin from embryonic chick neural retina inhibits retinal cell adhesion, *Dev. Biol.* **86**:250–255.

Marchase, R. B., Koro, L. A., Kelly, C. M., and McClay, D. R., 1982, A possible role for ligatin and the phosphoglycoproteins it binds in calcium-dependent retinal adhesion, *J. Cell. Biochem.* **18**:461–468.

McClay, D. R., and Marchase, R. B., 1982, Calcium-dependent and calcium-independent adhesive mechanisms are present during initial binding events of neural retinal cells, *J. Cell Biochem.* **18**:469–478.

McClay, D. R., Gooding, L. R., and Fransen, M. E., 1977, A requirement for trypsin-sensitive cell-surface components for cell–cell interactions of embryonic neural retina cells, *J. Cell. Biol.* **75**:56–66.

McGuire, E. J., and Burdick, C. L., 1976, Intercellular adhesive selectivity. I. An improved assay for the measurement of embryonic intercellular adhesion (liver and other tissues), *J. Cell Biol.* **68**:80–89.

Merrell, R., Gottlieb, D. I., and Glaser, L., 1975, Embryonal cell surface recognition. Extraction of an active plasma membrane component, *J. Biol. Chem.* **250**:5655–5659.

Meyer, R., 1978, Deflection of selected optic fibers into a denervated tectum in goldfish, *Brain Res.* **155**:213–227.

Moscona, A. A., 1956, Development of heterotypic combinations of dissociated embryonic chick cells, *Proc. Soc. Exp. Biol. Med.* **92**:410–416.

Moscona, A. A., 1957, The development *in vitro* of chimeric aggregates of dissociated embryonic chick and mouse cells, *Proc. Natl. Acad. Sci. USA* **43**:184–194.

Moscona, A. A., 1961a, Rotation mediated aggregation of dissociated cells, in: *Growth in Living Systems* (M. X. Zarrow, ed.), pp. 197–220, Basic Books, New York.

Moscona, A. A., 1962, Analysis of cell recombinations in experimental synthesis of tissues *in vitro*, *J. Cell Comp. Physiol. (Suppl. 1)* **60**:65–80.

Moscona, A. A., and Hausman, R. E., 1977, Biological and biochemical studies on embryonic cell–cell recognition, in: *Cell and Tissue Interactions* (J. W. Lash and M. M. Burger, eds.), pp. 173–185, Raven Press, New York.

Moyer, W. A., and Steinberg, M. S., 1976, Do rates of intercellular adhesion measure the cell affinities reflected in cell-sorting and tissue spreading configurations?, *Dev. Biol.* **52**:246–262.

Nielsen, L. D., Pitts, M., Grady, S. R., and McGuire, E. J., 1981, Cell–cell adhesion in the embryonic chick: Partial purification of liver adhesion molecules from liver membranes, *Dev. Biol.* **86**:315–326.

Nirenberg, M., Wilson, S., Higashida, H., Rotter, A., Kruger, K., Buis, N., Ray, R., Kenimer, J. G., and Adler, M., 1983, Modulation of synapse formation by cyclic adenosine monophosphate, *Science* **222**:794–799.

Ogou, S., Okada, T. S., and Takeichi, M., 1982, Cleavage stage mouse embryos share a common cell adhesion system with teratocarcinoma cells, *Dev. Biol.* **92**:521–528.

Phillips, H. M., 1969, Equilibrium measurements of embryonic cell adhesiveness: Physical formulation and testing of the differential adhesion hypothesis, Ph.D. thesis, The Johns Hopkins University, Baltimore.

Phillips, H. M., and Davis, G. S., 1978, Liquid-tissue mechanics in amphibian gastrulation: Germ layer assembly in *Rana pipiens*, *Am. Zool.* **18**:81–93.

Phillips, H. M., Wiseman, L. L., and Steinberg, M. S., 1977, Self vs. nonself in tissue assembly: Correlated changes in recognition behavior and tissue cohesiveness, *Dev. Biol.* **57**:150–159.

Prestige, M. C., and Willshaw, D. J., 1975, On a role for competition in the formation of patterned neural connections, *Proc. R. Soc. Lond. [Biol.]* **190**:77–98.

Roth, S., 1968, Studies on intercellular adhesive selectivity, *Dev. Biol.* **18**:602–631.

Roth, S., and Weston, J. A., 1967, The measurement of intercellular adhesion, *Proc. Natl. Acad. Sci. USA* **58**:974–980.

Roth, S., McGuire, E. J., and Roseman, S., 1971, An assay for intercellular adhesive specificity, *J. Cell Biol.* **51**:525–535.

Rothbard, J. B., Brackenbury, R., Cunningham, B. A., and Edelman, G. M., 1982, Differences in the carbohydrate structure of neural cell-adhesion molecules from adult and embryonic chicken brains, *J. Biol. Chem.* **257**:11064–11069.

Rutishauser, U., and Edelman, G. M., 1980, Effects of fasciculation on the outgrowth of neurites from spinal ganglia in culture, *J. Cell Biol.* **87**:370–378.

Rutishauser, U., Thiery, J.-P., Brackenbury, R., Sela, B.-A., and Edelman, G. M., 1976, Mechanisms of adhesion among cells from neural tissues of the chick embryo, *Proc. Natl. Acad. Sci. USA* **73**:577–581.

Rutishauser, U., Thiery, J.-P., Brackenbury, R., and Edelman, G. M., 1978a, Adhesion among neural cells of the chick embryo. III. Relationship of the surface molecule CAM to cell adhesion and the development of histotypic patterns, *J. Cell Biol.* **79**:371–381.

Rutishauser, U., Gall, W. E., and Edelman, G. M., 1978b, Adhesion among neural cells of the chick embryo. IV. Role of the cell surface molecule CAM in the formation of neurite bundles in cultures of spinal ganglia, *J. Cell Biol.* **79**:382–393.

Rutishauser, U., Hoffman, S., and Edelman, G. M., 1982, Binding properties of a cell adhesion molecule from neural tissue, *Proc. Natl. Acad. Sci. USA* **79**:685–689.

Rutz, R., and Lilien, J., 1979, Functional characterization of an adhesive component from the embryonic chick neural retina, *J. Cell Sci.* **26**:323–342.

Schmidt, J. T., 1978, Retinal fibers alter tectal positional markers during the expansion of the half-retinal projections in goldfish, *J. Comp. Neurol.* **177**:279–300.

Sperry, R. W., 1963, Chemoaffinity in the orderly growth of nerve fiber patterns and connections, *Proc. Natl. Acad. Sci. USA* **50**:703–710.

Springer, W. R., and Barondes, S. H., 1978, Direct measurement of species-specific cohesion in cellular slime molds, *J. Cell Biol.* **78**:937–942.

Steinberg, M. S., 1962a, On the mechanism of tissue reconstruction by dissociated cells. I. Population kinetics, differential adhesiveness, and the absence of directed migration, *Proc. Natl. Acad. Sci. USA* **48**:1577–1582.

Steinberg, M. S., 1962b, On the mechanism of tissue reconstruction by dissociated cells. II. Time course of events, *Science* **137**:762–763.

Steinberg, M. S., 1962c, On the mechanism of tissue reconstruction by dissociated cells. III. Free energy relationships and the reorganization of fused, heteronomic tissue fragments, *Proc. Natl. Acad. Sci. USA* **48**:1769–1776.

Steinberg, M. S., 1963, Reconstruction of tissues by dissociated cells, *Science* **141**:401–408.

Steinberg, M. S., 1964, The problem of adhesive selectivity in cellular interactions, in: *Cellular Membranes in Development* (M. Locke, ed.), pp. 321–366, Academic Press, New York.

Steinberg, M. S., 1970, Does differential adhesion govern self-assembly processes in histogenesis? Equilibrium configurations and the emergence of a hierarchy among populations of embryonic cells, *J. Exp. Zool.* **173**:395–434.

Steinberg, M. S., 1978, Cell–cell recognition in multicellular assembly: Levels of specificity. in: *Cell–Cell Recognition, Society for Experimental Biology Symposium* (A. S. G. Curtis, ed.), Vol. 32, pp. 25–49, Cambridge University Press, Cambridge.

Taghert, P. H., Bastiani, M. J., Ho, R. K., and Goodman, 1982, Guidance of pioneer growth cones: Filopodial contacts and coupling revealed with an antibody to lucifer yellow, *Dev. Biol.* **94**:391–399.

Takeichi, M., 1977, Functional correlation between cell adhesive properties and some cell surface proteins, *J. Cell. Biol.* **75**:464–474.

Takeichi, M., 1981, Identification of cell-to-cell adhesion molecules of Chinese hamster fibroblasts, in: *Cancer Cell Biology, Gann Monograph on Cancer Research*, No. 25 (T. Nagayo and W. Mori, eds.), pp. 3–8, Japan Scientific Societies Press, Tokyo.

Takeichi, M., Ozaki, H. S., Tokumaga, K., and Okada, T. S., 1979, Experimental manipulation of cell surface to affect cellular recognition mechanisms, *Dev. Biol.* **70**:195–205.

Takeichi, M., Atsumi, T., Yoshida, C., Uno, K., and Okada, T. S., 1981, Selective adhesion of embryonal carcinoma cells and differentiated cells by Ca^{2+}-dependent sites, *Dev. Biol.* **87**:340–350.

Thiery, J.-P., Brackenbury, R., Rutishauser, U., and Edelman, G. M., 1977, Adhesion among neural cells of the chick embryo. II. Purification and characterization of a cell adhesion molecule from neural retina, *J. Biol. Chem.* **252**:6841–6845.

Thiery, J.-P., and Duband, J. L., and Delouvee, A., 1982a, Pathways and mechanisms of avian trunk neural crest cell migration and localization, *Dev. Biol.* **93**:324–343.

Thiery, J.-P., Duband, J.-L., Rutishauser, U., and Edelman, G. M., 1982b, Cell adhesion molecules in early chicken embryogenesis, *Proc. Natl. Acad. Sci. USA* **79**:6737–6741.

Thomas, W. A., 1979, A behavioral and immunological analysis of intercellular adhesion, Ph.D. dissertation, Princeton University, Princeton, New Jersey.

Thomas, W. A., and Giudice, G., 1983, Fibronectin-binding proteins in the chick neural retina, *J. Cell Biol.* **97**:326a (abst. 1235).

Thomas, W. A., and Steinberg, M. S., 1980, A twelve-channel automatic device for continuous recording of cell aggregation by measurement of small-angle light-scattering, *J. Cell Sci.* **42**:1–18.

Thomas, W. A., and Steinberg, M. S., 1981, Two distinct adhesion mechanisms in embryonic neural retina cells. II. An immunological analysis, *Dev. Biol.* **81**:106–114.

Thomas, W. A., Edelman, B. A., Lobel, S. M., Breitbart, A. S., and Steinberg, M. S., 1981, Two chick embryonic adhesion systems: Molecular vs. tissue specificity, *J. Supramol. Struct.* **16**:15–27.

Thomas, W. A., Thomson, J., Magnani, J. L., and Steinberg, M. S., 1981, Two different adhesion mechanisms in embryonic neural retina cells. III. Functional specificity, *Dev. Biol.* **81**:379–385.

Townes, P. A., and Holtfreter, J., 1955, Directed movements and selective adhesion of embryonic amphibian cells, *J. Exp. Zool.* **128**:53–120.

Trinkaus, J. P., and Groves, P. W., 1955, Differentiation in culture of mixed aggregates of dissociated tissue cells, *Proc. Natl. Acad. Sci. USA* **41**:787–795.

Trisler, D. G., Schneider, M. D., and Nirenberg, M., 1981, A topographic gradient of molecules in retina can be used to identify neuron position, *Proc. Natl. Acad. Sci. USA* **78(4)**:2145–2149.

Underhill, C., and Dorfman, A., 1978, The role of hyaluronic acid in intercellular adhesion of cultured mouse cells, *Exp. Cell Res.* **117**:155–164.

Urushihara, H., and Takeichi, M., 1980, Cell–cell adhesion molecule: Identification of a glycoprotein relevant to the Ca^{2+}-independent aggregation of Chinese hamster fibroblasts, *Cell* **20**:363–371.

Urushihara, H., Ueda, M. J., Okada, T. S., and Takeichi, M., 1977, Calcium-dependent and independent adhesion of normal and transformed BHK cells, *Cell Struct. Funct.* **2**:289–296.

Urushihara, H., Ozaki, H. S., and Takeichi, M., 1979, Immunological detection of cell surface components related with aggregation of Chinese hamster and chick embryonic cells, *Dev. Biol.* **70**:206–216.

Walther, B. T., Ohman, R., and Roseman, S., 1973, A quantitative assay for intercellular adhesion, *Proc. Natl. Acad. Sci. USA* **70**:1569–1573.

Weiss, P., 1941, Nerve patterns: the mechanics of nerve growth, *Growth (Suppl.)* **51**:163–203.

Weiss, P., 1947, The problem of specificity in growth and development, *Yale J. Biol. Med.* **19**:235–278.

Whitelaw, V. A., and Cowan, J. D., 1981, Specificity and plasticity of retino-tectal connections: A computational model, *J. Neurosci.* **1**:1369–1387.

Willshaw, D. J., and von der Malsburg, C., 1979, A marker induction mechanism for the establishment of ordered neural mapping; its application to the retinotectal problem, *Philos. Trans. R. Soc. Lond. [Biol.]* **287**:203–243.

Wilson, H. V., 1907, On some phenomena of coalescence and regeneration in sponges, *J. Exp. Zool.* **5**:245–258.

Yoshida, C., and Takeichi, M., 1982, Teratocarcinoma cell adhesion: Identification of a cell surface protein involved in calcium-dependent cell aggregation, *Cell* **28**:217–224.

Chapter 10

Gap Junctions in Development

JEAN-PAUL REVEL

1. Introduction

The concept of a multicellular organism as an assemblage of cells that work together harmoniously for the good of the whole organism is well established. Over the past 15 years, however, it has become increasingly clear that the individual cells of multicellular animals are not as self-sufficient and independent of each other as once had been believed. Channels that exist between virtually all neighboring cells permit such widespread intercellular exchanges of low-molecular-weight materials that they must be considered syncytial with respect to small molecules. The specialized areas of the membrane that permit these exchanges are clusters of channels whose permeability is limited only by molecular size of the permeant (Simpson *et al.*, 1977; Flagg-Newton *et al.*, 1979), known collectively as *gap junctions*. Only in the case of molecules close to the exclusion limit can one detect selectivity (Brink and Dewey, 1980).

Gap junctions generally appear early in embryogenesis—because their distribution and permeability seem to change at developmentally interesting times, many theories have been proposed as to their possible role in differentiation. The goal of this chapter is to examine the behavior of junctions during development. It must be made clear from the outset, however, that recording the presence of junctions is not sufficient to define their role in differentiation. The channel is only one element of the system, since one must also consider the molecules that traverse them, carrying potentially permissive or instructive messages from one cell to the other. The nature of the message thus exchanged is of prime importance—gap junctions are found not only during development but throughout life. In the adult, (a nondeveloping organism, if such exists), gap junctions play roles in cell communication, metabolic cooperation, and homeostasis. In the embryo these or other junction-mediated exchanges presumably participate in controlling development.

JEAN-PAUL REVEL • Division of Biology, California Institute of Technology, Pasadena, California 91125.

2. Structure of Junctions

Gap junctions consist of collections of subunits or *connexons*. Connexons in one cell membrane associate with similar structures in the membrane of a neighboring cell to form a continuous aqueous path between the cell cytoplasms. Models based on X-ray diffraction work (Caspar *et al.*, 1977; Makowski *et al.*, 1977) and on analysis of low-dose electron micrographs (Unwin and Zampighi, 1980) show each connexon (in liver) to be hexameric, formed by the association of six protein molecules that delineate a central pore. The junctional protein has a molecular weight of about 25,000–30,000 (Duguid and Revel, 1976; Hertzberg and Gilula, 1979; Finbow *et al.*, 1980; Gros *et al.*, 1983). Connexons are grouped into close packed arrays. In some instances (developing heart for example) the junctions can have rather complex shapes (Griepp *et al.*, 1978; Gros *et al.*, 1978, 1979), but in general gap junctions have a sufficiently characteristic appearance that they are clearly recognizable morphologically at many developmental stages and in a very broad range of organisms.

In spite of a reasonable morphological homogeneity, the constituent proteins of gap junctions appear to differ (Hertzberg *et al.*, 1982; Nicholson *et al.*, 1983; Gros *et al.*, 1983; Nicholson *et al.*, 1985). Liver, heart, and lens junction proteins of the same organism show very different two-dimensional peptide maps. Proteins isolated from the livers (or lenses) of different species, however, appear to be very closely related. Few antibodies against gap junctions (Traub *et al.*, 1982; Hertzberg, 1980; Bok *et al.*, 1982; Benedetti *et al.*, 1981) have yet been prepared, but the early results do not show as marked a divergence between different junction proteins as seen by peptide mapping. This discrepancy is probably explained by the results of actual amino acid sequencing of the N-terminal portion of heart (Nicholson *et al.*, unpublished) and liver junction proteins (Nicholson *et al.*, 1981), which indicate that 40% of the residues are homologous. This level of conservation is sufficient to suggest both that gap junction proteins may form a family and that there are major differences between members of this family. It has long been intuitively assumed that in order to perform similar functions there must be sufficient conservation of the domains within the proteins to permit the formation of transmembrane channels and other appropriate structures. This does not require that the primary sequence of amino acids be conserved absolutely.

3. Study of Junctions in Embryonic Systems

Of the many cell interactions that take place during embryogenesis and development, only a fraction will be mediated *via* gap junctions, but it is not always known which are and which are not. Thus, the interesting ionic currents extensively studied by Jaffe (1982) could exert their effect by a junctional mechanism, but in fact the path of current is not known. Another class of interaction is exemplified by the observations of Moscona and Linser (1983) on the contact dependence of the induction of the enzyme glutamine synthetase. There is no actual evidence linking the induction of the enzyme to the presence of gap junctions; we only know that gap junctions are present in the developing retina, along with other types of cell contacts.

In those instances in which gap junctions appear implicated by their presence at a specific locale, at a critical time, even then one must beware of overinterpreting their significance. The presence of junctions is not sufficient by itself to indicate actual cell–cell coupling; it points only to the *possibility* of communication between neighboring cells. The experiments of the Peracchias (1980a,b) clearly show that junctions persist even when cells become temporarily uncoupled. These workers even argue that the packing of connexons changes in a characteristic manner upon uncoupling (Peracchia, 1977); several researchers have tried to use morphological criteria in deciding which junctions can and which cannot pass dyes or other molecules. Unwin and Ennis (1984) recently demonstrated that gap junctions can be found in two different states depending on whether or not calcium is present, corroborating the earlier suggestion by Makowski *et al.* (1977). Calcium is believed to be one of the mediators proposed to control junctional permeability (Rose *et al.*, 1977; Rose and Rick, 1978). However, not all investigators agree that morphological changes are a reliable index of uncoupling. Some of the morphological changes that take place on uncoupling could be so subtle as to be difficult to detect and might not be the same in different systems. Raviola *et al.* (1980) show that junction appearance may depend on preparative technique. S. B. Yancey (personal communication), using J. Heuser's rapid freezing apparatus, observed "crystallized" (i.e., "impermeable") junctions after rapid freezing, but they were "close packed" (i.e., "permeable") after glutaraldehyde fixation. Bennett and collaborators also recently emphasized that no alterations in connexon packing were seen in the systems they examined after uncoupling (Hanna *et al.*, 1984).

Conversely to the cases just discussed, cell coupling can be detected under circumstances in which junctions are essentially absent. It has become increasingly evident that discrepancies between conclusions can be drawn on the basis of physiological detection (electrotonic coupling, i.e., the exchange of ions or dye transfer between cells) and the morphological demonstration of gap junctions (Daniel *et al.*, 1976; D. J. Meyer *et al.*, 1981; Williams and DeHaan, 1981). Very small gap junctions, which could consist of very few particles or even of separate connexons, would be nearly impossible to identify morphologically in unequivocal fashion, but they could nevertheless be physiologically competent. Dye-coupling experiments generally show a much better correlation with the presence of morphologically recognizable junctions than does electrical coupling. However, technical problems such as leakage of dyes and quenching of fluorescence (Bennett *et al.*, 1978) sometimes complicate the detection of communicating junctions. It is therefore best to rely on data obtained by as many approaches as feasible. Different approaches are also needed because we do not know in what form developmental signals are that might be carried by gap junctions—they could be large molecules such as second messengers (T. S. Lawrence *et al.*, 1978), morphogens (Meinhardt, 1983; Crick, 1970), metabolites (Pitts, 1978), ionic currents (Jaffe, 1982) or still other forms, all requiring different techniques for their detection. To distinguish among the various ways to demonstrate cell–cell exchanges, we will use the terms "gap" *junctions* to denote detection by morphological means alone, and "electrotonic," "dye," or "metabolic" *coupling* to denote coupling by electrophysiological means, by exchange of dye molecules, or by exchange of detectable metabolically essential molecules, respectively.

4. Cell Communication in Embryonic Systems

4.1. Phylogenetic Distribution of Gap Junctions or Cell–Cell Coupling

Work on the phylogenetic distribution of junctions as well as their ontogeny has shown that they are essentially ubiquitous. Electrical coupling has been described between mechanically apposed cells of the sponge *Microciona* when the experiment was performed in the presence of aggregating factor (Loewenstein, 1967). Whether gap junctions are involved is not demonstrated (Revel and Goodenough, 1970; Lethias *et al.*, 1983). In freeze-cleaved samples, junctionlike clusters of particles have been seen, but images are not conclusive because the expected "steps" between P and F have not as yet been observed (M. Mann and J. P. Revel, unpublished observations). Gap junctions and electrotonic coupling, however, have been clearly demonstrated in other very primitive groups such as the Dicyemids and their embryos (Revel *et al.*, 1984), and the Coelenterates (Wood and Kuda, 1980; Hand and Gobel, 1972; Fraser and Bode, 1981).

Gap junctions are found in both the protostomes and deuterostomes. Of special interest among protostomes are the Arthropoda. Their cells are linked by structures resembling gap junctions, but the intramembrane particles typical of the freeze-fracture image remain with the E instead of the P face as they do in many other organisms (Epstein and Gilula, 1977; Lane and Skaer, 1980). The physiology of insect junctions has been particularly thoroughly studied. Insect junctions can transfer somewhat larger molecules than can junctions between mammalian cells (Flagg-Newton *et al.*, 1979).

An unresolved problem in the deuterostomes is that the existence of gap junction has been difficult to prove in both adult and embryonic echinoderms (Wood and Kuda, 1980; Spiegel and Howard, 1983). Gap junctions, however, are present in tunicates, where they seem inverted, like the arthropod junctions, in heart (Lorber and Rayns, 1977), but not in other tissues (Georges, 1979). Gap junctions are obviously present in higher deuterostomes as well.

4.2. Ontogenic Appearance of Gap Junctions and Cell Coupling

Gap junctions and/or coupling can be detected very early in development (see Bennett, 1973; Potter *et al.*, 1966; Lentz and Trinkaus, 1971; Tupper and Saunders, 1972; Ito and Loewenstein, 1969). A problem peculiar to developing systems is that caused by dividing or newly divided cells that might still be linked by the remains of intercellular bridges. These could permit the passage of ion currents and be permeable to low-molecular-weight materials. Such bridges can be detected by injecting dyes too large to pass through the gap junctions (Lo and Gilula, 1979). In the chick, gap junctions have been shown to be present between cells even in unincubated eggs (Bellairs *et al.*, 1975) and electrical coupling has been detected in early embryos (Sheridan, 1966). In vertebrates, they appear at about the 4–8-cell stage, coincident with or just before compaction (Ducibella *et al.*, 1975; Goodall and Johnson, 1982; Lo and Gilula, 1979; McLachlin *et al.*, 1983).

4.3. Types of Junctional Interaction in Development

Morphological changes and functional modulation of junction permeability have been reported often during differentiation and development. Junctions appear, reappear or else disappear altogether, or there are temporary changes or loss of junctional permeability. Examples are discussed below in an attempt to assess the possible role of these changes in development or differentiation.

4.3.1. Disappearance of Junctions

There are many instances in which gap junctions have been shown to disappear during the course of differentiation. Many investigators believe that, upon losing their connections with each other, cells are freed to follow different developmental pathways. The differentiation of sensory and support cells in chick otocysts (Ginzberg and Gilula, 1979) is a good example of junction disappearance at the time of differentiation. Before innervation, the cells are all linked to each other by junctional arrays (i.e., gap, tight, and adhaerens junctions) characteristic of all epithelia. Only gap junctions, however, disappear from sensory cells as they differentiate.

Another well-known example of junction loss that has been linked to a differentiative event is found in the case of the developing retina. At early stages, gap junctions are present between the developing retinal cells and connecting the retinal cells with the pigment epithelial cells. In *Xenopus*, Dixon and Cronly-Dillon (1972) suggested that the junctions may play an important role in retinal specification, disappearing at the time of specification. In fact, after treatment of *Xenopus* eyes with calcium ionophore, premature specification of the retina occurs (Jacobson, 1976). This could be interpreted as due to calcium-mediated uncoupling of gap junctions, thus preventing the cellular interactions necessary for the status quo. Observations made on the chick embryo retina (Hayes, 1977; Fujisawa et al., 1976) suggest, however, that loss of retinal gap junctions in these organisms is involved in the withdrawal of cells from the pool of dividing cells and their eventual differentiation, rather than directly with the specification of a particular retinotectal path.

The wealth of gap junctions found between follicular cells and ova (Anderson and Albertini, 1976) provides yet another opportunity to examine the close correlation between the disappearance of the junctions and a differentiative step (Coons and Espey, 1977). As shown by Gilula et al. (1978), electrotonic and dye coupling exist between the cumulus cells and oocyte before ovulation. The extent of coupling decreases progressively until no coupling can be detected in postovulatory specimens. Although there seems to be contradictory evidence (Eppig et al., 1983), the junctions may play the important role of keeping levels of cAMP in the oocyte high, thereby preventing termination of the meiotic division. Whatever the "signals" exchanged, gap junctions seem to play a role in the process (Eppig et al., 1983).

The disappearance of gap junctions in the examples above can be interpreted as enabling a new cell "species" to emerge by isolation from the communicating pool of cells. Although such a simple mechanism is appealing, causality has to be established, particularly because other changes, including changes in other junctions, have been noted in temporal coincidence with striking developmental or differentiation events (see Revel and Brown, 1976).

4.3.2. Modulation in the Number of Junctions

In many instances, the junctions do not actually disappear but rather seem to change in numbers (or percentage area of the membrane involved) or permeability. Whether this represents only a change in turnover rate or reflects a change in functional state because different junction proteins are being synthesized is not clear. It should be noted that junction proteins turn over very rapidly (Yancey et al., 1979; Fallon and Goodenough, 1981). Junction permeability probably can also be modulated without changing junction number.

Many instances have been documented in which electrotonic coupling was preserved, although there was a loss of dye coupling. Such changes might reflect a modulation of junctional permeability, but caution must be exercised, since there is the possibility of technical artifacts (Bennett et al., 1978). Low concentrations of dye are difficult to see, and the results are further distorted if dye leaks out of the cells. The use of lucifer dyes, introduced recently, seems to circumvent these problems (Bennett et al., 1978).

In the developing liver, R. A. Meyer and Overton (1983) documented stages during which junctions are abundant and others during which junctions are less so. In the regenerating liver, which can be considered a special case of "development," similar changes occur—the junctions dramatically decrease in number and size between 24 and 28 hr after hepatectomy, only to reappear thereafter (Yancey et al., 1979). The significance of this momentary eclipse is not clear. The change is seen in temporal coincidence with the first wave of mitosis in young animals (Yee and Revel, 1978) but not in older ones (unpublished observations). Although there is a reduction in the junctional complement by 100-fold, only subtle effects on cell–cell coupling occur as determined by electrophysiological means (D. J. Meyer et al., 1981). Dye coupling, however, is severely curtailed. Taken at face value, this observation could signify that electrical signaling does not play a major role in the differentiative events seen during regeneration but that the passage of low-molecular-weight materials is of importance, since it is curtailed. Clearly other interpretations are possible.

The disappearance of junctions as myoblasts fuse (Kalderon et al., 1977) during the formation of skeletal muscle cannot be seen in the same light. After all, the cells will soon form a true syncytium. Since it does not seem likely that the gap junctions are involved in the actual process of fusion, one must postulate that their formation, a necessary step in the process, may play a role in recognition or in triggering some of the following steps.

Although not strictly "developmental," the effect of vitamin A in increasing the number and size of gap junctions is both spectacular and confusing. The results presented by Elias et al. (1980, 1981) clearly show a very large increase in the number of junctions accompanying vitamin A treatment. Yet application of vitamin A and derivatives to cells in culture apparently decreases the extent of coupling. Until further work is done on this system, and correlation with the effects of vitamin A on development (cf. Maden, 1982) is attempted, it is difficult to reconcile these observations. Perhaps the proliferation of gap junctions is the result of a negative feedback of poor cell coupling. Could the assumption that morphologically recognizable gap junctions represent functional units be false?

4.3.3. Qualitative Modulation of Gap Junctions

Little is known of qualitative changes in junctions during development. The finding that gap junction proteins in different tissues are different entities clearly supports this possibility. Work on the junctional complement of the eye lens gives some insight into these possibilities.

In the adult eye there are structures between lens fibers that resemble gap junctions closely but (Bloemendal et al., 1972; Benedetti et al., 1974; Zampighi et al., 1982) are not typical. From a biochemical standpoint, lens junctions appear different from liver or heart (Hertzberg et al., 1982; Nicholson et al., 1983; Gros et al., 1983). The "lens junctions" have been claimed to be responsible for extensive cell–cell coupling between the cells (Goodenough, 1979; Goodenough et al., 1980; see also Paul and Goodenough, 1983). Perhaps they have a role in the nutrition of lens fiber cells, removed as the latter are from the direct blood supply. Early in development, however, typical gap junctions seem to be present between the cells of the developing organ (Schuetze and Goodenough, 1982). As cells from the anterior epithelium become lens fibers, they lose their typical junctional complements and acquire the "lens" junctional pattern. Concomitantly there seems to be a change in the sensitivity of the coupling to control by changes in the partial pressure of CO_2 (Schuetze and Goodenough, 1982). Rising pCO_2 values have been shown in other systems to uncouple cells (Turin and Warner, 1977) presumably by changing intracellular pH, which also affects cell–cell coupling (Spray et al., 1981; Turin and Warner, 1977). While the junctions between anterior epithelial cells show reduced permeability with increased pCO_2 mature fibers seem to be unaffected by this uncoupling agent (although they remain sensitive to others) (D. A. Goodenough, personal communication).

Although it is possible to interpret several of the instances of junction disappearance in terms of causality, other interpretations more simply account for the momentary junction disappearance seen in the liver or for the changes in junctional type seen in the case of the lens. One must keep in mind that junctional changes could represent one of the facets of the differentiative process rather than being causative.

4.3.4. Junctions as "Informational" Pathways

Many developmental systems show no obvious or dramatic changes in the junctional complement such as those just described. Gap junctions could nevertheless play a major role in controlling various developmental steps.

The role junctions might play in defining the compartmental boundaries observed in developing insect tissues (Warner and Lawrence, 1982; Weir and Lo, 1982) is particularly interesting. Meinhardt (1983) recently proposed a very ingenious mechanism by which boundaries resulting from the primary embryonic organization of a developing organism act as organizing regions for secondary embryonic fields, e.g., imaginal discs in insects. This would occur if two compartments had to cooperate for the production of a morphogenetic substance. A high concentration of morphogen would thus form at the boundary between the groups of cells involved. P. A. Lawrence and Green (1975) and Warner and Lawrence (1973) were early in showing that there were junctions and coupling between cells on opposite sides of a developmental boundary. These disappointing results seem to have been reversed by more recent experiments that suggest that communication boundaries can indeed be found if dye transfer instead of electrotonic coupling is used to test for their presence (Warner and Lawrence, 1982; Weir and Lo, 1982, 1984). To complicate the problem, it has not been established whether the boundaries seen are necessarily the same as the developmental boundaries (S. E. Fraser, 1984, personal communication; Weir and Lo, 1984). Another instance can be found in Hydra, where it has long been known (Schaller, 1973) that there are morphogens (M_r ~900) that influence the fate of regenerating portions of the organism. Although the presence of gap junctions is well documented in these organisms (Wood and Kuda, 1980), there has been some question as to whether these represent patent channels (De Laat et al., 1980). The recent demonstration by Fraser and Bode (1981) of both dye and electrical coupling in Hydra suggests that gap junctions may indeed play a role in allowing the formation of an intercellular gradient of morphogen, a potential means of specifying positional information in this (Wakeford, 1979; MacWilliams, 1983) and other systems (Summerbell *et al.*, 1973; Caveney, 1974; Tickle *et al.*, 1974).

The demonstration by Bennett and collaborators (Harris *et al.*, 1983) that changes in membrane potential can alter intercellular pathways in "permanent" fashion also offers the possibility that voltage-controlled changes in junction permeability could play an important role in defining patterns in developing systems.

4.3.5. Momentary Junction Formation (*en Passant*)

A particularly interesting type of interaction is one in which junctions form as cells meet during their migration to their ultimate positions in the adult organism. The case of the fusing myoblast has been discussed previously. Temporary junctions also seem to form during the differentiation of neuroblasts (Pannese *et al.*, 1977). It was noted early by LoPresti *et al.* (1974) that gap junctions are formed as a growth cone passes by other neurons. Similar behavior has been documented recently by Goodman and his collaborators (see Goodman *et al.*, 1982) in studies of pathfinding by neuronal growth cones in the grasshopper. Specific and stereotypic patterns of cell–cell coupling occur as the growth cones pass different landmarks (see Chapter 4, this volume) that can be interpreted to mean that neurons require intercellular communication mediated by cell coupling in order to navigate. Growth cones in filopodia contact many cells in their environment but become coupled only to a specific subset. It is interesting to contrast these findings with those made on fibroblasts in culture, in which contact inhibition of motion, which requires cell contacts, does not seem to be mediated by the presence of junctions.

5. Conclusions

The examples chosen to illustrate possible roles that gap junctions might play in controlling patterns are far from exhaustive. Nothing has been said of the patterns of junctions encountered in developing limbs (Kelley and Fallon, 1978) of the patterns of junction formation encountered during the formation of the neural tube (Revel and Brown, 1976), of the changes in the junctional complement that seem to occur in smooth muscle at the time of parturition (Garfield *et al.*, 1980) or epithelial–mesenchymal interactions, and so forth. The examples cited suffice, however, to make the point that no single mechanism is of unique importance in the embryo. In the embryo, as in the adult, junction formation seems to be part of the cell surface's repertoire. The circumstantial evidence that derives from the temporal correlation of junctional modulation with some change in behavioral pattern formation is strong enough to be able to make the general statement that gap junctions are most likely involved in playing a basic role in development. What this role is, however, must await the outcome of direct experiments in which junctional permeability is interfered with in known fashion. Specific reagents that will permit junctions to be turned off or on must be tested. To date, the most specific reagents of this sort are antibodies. First results (Warner *et al.*, 1984) suggest that it is indeed possible to describe drastic developmental consequences of shutting off junctions at critical times. Thus, a new era of our understanding of junction formation and development is about to begin.

ACKNOWLEDGMENTS. The work cited from the author's laboratory was supported by grants GM-06925 from the NIH RR-07003 from the BRSG and the Ruddock Foundation.

References

Anderson, E., and Albertini, D. F., 1976, Gap junctions between the oocyte and the companion follicle cells in the mammalian ovary, *J. Cell Biol.* **71:**680–686.

Bellairs, R., Breathnach, A. S., and Gross, M., 1975, Freeze fracture replication of junctional complexes in unincubated and incubated chick embryos, *Cell Tissue Res.* **162:**235–252.

Benedetti, E. L., Dunia, I., and Bloemendal, H., 1974, Development of junctions during differentiation of lens fibers, *Proc. Natl. Acad. Sci. USA* **71:**5073–5077.

Benedetti, E. L., Duria, I., Ramackers, F. C. S., and Kibbelaar, M. A., 1981, Lenticular plasma membranes and cytoskeleton, in: *Molecular and Cellular Biology of the Eye Lens* (H. Bloemendal, ed.), pp. 137–188, Wiley, New York.

Bennett, M. V. L., 1973, Function of electrotonic junctions in embryonic and adult tissues, *Fed. Proc.* **32:**65–75.

Bennett, M. V. L., Spira, M. E., and Spray, D. C., 1978, Permeability of gap junctions between embryonic cells of *Fundulus*: A reevaluation, *Dev. Biol.* **65:**114–125.

Bloemendal, H. A., Zweers, F., Vermorken, I., Dunia, I., and Benedetti, E. L., 1972, The plasma membranes of eye lens fibers. Biochemical and structural characterization, *Cell Diff.* **1:**91–106.

Bok, D., Dockstader, J., and Horwitz, J., 1982, Immunocytochemical localization of the lens main intrinsic polypeptide (MIP26) in communicating junctions, *J. Cell Biol.* **92:**213–220.

Brink, P. R., and Dewey, M. M., 1980, Evidence for fixed charge in the nexus, *Nature (Lond.)* **285:**101–103.

Caspar, D. L. D., Goodenough, D. A., Makowski, I., and Phillips, W. C., 1977, Gap junction structures. I. Correlated electron microscopy and X-ray diffraction, *J. Cell Biol.* **74:**605–628.

Caveney, S., 1974, Intercellular communication in a positional field: movement of small ions between insect epidermal cells, *Dev. Biol.* **40:**311–322.

Coons, L. W., and Espey, L. L., 1977, Quantitation of nexus junctions in the granulosa cell layer of rabbit ovarian follicles during ovulation, *J. Cell Biol.* **74:**321–325.

Crick, F., 1970, Diffusion in embryogenesis, *Nature (Lond.)* **225:**420–422.

Daniel, E. E., Daniel, V. P., Duchon, G., and Garfield, R. E., 1976, Is the nexus necessary for cell-to-cell coupling of smooth muscle?, *J. Membr. Biol.* **28:**207–239.

De Laat, S. W., Tertoolen, L. G., and Grimmlikhuijzen, C. J., 1980, No junctional communication between epithelial cells in *Hydra*, *Nature (Lond.)* **288:**711–713.

Dixon, J. S., and Cronly-Dillon, J. R., 1972, The fine structure of developing retina in *Xenopus laevis*, *J. Embryol. Exp. Morphol.* **28:**659–666.

Ducibella, J., Albertini, D., Anderson, E., and Biggers, J. D., 1975, The preimplantation mammalian embryo: Characterization of intercellular junctions and their appearance during development, *Dev. Biol.* **47:**231–250.

Duguid, J., and Revel, J. P., 1976, The protein components of the gap junction, *Cold Spring Harbor Symp. Quant. Biol.* **40:**45–47.

Elias, P. M., Grayson, S., Caldwell, T. M., and McNutt, N. S., 1980, Gap junction proliferation in retinoic acid-treated human basal cell carcinoma, *Lab. Invest.* **42:**469–474.

Elias, P. M., Grayson, S., Gross, E. G., Peck, G. L., and McNutt, N. S., 1981, Influence of topical and systemic retinoids on basal cell carcinoma membranes, *Cancer* **48**:932–938.

Eppig, J. J., Freter, R. R., Ward-Bailey, P. F., and Schultz, R. M., 1983, Inhibition of oocyte maturation in the mouse: Participation of cAMP, steroid hormones, and a putative maturation-inhibitory factor, *Dev. Biol.* **100**:39–49.

Epstein, M. L., and Gilula, N. B., 1977, A study of communication specificity between cells in culture, *J. Cell Biol.* **75**:769–787.

Fallon, R. F., and Goodenough, D. A., 1981, Five hour half-life of mouse liver gap junction protein, *J. Cell Biol.* **90**:521–525.

Finbow, M., Yancey, S. B., Johnson, R., and Revel, J. P., 1980, Independent lines of evidence suggesting a major junctional protein with a MW of 26000, *Proc. Natl. Acad. Sci. USA* **77**:970–974.

Flagg-Newton, J., Simpson, I., and Loewenstein, W. R., 1979, Permeability of the cell–cell membrane channels in mammalian cell junction, *Science* **205**:404–409.

Fraser, S. E., and Bode, H. R., 1981. Epithelial cells of Hydra are dye coupled, *Nature (Lond.)* **294**:356–358.

Fujisawa, H., Morioka, H., Watanabe, K., and Nakamura, M., 1976, A decay of gap junctions in association with cell differentiation of neural retina in chick embryo development, *J. Cell Sci.* **22**:585–596.

Garfield, R. E., Merrett, D., and Grover, A. K., 1980, Gap junction formation and regulation in myometrium, *Am. J. Physiol.* **239**:C217–C228.

Georges, D., 1979, Gap and tight junctions in tunicates, *Tissue Cell* **11**:781–792.

Gilula, N. B., Epstein, M. L., and Beers, W. H., 1978, Cell-to-cell communication and ovulation. A study of the cumulus–oocyte complex, *J. Cell Biol.* **78**:58–75.

Ginzburg, R. D., and Gilula, N. B., 1979, Modulation of cell junctions during differentiation of the chicken otocyst sensory epithelium, *Dev. Biol.* **68**:110–129.

Goodall, H., and Johnson, M. H., 1982, Use of carboxyfluorescein diacetate to study the formation of permeable channels between mouse blastomeres, *Nature (Lond.)* **295**:524–526.

Goodenough, D. A., 1979, Lens gap junctions: A structural hypothesis for nonregulated low-resistance intercellular pathways, *Invest. Ophthalmol.* **18**:1104–1122.

Goodenough, D. A., Dick, J. S., and Lyons, J. E., 1980, Lens metabolic cooperation: A study of mouse lens transport and permeability visualized with freeze-substitution autoradiography and electron microscopy, *J. Cell Biol.* **86**:576–589.

Goodman, C. S., Raper, J. A., Ho, R. K., and Chang, S., 1982, Path finding by neuronal growth cones in grasshopper embryos, in: *Developmental Order: Its Origin and Regulation* (S. S. Subtelny and P. B. Green, eds.), pp. 275–316, Alan R. Liss, New York.

Griepp, E. B., Peacock, J., Bernfield, M., and Revel, J. P., 1978, Morphological and functional correlates of synchronous beating between embryonic heart cell aggregates and layers, *Exp. Cell Res.* **113**:273–282.

Gros, D., Mocquard, J. P., Challice, C. E., and Schrevel, J., 1978, Formation and growth of gap junctions in mouse myocardium during ontogenesis: A freeze cleavage study, *J. Cell Sci.* **30**:45–61.

Gros, D., Macquard, J. P., Challice, C. E., and Schrevel, J., 1979, Formation and growth of gap junctions in mouse myocardium during ontogenesis: Quantitative data and their implications on the development of intercellular communication, *J. Mol. Cell Cardiol.* **11**:543–554.

Gros, D. B., Nicholson, B. J., and Revel, J. P., 1983, Comparative analysis of the gap junction protein from rat heart and liver: Is there a tissue specificity of gap junctions? *Cell* **35**:539–549.

Hand, A. R., and Gobel, S., 1972, The structural organization of the septate and gap junctions of the Hydra, *J. Cell Biol.* **52**:397–408.

Hanna, R. B., Pappas, G. D., and Bennett, M. V. L., 1984, The fine structure of identified electrotonic synapses following increased coupling resistance, *Cell Tissue Res.* **235**:243–249.

Harris, A. L., Spray, D. C., and Bennett, M. V. L., 1983, Control of intercellular communication by voltage dependent gap junction conductance, *J. Neurosci.* **3**:79–100.

Hayes, B. P., 1977, Intercellular gap junctions in the developing retina and pigment epithelium of the chick, *Anat. Embryol. (Berl.)* **151**:325–333.

Hertzberg, E. L., 1980, Biochemical and immunological approaches to the study of gap junctional communication, *In Vitro* **16**:1057–1067.

Hertzberg, E. L., and Gilula, N. B., 1979, Isolation and characterization of gap junctions from rat liver, *J. Biol. Chem.* **254**:2138–2147.

Hertzberg, E. L., Anderson, D. J., Friedlander, M., and Gilula, N. B., 1982, Comparative analysis of the major polypeptides from liver gap junctions and lens fiber junctions, *J. Cell Biol.* **92**:53–59.

Ito, S., and Loewenstein, W. R., 1969, Ionic communication between early embryonic cells, *Dev. Biol.* **19**:228–243.

Jacobson, M., 1976, Premature specification of the retina in embryonic *Xenopus* eyes treated with ionophore X537A, *Science* **191**:288–289.

Jaffe, L. F., 1982, Developmental currents, voltages and gradients, in: *Developmental Order: Its Origin and Regulation* (S. S. Subtelny and P. B. Green, eds.), pp. 183–215, Alan R. Liss, New York.

Kalderon, N., Epstein, M. L., and Gilula, N. B., 1977, Cell to cell communication and myogenesis, *J. Cell Biol.* **75**:788–806.

Kelley, R. O., and Fallon, J. F., 1978, Identification and distribution of gap junctions in the mesoderm of the developing chick limb bud, *J. Embryol. Exp. Morphol.* **46**:99–110.

Lane, N., and Skaer, H., H. Le B., 1980, Intercellular junctions in insect tissues, *Adv. Insect. Physiol.* **15**:35–213.

Lawrence, P. A., and Green, S. M., 1975, The anatomy of a compartment border. The intersegmental boundary in *Oncopeltus*, *J. Cell Biol.* **65**:373–383.

Lawrence, T. S., Beers, W. H., and Gilula, N. B., 1978, Transmission of hormonal stimulation by cell-to-cell communication, *Nature (Lond.)* **272**:501–506.

Lentz, T. L., and Trinkaus, J. P., 1971, Differentiation of the junctional complex of surface cells in the developing *Fundulus* blastoderm, *J. Cell Biol.* **48**:455–472.

Lethias, C., Garrone, R., and Mazzorana, M., 1983, Fine structure of sponge cell membranes: Comparative study with freeze fracture and conventional thin section methods, *Tissue Cell* **15**:523–535.

Lo, C. W., and Gilula, N. B., 1979, Gap junctional communication in the preimplantation mouse embryo, *Cell* **18**:399–409.

Loewenstein, W. R., 1967, On the genesis of cellular communication, *Dev. Biol.* **15**:502–520.

LoPresti, V., Macagno, E. R., and Levinthal, C., 1974, Structure and development of neuronal connections in isogenic organisms: Transient gap junctions between growing optic axons and lamina neuroblasts, *Proc. Natl. Acad. Sci. USA* **71**:1098–1102.

Lorber, V., and Rayns, D. G., 1977, Fine structure of the gap junction in the tunicate heart, *Cell Tissue Res.* **179**:169–175.

MacWilliams, H. K., 1983, *Hydra* transplantation phenomena and the mechanism of *Hydra* head regeneration, *Dev. Biol.* **96**:239–257.

Maden, M., 1982, Vitamin A and pattern formation in the regenerating limb, *Nature (Lond.)* **295**:672–675.

Makowski, L., Caspar, D. L. D., Phillips, W. C., and Goodenough, D. A., 1977, Gap junction structures. II. Analysis of the X-ray diffraction data, *J. Cell Biol.* **74**:629–645.

McLachlin, J., Caveney, S., and Kidder, G. M., 1983, Control of gap junction formation in early mouse embryos, *Dev. Biol.* **98**:155–164.

Meinhardt, H., 1983, Cell determination boundaries as organizing regions for secondary embryonic fields, *Dev. Biol.* **96**:375–385.

Meyer, D. J., Yancey, S. B., and Revel, J. P., 1981, Intercellular communication in normal and regenerating rat liver: A quantitative analysis, *J. Cell Biol.* **91**:505–523.

Meyer, R. A., and Overton, J., 1983, Changes in intercellular junctions. I. Embryonic chick liver development, *Dev. Biol.* **99**:172–180.

Moscona, A. A., and Linser, P., 1983, Developmental and experimental changes in retinal glia cells: Cell interactions and control of phenotype expression and stability, *Curr. Topics Dev. Biol.* **18**:155–188.

Nicholson, B. J., Hunkapiller, M. W., Grim, L. B., Hood, L. E., and Revel, J.-P., 1981, The rat liver gap junction protein: Properties and partial sequence, *Proc. Natl. Acad. Sci. USA* **78**:7594–7598.

Nicholson, B. J., Gros, D. B., Kent, S. B. H., Hood, L. E., and Revel, J.-P., 1985, The M_r 28,000 gap junction proteins from rat heart and liver are different but related, *J. Biol. Chem.* **260**:6514–6517.

Nicholson, B. J., Takemoto, L. J., Hunkapiller, M. W., Hood, L. E., and Revel, J.-P., 1983, Differences between the liver gap junction protein and lens MIP26 from rat: Implications for tissue specificity of gap junctions, *Cell* **32**:967–978.

Pannese, E., Luciano, L., Iurato, S., and Reale, E., 1977, Intercellular junctions and other membrane specializations in developing spinal ganglia, *J. Ultrastruct. Res.* **60**:169–180.

Paul, D. L., and Goodenough, D. A., 1983, Preparation, characterization and localization of antisera against bovine MIP26, an integral membrane protein of the lens fiber, *J. Cell Biol.* **96**:625–632.

Peracchia, C., 1977, Gap junction. Structural changes after uncoupling, *J. Cell Biol.* **72**:628–645.

Peracchia, C., and Peracchia, L. L., 1980a, Gap junction dynamics: reversible effects of divalent cations, *J. Cell Biol.* **87**:708–718.

Peracchia, C., and Peracchia, L. L., 1980b, Gap junction dynamics: Reversible effects of hydrogen ions, *J. Cell Biol.* **87**:719–727.

Pitts, J. D., 1978, Junctional communication and cellular growth control, in: *Intercellular Junctions and Synapses*, (J. Feldman, N. B. Gilula, and J. D. Pitts, eds.), pp. 68–80, Chapman and Hall, London.

Potter, D. D., Furshpan, E. J., and Lennox, E. S., 1966, Connections between cells of the developing squid as revealed by electrophysiological methods, *Proc. Natl. Acad. Sci. USA* **55**:328–335.

Raviola, E., Goodenough, D. A., and Raviola, G., 1980, Structure of rapidly frozen junctions, *J. Cell Biol.* **87**:273–279.

Revel, J.-P., and Brown, S. S., 1976, Cell junctions in development with particular reference to the neural tube, *Cold Spring Harbor Symp. Quant. Biol.* **40**:443–455.

Revel, J.-P., and Goodenough, D. A., 1970, Cell coats and intercellular matrix, in: *Chemistry and Molecular Biology of the Intercellular Matrix* (E. A. Balasz, ed.), pp. 1361–1380, Academic Press, New York.

Revel, J.-P., Levy, L., and Wang, C., 1984, Electron microscope study of junctions in the mesozoa, *Cell Tissue Res.* (in press).

Rose, B., and Rick, R., 1978, Intracellular pH, intracellular free Ca, and junctional cell–cell coupling, *J. Membr. Biol.* **44**:377–415.

Rose, B., Simpson, I., and Loewenstein, W. R., 1977, Calcium ion produces graded changes in permeability of membrane channels in cell junction, *Nature (Lond.)* **267**:625–627.

Schaller, H. C., 1973, Isolation and characterization of a low molecular weight substance activating head and bud formation in *Hydra*, *J. Embryol. Exp. Morphol.* **29**:27–38.

Schuetze, S. M., and Goodenough, D. A., 1982, Dye transfer between cells of embryonic chick lens becomes less sensitive to CO_2 treatment with development, *J. Cell Biol.* **92**:694–705.

Sheridan, J. D., 1966, Electrophysiological studies of special connections between cells in the early chick embryo, *J. Cell Biol.* **31**:C1–C5.

Simpson, I., Rose, B., and Loewenstein, W. R., 1977, Size limit of molecules permeating the junctional membrane channels, *Science* **195**:294–296.

Spiegel, E., and Howard, L., 1983, Development of cell junctions in sea urchin embryos, *J. Cell Sci.* **62**:27–48.

Spray, D. C., Harris, A. L., and Bennett, M. V. L., 1979, Voltage dependence of junctional conductance in early amphibian embryos, *Science* **204**:432–434.

Spray, D. C., Harris, A. L., and Bennett, M. V. L., 1981, Gap junctional conductance is a simple and sensitive function of intracellular pH, *Science* **211**:712–715.

Summerbell, D., Lewis, J. H., and Wolpert, L., 1973, Positional information in chick limb morphogenesis, *Nature (Lond.)* **244**:492–496.

Tickle, C., Summerbell, D., and Wolpert, L., 1975, Positional signaling and specification of digits in chick limb morphogenesis, *Nature (Lond.)* **254**:199–202.

Traub, O., Janssen-Timmen, U., Druge, P. M., Dermietzel, R., and Willecke, K., 1982, Immunological properties of gap junction protein from mouse liver, *J. Cell. Biochem.* **19**:27–44.

Tupper, J. T., and Saunders, J. W., 1972, Intercellular permeability in the early *Asterias* embryo, *Dev. Biol.* **27**:446–554.

Turin, L., and Warner, A. E., 1977, Carbon dioxide reversibly abolishes ionic communication between cells of early amphibian embryo, *Nature (Lond.)* **270**:56–57.

Unwin, P. N. T., and Ennis, P. D., 1984, Two configurations of a channel-forming membrane protein, *Nature (Lond.)* **307**:609–613.

Unwin, P. N. T., and Zampighi, G., 1980, Structure of the junctions between communicating cells, *Nature (Lond.)* **283**:545–549.

Wakeford, R. J., 1979, Cell contact and positional communication in *Hydra, J. Embryol. Exp. Morphol.* **54**:171–183.

Warner, A. E., and Laurence, P. A., 1973, Electrical coupling across developmental boundaries in insect epidermis, *Nature* **245**:47–49.

Warner, A. E., and Lawrence, P. A., 1982, Permeability of gap junctions at the segmental border of the insect epidermis, *Cell* **28**:243–252.

Warner, A. E., Guthrie, S., and Gilula, N. B., 1984, Antibodies to gap junctional protein selectively disrupt junctional communication in the early amphibian embryo, *Nature* **311**:127–131.

Weir, M. P., and Lo, C. W., 1982, Gap junction communication compartments in the *Drosophila* wing disc, *Proc. Natl. Acad. Sci. USA* **79**:3232–3235.

Weir, M. P., and Lo, C. W., 1984, Gap-junctional communication compartments in the *Drosophila* wing imaginal disk, *Dev. Biol.* **102**:130–146.

Williams, E. H., and De Haan, R. L., 1981, Electrical coupling among heart cells in the absence of ultrastructurally defined gap junctions, *J. Membr. Biol.* **60**:237–248.

Wood, R. L., and Kuda, A. M., 1980, Formation of junctions in regenerating *Hydra:* Gap junctions, *J. Ultrastruct. Res.* **73**:350–360.

Yancey, S. B., Easter, D., and Revel, J.-P., 1979, Cytological changes in gap junctions during liver regeneration, *J. Ultrastruct. Res.* **67**:229–242.

Yancey, S. B., Nicholson, B. J., and Revel, J.-P., 1981, The dynamic state of liver gap junctions, *J. Supramol. Struct. Cell Biochem.* **16**:221–232.

Yee, A. G., and Revel, J.-P., 1978, Loss and reappearance of gap junctions in regenerating liver, *J. Cell Biol.* **78**:544–564.

Zampighi, G., Simon, S., Robertson, J. D., McIntosh, T., and Costello, M., 1982, On the organization of isolated bovine lens fiber junctions, *J. Cell Biol.* **93**:175–189.

Chapter 11

Cell–Cell and Cell–Matrix Interactions in the Morphogenesis of Skeletal Muscle

DAVID C. TURNER

> Developmental biology is the search for understanding how structure arises in biological systems, it being assumed as axiomatic that structure and function are inexorably linked. In skeletal muscle this relationship need not be assumed; it is nowhere more obvious.
>
> Irwin R. Konigsberg (1965)

There is a tendency, widespread but by no means universal, to equate developmental biology with the study of cell differentiation. An emphasis on the problem of differentiation is understandable because it has been the area of greatest progress. Recent advances in molecular biology make it likely that an understanding of the molecular mechanisms responsible for differential gene expression in eukaryotic development will be achieved fairly soon. And yet, as the above quotation reminds us, the ultimate concern of the developmental biologist is to comprehend morphogenesis.

In a fundamental sense, the goal of the molecular biologist is the same. He or she ultimately wants to understand what Monod (1970) called *ontogénèse moléculaire*—the processes of self-assembly and catalyzed assembly by which structures such as viruses and organelles are formed, and then, successively, cells, tissues, organs, and whole organisms.

For the present, however, there remain valid differences in outlook among students of developmental problems. Just as it is possible to consider a topic such as skeletal myogenesis primarily from the standpoint of cell differentiation, so too can it be viewed from the standpoint of morphogenesis. This chapter takes the latter approach, as indeed I have been trying to do in my research.

DAVID C. TURNER • Department of Biochemistry, State University of New York Upstate Medical Center, Syracuse, New York 13210.

The chapter begins with an overview of skeletal myogenesis and of the problems of muscle morphogenesis. The remainder of the discussion is devoted to experimental approaches to certain of these problems, with most of the examples chosen from work performed in my own laboratory. Fibronectin-mediated attachment of muscle cells is given particular attention.

1. Introduction to Skeletal Myogenesis

I believe that muscle cell cultures will be as important in the study of muscle morphogenesis as they have been in the study of muscle differentiation. Accordingly, this chapter begins with a brief and selective review of skeletal myogenesis as obtained in vitro, followed by a comparison of myogenesis in vitro and in the embryo. Against this background of experimental findings and limitations, this introductory section concludes by attempting to define the problems of muscle morphogenesis more closely.

1.1. Skeletal Myogenesis in Primary Cultures

The standard procedures that have been in use for more than two decades (Hauschka, 1972; Konigsberg, 1979) involve the dissociation of pieces of muscle from avian or mammalian embryos by a combination of mechanical disruption and mild protease treatment in the absence of divalent cations, followed by filtration to yield a suspension of single cells. The cells are plated in an appropriate medium (any of several basal media consisting of salts, glucose, amino acids, and vitamins, supplemented with serum and often embryo extract) and on a suitable surface (usually "tissue culture"-quality polystyrene dishes coated with collagen or gelatin). At the outset, most of the cells in such cultures are proliferative myogenic precursor cells called myoblasts,* although some myocytes and nonmuscle cells (mainly fibroblasts) are invariably included. No nerve cells are present. Depending on the initial cell density and the medium composition, the spindle-shaped, refractile myoblasts will undergo one or more rounds of cell division before giving rise to nondividing terminally differentiating myocytes. Like the myoblasts, myocytes are bipolar and uninucleate. Unlike them, myocytes promptly fuse with one another to form multinucleate myotubes or primitive muscle fibers. Initiation of differentiation can be roughly synchronized, so that most myocyte fusion occurs within half a day or so. In well-fused cultures, the swirling pattern of branched anastomosing myotubes covers the entire surface; 80% or more of all nuclei are in myotubes. The entire

---

*The terminology used here (proliferative myoblasts, nondividing myocytes) differs from the one I have used heretofore (see Turner, 1978). It has the advantage that the "blast" is a committed, but still proliferative, precursor, like other similarly named embryonic cell types (e.g., chondroblast, erythroblast, neuroblast). It also appears to be gaining wide acceptance.

process of cytodifferentiation requires several days, with most of the obvious changes occurring after the syncytia have formed.

The cytodifferentiation of cultured myotubes is impressive (for review, see Holtzer, 1970; Hauschka, 1972; Buckingham, 1977; Pearson, 1980). The skeletal muscle isoforms of myosin, actin, and other proteins of the contractile apparatus are synthesized at high rates and are assembled into myofibrils. Other muscle-specific proteins, e.g., acetylcholine (ACh) receptors, acetylcholinesterase (AChE), and isoenzymes of creatine kinase and aldolase, are rapidly synthesized and accumulated. Myofibrils come to be arrayed in register, giving the entire myotube the cross-striated appearance typical of muscle fibers, and the cells begin to contract spontaneously.

Remarkable as this *in vitro* cytodifferentiation is, the end result is not equivalent to the situation in fully differentiated adult muscle. None of the myotubes becomes specialized into one of the recognizable skeletal muscle fiber types, largely—but probably not entirely—because of a lack of proper innervation. Moreover, certain (apparently nerve-independent) differentiative changes fail to go to completion *in vitro*. An example is the creatine kinase isoenzyme transition from the B to the M form; even after many days, cultured myotubes continue to synthesize some creatine kinase B along with greatly increased amounts of M (Caravatti et al., 1979).

As an alternative to the primary culture system, established lines of myogenic cells from rat and mouse are available (Pearson, 1980; Linkhart et al., 1982). Such lines can be induced to differentiate by mitogen depletion, but cell morphology, response to growth factors, kinetics of cell fusion, and biochemical changes vary greatly among the lines and often differ significantly from what is seen in primary cultures. Despite the advantages of established lines (cell homogeneity, ability to select variants), results with established lines must be interpreted with caution. For example, repeated passaging may select for variants that adhere well to the culture substrate. Consequently, the adhesive interactions of such cells with other cells and with extracellular matrices may differ significantly from those of primary cells. The work described from my own laboratory has been with primary cultures of cells from chicken embryos.

Much of the myogenesis research over the past 25 years or so has been focused on the regulation of gene expression during terminal differentiation *in vitro*. Two prominent areas of investigation have been the relationship between proliferation and differentiation and the possible role of myocyte fusion in triggering terminal differentiation (for review, see Holtzer, 1970; Buckingham, 1977; Bischoff, 1978; see also the many contributions in Pearson and Epstein, 1982, and in Eppenberger and Perriard, 1984). All along, the newest techniques in eukaryotic molecular biology have been applied to these problems. It is now well established that, as muscle cells begin to differentiate, messages specifying the sequences of a variety of muscle-specific proteins are transcribed at greatly increased rates, whereas the transcription of other messages is reduced. Regulation at post-transcriptional levels appears to be of lesser importance in this system. The search is now on for regulatory elements responsible for the coordinated changes in transcription. Skeletal muscle is likely to be one of the first vertebrate cell types for which the molecular mechanisms of differential gene expression in development are worked out. The availability of a suitable cell culture system will have contributed significantly to this outcome.

1.2. Skeletal Myogenesis in the Vertebrate Embryo

Comparative developmental anatomy and careful histology had, by the end of the last century, largely established that the skeletal muscles of the trunk and extremities are derived from cells that migrate (or are translocated) from the somites into the flanks and limb buds (Field, 1894; Fischel, 1895). Experiments by later embryologists were interpreted as contradicting this view; for more than half a century it was taught that the muscles of the limbs and ventral body wall arise from lateral plate mesoderm. One prominent model held that the mesenchyme of the limb bud was initially undetermined and that local environmental influences caused some of the cells to become committed to a myogenic, and others to a chondrogenic, lineage (Zwilling, 1968). Recent chick–quail grafting experiments have shown, however, that all the skeletal muscle fibers of the trunk and extremities are derived from the somites (reviewed in Shellswell, 1980; Chiquet et al., 1981; Christ et al., 1983; Kieny, 1983). Unlike the muscle cells proper, the cells of the muscle sheaths (epimysium, perimysium, and endomysium) are derived from the lateral plate.

In the limb bud, myogenic cells first appear in two premuscle masses (Fig. 4, left). As the limb develops, these cells are distributed, by successive splittings of the premuscle masses, into the anlagen of the many limb muscles (Shellswell, 1980; Kieny, 1983). As late arrivals, myogenic cells do not participate in laying down the primary limb pattern. Presumably, the spatial arrangement of muscle cells is determined by their interactions with tendons and other elements of connective tissue (Shellswell, 1980).

Muscle differentiation occurs at different times in different parts of the embryo. By the third day of embryonic development in the chicken, the anterior somites contain rows of cross-striated differentiated cells (Holtzer, 1970) (see Fig. 1). Interestingly, these cells differentiate as uninucleate myocytes. The extracellular matrix of the myotome presumably prevents intimate cell–cell contacts required for fusion, as somite cells are able to fuse when placed in culture (Holtzer, 1970). The usual sources of chicken myogenic cells for primary cultures are the leg and breast muscles of older (10–12-day) embryos that already contain, in addition to myoblasts, many myocytes and small myotubes. Subsequent fusion in these tissues results in long, thin muscle fibers arrayed in parallel. Branching is rare. Oriented elements of the extracellular matrix may serve to align the cells and to prevent fusion events that would crosslink adjacent myotubes.

Even before the premuscle masses of the limb buds begin to split up they contain, in addition to proliferating myogenic precursor cells, significant num-

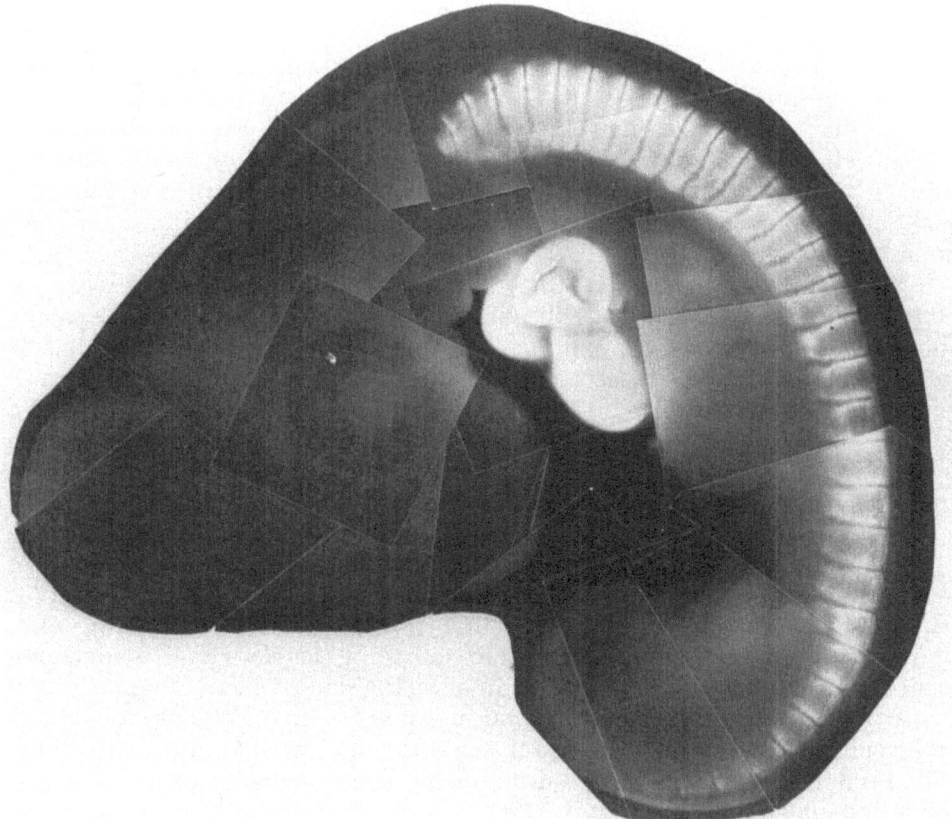

Figure 1. Whole-mount immunofluoresence histochemistry. A chicken embryo (3.5 days, Hamburger-Hamilton stage 23) was fixed and permeabilized for 20 hr at 0°C with an aged mixture of 10 parts 70% ethanol, 2 parts glacial acetic acid, and 1 part formalin (Turner *et al.*, 1976). After washing for 1 day with two changes of phosphate-buffered saline, the embryo was reacted with a 1 : 2 dilution of hybridoma culture supernate containing the anti-myosin antibody MF 20 (Bader *et al.*, 1982). The embryo was then washed for several days with phosphate-buffered saline, reacted with rhodamine-coupled goat anti-mouse IgG, washed for several more days, and photographed. Only the myocytes of myotome and heart show positive staining; if there are any terminally differentiated myocytes in the wing bud at this stage, they cannot be detected.

bers of terminally differentiating cells (Chiquet *et al.*, 1981). Analysis of clones generated *in vitro* by limb bud cells indicates several distinct cell types among the precursors, and even distinct myogenic lineages (Rutz and Hauschka, 1982), but definitive information about early commitment events in myogenesis is lacking. Cellular heterogeneity within the premuscle masses is a problem. When considering morphogenetically important interactions, such as those with extracellular matrices or with nerves and tendons, we do not know which myogenic cells (i.e., which developmental stages) are participating.

1.3. Problems of Muscle Morphogenesis

Ultimately, we want to understand how individual, functional muscles of the right size and shape are formed—at the right times, in the right places, and with all the right connections to nerves, tendons, and so forth. Detailed knowledge of cell interactions will be an important part of such an understanding. We need to know how, at various stages of development, myogenic cells interact with one another, with cells of other lineages, and with extracellular matrices deposited by such cells.

Some of the relevant interactions have already been alluded to. They include those responsible for (1) the translocation of myogenic cells from the somite to other parts of the embryo and the later subdivision of premuscle masses; (2) the mutual adherence and fusion of myogenic cells; (3) the organization and alignment of muscle cells within the growing tissue; and (4) the formation of proper connections to nerve and glia cells, connective tissue (sheath and tendon) cells, and blood vessels. These morphogenetically relevant cell–cell and cell–matrix interactions all occur at the cell surface.

Having identified the cellular interactions of interest, the task becomes one of understanding them in molecular terms. The questions are the same as in many other developmental systems. What cell-surface and matrix molecules participate in these morphogenetic events? How do their structures, concentrations, and locations change during the course of development? Which molecular interactions—with molecules in or on the same cell, on other cells, and in matrices—are significant for morphogenesis? How do molecular interactions at the cell surface mediate between the exterior and interior of the cell? Which cells produce which molecules of the extracellular matrix, and how are matrices assembled and modified?

2. Experimental Analysis of Cell–Substratum Attachment in Muscle Cell Cultures

Four morphogenetic processes have been studied most intensively in culture: (1) myocyte attachment to, and elongation upon, acellular substrata; (2) the mutual adhesion of myogenic cells; (3) myocyte fusion; and (4) formation of neuromuscular junctions. The following discussion considers the first of these in detail, deals briefly with the second and the third, and does not include the fourth process. This deliberate imbalance, reflecting the emphases in my own research, has obvious drawbacks. I believe, however, that the lessons to be learned from a more thoroughgoing analysis of a limited set of development interactions can, at least partially, justify this approach.

Myocyte attachment is an adhesive interaction. The central importance of adhesive interactions for development is now widely appreciated (see Steinberg and Poole, 1982, and Chapters 1, 4, 8, 9, and 12–16 in this volume).

2.1. Mediation by Fibronectin of Myocyte Attachment

During the mid-1970s my student, Emilio Puri, began experimenting with simplified media for chick myogenic cells (Fig. 2). He found that when serum was omitted, myocytes accumulated in suspension, while fibroblasts remained attached to the dish. When the myocytes were decanted into gelatinized dishes, few of them attached to the substratum. Upon the addition of serum, however, the myocytes attached, elongated, and fused to form myotubes. This culture system (Puri and Turner, 1978) has served as the basis of our assay for exogenous factors that promote myocyte attachment.

Three advantages of this system may be noted. First, the cell suspension used is almost free of fibroblasts, which adhere well in the absence of exogenous factors and which had long been suspected of secreting substances that promote myocyte attachment (Hauschka, 1972). Second, at the time they are used, the cells have had 2 days to recover from any damage suffered in the initial tissue disruption (although we have no proof that repair is complete). Third, most of the suspended cells are postmitotic myocytes; there are few proliferative myoblasts, but some myocytes do aggregate and fuse in suspension to form paucinucleate "myoballs."

Another student, Matthias Chiquet, purified the attachment-promoting factor from horse serum and identified it as the glycoprotein now known as fibronectin (FN). Chiquet *et al.* (1979) showed that (1) addition to the myocyte suspension of FN purified from horse serum, human plasma, or culture medi-

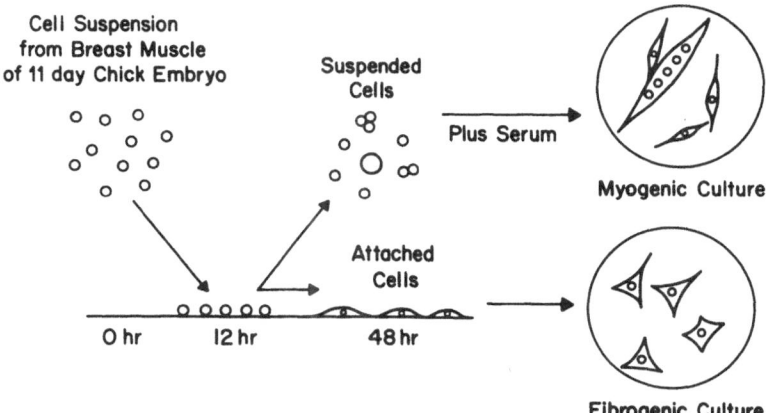

Figure 2. Suspension cultures in serum-free medium (Puri and Turner, 1978). Most of the freshly trypsinized cells from embryonic muscle stick to the plastic dish when first plated in serum-free medium. During the course of the next day or two, most myogenic cells detach, while fibroblasts remain attached. When the suspended cells are poured off and subcultured in the presence of fibronectin (FN), they reattach. Left in suspension culture, they aggregate and fuse to form *myoballs*.

um conditioned by chick fibroblasts promoted attachment and elongation in a concentration-dependent manner (Fig. 3); (2) gelatinized dishes pretreated with purified FN also promoted attachment and elongation; (3) rabbit antibodies against human or horse FN blocked FN-mediated attachment; and (4) serum selectively depleted of FN by affinity chromatography on gelatin-agarose or by immunoprecipitation was ineffective in promoting attachment. Thus we were able to conclude that FN is the serum component responsible for mediating myocyte attachment to gelatin. The existence of a collagen-binding, attachment-mediating substance in serum had been demonstrated some years earlier in the laboratory of Hauschka (1972), but the factor had not been purified or otherwise identified.

We did observe that sustained attachment (over many hours) was better when the cultures were supplemented with FN-depleted serum in addition to FN (Chiquet *et al.*, 1979). Inasmuch as serum contains a variety of substances that may have a general stimulatory effect on muscle cell metabolism, the observed enhancement could be indirect, with no particular significance for attachment. Serum is, however, known to contain attachment-promoting proteins other than FN (see, e.g., Barnes *et al.*, 1983). It may be that one of these has a secondary role in enhancing myocyte attachment itself.

Chiquet also investigated whether myocytes depend on exogenous FN for attachment because they themselves synthesize none or too little of this protein. The answer was clear (Chiquet *et al.*, 1981). Under conditions in which fibroblasts synthesize FN in copious amounts and release it into the medium, myocytes and myotubes synthesize and release much less. Myocytes and myotubes also deposit much less insoluble FN in a pericellular matrix, as is particularly evident in older primary cultures. After a week or so, fibroblasts have typically proliferated to fill the spaces between myotubes. When FN in such cultures is visualized by indirect immunofluorescence, a bright, reticular matrix covers all areas populated by fibroblasts. In striking contrast, the myotubes appear black.

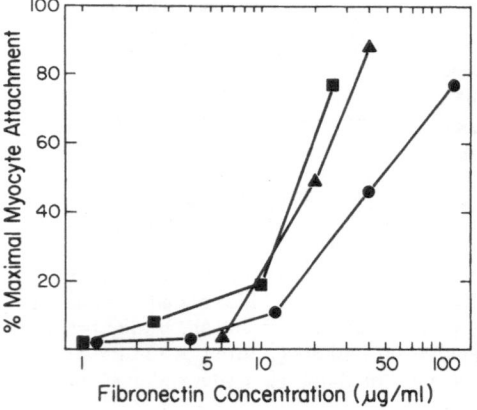

Figure 3. Dependence of myocyte attachment on the concentration of fibronectin (FN). FNs purified from human plasma (■), from horse serum (●), and from medium conditioned by chicken fibroblasts (▲) were added to aliquots of cell suspension (to the final concentrations indicated). Attached cells were counted at 20 hr; attachment is expressed as a percentage of that obtained in the presence of 3% horse serum. Similar curves were obtained when attachment was scored at 2–3 hr. (Adapted from Chiquet *et al.*, 1979.)

2.2. Mode of Action of Fibronectin

FN molecules consist of two subunits, 220,000–235,000-M_r in size, linked covalently by disulfide bridges located near the carboxy termini (for recent reviews of FN structure and functions, see Hynes and Yamada, 1982; Yamada, 1983). FN occurs in two forms that are structurally similar but that differ in some (but not all) biological activities. The more soluble form is called plasma FN. It is found in plasma and other body fluids. Cellular FN has a much greater tendency to self-associate. It is found in extracellular matrices, on cell surfaces, and in conditioned media. The two forms appear to be equally effective in promoting attachment of a variety of cell types. Cellular FN is far more potent in agglutinating erythrocytes and in restoring normal morphology to neoplastically transformed cells.

FNs are known to bind to an almost bewildering number of other molecules, including collagens, glycosaminoglycans and proteoglycans, fibrin, actin, and DNA; they also bind to at least one "receptor" on cell surfaces that is evidently distinct from any of these, as well as to other FN molecules. Binding to one macromolecular ligand can, in some cases, influence binding to another. FN, collagens, hyaluronic acid, and proteoglycans are all found in the extracellular matrices of the embryo, carrying at least the potential for multiple molecular interactions that might modulate the rate or strength of FN-mediated attachment *in vivo*.

FNs can be dissected by limited proteolysis into fragments that retain one or more of the activities of the intact molecule. FNs evidently consist of multiple compact domains relatively resistant to proteolytic attack, strung together by open stretches of polypeptide at which cleavage occurs readily. Obviously, a set of fragments, each capable of a limited range of intermolecular interactions, would have advantages for the study of the mode of action of FN. In setting out to obtain such fragments, we were particularly interested in fragments capable of promoting myocyte attachment and elongation.

In a key first step, another student, Ruth Ehrismann, showed that dishes coated directly with purified FN promote myocyte attachment about as well as dishes coated first with gelatin and then with FN. She then isolated several proteolytic fragments of horse serum FN and characterized them, with respect to both ability to bind to gelatin or heparin and myocyte attachment activity (Ehrismann *et al.*, 1981, 1982). She was able to obtain a fragment of about 105,000 daltons that promoted attachment but lacked binding sites for heparin and gelatin. Further digestion with a different protease produced a slightly smaller fragment that lacked attachment-promoting activity.

The results of the fragment studies, when taken together with certain information from other laboratories, allowed us to propose a model for the arrangement of functional domains in horse serum FN (Ehrismann *et al.*, 1982). In its general features and in many details, this model agrees well with models proposed for FNs from other sources. The region of the molecule active in promoting myocyte attachment appears to be the same one identified as binding to other cell types. The attachment-promoting activity of human FN has since been shown to reside in a sequence of 3 or 4 amino acids (Pierschbacher and Ruoslahti, 1984).

2.3. Collagen in Myogenesis: A Reappraisal

Hauschka and Konigsberg (1966) showed that, compared with uncoated surfaces or to surfaces coated with various other substances, dishes coated with collagen are an excellent substratum for supporting myogenesis in culture. The effects are so dramatic, especially when cells are plated at very low density, that it became commonplace to speak of a requirement for collagen. A brief review of this subject is given elsewhere (Turner and Carbonetto, 1984).

Hauschka and co-workers later showed that (1) the effects of collagen on myogenesis *in vitro* could all be traced to increased cell attachment, (2) gelatin is as effective as undenatured collagen, (3) only certain CNBr fragments of collagen are effective when used to coat substrata and (4) myocyte attachment is mediated by a collagen-binding protein found in serum (reviewed in Hauschka, 1972).

We now know that exogenous collagen is not required to support myocyte attachment *in vitro*. Purified FN and FN fragments are active in the absence of added collagen. Surfaces coated with gelatin (Hauschka, 1972; Chiquet *et al.*, 1979) or native collagen (D. Turner, unpublished observations) are effective only if FN is contained in the medium. They work because they collect FN so efficiently; Hauschka's active CNBr fragments are the same ones now known to bind FN (Kleinman *et al.*, 1978). As far as I know, there is no effect of collagen on myogenesis *in vitro* that cannot be rationalized in these terms.

Our working assumption (see Section 3) is that, in the embryo as well as in culture, muscle cells attach to extracellular matrices only *via* FN molecules—that these cells do not adhere directly to collagens, glycosaminoglycans, or proteoglycans. Collagen might nevertheless be required for myogenesis *in vivo*. It could be that, in constructing a matrix to which myogenic cells must attach, collagen fibrils are always laid down first, with FN binding secondarily. In such a case, it would be hard to argue that collagen is not essential. On the other hand, it may be that a FN network forms by FN self-association independently of or prior to collagen deposition, and that it is predominantly this matrix to which muscle-forming cells attach. More experiments are needed to settle this question.

3. Effects of a Fibronectin-Containing Matrix on Muscle Morphogenesis

One interesting generalization that has emerged during the past few years is that migratory cells fail to elaborate a FN-containing pericellular matrix (reviewed by Chiquet *et al.*, 1981; Turner *et al.*, 1983). They synthesize, at best, little FN, yet their attachment is mediated by FN. This is true for neural crest cells and primordial germ cells as well as for myogenic cells. It also holds for nerve cells that extend axons into surrounding tissues and for many types of malignant cells which invade and metastasize. We have suggested (Chiquet *et al.*, 1981) that these cells have the ability (1) to wander as single cells because they do not become trapped in an FN-containing matrix of their own making, and (2) to sense and respond to a differential distribution of FN in their surroundings because they are not constantly depositing their own FN.

For primordial germ cells (Heasman *et al.*, 1981) and for neural crest cells (Newgreen *et al.*, 1982; Rovasio *et al.*, 1983) there is impressive evidence that migration is guided by oriented FN-containing matrices. For myogenic cells the evidence so far is less compelling. What we have seen by immunohistochemistry on cryosections of limb buds (Chiquet *et al.*, 1981) is that there is much less matrix FN in the premuscle masses than in the surrounding mesenchyme (Figs. 4–6). Where the premuscle masses are beginning to split, a FN matrix is present in the clefts (Fig. 5). Within the premuscle masses, myocytes and myotubes are oriented in the long (proximodistal) axis of the limb (Fig. 6). As continued growth further extends the limb, the FN-containing latticework is stretched

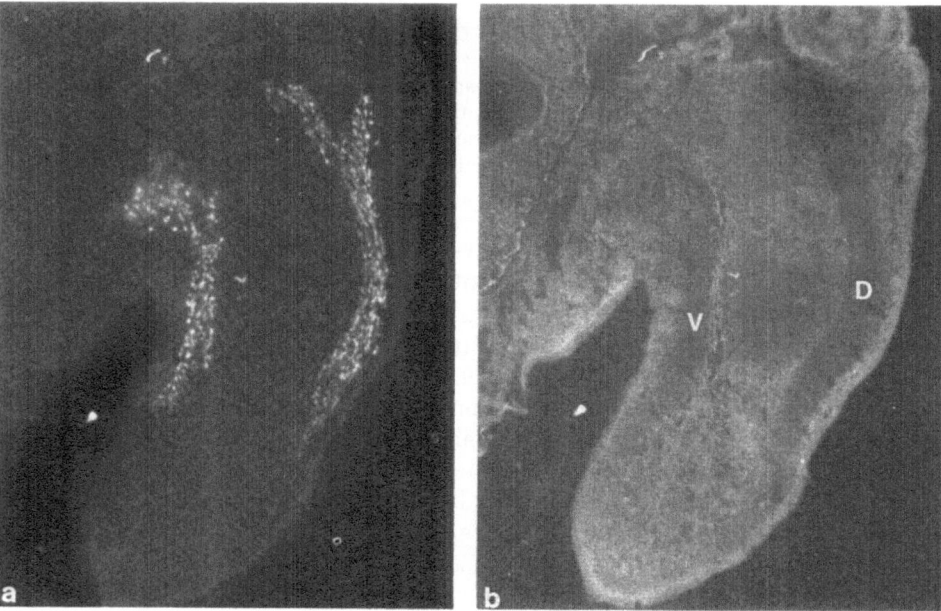

Figure 4. Double-antibody immunofluorescent staining of a cryosection through the wing bud of a 5-day chick embryo. (Left) Reaction with mouse monoclonal anti-myosin (see legend to Fig. 1) shows differentiating myocytes in the premuscle masses. (Right) Reaction with rabbit anti-FN (visualized with fluorescein-labeled goat antibodies against rabbit IgG) is noticeably weaker in the premuscle masses (D, dorsal premuscle mass; V, ventral premuscle mass) than in the distal limb mesenchyme into which myocytes have not penetrated.

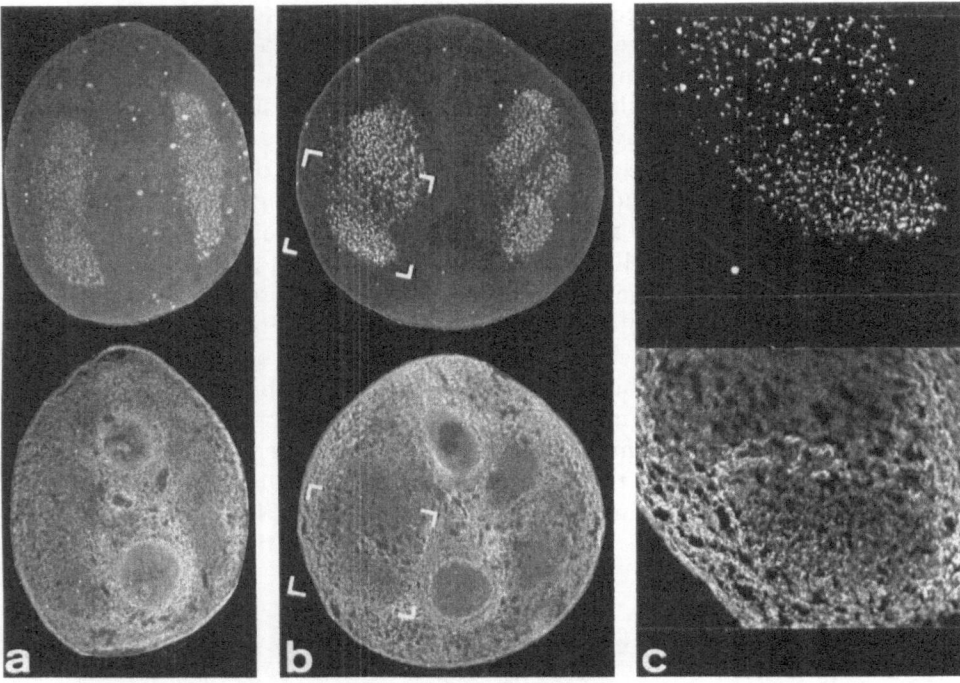

Figure 5. Distribution of fibronectin (FN) during splitting of the premuscle masses of stage 27 (a) and stage 28 (b) chick embryos. (Top) Staining with antibodies against myomesin (a myofibrillar protein). (Bottom) Next section in the series, reacted with anti-FN. Dorsal side is on the right. (c) Enlargement of area in (b). (Reproduced with permission from Chiquet *et al.*, 1981.)

along the proximodistal axis. The observed alignment of muscle cells presumably occurs as they attach to the FN-containing matrix.

But what about the earlier migration of myogenic cells from the somite to the limb? Are the cells guided by an oriented FN-containing matrix? We do not know. We have been unable to label selectively the myogenic cells as they first emigrate from the somites, probably because we lack an antibody against a marker antigen that is both particularly abundant in myoblasts (or earlier muscle precursors) and not redistributed or destroyed in our immunohistochemical procedures. Until we can use double-label immunofluorescence in the same section to identify both the migratory cells and the FN with which they may interact, we cannot answer the question.

Meanwhile, it should be noted that what has been called cell "migration" in this and other systems may not involve active cell migration over long distances. Rather, expansion of proliferating cells into nearby areas along paths of least resistance may lead to early and firm attachment to matrix elements, with most of the translocation being brought about by later relocation of the matrix (see Chapter 16, this volume). In his analysis of neurogenesis, Weiss (1941) stressed that "pioneering fibers" establish permanent contacts to target cells early on, when the targets are still nearby (see Chapter 14, this volume). The terminally differentiating myocytes and small myotubes present fairly early in premuscle masses might have a similar "pioneering" function. Whereas pioneering nerve fibers were thought by Weiss to provide tracks to which later-emerging fibers could adhere, pioneering myocytes might function to keep channels open (i.e., relatively free of FN matrix), thus allowing later-arriving myogenic cells to follow.

3.1. A Model System for Studying Morphogenetic Interactions *in Vitro*

We have developed a procedure for making oriented deposits of FN (Turner *et al.*, 1983). A solution of FN in urea is allowed to dry in a culture dish. As urea crystals form, FN is precipitated between them. The pattern of crystallization determines the pattern of FN deposition. Washing the plate redissolves the urea, leaving the FN behind. Often, parallel tracks cover large areas of the dish.

Myoblasts and myocytes attach preferentially to the threadlike FN deposits. These bipolar cells align themselves with the tracks. Most fusion occurs end-to-end among myocytes and myotubes aligned along the same track. As a

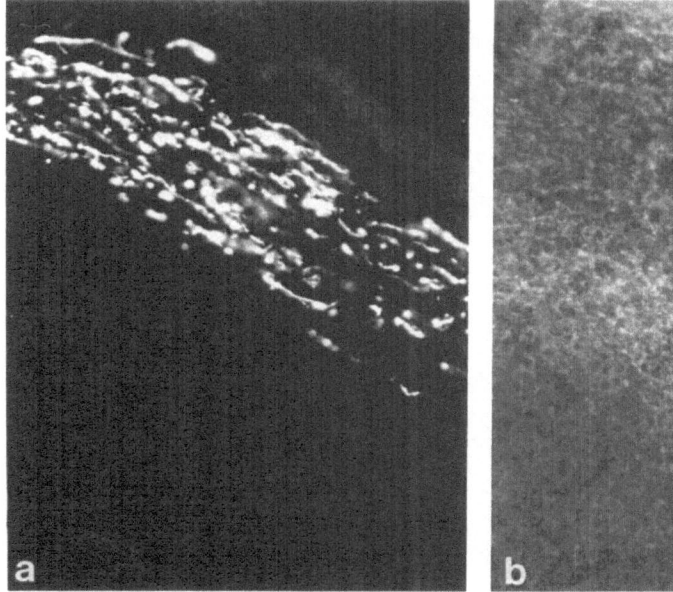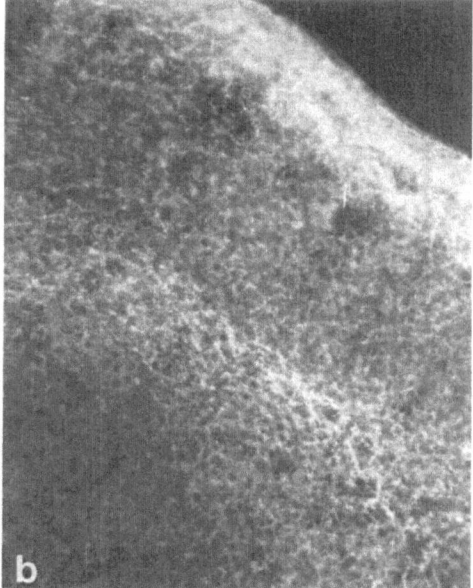

Figure 6. Detail of dorsal premuscle mass, 5-day embryo. (Left) Fluorescent antibody staining with anti-myosin (see Fig. 1). Myocytes and myotubes are aligned with the proximodistal axis of the limb. (Right) Staining of the same section with anti-FN. A prominent fibronectin (FN)-containing matrix lines the area occupied by the myogenic cells.

result, myotubes are generally long, thin, unbranched, and arrayed in parallel (Fig. 7). Myogenic clones on these artificial matrices tend to be highly elongated (Fig. 8), in striking contrast to the roughly circular clones that form on uniform substrata. Where the orientation of the deposits changes, the direction of outgrowth changes also. From our results with this model system (Turner *et al.*, 1983) there can be no doubt that the spatial distribution of FN molecules can guide the contacts of myogenic cells with one another and so the form of differentiated muscle cells.

Oriented deposits of collagen can—provided serum or some other source of FN is present—also guide myotube formation (John and Lawson, 1980) (see Fig. 9). I would predict that oriented fibrin deposits would also be effective, but again only in the presence of FN. From what we know now, all such substrate guidance of myocytes will depend on differential attachment, with more stable attachments occurring at substrate sites where the density of bound FN is greater. The importance of FN is that it is the molecule that binds directly to the cell surface, mediating attachment to any of a variety of collagens and, possibly, to other matrix molecules as well.

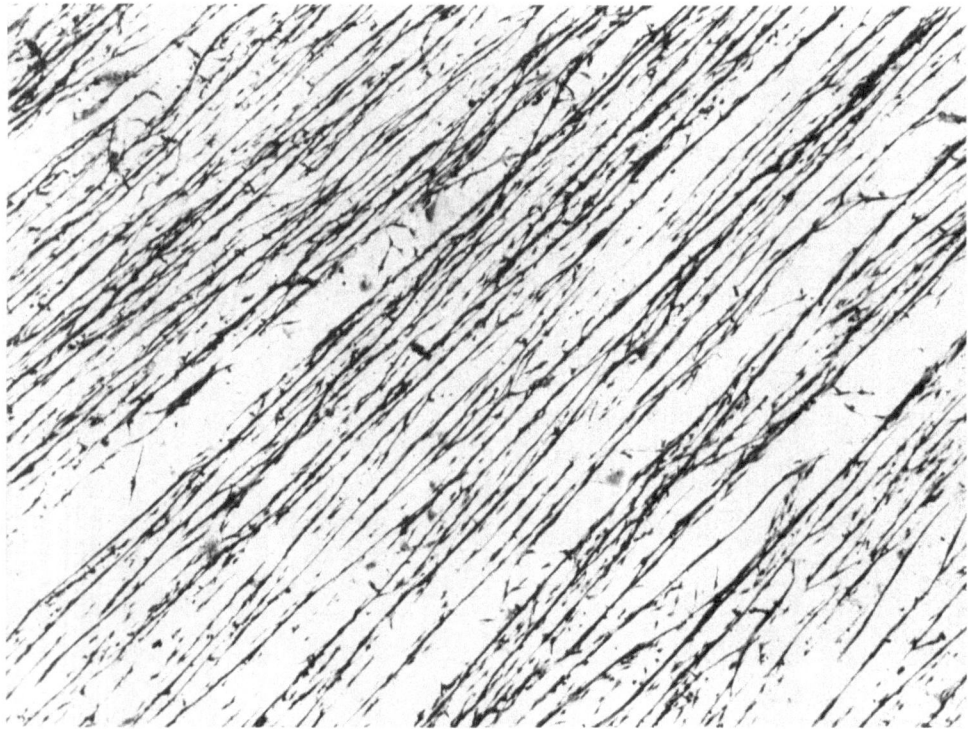

Figure 7. Myogenic cells cultured on oriented fibronectin (FN) deposits give rise to thin myotubes arrayed in parallel, with little anastomosing.

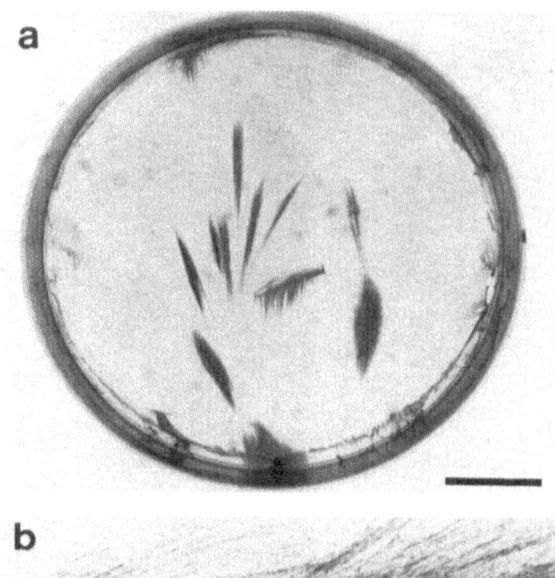

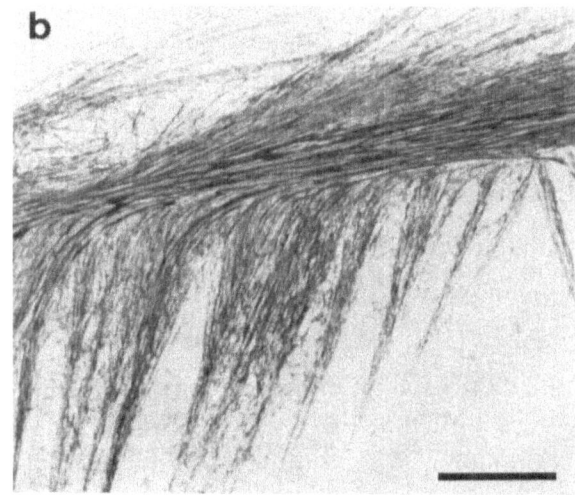

Figure 8. Clonal culture on oriented fibronectin (FN) matrix. (a) Of the 10 colonies not in contact with the sides of the dish, all are myogenic except for the two lens-shaped ones in the lower part of the figure. Several of the myogenic clones appear frayed. (b) Detail shows this is because cells migrating in one direction encountered FN tracks with a different orientation and turned onto them. Scale bars: (a) 1 cm; (b) 1 mm. (Reproduced with permission from Turner et al., 1983.)

In work now in progress, we have been experimenting with oriented matrices formed by depositing FN fragments and by codepositing FN with glycosaminoglycans. We have also been examining the behavior of a variety of cell types on oriented FN. Nerve fibers grow out along the FN threads (Turner and Carbonetto, 1984). Pleomorphic cells derived from limb buds adopt a bipolar shape as they align with, and move along, the FN tracks. The morphology and direction of migration of certain other cell types is less affected. Fibroblasts, for example, cross the FN tracks readily, probably because of a combination of two factors: attachment *via* a mechanism independent of FN and deposition of new FN that confounds the directionality of the artificial matrix (Turner et al.,

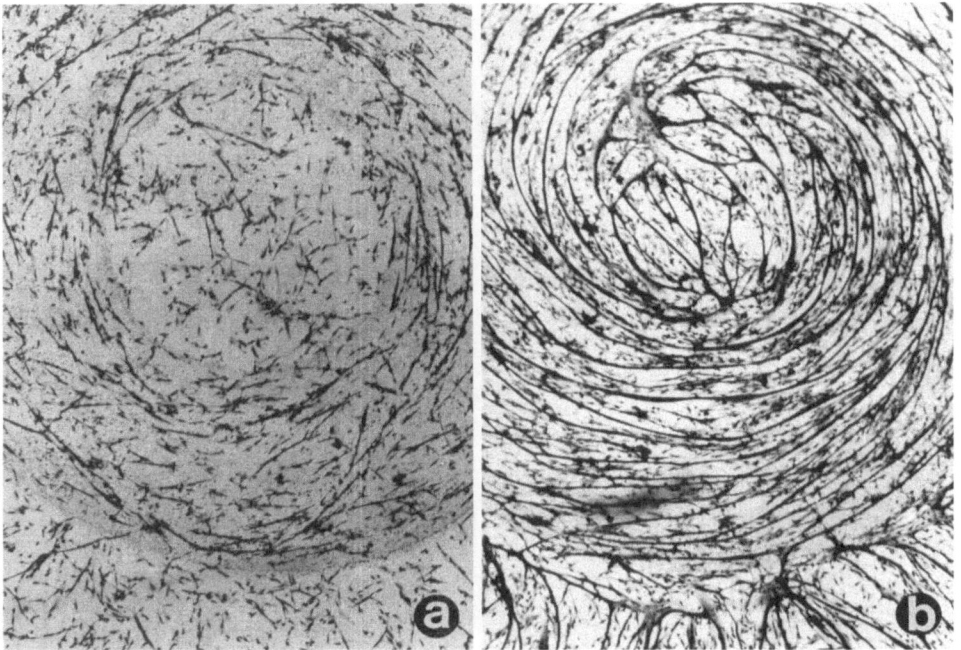

Figure 9. Demonstration that oriented collagen fibrils, deposited in a circular pattern according to John and Lawson (1980), can guide myotube formation. Chick myogenic cells cultured in fibronectin (FN)-containing medium (to permit attachment) were fixed and Giemsa stained after 1.5 days (a) and 3 days (b).

1983). I believe the oriented FN deposits hold considerable promise for further studies of the regulation of cell shape.

4. Effects of Fibronectin on Other Events in Myogenesis

After conflicting claims had appeared that FN influences the process of myocyte fusion in cultures of established myogenic lines (Chen *et al.*, 1978; Podleski *et al.*, 1979), we decided to investigate this point in chicken myogenic cultures. In cultures of attached cells, the rate and extent of cell fusion depend on cell attachment and elongation, which are FN mediated. Moreover, cell migration, which precedes fusion of attached myocytes, is also stimulated by FN (Ali and Hynes, 1978). We concluded that the issue would be hard to settle in conventional cultures and turned instead to the FN-free cultures mentioned earlier, in which suspended myocytes aggregate and fuse (Puri and Turner, 1978).

We first showed that the fusion of suspended cells could be quantitated and then demonstrated that exogenous FN has no effect on the extent of fusion (Puri *et al.*, 1979). Polyclonal rabbit antibodies against FN, in concentrations sufficient to block FN-mediated attachment, also had no effect (Puri *et al.*, 1979). Since the amount of cell-surface FN on myocytes is so low as to be virtually undetectable by immunofluorescence, we interpreted this result as indicating that myocyte fusion cannot be mediated by endogenous FN.

It is also unlikely that FN has any effect on the mutual recognition and adhesion of myogenic cells, but for technical reasons this conclusion is less firm (Puri *et al.*, 1979).

On the basis of the present evidence there is no reason to think that FN has some separate and distinct function in myocyte differentiation. We have argued (Chiquet *et al.*, 1981) that all known effects of FN on myogenesis can be plausibly traced to its primary effect in promoting attachment and elongation.

5. Perspectives

Our inquiry began rather broadly, with a panoramic view of morphogenetic interactions in myogenesis. Very soon, however, it devolved into a consideration of the functions of a single molecule. Narrowing of focus is an inevitable part of research, even in the very largest research groups, yet I confess to a certain uneasiness about doing developmental biology one molecule at a time. Can such an approach be justified? I would say yes—and no.

Yes—in the short term. As I believe my account of our research shows, it helps enormously to be concentrating on one molecule (or a small group of related molecules). One learns how to handle the molecule(s) and how to modify molecular structure to suit one's needs. One raises specific antibodies, characterizes them thoroughly, and then uses them to purify, localize, and quantitate the molecule(s) in a variety of applications. In short, one acquires that intimate familiarity needed for research in depth.

And *no*—certainly not over the long term. It is not going to be satisfying to have a number of parallel "stories" of muscle morphogenesis, each with a single major character, its own "molecular hero." Let me give one example. Toole and Underhill (1983) have been investigating the functions of hyaluronate and hyaluronidase in several morphogenetic processes, including muscle morphogenesis. They have shown that a hyaluronate-containing matrix inhibits myocyte contact and thus myocyte fusion. We know that FN can bind to hyaluronate. We know that the two molecules must coexist in some extracellular matrices. To reach an adequate understanding of muscle morphogenesis, we will have to perform experiments involving both molecules in an effort to reconcile what is known about them separately. And we will not be able to stop there. Cell-surface molecules (including the FN receptor and molecules involved in cell–cell adhesion) and other matrix molecules (including proteoglycans, collagens, and laminin) will have to be integrated into one grand story. It is something to look forward to.

ACKNOWLEDGMENTS. Many colleagues in both Zurich and Syracuse have participated in the research described here; their names appear in the references. It is a pleasure to acknowledge their essential contributions. I thank Dr. Donald Fischman for providing the antibody against myosin. Our research has been supported by the Swiss National Science Foundation, by the Muscular Dystrophy Association, and by NIH (R01 HD 17060).

References

Ali, I. U., and Hynes, R. O., 1978, Effects of LETS glycoprotein on cell motility, *Cell* **14**:439–446.

Bader, D., Masaki, T., and Fischman, D. A., 1982, Immunochemical analysis of myosin heavy chain during avian myogenesis *in vivo* and *in vitro*, *J. Cell Biol.* **95**:763–770.

Barnes, D. W., Silnutzer, J., See, C., and Shaffer, M., 1983, Characterization of human serum spreading factor with monoclonal antibody, *Proc. Natl. Acad. Sci. USA* **80**:1362–1366.

Bischoff, R., 1978, Myoblast fusion, in: *Membrane Fusion* (G. Poste and G. L. Nicolson, eds.), pp. 128–179, Elsevier/North-Holland, Amsterdam.

Buckingham, M. E., 1977, Muscle protein synthesis and its control during differentiation of skeletal muscle cells in vitro, in: *Biochemistry of Cell Differentiation II*, Vol. 15 (J. Paul, ed.), pp. 269–332, University Park Press, Baltimore.

Caravatti, M., Perriard, J. C., and Eppenberger, H. M., 1979, Developmental regulation of creatine kinase isozymes in myogenic cell cultures from chicken: Biosynthesis of creatine kinase subunits M and B, *J. Biol. Chem.* **254**:1388–1394.

Chen, L. B., Murray, A., Segal, R. A., Bushnell, A., and Walsh, M. L., 1978, Studies on intercellular LETS glycoprotein matrices, *Cell* **14**:377–391.

Chiquet, M., Puri, E. C., and Turner, D. C., 1979, Fibronectin mediates attachment of chick myoblasts to a gelatin-coated substratum, *J. Biol. Chem.* **254**:5475–5482.

Chiquet, M., Eppenberger, H. M., and Turner, D. C., 1981, Muscle morphogenesis: Evidence for an organizing function of exogenous fibronectin, *Dev. Biol* **88**:220–235.

Christ, B., Jacob, H. J., Jacob, M., and Wachtler, F., 1983, On the origin, distribution, and determination of avian limb mesenchymal cells, in: *Limb Development and Regeneration*, Part B (R. O. Kelley, P. F. Goetinck, and J. A. MacCabe, eds.), pp. 281–292, Alan R. Liss, New York.

Ehrismann, R., Chiquet, M., and Turner, D. C., 1981, Mode of action of fibronectin in promoting chicken myoblast attachment. A M_r = 60,000 gelatin-binding fragment also binds native fibronectin, *J. Biol. Chem.* **256**:4056–4062.

Ehrismann, R., Roth, D., Eppenberger, H. M., and Turner, D. C., 1982, Arrangement of attachment-promoting, self-association, and heparin-binding sites in horse serum fibronectin, *J. Biol. Chem.* **257**:7381–7387.

Eppenberger, H. M., and Perriard, J. C. (eds.), 1984, *Developmental Processes in Normal and Diseased Muscle*, S. Karger, Basel, Switzerland.

Field, H. H., 1894, Die Vornierenkapsel, ventrale Muskulatur und Extremitaetenanlagen bei den Amphibien, *Anat. Anz.* **9**:713–714.

Fischel, A., 1895, Zur Entwicklung der ventralen Rumpf- und der Extremitaetenmuskulatur der Voegel und Saeugethiere, *Morphol. Jahrb.* **23**:544–561.

Hauschka, S. D., 1972, Cultivation of muscle tissue, in: *Growth, Nutrition and Metabolism of Cells in Culture*, Vol. 2 (G. H. Rothblat and V. J. Cristofalo, eds.), pp. 67–130, Academic Press, New York.

Hauschka, S. D., and Konigsberg, I. R., 1966, The influence of collagen on the development of muscle colonies, *Proc. Natl. Acad. Sci. USA* **55**:119–126.

Heasman, J., Hynes, R. O., Swan, A. P., Thomas, V., and Wylie, C. C., 1981, Primordial germ cells of *Xenopus* embryos: The role of fibronectin in their adhesion during migration, *Cell* **27**:437–447.

Holtzer, H., 1970, Myogenesis, in: *Cell Differentiation* (O. Schjeide and J. de Vellis, eds.), pp. 476–503, Van Nostrand Reinhold, New York.

Hynes, R. O., and Yamada, K. M., 1982, Fibronectins: Multifunctional modular glycoproteins, *J. Cell. Biol.* **95**:369–377.

John, H. A., and Lawson, H., 1980, The effect of different collagen types used as substrata on myogenesis in tissue culture, *Cell Biol. Int. Rep.* **4**:841–849.

Kieny, M. A., 1983, Cell and tissue interactions in the organogenesis of the avian limb musculature, in: *Limb Development and Regeneration, Part B* (R. O. Kelley, P. F. Goetinck, and J. A. MacCabe, eds.), pp. 293–302, Alan R. Liss, New York.

Kleinman, H. K., McGoodwin, E. C., Martin, G. R., Klebe, R. J., Fietzek, P. P., and Wooley, D. E., 1978, Localization of the binding sites for cell attachment in the α_1(I) chain of collagen, *J. Biol. Chem.* **253**:5642–5646.

Konigsberg, I. R., 1965, Aspects of cytodifferentiation of skeletal muscle, in: *Organogenesis* (R. L. DeHaan and H. Ursprung, eds.), pp. 337–358, Holt, Rinehart and Winston, New York.

Konigsberg, I. R., 1979, Skeletal myoblasts in culture, *Methods Enzymol.* **58**:511–527.

Linkhart, T. A., Clegg, C. H., Lim, R. W., Merrill, G. F., Chamberlain, J. S., and Hauschka, S. D., 1982, Control of mouse myoblast commitment to terminal differentiation by mitogens, in: *Muscle Development: Molecular and Cellular Control* (M. L. Pearson and H. F. Epstein, eds.), pp. 377–382, Cold Spring Harbor Laboratory, Cold Spring Harbor, New York.

Monod, J., 1970, *Le Hasard et la Nécessité: Essai sur la Philosophie Naturelle de la Biologie Moderne,* Editions du Seuil, Paris.

Newgreen, D. F., Gibbons, I. L., Sauter, J., Wallenfels, B., and Wutz, R., 1982, Ultrastructural and tissue-culture studies on the role of fibronectin, collagen, and glycosoaminoglycans in the migration of neural crest cells in the fowl embryo, *Cell Tissue Res.* **221**:521–549.

Pearson, M. L., 1980, Muscle differentiation in cell culture: A problem in somatic and molecular genetics, in: *The Molecular Genetics of Development* (T. Leighton and W. F. Loomis, eds.), pp. 361–418, Academic Press, New York.

Pearson, M. L., and Epstein, H. F. (eds.), 1982, *Muscle Development,* Cold Spring Harbor Laboratory, Cold Spring Harbor, New York.

Pierschbacher, M., Hayman, E. G., and Ruoslahti, E., 1984, Variants of the cell recognition site of fibronectin that retain attachment-promoting activity, *Proc. Natl. Acad. Sci. USA* **81**:5985–5988.

Podleski, T. R., Greenberg, I., Schlessinger, J., and Yamada, K. M., 1979, Fibronectin delays the fusion of L-6 myoblasts, *Exp. Cell Res.* **122**:317–326.

Puri, E. C., and Turner, D. C., 1978, Serum-free medium allows chicken myogenic cells to be cultivated in suspension and separated from attached fibroblasts, *Exp. Cell Res.* **115**:159–173.

Puri, E. C., Chiquet, M., and Turner, D. C., 1979, Fibronectin-independent myoblast fusion in suspension cultures, *Bioch. Biophys. Res. Commun.* **90**:883–889.

Rovasio, R. A., Delouvée, A., Yamada, K. M., Timpl, R., and Thiery, J. P., 1983, Neural crest cell migration: Requirements for exogenous fibronectin and high cell density, *J. Cell Biol.* **96**:462–473.

Rutz, R., and Hauschka, S., 1982, Clonal analysis of vertebrate myogenesis. VII. Heritability of muscle colony type through sequential subclonal passages *in vitro, Dev. Biol.* **91**:103–110.

Shellswell, G. B., 1980, Cellular events in the early development of skeletal muscles, in: *Development in Mammals,* Vol. 4 (M. H. Johnson, ed.), pp. 137–159, Elsevier, Amsterdam.

Steinberg, M. S., and Poole, T. J., 1982, Cellular adhesive differentials as determinants of morphogenetic movements and organ segregation, in: *Developmental Order: Its Origin and Regulation* (S. Subtelny and P. B. Green, eds.), pp. 351–378, Alan R. Liss, New York.

Toole, B. P., and Underhill, C. B., 1983, Regulation of morphogenesis by the pericellular matrix, in: *Cell Interactions and Development: Molecular Mechanisms* (K. M. Yamada, ed.), pp. 203–230, Wiley, New York.

Turner, D. C., 1978, Differentiation in cultures of chick skeletal muscle cells. The postmitotic fusion-capable myoblast as a distinct cell type, *Differentiation* **10**:81–93.

Turner, D. C., and Carbonetto, S. T., 1984, Model systems for studying the functions of extracellular matrix molecules in muscle development, *Exp. Biol. Med.* **9**:72–79.

Turner, D. C., Gmür, R., Lebherz, H. G., Siegrist, M., Wallimann, T., and Eppenberger, H. M., 1976, Differentiation in cultures of chick skeletal muscle cells. II. Phosphorylase histochemistry and fluorescent antibody staining for creatine kinase and aldolase, *Dev. Biol.* **48**:284–307.

Turner, D. C., Lawton, J., Dollenmeier, P., Ehrismann, R., and Chiquet, M., 1983, Guidance of myogenic cell migration by oriented deposits of fibronectin, *Dev. Biol.* **95:**497–504.

Weiss, P., 1941, Nerve patterns: The mechanics of nerve growth, *Growth* **5:**163–203.

Yamada, K. M., 1983, Fibronectin in cell interactions, in: *Cell Interactions and Development: Molecular Mechanisms* (K. M. Yamada, ed.), pp. 231–249, Wiley, New York.

Zwilling, E., 1968, Morphogenetic phases in development, *Dev. Biol.* **2:**184–207.

Chapter 12

Extracellular Matrix Involvement in Epithelial Branching Morphogenesis

BRIAN S. SPOONER, HOLLY A. THOMPSON-PLETSCHER,
BRAD STOKES, and KENNETH E. BASSETT

1. Introduction

Branching morphogenesis is a major developmental process used by a large array of embryonic organs. Examples of organs whose development involves branching morphogenesis include submandibular, sublingual, and parotid salivary glands, mammary glands, lungs, pancreas, and kidneys. In each case, an epithelium undergoes repeated branching activity, in concert with mitotic growth, that results in the formation of an organ-specific pattern of clefts or branch points and the generation of large amounts of epithelial surface area for the available amount of organ space. The advantage would appear to be that epithelial surface area for such functions as secretory activity and gas exchange is maximized, resulting in highly efficient compact organs.

Branching morphogenesis has been shown to take place in response to epithelial–mesenchymal tissue interactions, in which the mesenchymal component dictates the branching pattern of the epithelium (see Grobstein, 1967; Spooner, 1974, for review). These epithelial–mesenchymal tissue interactions are more properly referred to as morphogenetic tissue interactions that affect the mitotic and cell shape behavior of the epithelium to produce the appropriate branching pattern. Older views that such interactions regulate cytodifferentiation and that cytodifferentiation and morphogenesis are coupled events are inconsistent with demonstrations that cytodifferentiation can take place in the absence of morphogenesis (Spooner et al., 1977) and that morphogenesis can be altered experimentally without concomitant alteration of cytodifferentiation (Sakakura et al., 1976). The fact that branching morphogenesis requires epithelial–mesenchymal tissue interactions draws attention to the extracellular matrix at the epithelial–mesenchymal tissue interface as the site of significant

BRIAN S. SPOONER, BRAD STOKES, and KENNETH E. BASSETT • Division of Biology, Kansas State University, Manhattan, Kansas 66506.   HOLLY A. THOMPSON-PLETSCHER • Department of Chemistry, University of Montana, Missoula, Montana 59812.

communication between the interacting tissues. Grobstein and co-workers recognized the fundamental involvement of the extracellular matrix and the necessity for characterizing its macromolecular components many years ago (Grobstein and Cohen, 1965; Kallman and Grobstein, 1965, 1966), but only during the past decade has major progress been made.

This chapter reviews the evidence we have generated over the past few years on extracellular matrix involvement in epithelial branching morphogenesis. Our studies have focused on the embryonic lung and, particularly, on the embryonic submandibular salivary gland. A major advantage is that the lung, isolated from 11-day mouse embryos, and the salivary gland, isolated from 13-day mouse embryos, both undergo spectacular degrees of branching morphogenesis in culture that are readily monitored in the living state (Fig. 1). Both systems require epithelial–mesenchymal interactions, have highly specific mesenchymal requirements (Alescio and Cassini, 1962; Tadarera, 1967; Grobstein, 1953; Spooner and Wessells, 1970), and produce distinctive branching patterns (Fig. 1). In addition, we know that the immediate cause of branch point (cleft) formation in the salivary epithelium is a change in shape by a discrete group of cells in the continuous epithelium, generated by actin-containing arrays of cytoplasmic microfilaments (Spooner and Wessels, 1972; Spooner et al., 1973, 1978; Spooner, 1973, 1974, 1975). Thus, we have some insight into what epithelial cell activities need to be controlled, if branching morphogenesis is to take place. We have used an approach to investigation of extracellular matrix that asks whether specific interference with synthesis, secretion, or deposition of particular classes of extracellular matrix macromolecules interferes with branching morphogenesis. Such a perturbation approach can be of enormous value if care is taken to avoid misinterpretation due to nonspecific effects; it provides experimental insight into matrix requirements that is not possible by descriptive correlations, be they biochemical, immunological, or anatomical. The results demonstrate an essential involvement of collagen and proteoglycans in epithelial branching morphogenesis.

2. Collagen in Branching Morphogenesis of Embryonic Salivary Gland and Lung

Collagen is a major macromolecular component of extracellular spaces in embryonic organ systems, recognizable biochemically, by electron microscopy, and by immunolabeling. In branching organs, such as the salivary glands and lungs, interstitial (presumably types I and/or III) collagen fibrils are present at the epithelial–mesenchymal interface, where they have been proposed to mediate epithelial branching by differential deposition patterns and differential stabilization of the epithelial surface (Grobstein and Cohen, 1965; Wessells and Cohen, 1968; Wessells, 1970). Since the bulk of such collagen is produced by the mesenchyme (Bernfield, 1970), it may represent one essential product of morphogenetic tissue interactions. The epithelially derived basal lamina of these organs most probably contains basement membrane collagen (presum-

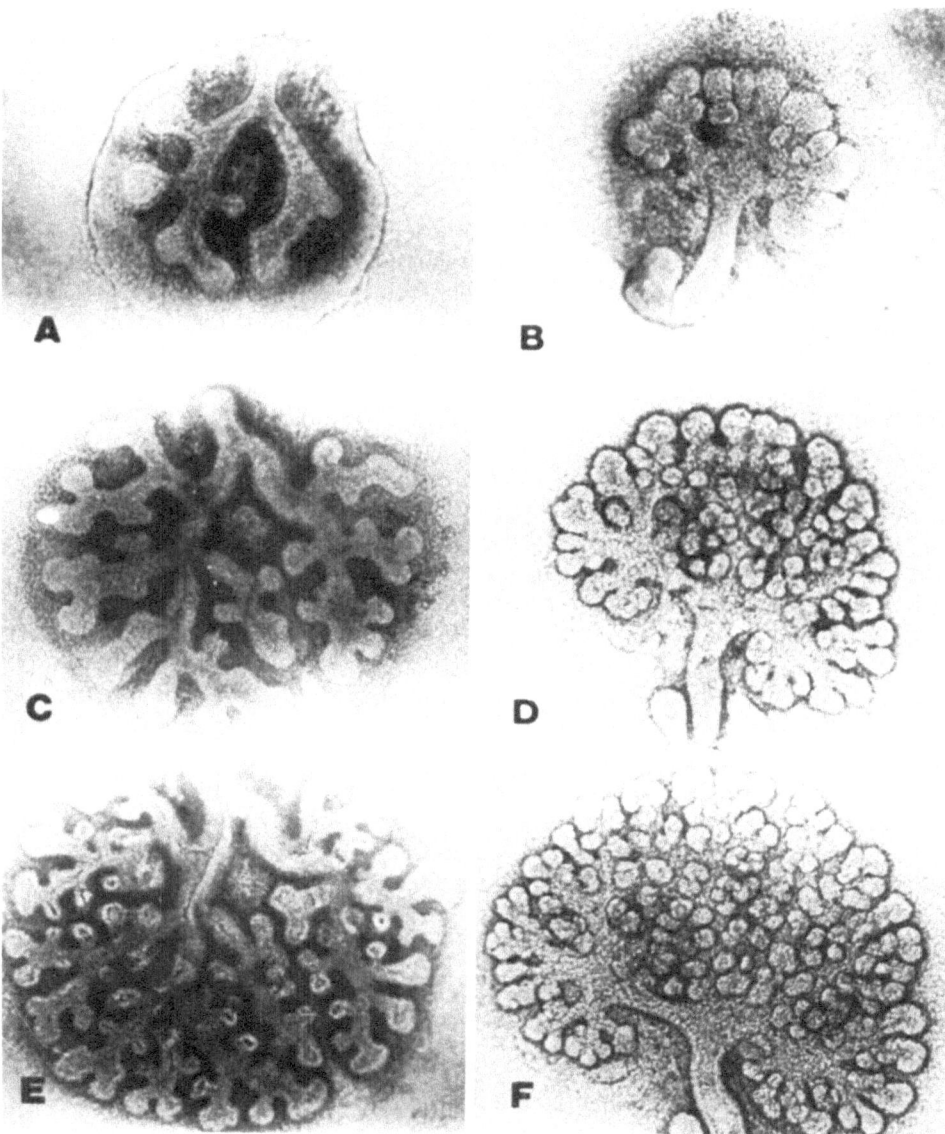

Figure 1. Living lung (A) and salivary gland (B) rudiments, isolated from mouse embryos at 11 and 13 days of gestation, respectively, and shown after 24 hr of organ culture. The same lung rudiment is shown at 48 hr (C) and 72 hr (E) of culture, and the same salivary rudiment is shown at 48 hr (D) and 72 hr (F) of culture. Both rudiments consist of an epithelium surrounded by mesenchyme, and undergo spectacular degrees of branching morphogenesis during the 3-day culture period. Epithelial growth and repetitive branch point (or cleft) formation result in an expanding epithelial "tree" with increasing branching complexity. The branching patterns are unique to lung and salivary, respectively, and are dependent on the presence of the mesenchyme. (From Spooner and Faubion, 1980.)

ably type IV); since the basal lamina is essential to ongoing branching (Bern-field *et al.*, 1972, 1973; Cohn *et al.*, 1977), it is likely that such collagen is essential for basal lamina structure and/or function and, thus, branching mor-phogenesis. Although collagen is present at the right time and in the right place to function in the branching process, there is little experimental evidence dem-onstrating that its presence is essential. Early studies showing that collagenase treatment interrupts salivary (Grobstein and Cohen, 1965) and lung (Wessells and Cohen, 1968) morphogenesis are difficult to interpret because of con-taminating protease and glycosaminoglycan-degradative activities (Bernfield *et al.*, 1972). We have therefore sought to investigate possible collagen involve-ment in branching morphogenesis by determining whether compounds that interfere with collagen synthesis, secretion, or extracellular crosslinking con-comitantly interfere with branching morphogenesis.

Common features of the various collagens include high glycine (approx-imately one-third of the residues) and proline (approximately one-fifth of the residues) contents in the α-chains, hydroxylation of proline (about 50%) and lysine residues in peptide linkage, and the unique triple helical structure of the assembled and processed molecule (see Bornstein and Traub, 1979). Under-hydroxylated collagen does not form the native triple helix and is not secreted, a condition that can be achieved experimentally by a variety of proline ana-logues or by inhibition of prolyl hydroxylase. We have exploited the proline analogue L-azetidine-2-carboxylic acid (LACA) and the ferrous ion chelator α,α'-dipyridyl as tools to assess collagen involvement in branching mor-phogenesis. LACA substitutes for proline during α-chain synthesis (Uitto and Prockop, 1974), with even minor degrees of substitution (i.e., a few percent) resulting in underhydroxylation of the proline residues that are incorporated, poor triple-helix formation, and reduced secretion (Takeuchi and Prockop, 1969; Takeuchi *et al.*, 1969; Lane *et al.*, 1971; Uitto and Prockop, 1974; Uitto *et al.*, 1976). On the other hand, whereas α,α'-dipyridyl also inhibits hydroxyla-tion, triple-helix formation, and subsequent secretion, it acts at the level of prolyl hydroxylase (Switzer and Summer, 1973). LACA does not directly affect that enzyme (Uitto and Prockop, 1974), but instead appears to alter polypeptide structure sufficiently so that the proline residues present are not recognized as an appropriate hydroxylation substrate.

2.1. LACA and α,α'-Dipyridyl Effects on Branching Morphogenesis

Treatment of embryonic lung and salivary gland cultures with LACA or α,α'-dipyridyl results in inhibition of morphogenesis (Spooner and Faubion, 1980). The inhibitory effects of α,α'-dipyridyl are observable within 24 hr, at concentrations of 5–15 µg/ml, for both types of rudiments. Branch points or clefts in the epithelium at the time of explantation remain, but no additional branching occurs. Growth of such rudiments is also substantially retarded rela-tive to controls. Treatment with LACA also inhibits branching (see also Al-escio, 1973), but the effect is not apparent until 48 hr, perhaps consistent with the fact that it must be incorporated into α-chains during synthesis, while α,α'-dipyridyl can affect hydroxylation of already synthesized α-chains. The minimum effective dose of LACA is 60 µg/ml for lung cultures and 40 µg/ml for salivary cultures. Branching ceases at 48 hr of culture in otherwise healthy-appearing rudiments, resulting in a dramatic difference between control and LACA-treated lungs (Fig. 2) and salivary glands (Fig. 3) at 72 hr of culture. Thus, while controls double (lung) or triple (salivary) their branches between 48 and 72 hr of culture, LACA-treated rudiments exhibit no new branches.

The dramatic effect of LACA on branching morphogenesis is not accompanied by an equivalent effect on growth, although a slight degree of growth inhibition is observed. Growth inhibition, monitored by LACA effects on gener-

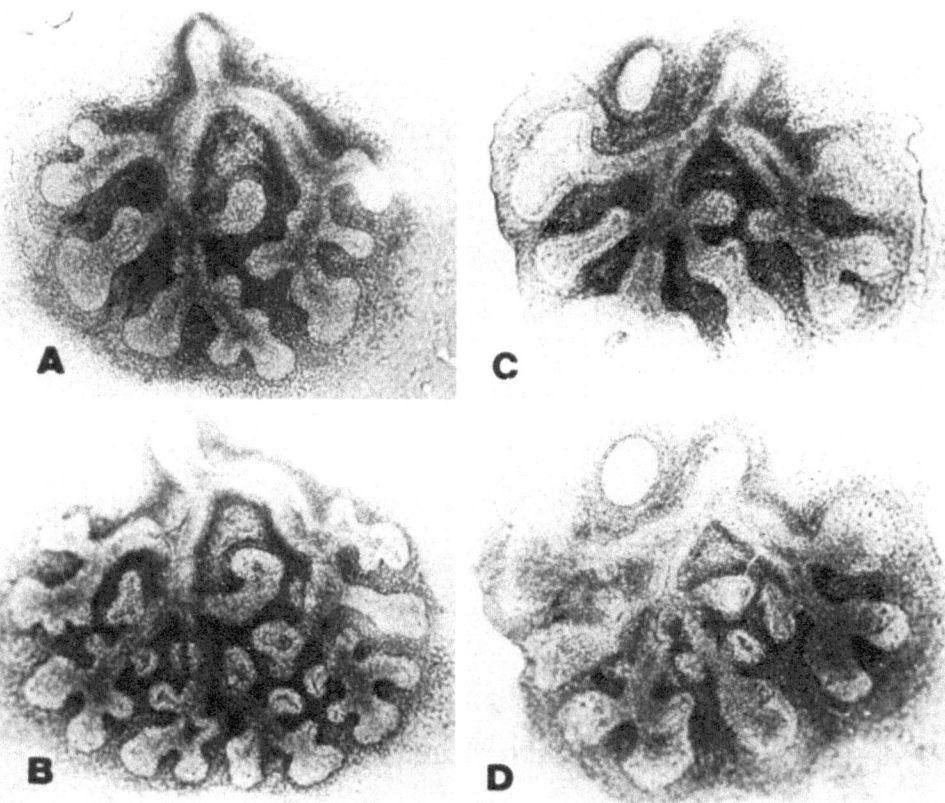

Figure 2. LACA effects on lung morphogenesis. Control and LACA-treated (100 µg/ml) living rudiments from sibling embryos are shown at 48 and 72 hr of culture. While the control lung continues branching morphogenesis between 48 hr (A) and 72 hr (B) of culture, doubling the branches present (a 100% increase in 24 hr), the LACA-treated lung exhibits no increase in branching complexity between 48 hr (C) and 72 hr (D) of culture (a 0% increase in 24 hr). Note that epithelial growth, compared with controls, is not obviously retarded, in spite of a complete cessation of epithelial branching morphogenesis. (From Spooner and Faubion, 1980.)

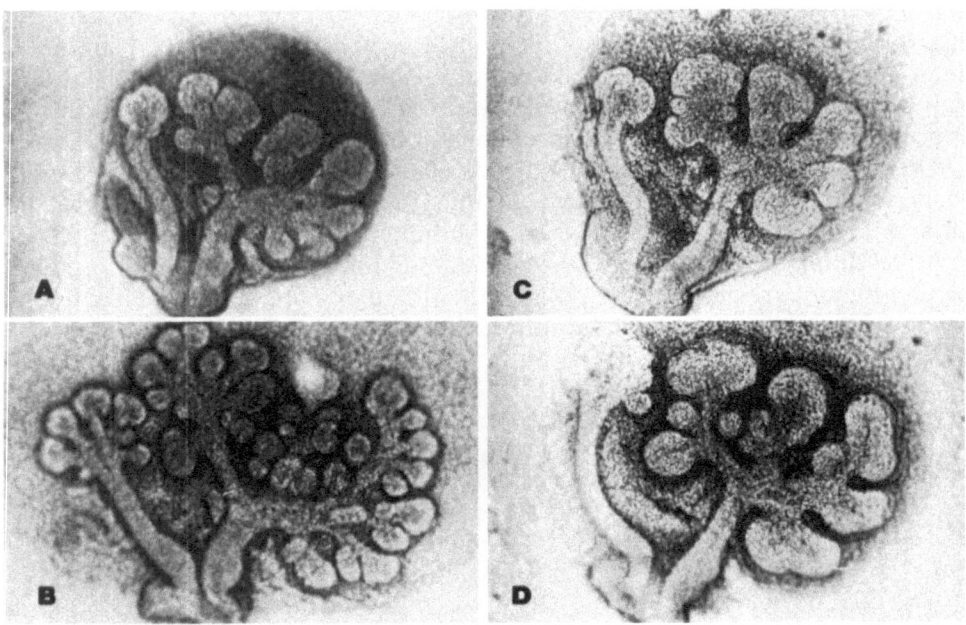

Figure 3. LACA effects on salivary morphogenesis. A living control rudiment is shown at 48 hr (A) and 72 hr (B) of culture. During the 24-hr period, the number of branches has approximately tripled (a 300% increase). A living LACA-treated (40 μg/ml) rudiment (isolated from the same embryo as the control rudiment) is also shown at 48 hr (C) and 72 hr (D) of culture. Epithelial branching morphogenesis is strikingly inhibited in the presence of LACA. Although there are no new branch points or clefts in the epithelium, existing clefts (branch points) have continued to deepen and epithelial growth has continued during this 24-hr period. (From Spooner and Faubion, 1980.)

al protein synthesis, total protein accumulation, and epithelial expansion, is maximally 20–25%, whereas branching activity is inhibited 85–100%. It appears unlikely that such growth inhibition can account for inhibition of branching morphogenesis, since experimental inhibition of general protein synthesis by some 50% with cycloheximide fails to block branching morphogenesis (see Section 3).

2.2. LACA Effects on Extracellular Collagen Presence

The experiments reviewed above show that the proline analogue LACA blocks branching morphogenesis of lungs and salivary glands. To assess whether such inhibition correlated with collagen presence or distribution in the extracellular spaces at the epithelial–mesenchymal tissue interface, electron microscopic examinations were made on control and LACA-treated rudiments. The observations are equivalent for lung and salivary gland cultures. Both epithelial and mesenchymal tissue appear healthy and devoid of obvious necrotic regions or organelle deficiencies or abnormalities in the presence of LACA. In addition, the epithelium contains typical basal arrays of cytoplasmic microfilaments and is covered by a basal lamina. The difference between control and LACA preparations is in the presence of collagen. In sharp contrast to control rudiments, LACA-treated rudiments are characteristically devoid of fibrillar collagen both in the spaces between mesenchymal cells and at the epithelial–mesenchymal interface (Spooner and Faubion, 1980). While controls exhibit typical banded collagen fibrils in close association with the epithelial basal lamina and among the interfacial mesenchymal cells, only the basal lamina and occasional wisps of fibrous material are observed in LACA-treated cultures (Fig. 4).

Thus, LACA inhibition of branching morphogenesis correlates with a dramatic decrease in the presence of collagen in the extracellular matrix. It is of interest that the transmission electron microscopic (TEM) appearance of the basal lamina is unperturbed by LACA, since collagen is a typical component of basal laminae. Whether basal lamina structural stability means that collagen is not required for a structurally intact (but possibly functionally deficient) basal lamina is not clear. It is worthy of note that efforts to demonstrate collagen chemically in the salivary basal lamina have not been successful (Banerjee et al., 1977). On the other hand, it may be that basement membrane (i.e., basal lamina) collagens are not affected by LACA. Experiments with other systems have shown that α,α'-dipyridyl does not inhibit basement membrane collagen secretion (Crouch and Bornstein, 1979), even though it blocks interstitial collagen secretion. Whether the secreted material forms a functional basal lamina and whether LACA (and other proline analogues) also fails to inhibit basement membrane collagen secretion is unknown. What is clear is that inhibition of branching morphogenesis by LACA correlates with a sharp depression of interstitial collagen presence in the extracellular matrix.

2.3. LACA Effects on Collagen Synthesis

To assess whether the effects of LACA on extracellular collagen presence and branching morphogenesis could in fact be related to effects at the level of collagen synthesis, we monitored collagen synthesis in control and LACA-treated cultures. Cultures were metabolically labeled with [³H]proline, the hot trichloroacetic acid (TCA)-soluble collagen fraction was isolated and acid hydrolyzed, and the material was analyzed for radioactive proline and hydroxyproline (Spooner and Faubion, 1980). Radioactive proline content in the collagen fraction of 72-hr LACA cultures was reduced 25–30% relative to control cultures. However, radioactive hydroxyproline was reduced 70% (lungs) to 80% (salivaries) in the same preparations (Fig. 5). On the basis of [³H]proline incorporation, LACA inhibits collagen synthesis only slightly, if at all (the 25–30% reduction could represent LACA substitution for labeled proline). However, the "collagen" that was made is not normal with respect to radioactive hydroxyproline content. Thus, LACA inhibits normal or native collagen syn-

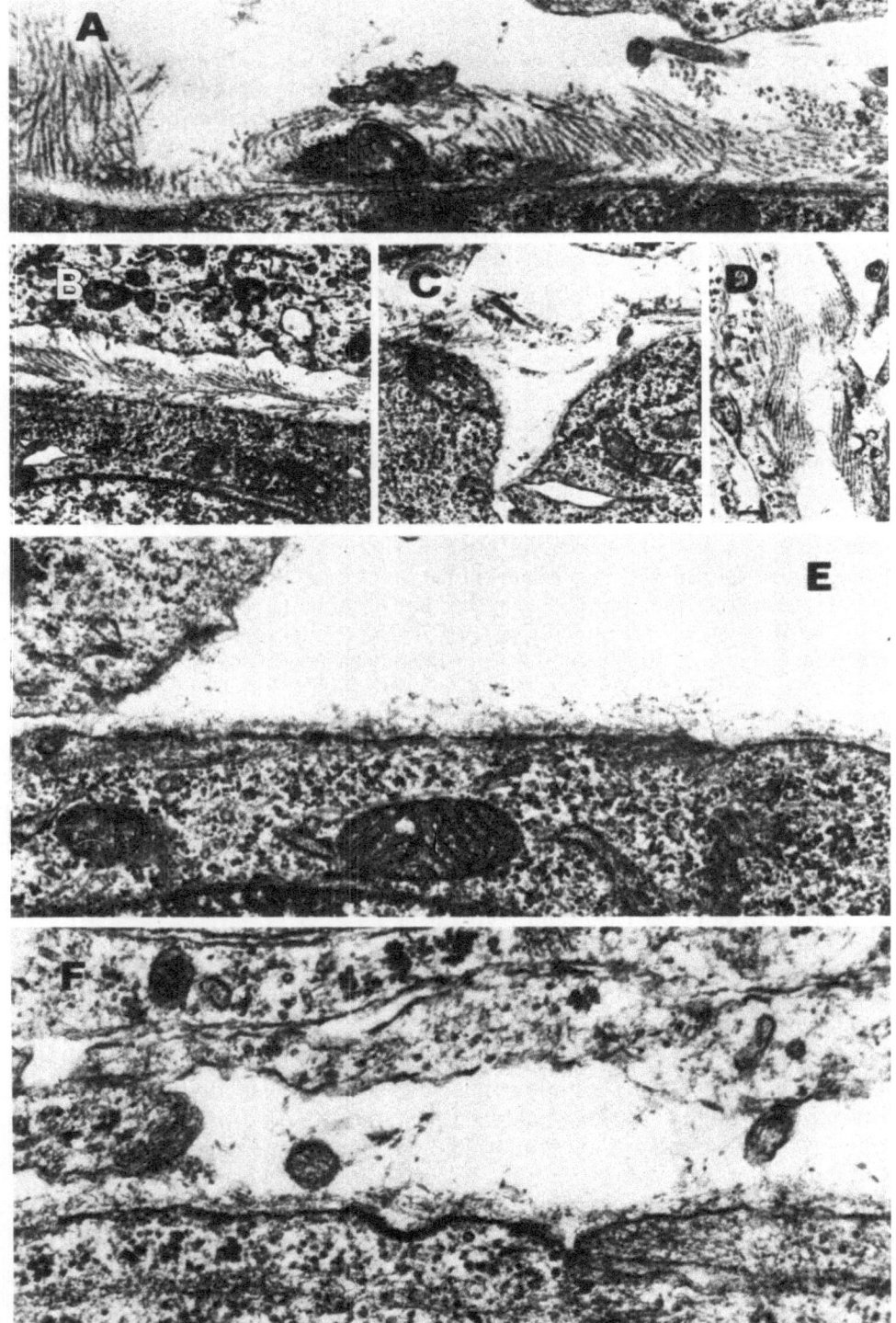

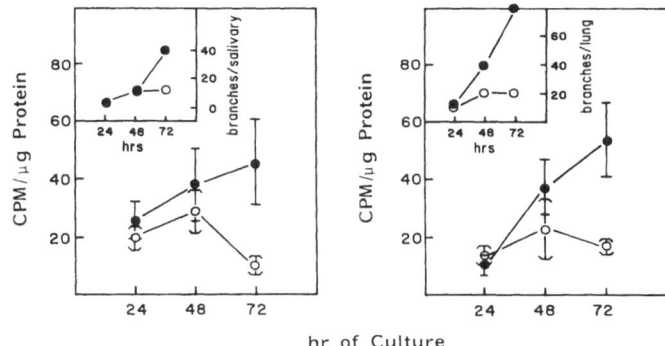

Figure 5. Collagen synthesis in control and LACA-treated lung and salivary rudiments. Native hydroxylated collagen synthesis was monitored, over time in culture, by radioactive hydroxyproline synthesis. Cultures were labeled with [³H]proline, the collagen fraction was isolated, and after acid hydrolysis of the collagen fraction radioactive proline and hydroxyproline were resolved by ion-exchange chromatography. The striking correlation between LACA inhibition of native collagen (i.e., hydroxyproline) synthesis and inhibition of branching morphogenesis is illustrated by the insets, for both lung (left) and salivary (right) cultures. (●) Control cultures; (○) LACA-treated cultures. (From Spooner and Faubion, 1980.)

thesis dramatically, by some 70–80%. The striking correlation between LACA effects on native collagen synthesis and branching morphogenesis is readily observed in Figure 5, in which the insets show the effects on branching.

2.4. Summary

The experiments reviewed above demonstrate that two different inhibitors of native collagen synthesis and secretion, α,α'-dipyridyl and the proline analogue LACA, both dramatically inhibit the branching morphogenesis of embryonic lung and salivary glands in culture. In the case of LACA, the biochemical and electron microscopic correlations suggest that synthesis of analogue-containing nonsecretable collagen results in a paucity of extracellular matrix collagen and cessation of epithelial branching morphogenesis. The interpretation is that collagen presence in the matrix is required for ongoing branching. That requirement is at present interpreted to mean interstitial (types I and/or III)

Figure 4. LACA effects on collagen presence at the epithelial–mesenchymal interface. Electron micrographs of control rudiments (A,B,C,D) and LACA-treated rudiments (E,F). Interfacial collagen fibrils are found in close association to the epithelial basal lamina (A,B), as well as between mesenchymal cells (D), in controls. Collagen fibrils do not follow the epithelial basal lamina down into small epithelial clefts (C). Collagen fibrils are not observed in LACA-treated rudiments, either in association with the basal lamina (E,F) or among the mesenchymal cells (F). The absence of collagen in the extracellular matrix is the only electron microscopically observed difference between control and treated rudiments. (A,E,F from Spooner and Faubion, 1980.)

collagens, as the experiments cannot comment on possible basement membrane collagen involvement. It is interesting that β-aminopropionitrile, an inhibitor of extracellular collagen crosslinking (Pinnell and Martin, 1968), has no effect on branching morphogenesis (Spooner and Faubion, 1980), suggesting (but not proving) that although collagen is required, it need not be crosslinked. The possibility that cytotoxicity accounts for inhibition of branching is not supported by the evidence: Ultrastructure of LACA-treated rudiments appears normal except for absence of collagen fibrils. The small effect of LACA on general protein synthesis and growth is unlikely to account for the effects of LACA on branching, as morphogenetic branching is essentially unaffected when protein synthesis is experimentally inhibited by more than 50% with cycloheximide (see Section 3). The possible role(s) of interstitial collagen in branching morphogenesis in conjunction with other extracellular matrix molecules are discussed in Section 7.

3. Glycoconjugates in Branching Morphogenesis

The extracellular matrix is endowed with a rich array of glycosylated molecules, ranging from collagens (with a few sugar residues) to classic glycoproteins (with variable numbers and lengths of sugar side chains) to proteoglycans (with large glycosaminoglycan side chains) to hyaluronic acid (a large glycosaminoglycan with little or no covalently linked protein). One compound whose biosynthetic position is central to formation of most of the other sugars present in these various classes of molecules is the aminosugar [UDP]glucosamine. Interference with cellular biosynthesis of [UDP]glucosamine therefore has major inhibitory effects on glycoconjugate synthesis in general.

A key step in biosynthesis of [UDP]glucosamine is the conversion of fructose 6-phosphate to glucosamine 6-phosphate, a reaction catalyzed by an aminotransferase enzyme, with glutamine serving as the amino donor. This reaction can be inhibited by a number of glutamine analogues, including 6-diazo-5-oxo-L-norleucine (DON) and azaserine (Ghosh et al., 1960; Bates et al., 1966; Hill and Bennett, 1969). Such analogues have been used to inhibit glycoprotein, proteoglycan, and glycosaminoglycan synthesis in a number of systems, to relate such molecules to various biological activities (Aydelotte and Kochhar, 1975; Coulombre and Coulombre, 1975; Pratt and Greene, 1976; Greene and Pratt, 1977; Ekblom et al., 1979; Hurmerinta et al., 1979; Linsenmayer and Kochhar, 1979; Hurmerinta and Thesleff, 1982; Hall et al., 1982). Because these analogues have several side effects, it is common to ask whether the observed biological effects are actually due to inhibition of glucosamine biosynthesis by determining whether such effects are relieved by exogenous glucosamine (i.e., by bypassing the enzymatic block). In addition, since an equivalent aminotransferase is involved in purine biosynthesis (Hill and Bennett, 1969), it is useful to control for analogue effects at that level. We have used both DON and azaserine to determine whether glycoconjugate synthesis is required during branching morphogenesis.

3.1. DON and Azaserine Effects on Salivary Branching

Branching morphogenesis of embryonic salivary rudiments in culture is inhibited by both DON and azaserine. Both glutamine analogues exhibit dose-dependent inhibitory effects, apparent at both 48 and 72 hr of culture (Fig. 6). Increasing concentrations of azaserine produce increasing degrees of branching inhibition, relative to control cultures, with branching activity being reduced by about 50% at a concentration of 5 μg/ml and ceasing at 10 μg/ml. The DON effect is dose dependent as well, but this analogue is more potent, with all branching activity ceasing at a concentration of 1 μg/ml. The more potent activity of DON is consistent with its greater structural similarity to glutamine. We chose an azaserine concentration of 5 μg/ml and a DON concentration of 1 μg/ml for additional experiments. At these concentrations, branching activity is dramatically inhibited (Figs. 6 and 7), while the rudiments appear healthy and only slightly growth inhibited (Fig. 7).

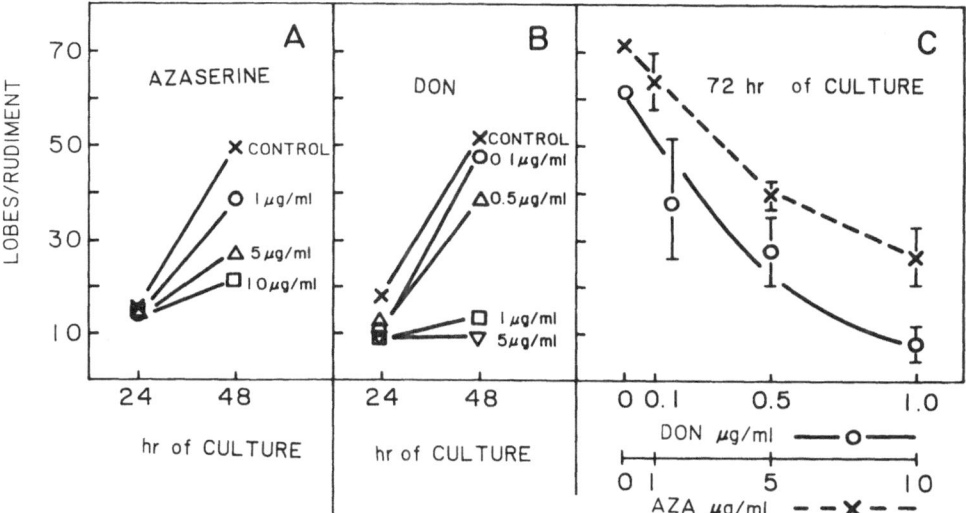

Figure 6. Dose-dependent inhibition of salivary branching by the glutamine analogues DON and azaserine. The effects of increasing analogue concentrations on branching morphogenesis over the first 48 hr of culture are shown for azaserine (A) and for DON (B), and the effects of both analogues at 72 hr of culture are shown in (C). Branching activity is progressively inhibited, at both 48 and 72 hr, at progressively higher analogue concentrations, with DON exhibiting approximately five to ten times more potent effect.

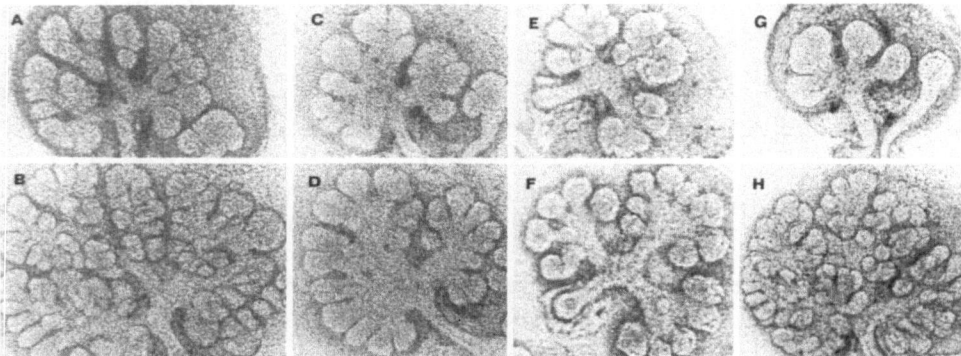

Figure 7. Branching morphogenesis in azaserine, DON, and cycloheximide-treated salivary rudiments. The control increases its branches by 300% during the 24-hr (A) to 48-hr (B) period. By contrast, the azaserine-treated (5 μg/ml) rudiment exhibits only a 65% increase during the 24-hr (C) to 48-hr (D) period, and the DON-treated (1 μg/ml) rudiment completely fails to continue branching during the same 24-hr (E) to 48-hr (F) period. Continuous culture in the presence of cycloheximide (0.01 μg/ml), however, results in at least a 300% increase in branches during the 24-hr (G) to 48-hr (H) period, equivalent to control branching activity, even though protein synthesis is inhibited by 50% in such cultures (see Fig. 9) and growth is reduced to the same degree as that observed with DON and azaserine. (Compare D, F, H with the control B) at 48 hr.

3.2. DON and Azaserine Effects on GAG Synthesis

A primary purpose in initiating the DON and azaserine experiments was to determine whether experimental inhibition of glycosaminoglycan (GAG) biosynthesis would result in inhibition of morphogenesis. Such a perturbation-based correlation could serve to complement existing data showing that the basal surface of the salivary epithelium was rich in GAG (Bernfield and Banerjee, 1972) and that removal of this material would interrupt morphogenesis (Bernfield *et al.*, 1972). We therefore assessed total sulfated GAG synthesis in control, azaserine (5 μg/ml), and DON (1 μg/ml) cultures.

Cultures were metabolically labeled with [^{35}S]sulfate for 10 hr (after 48 hr of culture in nonradioactive medium), and the medium and tissue were harvested, pooled, and digested with pronase (see Thompson and Spooner, 1982). Digests were chromatographed on Sephadex G-50 columns, and the radioactivity in the excluded fractions was determined. This technique separates GAG from free sulfate and sulfated glycopeptides (which are included in the column), since the GAG chains are sufficiently large to be excluded on G-50, even though their core proteins have been digested.

The results of this analysis and the correlation with branching morphogenesis are shown in Figure 8. The stepwise decrease in branching activity from control to azaserine to DON cultures is positively correlated with analogue depressions of GAG synthesis. It should be pointed out that the data do not prove that glutamine analogue inhibition of sulfated GAG (i.e., proteoglycan) synthesis is the cause of branching inhibition, although they are consistent with that possibility. More definitive data on proteoglycan involvement in branching morphogenesis are considered in Section 4. DON and azaserine inhibition of branching might instead reflect a nonsulfated GAG (i.e., hyaluronic acid) or glycoprotein requirement, effects on purine synthesis, or possibly a nonspecific effect on general protein synthesis.

3.3. Effects of Protein Synthesis on Salivary Branching

It was clear from observation of control and glutamine analogue-treated cultures that these compounds were causing a slight inhibition of growth, in addition to a major inhibition of branching (see Fig. 7). We therefore analyzed azaserine-treated cultures further to determine whether general protein synthesis was perturbed. On the basis of [³H]leucine incorporation into TCA-precipitable material, protein synthesis by 48-hr salivary rudiments was inhibited 45–50% by 5 μg/ml azaserine, relative to control rudiments. To determine whether such inhibition of protein synthesis was, in itself, sufficient to account for inhibition of branching morphogenesis, we employed the protein synthesis inhibitor cycloheximide. Dose-response studies demonstrated that a 0.01 μg/ml concentration of cycloheximide inhibited protein synthesis in 48-hr salivary cultures by 50–55%, that is, a degree of inhibition slightly greater than that produced by 5 μg/ml azaserine.

We therefore compared salivary rudiments cultured for 48 hr in azaserine (5 μg/ml) with ones cultured for 48 hr in cycloheximide (0.01 μg/ml) for effects

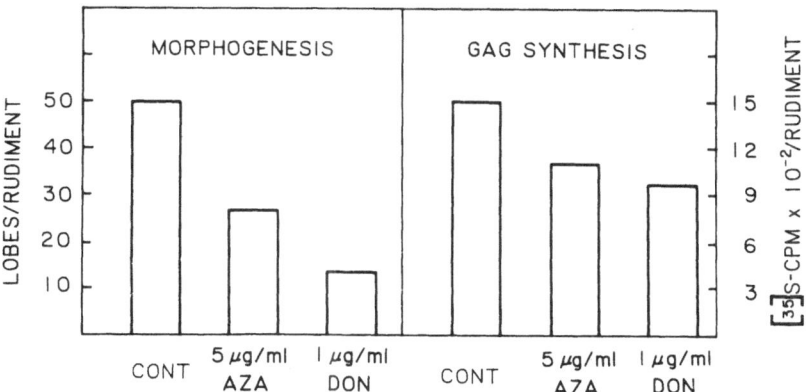

Figure 8. Correlation between DON and azaserine effects on salivary morphogenesis and glycosaminoglycan biosynthesis. Comparison of branching activity during 48 hr of culture by control, azaserine (5 μg/ml), and DON (1 μg/ml)-treated rudiments, showing that inhibition by both glutamine analogues (with DON exhibiting the more potent effect) correlates with total sulfated GAG biosynthesis in identically treated rudiments. As with branching, both glutamine analogues inhibit GAG synthesis, with DON exhibiting the stronger inhibitory effect.

on protein synthesis, i.e., [³H]leucine incorporation into TCA-precipitable material, total protein content (determined by micro-Lowry), and branching morphogenesis. The results are shown in Figure 9. Azaserine inhibits protein synthesis by 50%, total protein by 20%, and branching morphogenesis by 50%. By contrast, cycloheximide has slightly greater inhibitory effects on protein synthesis (55% inhibition) and total protein (25% inhibition) but has no significant effect on branching morphogenesis (branching activity is about 90% of control values). The remarkable degree of branching activity that occurs in these cycloheximide-treated cultures is represented in Figure 7. These experiments clearly demonstrate that general inhibition of protein synthesis, a side effect of these glutamine analogues and, less dramatically, of LACA (Section 2) cannot account for their striking inhibition of branching morphogenesis.

3.4. DON and Azaserine Effects on Lung Branching

DON and azaserine also inhibit branching morphogenesis of cultured embryonic lung rudiments. The inhibitory effects of these analogues is slightly delayed in the lung system, with pronounced inhibition apparent in 72-hr cultures. Inhibition of branching is dose dependent (Fig. 10) and, as in the salivary system, DON is a more potent inhibitor than is azaserine. Again, since DON is structurally more closely related to glutamine than azaserine, it should be a more potent inhibitor. Branching activity is inhibited by 85% at a DON concentration of 1 µg/ml and by 50% at a concentration of 10 µg/ml azaserine. Although these results are consistent with those obtained in the salivary system, we have not conducted additional glutamine analogue experiments with

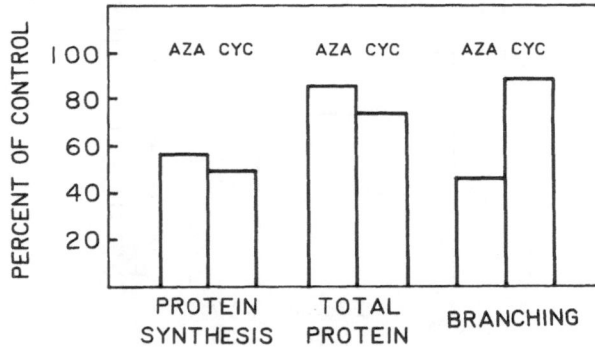

Figure 9. Comparative effects of continuous treatment with azaserine (5 µg/ml) and cycloheximide (0.01 µg/ml) on protein synthesis, total protein content, and branching morphogenesis in 48-hr cultures, relative to control culture values. Although azaserine and ⋅ cycloheximide have equivalent effects on protein synthesis (about 50% inhibition) and total protein content (about 20% inhibition), with cycloheximide producing slightly greater inhibition of each parameter, they are sharply distinguishable in their effects on branching. Azaserine inhibits branching by about 50%, while cycloheximide-treated rudiments exhibit branching activity about 90% of control values (see also Figure 7). Since cycloheximide-treated rudiments actively branch, even though protein synthesis is depressed by 55%, the general protein synthesis inhibition activity of azaserine (about 50%) cannot be the reason that azaserine inhibits salivary branching morphogenesis.

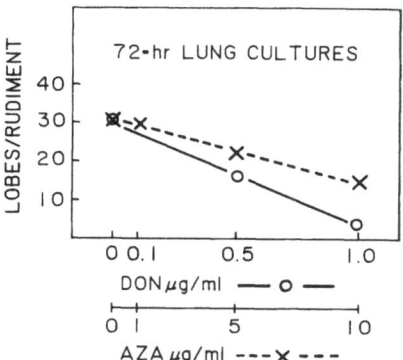

Figure 10. DON and azaserine inhibit branching morphogenesis of lung rudiments in a dose-dependent manner. The data show that increasing concentrations of the glutamine analogues progressively inhibit the number of branches formed over a 72-hr period in culture, with DON being a more potent inhibitor than azaserine.

embryonic lungs; consequently, we do not know whether comparable GAG synthesis correlations exist.

3.5. Cycloheximide Effects on Lung Branching and Protein Synthesis

The remarkable observation, reported above, that branching morphogenesis of cultured salivary rudiments continues unabated when protein synthesis is inhibited by more than 50% prompted us to ask whether lung branching activity is similarly insensitive to such protein synthesis perturbation. Lung rudiments were cultured for 72 hr in the continuous presence of various concentrations of cycloheximide and then monitored for protein synthesis (as above) and branching morphogenesis. The results are shown in Figure 11. Increasing concentrations of cycloheximide produce increasing degrees of protein synthesis inhibition but have no effect on branching morphogenesis. Thus, the embryonic lung, like the salivary rudiment, can suffer inhibition of general

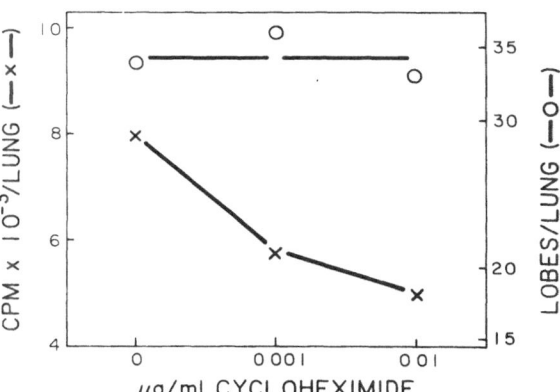

Figure 11. Cycloheximide inhibits protein synthesis, but not branching morphogenesis, of cultured lung rudiments. The data show that, at 72 hr of culture, branching activity is unchanged by increasing concentrations of cycloheximide, while protein synthesis is progressively inhibited.

protein synthesis by at least 40% without effect on its ability to continue branching morphogenesis.

3.6. Summary

The glutamine analogues DON and azaserine both exhibit a dose-dependent ability to inhibit branching morphogenesis of cultured lungs and salivary glands. At doses that inhibit branching activity dramatically, total sulfated GAG synthesis by salivary rudiments is depressed. Although the result is consistent with the idea that such synthesis is required for branching morphogenesis, the data do not distinguish between sulfated GAG synthesis and possible requisite synthesis of hyaluronic acid and other glycoconjugates. Thus, the inhibition of branching morphogenesis by these compounds is most cautiously interpreted as implying a necessary degree of glycoconjugate biosynthesis. Even this interpretation, however, requires demonstration that DON and azaserine inhibition of branching does not result from side effects that are independent of glycoconjugate inhibition (e.g., inhibition of purine biosynthesis).

Attempts to demonstrate specificity by showing that exogenous glucosamine bypasses the inhibitory block and permits branching to continue have not succeeded. Part of the problem stems from the fact that glucosamine is itself toxic at high concentrations (Kim and Conrad, 1974). Thus, glucosamine is observed to provide only partial relief of glutamine analogue inhibition in several systems (e.g., Spooner and Conrad, 1975). However, we were also unable to demonstrate that inhibition of branching resulted from analogue inhibition of purine synthesis, by adding exogenous aminoimidazole carboxamide to bypass the enzymatic block. Again, part of the problem is that such substances may themselves be toxic (Ishii and Green, 1973). A second and far more successful approach to eliminating a drug side effect as the cause of branching inhibition stems from our observation that, in addition to inhibiting branching and GAG synthesis, these glutamine analogues are potent inhibitors of general protein synthesis. Thus, azaserine (5 μg/ml) inhibits protein synthesis by 50%. We therefore asked whether inhibition of protein synthesis was itself sufficient to inhibit branching morphogenesis. The remarkable observation was that cycloheximide concentrations that inhibit protein synthesis to the same (actually somewhat greater) degree as azaserine failed to inhibit branching. Therefore, inhibition of protein synthesis is not the reason DON and azaserine inhibit branching of lungs and salivary glands. That inhibition may be due to inhibition of glycoconjugate biosynthesis.

The cycloheximide experiments, in addition to serving as controls for the glutamine analogs, provide important information on growth requirements in branching morphogenesis. The data demonstrate that branching morphogenesis is unperturbed when protein synthesis is inhibited in excess of 50% and total protein content is reduced by 25% by culture in the continuous presence of cycloheximide. Thus, control levels of protein synthesis and growth can be significantly impaired without effect on epithelial branching activity.

4. Proteoglycans in Branching Morphogenesis

Proteoglycans constitute a major macromolecular component of extracellular matrices, whose presence, distribution, and type may be crucial to branching morphogenesis. Proteoglycans are large molecules composed of a core protein and covalently attached glycosaminoglycan chains. The molecules may have a few or many GAG chains per core protein, and a single proteoglycan may contain more than one type of GAG chain. Proteoglycans are present in the interstitial matrix, in association with collagen, hyaluronic acid, and fibronectin, and are found as components of the epithelial basal lamina. In addition, they are found as components of cell surfaces, possibly as integral plasma membrane molecules.

The demonstration that GAG and proteoglycan are major components of the extracellular matrix associated with the basal epithelial surface of the embryonic salivary rudiment and with that of other embryonic organ rudiments was initially made by Bernfield and collaborators (Bernfield and Banerjee, 1972; Bernfield et al., 1972, 1973). It should be noted that GAG presence, other than hyaluronic acid, is evidence of proteoglycan presence, since the sulfated GAGs normally exist only as proteoglycans. The Bernfield laboratory has further shown that (1) proteoglycans are major components of the epithelially derived basal lamina of the embryonic salivary (Cohn et al., 1977; Banerjee et al., 1977), (2) removal of the GAG-rich basal lamina interrupts morphogenesis (Bernfield et al., 1972; Banerjee et al., 1977), and (3) mesenchyme-mediated turnover of lamina GAG is a normal morphogenetic event (Bernfield and Banerjee, 1982; Smith and Bernfield, 1982).

We have sought to determine requisite GAG and proteoglycan synthesis in salivary branching by interference with normal synthesis. As described in Section 3, inhibition of GAG synthesis and branching by glutamine analogues are suggestive, but do not distinguish proteoglycans from other glycoconjugates. A more specific approach has been possible by the use of β-xylosides, which stimulate synthesis of GAG chains not attached to protein (free GAG) and can inhibit proteoglycan synthesis. During normal proteoglycan biosynthesis, synthesis of core protein is followed by xylosylation of certain serine residues in the core protein. The xylosylated core protein then serves as substrate for a galactosyl transferase. A second galactosyl transferase reaction completes the linkage region; monosaccharides are then sequentially added to produce the appropriate GAG chain, which is then sulfated. The Ser-Xyl-Gal-Gal linkage region is identical for chondroitin 4-sulfate, chondroitin 6-sulfate, dermatan sulfate, and heparan sulfate, even though these GAGs differ in their characteristic disaccharides, position or mode of sulfation, and may exist on different core proteins.

Certain exogenously supplied xylose derivatives, β-D-xylosides, will compete with xylosylated core protein for the first galactosyl transferase. Following addition of a galactose to the xylose derivative, the second galactose is added; monosaccharides are then sequentially added, sulfation takes place, and core protein-free GAG chains are produced. Large amounts of free GAG can be synthesized in this manner (Schwartz *et al.*, 1974; Fukunaga *et al.*, 1975; Galligani *et al.*, 1975) and, depending on the xyloside concentration, normal proteoglycan synthesis may be inhibited (Schwartz, 1977; Lohmander *et al.*, 1979). The galactosyl transferase is specific for the β-anomer of the xyloside substrate (Schwartz *et al.*, 1974; Galligani *et al.*, 1975), so the α-xylosides provide an excellent control for nonspecific biological effects. We have used β-xylosides, specifically p-nitrophenyl-β-D-xylopyranoside, as a tool to investigate GAG and proteoglycan synthetic requirements in branching morphogenesis of the embryonic salivary rudiment (Thompson and Spooner, 1982, 1983).

4.1. β-Xyloside Effects on Salivary Branching Morphogenesis

β-Xylosides inhibit branching morphogenesis in a dose-dependent manner (see Fig. 14), with branching activity virtually ceasing at 48 hr of culture at a 0.5 mM concentration. However, even at a 1.0 mM β-xyloside concentration, the inhibitory effect is reversible. Thus, removal of the β-xyloside-containing medium at 48 hr of culture, and replacement with normal medium, results in resumption of branching morphogenesis within an additional 24 hr of culture (Fig. 12). The inhibitory effect is specific to β-xyloside, since branching morphogenesis is completely normal in the presence of 1.0 mM α-xyloside (Thompson and Spooner, 1982). Consistent with the presumed specificity of the β-xyloside effect, we find little effect on general protein synthesis or total protein content in the treated rudiments (the values are 85–90% of control values), and transmission electron microscopy (TEM) shows a normal appearing epithelium and mesenchyme, with typical interstitial collagen fibrils and an intact basal lamina.

4.2. β-Xyloside Effects on GAG and Proteoglycan Synthesis

When β-xyloside-treated cultures are metabolically labeled with [^{35}S]sulfate and analyzed for total GAG synthesis, a threefold stimulation is observed. This material includes both proteoglycan-associated GAG and free GAG, since the assay monitors G-50 excludable radioactivity of pronase digested samples. The material is most certainly GAG, since it is degradable by testicular hyaluronidase (which acts on chondroitin 4-sulfate, chondroitin 6-sulfate, and, variably, on dermatan sulfate) and nitrous acid (which degrades heparan sulfate). The β-xyloside effect is not simply an increased level of sulfation, since a similar stimulation is observed even when [^{3}H]glucosamine is substituted for [^{35}S]sulfate as the metabolic label. Under such conditions, the

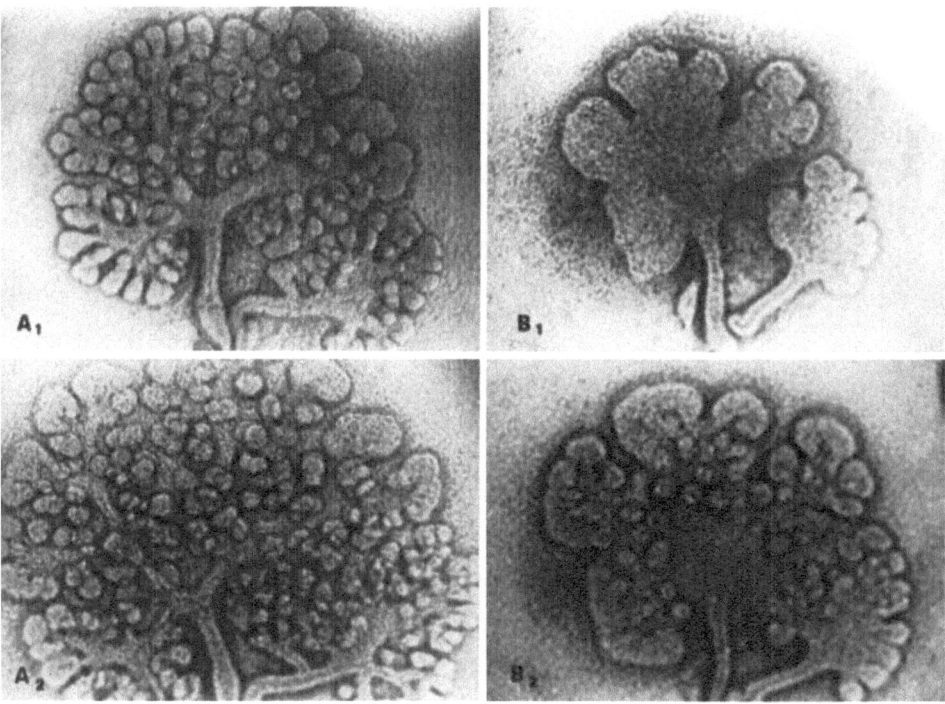

Figure 12. β-Xyloside effects on branching morphogenesis of cultured salivary rudiments. A control rudiment is shown at 48 hr of culture (A_1) and the same rudiment is shown at 72 hr of culture (A_2). A β-xyloside-treated (1mM) rudiment is shown (B_1), with severely inhibited branching activity (compare with A_1). The same β-xyloside-treated rudiment is shown in B_2, 24 hr after removal of the β-xyloside medium and reincubation in control medium. Recovery of branching activity is evident (compare B_1 and B_2). Thus, β-xyloside drastically inhibits branching morphogenesis (see also Figure 14), and the effect is reversible. (From Thompson and Spooner, 1982.)

GAG stimulation is about twofold, rather than threefold, a result accounted for by label incorporation into hyaluronic acid, a nonsulfated GAG, which does not share the xylose linkage region and the synthesis of which is not stimulated by β-xylosides. Two indirect observations suggested that the increased GAG synthesized was free GAG (i.e., initiated on xyloside rather than on xylosylated core protein). First, when medium and tissue were analyzed independently, the bulk of the excess GAG (i.e. 3X control amounts) was found in the medium, consistent with reports that core protein-free GAG diffuses from tissue into the medium (Lohmander *et al.*, 1979). Second, the huge stimulation of sulfated GAG synthesis was still observed when protein synthesis was inhibited by more than 95% with cycloheximide. Under such conditions, core protein synthesis should be reduced along with other proteins, meaning that the β-xyloside stimulation of GAG synthesis was as core-free GAG. In fact, total GAG synthesis in these cycloheximide concentrations, in the absence of β-xyloside, was inhibited by 75%, consistent with the conclusion that free GAG is synthesized in the presence of β-xyloside.

To better ascertain the effects of β-xylosides on GAG and proteoglycan synthesis, we labeled 48-hr control and 0.5 mM β-xyloside-treated salivary cultures with [³⁵S]sulfate, extracted total GAG and proteoglycan, and resolved the material by column chromatography on Sepharose CL-4B, under dis-

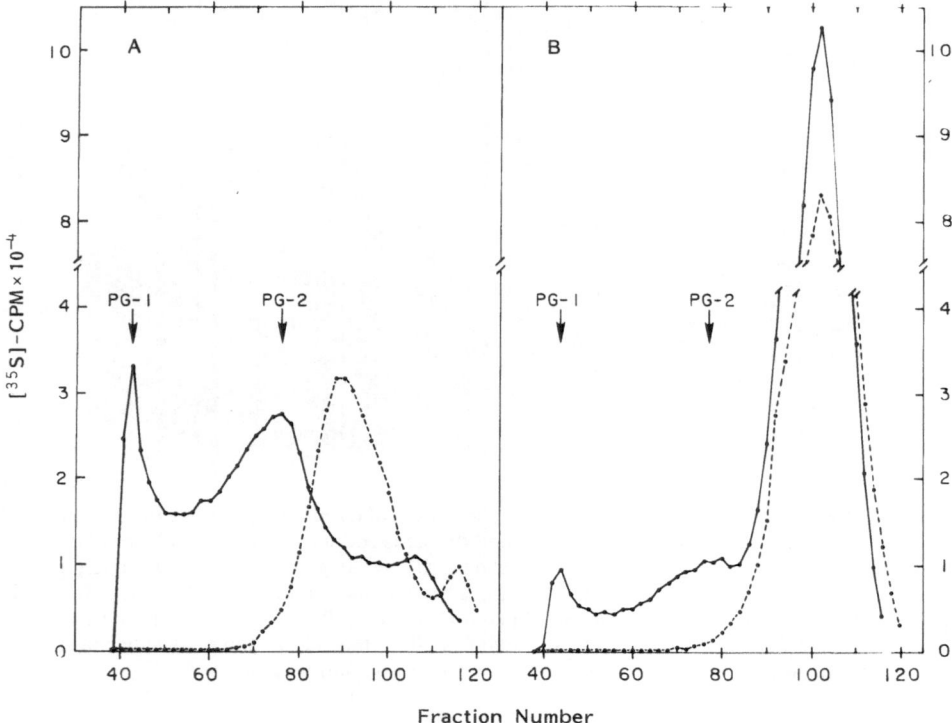

Figure 13. β-Xyloside effects on proteoglycan and GAG synthesis by salivary gland rudiments in culture. Salivary rudiments were cultured for 48 hr in control or β-xyloside (0.5 mM) medium, and then labeled with [³⁵S]sulfate. Medium and tissue were extracted and chromatographed on Sepharose CL-4B columns described in Thompson and Spooner, 1983. Radioactive material from controls (A) migrated predominantly as two size classes of proteoglycan (PG-1 and PG-2), which shifted to a lower molecular weight size class following digestion with pronase (broken line). GAG composition of PG-1 was predominantly chondroitin/dermatan sulfate (90%), while PG-2 was 75% chondroitin/dermatan sulfate and 25% heparan sulfate, on the basis of chondroitin ABC lyase and nitrous acid degradation analyses. Peak 3 of controls was small, insensitive to pronase (data not shown), and about 85% heparan sulfate. β-Xyloside-treated cultures (B) contained both PG-1 and PG-2 proteoglycan size classes and a huge peak 3. The proteoglycan peaks both shifted to the peak 3-position following pronase digestion (broken line), and had the same GAG compositions as control PG-1 and PG-2. However, β-xyloside PG-1 plus PG-2 was only 50% of control PG-1 plus PG-2 radioactivity. Peak 3 radioactivity was insensitive to pronase, was about 98% chondroitin/dermatan sulfates, and accounts for the three-fold stimulation of total GAG by β-xyloside. Thus, β-xylosides (0.5 mM) inhibit proteoglycan synthesis by 50% and stimulate free GAG synthesis in cultured salivary rudiments.

sociative conditions in the presence of guanidine HCl (Thompson and Spooner, 1983). Control cultures synthesize two major size classes of sulfated material (Fig. 13). Both peaks shift to a common lower molecular size peak after digestion by pronase (Fig. 13), and the radioactivity in both peaks is susceptible to chondroitin ABC lyase and nitrous acid treatment. Therefore, the two peaks appear to represent two different-size classes of proteoglycan that we have designated PG-1 and PG-2. Controls also contain a third very small retarded peak, the position of which closely coincides with that of pronase-digested PG-1 and PG-2, that represents free GAG. Although PG-1 contains 90% chondroitin sulfate GAG and PG-2 contains about 25% heparan sulfate and about 75% chondroitin sulfate, peak 3 is almost exclusively heparan sulfate, and its position is not altered by pronase digestion.

β-Xyloside-treated cultures contain both PG-1 and PG-2 peaks, as well as an enormous peak 3 (Fig. 13). Pronase treatment shifts PG-1 and PG-2 to the position of peak 3 but does not affect peak 3. Thus, peak 3 represents the large amount of free GAG synthesized in the presence of β-xylosides. Degradation analyses show that peak 3 is almost exclusively (98%) chondroitin sulfate. Thus, in the salivary system, β-xylosides stimulate free chondroitin (and/or dermatan) sulfate synthesis but do not stimulate free heparan sulfate synthesis, even though the linkage region is identical. In addition to free GAG synthesis, β-xylosides inhibit proteoglycan synthesis. Thus, PG-1 plus PG-2 amounts to about 50% of the PG-1 plus PG-2 total synthesized by control cultures. The effect appears to be quantitative rather than qualitative, since the GAG composition of PG-1 and PG-2 is unaltered by β-xylosides, and their relative sizes, as judged by chromatography on Sepharose CL-4B, are the same as those of control PG-1 and PG-2. The data demonstrate the β-xyloside inhibition of branching morphogenesis could result from production and accumulation of large amounts of free GAG or from inhibition of synthesis of proteoglycan.

4.3. Inhibition of Branching Correlates with Inhibition of Proteoglycan Synthesis

The data reviewed in Section 4.2 show that perturbation of normal proteoglycan synthesis reversibly inhibits branching morphogenesis but do not distinguish between inhibition of proteoglycan synthesis and the stimulation of free GAG synthesis as the basis for that effect. We have conducted two kinds of experiments, designed to achieve such a distinction. In the first case, we have asked whether the presence of large amounts of free GAG that accumulate in the medium of β-xyloside-treated cultures is sufficient to inhibit branching. The peak 3 material from β-xyloside cultures (i.e., free chondroitin sulfate) was isolated by Sepharose CL-4B chromatography and added to control cultures at two and five times the concentration found in β-xyloside cultures. Branching morphogenesis was completely unperturbed by these free GAG concentrations, which exceeded those in β-xyloside cultures, where branching is blocked. We also tested the effects of purified commercial chondroitin sulfate on similar cultures and, again, morphogenesis continued unabated. These data do not support the idea that inhibition of branching morphogenesis in the presence of β-xylosides results from the large amount of free GAG that is produced.

A second, and more definitive, approach to distinguish the basis for inhibition of branching has been a detailed comparison of branching versus proteoglycan and free GAG synthesis, at a series of β-xyloside concentrations. Comparisons were made at 0.0, 0.1, 0.2, 0.3, and 0.4 mM β-xyloside concentrations. Salivary morphogenesis is progressively inhibited at increasing β-xyloside concentrations (Fig. 14). Effects on morphogenesis of both submandibular and sublingual gland rudiments are shown, since both rudiments are included in the biochemical determinations, and both exhibit dose-dependent β-xyloside inhibition of branching. Proteoglycan synthesis is also progressively inhibited at increasing β-xyloside concentrations (Fig. 14). However, free GAG synthesis is already maximal at the lowest β-xyloside concentrations (Thompson and Spooner, 1983). Thus, there is a direct correlation between β-xyloside concentration, inhibition of branching morphogenesis, and inhibition of proteoglycan synthesis, but there is no correlation with free GAG synthesis. The

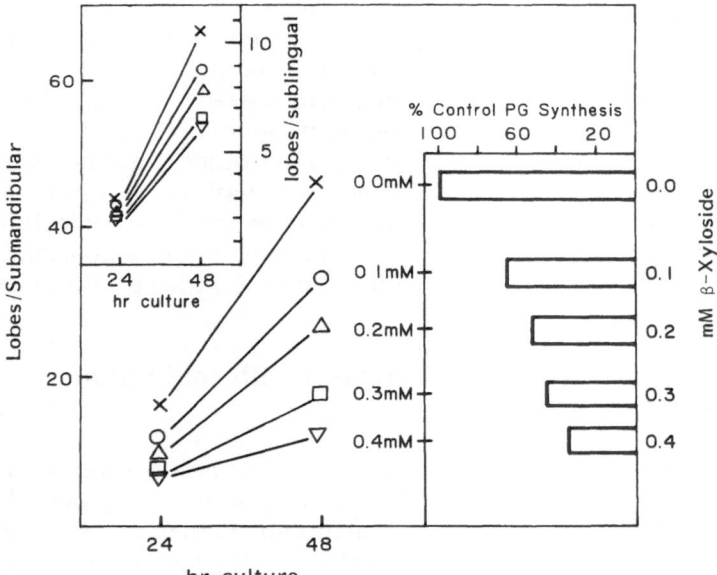

Figure 14. β-Xyloside inhibition of salivary branching morphogenesis correlates with inhibition of proteoglycan biosynthesis. Submandibular gland branching activity is progressively inhibited by increasing concentrations of β-xyloside, over the range of 0, 0.1, 0.2, 0.3, and 0.4 mM. Sublingual rudiment branching also exhibits dose-dependent inhibition over the same concentration range (upper left inset). A parallel dose-dependent inhibition of proteoglycan biosynthesis (right inset) shows a direct correlation between β-xyloside effects on branching morphogenesis and on proteoglycan synthesis. In contrast, core protein-free GAG synthesis does not correlate with inhibition of branching at these same xyloside concentrations (Thompson and Spooner, 1983). (Data, in part, reexpressed from Thompson and Spooner, 1983.)

interpretation of these data is that β-xyloside inhibition of branching morphogenesis results from inhibition of proteoglycan synthesis.

4.4. β-Xyloside Effects on Proteoglycan Deposition and Turnover at the Basal Epithelial Surface

The β-xyloside experiments reviewed in Section 4.3 demonstrate, for the first time, that ongoing proteoglycan synthesis is required for ongoing branching morphogenesis of salivary epithelium. However, the biochemical determinations were made on intact rudiments and do not comment on whether the requisite proteoglycan synthesis is related to the basal lamina (epithelially derived) proteoglycan described by Bernfield (1981) or whether that specific proteoglycan compartment is unaltered and our results derive from inhibition of mesenchymally produced interstitial proteoglycan. If β-xyloside inhibition of proteoglycan synthesis is demonstrable at the level of the basal lamina, a crucial link to data suggesting that basal lamina proteoglycan deposition and turnover are necessary for branching morphogenesis (Bernfield and Banerjee, 1982) would be established. We have therefore investigated that possibility.

Embryonic salivary rudiments were cultured in control or 0.5 mM β-xyloside medium for 48 hr and then pulse-labeled for 2 hr with [^{35}S]sulfate. After labeling, the medium was removed; the rudiments were rinsed with nonradioactive medium, fixed, embedded, sectioned, and processed for light microscopic autoradiography. Controls exhibit intense radioactivity at the basal surface of the epithelium, which sharply defines the epithelial contour (Fig. 15). By contrast, much less radioactivity is present at the basal epithelial surface of rudiments treated with β-xyloside (Fig. 15), even though the medium from such cultures contains four times more total GAG than does control medium. The treated rudiments do possess sufficient label to define the surface of the epithelium, but the reduced intensity is in sharp contrast to control labeling. The radioactivity in both control and β-xyloside-treated preparations is sensitive to testicular hyaluronidase and chondroitin ABC lyase treatment of sections before autoradiography (Fig. 16), demonstrating that the autoradiograms are resolving sulfated GAG and that the predominant material labeled at the level of the basal lamina is chondroitin sulfate proteoglycan. Thus, inhibition of proteoglycan biosynthesis by treatment with β-xyloside is reflected in reduced amounts of newly synthesized proteoglycan at the basal surface of the epithelium, the level of the basal lamina.

It is possible that the reduced amounts of proteoglycan in the basal lamina are the immediate reason that the epithelium fails to branch. Thus, the immediate epithelial environment is wrong and inconsistent with branching morphogenesis. It is also possible that the defect responsible for failure to branch is at the level of subsequent processing of basal lamina proteoglycan. Bernfield has shown that basal lamina proteoglycan undergoes a distinct pattern of mesenchyme-dependent degradation and has proposed that the turnover is essential to epithelial growth, expansion, and repetitive branching (Bernfield and

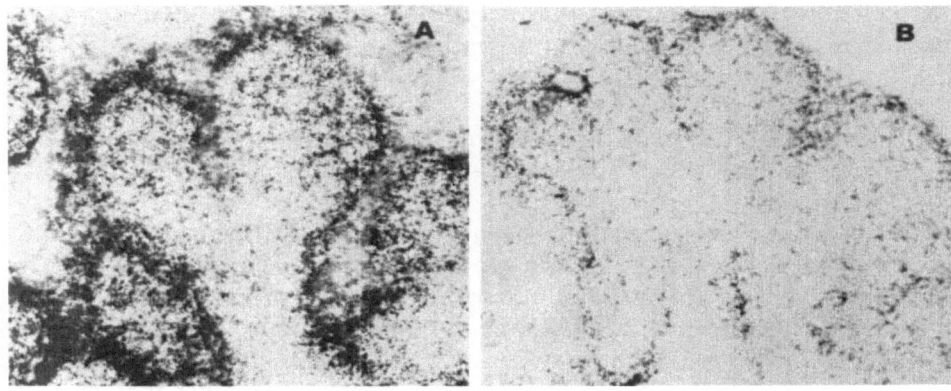

Figure 15. β-Xyloside-treated rudiments localize reduced amounts of radiolabeled proteoglycan to the basal lamina. Salivary rudiments were cultured 50 hr in control or β-xyloside (0.5 mM) medium, and metabolically labeled with [^{35}S]sulfate for the final 2 hr of culture. After rinsing, fixation, embedding, and sectioning, sections were coated with emulsion, exposed for 5 days, and autoradiograms developed. Light microscopic autoradiography reveals intense radioactivity at the epithelial basal surface of controls (A) that defines the epithelial contour, while β-xyloside-treated rudiments exhibit dramatically less radioactivity (B), consistent with the β-xyloside inhibition of proteoglycan biosynthesis. ((B) From Spooner *et al.*, 1985.)

Banerjee, 1982; Smith and Bernfield, 1982). Thus, it could be that reduced amounts of basal lamina proteoglycan either fail to activate the mesenchymal degradative program or that the degradative activity is random and unpatterned, and that this is the reason that epithelial branching stops. To assess this possibility, control and β-xyloside rudiments were cultured and labeled as

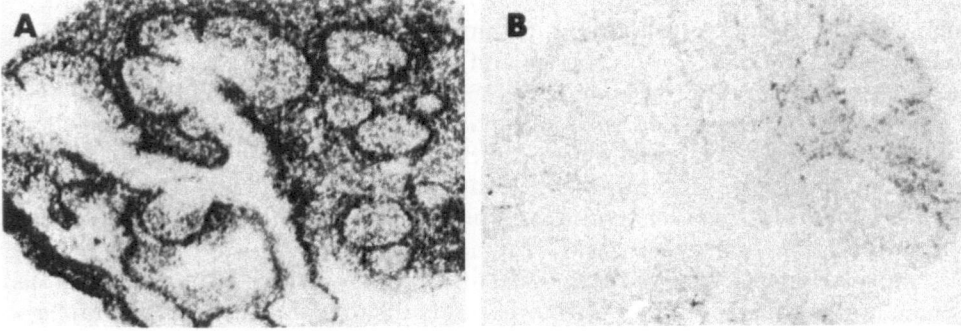

Figure 16. Autoradiogram of control salivary culture, incubated, labeled, and processed as in Figure 15, except that sections were incubated overnight in buffer alone (A) or chondroitin ABC lyase (B) before autoradiography. Enzyme treatment eliminates the bulk of the radioactivity (compare A and B). Equivalent results are obtained with testicular hyaluronidase. Similar results are observed with β-xyloside preparations. Thus, the bulk of the localized radioactivity in control and xyloside cultures is chondroitin sulfate. (From Spooner *et al.*, 1985.)

above and then fixed and processed for autoradiography either immediately or after 2, 4, 6, or 8 hr of postlabeling chase in nonradioactive medium. The autoradiograms revealed no differences in the time course or pattern of turnover of the basal epithelial surface-associated radioactivity (Fig. 17). In both cases, label was initially lost from the tips of lobes and was most stable in the clefts between lobes; by 6–8 hr of chase, the epithelial surface was poorly defined in both control and β-xyloside-treated rudiments. Thus, even though substantially less newly synthesized proteoglycan is deposited, its subsequent processing appears to be normal.

Thus, β-xylosides inhibit proteoglycan biosynthesis, resulting in sharply

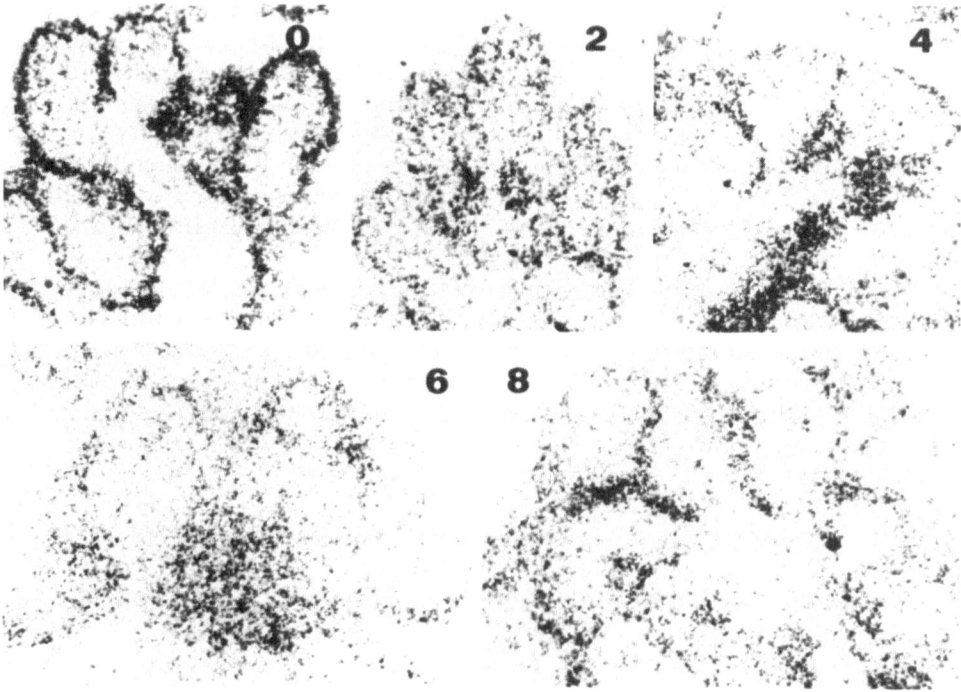

Figure 17. Basal lamina proteoglycan turnover is not affected by β-xyloside. Salivary rudiments were cultured in control or β-xyloside-containing medium, pulse- labeled with [^{35}S]sulfate for 2 hr, and processed for autoradiography, as in Figure 15, except that rudiments were fixed immediately or chased with nonradioactive medium for 2, 4, 6, or 8 hr before fixation. β-Xyloside (0.5 mM)-treated rudiments exhibit basal surface-associated turnover of radioactivity, with a time course and pattern identical to controls. The well-defined epithelial contour observed in autoradiograms of pulse-labeled unchased rudiments (0), is progressively lost with increasing chase times from 2 hr (2) to 4 hr (4) to 6 hr (6) to 8 hr (8), resulting in a poorly defined epithelial surface. The pattern of label disappearance is that label is first lost from the tips of lobes and more slowly lost from clefts or branch points between lobes. By 8 hr of chase (8), lobular tips are poorly resolvable, while radioactivity is still present in clefts. Although the amount of label present immediately after labeling is greater in controls, label turnover in controls is identical to that observed in β-xyloside cultures. (From Spooner et al., 1985.)

reduced proteoglycan deposition at the level of the basal lamina, and branching morphogenesis stops.

4.5. Summary

β-Xylosides are remarkably useful tools to probe proteoglycan synthesis involvement in developmental events. In the salivary system, they reversibly inhibit branching morphogenesis in a dose-dependent manner, whereas control α-xylosides have no effect (Thompson and Spooner, 1982). At 0.5-mM concentrations, β-xyloside causes a 50% inhibition of proteoglycan synthesis and a 300% stimulation of total GAG synthesis (Thompson and Spooner, 1983). Free GAG synthesis does not appear to cause branching inhibition, but inhibition of proteoglycan synthesis directly correlates with inhibition of branching morphogenesis. The effect appears to be quantitative, that is, a function simply of the amount of proteoglycan inhibition, and not a qualitative change in the proteoglycans, since they are unaltered in terms of size class and GAG composition. The inhibition of proteoglycan synthesis, seen biochemically, is resolvable at the level of the basal lamina, where reduced amounts are deposited. However, proteoglycan turnover patterns are identical in control and β-xyloside-treated cultures (Spooner et al., 1985.)

The interpretation of these data is that β-xylosides inhibit proteoglycan synthesis, leading to reduced proteoglycan deposition in the basal lamina, which results in inhibition of branching morphogenesis. Although the data demonstrate involvement, they do not elucidate the exact role of proteoglycans in morphogenesis. Possible roles, in association with other extracellular matrix components, are considered in Section 7.

5. Tunicamycin Effects on Branching Morphogenesis

The data discussed thus far have principally focused on collagen and proteoglycan involvement in branching morphogenesis. However, it is highly likely that glycoproteins of the extracellular matrix have important roles as well. In particular, it would be reasonable to suspect that fibronectin and laminin might be involved. Although the glycoconjugate inhibition approach, using DON and azaserine (described in Section 3), would include glycoprotein inhibition (the β-xyloside approach would not), a glycoprotein synthesis perturbation tool of higher specificity is the antibiotic tunicamycin. Tunicamycin inhibits synthesis of glycoconjugates whose construction involves glycosylation of the lipid-linked intermediate dolichol phosphate (Mahoney and Duksin, 1979), and then covalent attachment to protein (in the case of glycoproteins). Tunicamycin will therefore inhibit asparagine-linked glycoprotein (Hart and Lennarz, 1978; Duksin et al., 1982; Hart, 1982) and glycolipid synthesis but not proteoglycan synthesis. A recognized problem is that tunicamycin preparations also possess significant protein synthesis inhibitory activity (Mahoney and Duksin, 1979).

We do not yet have any information on glycoprotein synthesis in branching morphogenesis. However, we have assessed the effects of tunicamycin on branching morphogenesis of the salivary rudiment and have made an observation worth noting here. Since tunicamycin severely retards growth in these cultures, and is obviously toxic in the microgram per milliliter range, we cultured rudiments in progressively lower concentrations, with the goal of determining a concentration that inhibited branching activity without significant effects on growth. In fact, we obtained just the opposite result. As shown in Figure 18, at a tunicamycin concentration of 25 ng/ml, treated rudiments are clearly growth inhibited but continue to branch on pace with controls. However, since the epithelial mass is so much smaller than in controls, the crowding of so many branches produces many tiny branches. As yet, we have no idea whether glycoprotein synthesis is normal or abnormal in these cultures. The important point is that branching morphogenesis is independent of the normal pattern of growth, whatever the biochemical basis for the growth inhibition may be. This observation complements and extends the cycloheximide results presented in Section 3. The data are not compatible with models of morphogenesis that include substantial degrees of ongoing growth as being essential for ongoing branching activity. They do not comment on models requiring increased cell numbers, as long as such increases are not coupled to required substantial increases in tissue mass (growth).

6. Anionic Sites in the Salivary Basal Lamina

In recent years, much attention has focused on basement membrane structure and function, with particular interest in proteoglycan presence and distribution. A major approach in such studies has been the use of electron-dense cationic substances (e.g., ruthenium red) to localize anionic sites. Such sites are generally interpreted to represent proteoglycan (particularly heparan sulfate proteoglycan), and are sometimes documented as proteoglycan by sensitivity of such sites to appropriate GAGase digestion. The basis for the anionic characteristics of proteoglycans is the high net negative charge of the GAG chains, resulting chiefly from their sulfate groups, and free carboxyl groups on their uronic acid residues. The free carboxyl contribution can be substantial, since, for example, hyaluronic acid is highly anionic but nonsulfated. Free carboxyl groups, and certainly sialic acid residues, on glycoproteins lend anionic properties to these molecules as well. For light microscopic assessment of proteoglycan distribution with such dyes as Alcian blue, sections are stained at low pH ($\sim$2–3) or at critical $MgCl_2$ concentrations ($\leqslant$0.3 M), where only the sulfate groups bind the dye. However, since binding of cationic probes such as ruthenium red and polyethyleneimine for EM studies is carried out on intact tissue, before embedding and sectioning, usually at $\geqslant$pH 5 (e.g., Vaccaro and

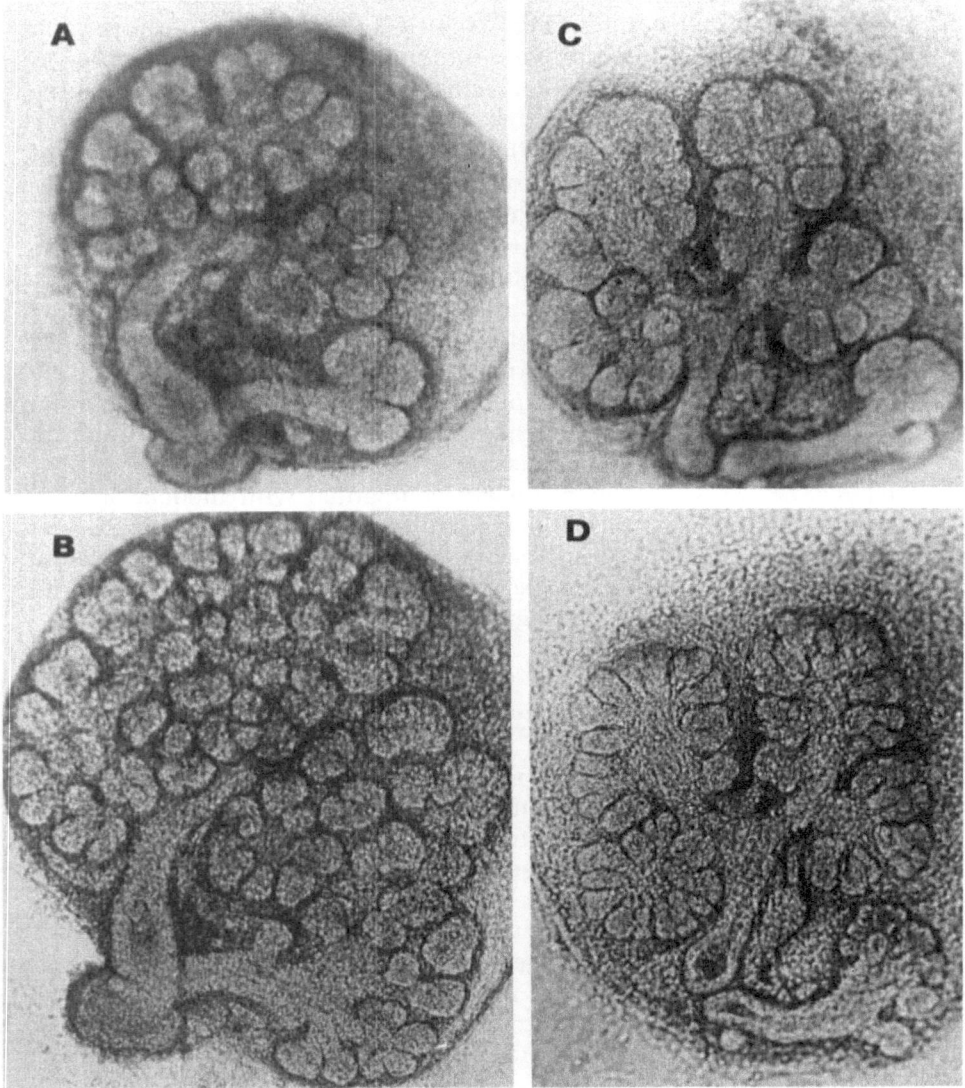

Figure 18. Tunicamycin effects on branching morphogenesis. Salivary rudiments were cultured in control (A) or tunicamycin (25 ng/ml)-containing (C) medium for 24 hr. The same rudiments are shown in B (control) and D (tunicamycin) at 48 hr of culture. Although tunicamycin severely inhibits rudiment growth (compare B and D), branching morphogenesis continues at a normal rate in tunicamycin.

Brody, 1981; Essner and Pino, 1982), interpretation of anionic sites as equivalent to proteoglycan should be viewed with caution. Greater care in binding has been possible where selected populations of cationized ferritin have been used (Farquhar, 1978; Simionescu et al., 1982).

Electron microscopic anionic site localization studies using ruthenium red have been reported for the embryonic lung (Grant et al., 1983) and for the embryonic salivary (Cohn et al., 1977). The lung studies were conducted on late embryonic tissue, not comparable to the earlier intense bronchial branching stages used in our lung studies, but discrete basal lamina anionic sites, sensitive to hyaluronidase and chondroitinase, were well resolved. The salivary studies used tannic acid analysis as well as ruthenium red and resolved some aspects of basal lamina substructure that permitted the proposal of a model of its organization.

We have conducted standard TEM analysis, as well as tannic acid and ruthenium red analyses on salivary rudiments, with results that essentially agree with those reported by Cohn et al. (1977). Anionic site localization by ruthenium red, in the salivary system, is really not very satisfactory, for discrete electron-dense deposits are not well resolved. We have had much greater success with an alternative cationic probe, polyethyleneimine (Schurer et al., 1978; Essner and Pino, 1982). Polyethyleneimine binding resolves anionic sites as discrete well-resolved electron-dense "dots." A regularly spaced double row of anionic sites is associated with the salivary basal lamina (Fig. 19). One row of sites is associated with the inner surface of the lamina densa, predominantly in the lamina rara, between the lamina densa and the epithelial plasmalemma. A second row of sites is associated with the outer surface of the lamina densa, i.e., on the extracellular matrix side of the lamina densa, toward the mesenchyme. Thus, the basal lamina appears as a continuous lamina densa "line" (in section), with regular arrays of anionic sites associated with both the inner and outer surfaces of the lamina densa. The "extended matrix" immediately peripheral to the basal lamina is also rich in anionic sites and appears as regular arrays of anionic sites interconnected by strands of fibrous material (Fig. 19). Finally, highly regular periodic arrays of anionic sites are associated with bundles of interstitial collagen (Fig. 19). Thus, localization of anionic site distribution in the extracellular matrix of the embryonic salivary rudiment, with polyethyleneimine, reveals regular arrays of such sites throughout the matrix, in the basal lamina and the interstitial matrix.

Whether the anionic sites resolved with polyethyleneimine are equivalent to proteoglycan or not is not yet clear. Some insight into their nature should be achievable by analysis of sensitivity to various enzymes, although susceptibility to enzymes may mean that the site is either a particular compound or that it is bound to that compound. The exciting potential of this impressive localization, however, lies in the opportunity to evaluate possible changes in anionic site distribution that might result from experimental alterations of the extracellular matrix—including those involving collagen and proteoglycan—that cause alterations in branching morphogenesis of the salivary epithelium.

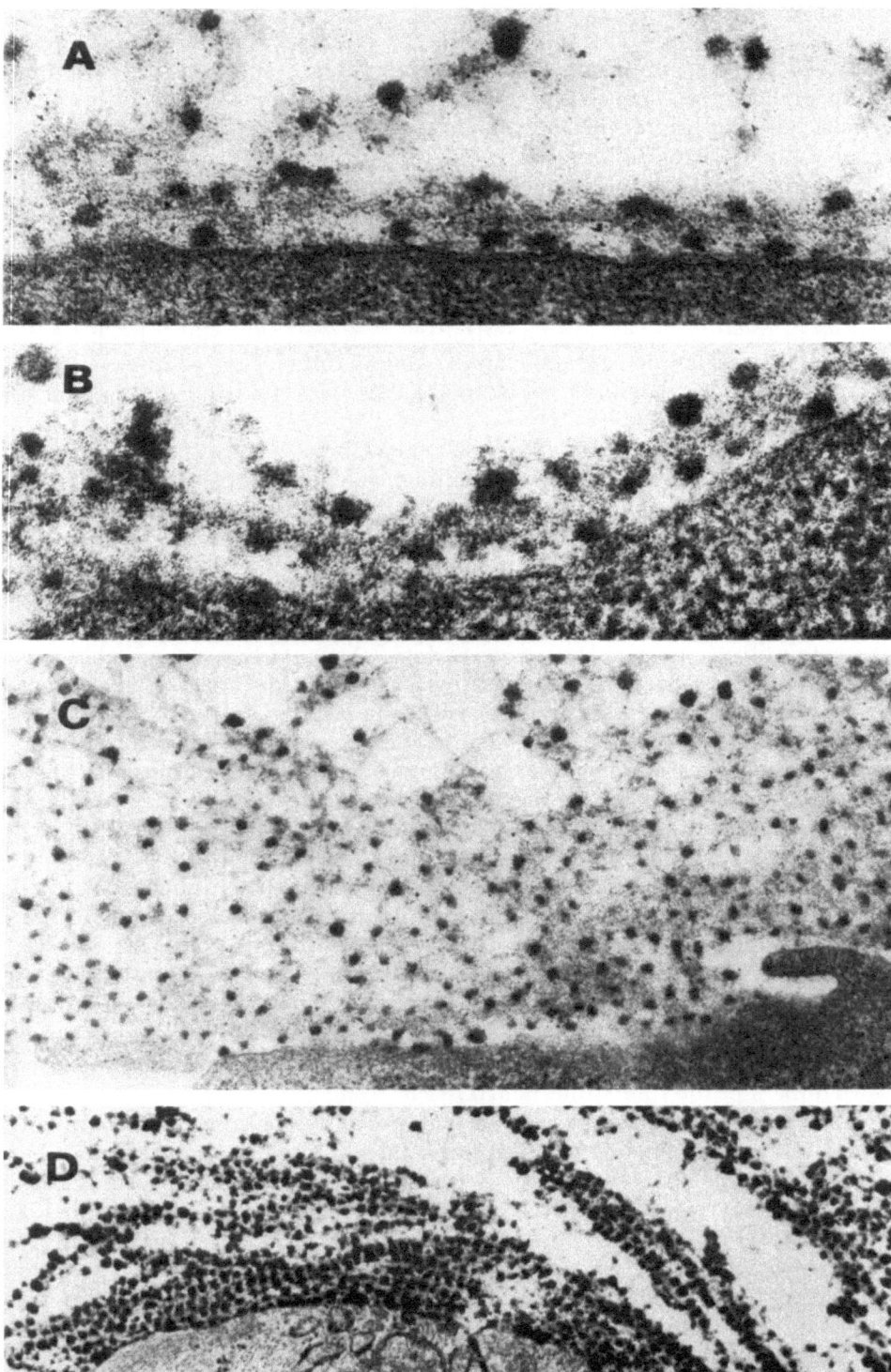

7. Conclusions

The data described in this chapter lead to two conclusions regarding the extracellular matrix requirements for epithelial branching morphogenesis. The first major conclusion is that interstitial collagen is intimately involved in branching, a conclusion supported by experimental demonstration that branching ceases under conditions that eliminate interstitial collagen from the extracellular matrix. The second major conclusion is that proteoglycans are also intimately involved in branching, for inhibition of proteoglycan synthesis also results in cessation of branching morphogenesis. Although both collagen and proteoglycans have been recognized for years as major components of the extracellular matrix of embryonic organs participating in branching morphogenesis, the extent of their involvement in that process has not been clear. Earlier studies have provided intriguing correlations that suggested a fundamental involvement. For example, differential patterns of collagen localization relative to branching and nonbranching regions of the epithelium have been observed (Wessells, 1970). In the case of proteoglycans, it has been shown that the epithelial basal lamina is rich in proteoglycans (Cohn et al., 1977), that the proteoglycan-rich basal lamina is required for branching (Banerjee et al., 1977), and that differential turnover of basal lamina proteoglycan is correlated with regions of branch point formation (Bernfield and Banerjee, 1982; Smith and Bernfield, 1982). However, whether collagen presence was actually required and whether the basal lamina requirement reflected a necessity for proteoglycan was unknown. Thus, the demonstration that collagen and proteoglycans are in fact required constitutes a fundamental advance in our understanding of epithelial branching morphogenesis.

The fact that collagen and proteoglycans are necessary for branching morphogenesis establishes an essential involvement but does not tell us how they are involved. What are the respective roles of collagen and proteoglycans? Are they interrelated roles? Do they interface in fundamental ways with possible glycoprotein roles in the extracellular matrix regulation of branching morphogenesis? These are the questions that now need to be addressed. It is useful to consider some possible functions that may suggest experimental approaches leading to their verification or rejection. For example, the role of collagen might be to regulate epithelial proteoglycan synthesis/secretion, deposition, or turnover. It has been reported that collagen can stimulate glycosaminoglycan synthesis by the corneal epithelium (Meier and Hay, 1974) and that proteoglycan synthesized by mammary cell lines is not localized at the tissue layer in the absence of collagen (David and Bernfield, 1979). Collagen might also be respon-

Figure 19. Electron microscopic resolution of anionic site distribution in the extracellular matrix of 48 hr salivary rudiments with polyethyleneimine. Anionic sites are present in the basal lamina (A,B), associated with both the inner and outer surfaces of the lamina densa. In addition, the extended matrix is rich in anionic sites (C) and a regular array of anionic sites is associated with interstitial collagen arrays (D). Polyethyleneimine labeling was carried out in Pipes buffer, before fixation and processing for electron microscopic evaluation.

sible for stabilizing the epithelium by differential distribution that either promotes or prohibits mesenchyme-mediated turnover of basal lamina proteoglycan (Bernfield and Banerjee, 1982; Smith and Bernfield, 1982) and produces a unique branching pattern. Some of these possibilities appear to be inconsistent with the demonstration that the salivary epithelium alone—without its investing mesenchyme and interstitial collagen—can regenerate a proteoglycan-rich basal lamina (Banerjee *et al.*, 1977). However, it is significant that such epithelia do not resume branching activity until they are recombined with mesenchyme (Bernfield *et al.*, 1972), which will provide interstitial collagen. Collagen might also act independently of proteoglycan, alone or in concert with matrix glycoproteins, to promote branch point formation in the epithelium *via* (1) selective mitogenic activity that causes some essential increase in epithelial cell numbers, or (2) a transmembrane action that activates cellular contractile activity for essential epithelial cell shape changes. An effect on epithelial growth would have to be one on necessary increases in cell numbers rather than on tissue mass, however, since inhibition of protein synthesis and rudiment growth does not perturb branching morphogenesis.

In a reciprocal sense, proteoglycan might function to stimulate collagen synthesis or deposition in the extracellular matrix. In addition to the possibility that basal lamina proteoglycan might be related to the arrays of cross-banded collagen closely associated with the basal lamina at the epithelial–mesenchymal interface, our anionic site distribution studies demonstrate extensive associations of such sites with interstitial collagen, in general. If these anionic sites do represent proteoglycan as generally presumed, they could be crucial to the collagen deposition patterns observed. In the case of the basal lamina proteoglycans, it is of further interest to speculate on how they might function in directly regulating epithelial branching. It is particularly intriguing to consider the possible significance of their anionic character to control of epithelial cell shape changes. Thus, basal lamina proteoglycan could function to bind large amounts of extracellular calcium ion near the epithelial surface. If such calcium were then released to enter the epithelial cells, it could serve to activate actin–myosin interactions that change cell shape and produce branch points in the epithelium (Spooner, 1975).

It is clear that branch points are generated by changes in cell shape that are produced by actin-containing arrays of microfilaments, and there is suggestive evidence that extracellular calcium ions are involved in these changes in shape (Ash *et al.*, 1973; Spooner, 1973). If this concept is correct, we still need to explain how or why calcium ion is released from the proteoglycan and enters the epithelial cell cytoplasm. The epithelial cell membrane would be permeable to calcium if that membrane were depolarized. If the depolarization stimulus came from the mesenchyme, it might account for a part of the mesenchyme role in the tissue interaction. Under such circumstances, it is possible that the resulting potassium ion efflux would result in an ion-exchange reaction releasing calcium from the proteoglycan to diffuse into the epithelial cells. However, if the epithelial cells are ionically coupled (although that is not known for these embryonic organs), the entire epithelium would depolarize and there would be no apparent basis for differential local increase in calcium availability to initiate contractility and branch point formation.

There are two theoretically possible solutions to this potential problem. First, proteoglycan (and GAG, if hyaluronic acid is involved) distribution in the basal lamina may be quantitatively or qualitatively asymmetric, with respect to calcium binding capacity, so that even if calcium is released and enters all the epithelial cells, only some cells will receive sufficient calcium to activate contractility and initiate branch point formation. The asymmetry could be in proteoglycan types, amounts, or degrees of sulfation. We do know that the bulk of the proteoglycan in the salivary system is chondroitin sulfate proteoglycan, but heparan sulfate proteoglycan is also present, and an assessment of their relative distribution along the basal lamina should be made. A second possibility that could result in higher calcium availability at regions where new branch points are to form is that the high rate of proteoglycan turnover at those sites results in locally higher free calcium ion concentrations, without the need for ion exchange release. As the GAG chains of the proteoglycans are degraded by the mesenchyme, reduced calcium binding capacity would be present. That is, the degradation itself could serve to release calcium from the GAG chains. There is a correlation between the regions of most rapid proteoglycan (GAG) turnover and the sites of new branch point formation, so that this possible mechanism might operate whether or not asymmetries in proteoglycan distribution were found.

Whether any of these possible roles for collagen and proteoglycan action are correct is of little consequence, if their consideration stimulates experimental analysis of the extracellular matrix regulation of branching morphogenesis. We know that both classes of matrix molecules are essential and anticipate that future investigation will reveal their function, as well as those for matrix glycoproteins, that will produce sharper focus and an ever-emerging understanding of the role of the extracellular matrix in epithelial branching morphogenesis.

ACKNOWLEDGMENTS.   We thank Gary W. Conrad for criticism, Mark Sullins for assistance, and he and Brian S. Spooner, Jr., for darkroom work. DON (NSC-7365) was supplied by the Laboratory of Toxicology, NCI. Original studies were supported by NIH grant HL25910 to B.S.S.

References

Alescio, T., 1973, Effect of a proline analog, azetidine-2-carboxylic acid, on the morphogenesis *in vitro* of mouse embryonic lung, *J. Embryol. Exp. Morphol.* **29:**493–514.

Alescio, T., and Cassini, A., 1962, Induction *in vitro* of tracheal buds by pulmonary mesenchyme grafted on tracheal epithelium, *J. Exp. Zool.* **150:**83–94.

Ash, J. F., Spooner, B. S., and Wessells, N. K., 1973, Effects of papaverine and calcium-free medium on salivary gland morphogenesis, *Dev. Biol.* **33:**463–469.

Aydelotte, M., and Kochhar, D., 1975, Influence of 6-diazo-5-oxo-L-norleucine (DON), a glutamine analog, on cartilagenous differentiation in mouse limb buds *in vitro*, *Differentiation* **4:**73–80.

Banerjee, S. D., Cohn, R. H., and Bernfield, M. R., 1977, Basal lamina of embryonic salivary epithelia: Production by the epithelium and role in maintaining lobular morphology, *J. Cell Biol.* **73**:445–463.

Bates, C. J., Adams, W. R., and Handschumacher, R. E., 1966, Control of formation of uridine diphospho-N-acetyl-hexosamine and glycoprotein synthesis in rat liver, *J. Biol. Chem.* **241**:1705–1712.

Bernfield, M. R., 1970, Collagen synthesis during epitheliomesenchymal interactions, *Dev. Biol.* **22**:213–231.

Bernfield, M. R.,1981, Organization and remodeling of the extracellular matrix in morphogenesis, in: *Morphogenesis and Pattern Formation: Implications for Normal and Abnormal Development* (L. L. Brinkley, B. M. Carlson, and T. G. Connelly, eds.), pp. 139–162, Raven Press, New York.

Bernfield, M. R., and Banerjee, S. D., 1972, Acid mucopolysaccharide (glycosaminoglycan) at the epithelial–mesenchymal interface of mouse embryo salivary glands, *J. Cell Biol.* **52**:664–673.

Bernfield, M. R., and Banerjee, S. D., 1982, The turnover of basal lamina glycosaminoglycan correlates with epithelial morphogenesis, *Dev. Biol.* **90**:291–305.

Bernfield, M. R., Banerjee, S. D., and Cohn, R. H., 1972, Dependence of salivary epithelial morphology and branching morphogenesis upon acid mucopolysaccharide-protein (proteoglycan) at the epithelial surface, *J. Cell Biol.* **52**:674–689.

Bernfield, M. R., Cohn, R. H., and Banerjee, S. D., 1973, Glycosaminoglycans and epithelial organ formation, *Am. Zool.* **13**:1067–1083.

Bornstein, P., and Traub, W., 1979, The chemistry and biology of collagen, in: *The Proteins*, Vol. IV (H. Neurath and R. L. Hall, eds.), pp. 411–432, Academic Press, New York.

Cohn, R. H., Banerjee, S. D., and Bernfield, M. R., 1977, Basal lamina of embryonic salivary epithelia. Nature of glycosaminoglycan and organization of extracellular materials, *J. Cell Biol.* **73**:464–478.

Coulombre, J., and Coulombre, A., 1975, Corneal development. V. Treatment of five-day-old embryos of domestic fowl with 6-diazo-5-oxo-L-norleucine (DON), *Dev. Biol.* **45**:291–303.

Crouch, E., and Bornstein, P., 1979, Characterization of a type IV procollagen synthesized by human amniotic fluid cells in culture, *J. Biol. Chem.* **254**:4197–4204.

David, G., and Bernfield, M. R., 1979, Collagen reduces glycosaminoglycan degradation by cultured mammary epithelial cells: A possible mechanism for basal lamina formation, *Proc. Natl. Acad. Sci. USA* **76**:786–790.

Duksin, D., Seiberg, M., and Mahoney, W. C., 1982, Inhibition of protein glycosylation and selective toxicity toward virally transformed fibroblasts caused by B_3-tunicamycin, *Eur. J. Biochem.* **129**:77–80.

Ekblom, P., Lash, J. W., Lehtonen, E., Nordling, S., and Saxén, L., 1979, Inhibition of morphogenetic cell interactions by 6-diazo-5-oxo-L-norleucine (DON), *Exp. Cell Res.* **121**:121–126.

Essner, E., and Pino, R. M., 1982, Distribution of anionic sites in Bruch's membrane of the rabbit eye, *Eur. J. Cell Biol.* **27**:251–255.

Farquhar, M. G., 1978, Recovery of surface membrane in anterior pituitary cells. Variations in traffic detected with anionic and cationic ferritin, *J. Cell Biol.* **77**:R35–R42.

Fukunaga, Y., Sobue, M., Suzuki, N., Kushida, H., Suzuki, S., and Suzuki, S., 1975, Synthesis of fluorogenic mucopolysaccharide by chondrocytes in cell culture with 4-methylumbelliferyl-β-D-xyloside, *Biochim. Biophys. Acta* **381**:443–447.

Galligani, L., Hopwood, J., Schwartz, N. B., and Dorfman, A., 1975, Stimulation of synthesis of free chondroitin sulfate chains by β-D-xylosides in cultured cells, *J. Biol. Chem.* **250**:5400–5406.

Ghosh, S., Blumenthal, H. J., Davidson, E., and Roseman, S., 1960, Glucosamine metabolism. V. Enzymatic synthesis of glucosamine-6-phosphate, *J. Biol. Chem.* **235**:1265–1273.

Grant, M. M., Cutts, N. R., and Brody, J. S., 1983, Alterations in lung basement membrane during fetal growth and type 2 cell development, *Dev. Biol.* **97**:173–183.

Greene, R. M., and Pratt, R. M., 1977, Inhibition by diazo-oxo-norleucine (DON) of rat palatal glycoprotein synthesis and epithelial cell adhesion in vitro, *Exp. Cell Res.* **105**:27–37.

Grobstein, C., 1953, Epithelio-mesenchymal specificity in the morphogenesis of mouse submandibular rudiments *in vitro*, *J. Exp. Zool.* **124**:383–404.

Grobstein, C., 1967, Mechanism of organogenetic tissue interaction, *Natl. Cancer. Inst. Monogr.* **26**:279–299.

Grobstein, C., and Cohen, J., 1965, Collagenase: Effect on the morphogenesis of embryonic salivary epithelium in vitro, Science **150**:626–628.

Hall, H. G., Farson, D. A., and Bissell, M. J., 1982, Lumen formation by epithelial cell lines in response to collagen overlay: A morphogenetic model in culture, Proc. Natl. Acad. Sci. USA **79**:4672–4676.

Hart, G. W., 1982, The role of asparagine-linked oligosaccharides in cellular recognition by thymic lymphocytes, J. Biol. Chem. **257**:151–158.

Hart, G. W., and Lennarz, W. J., 1978, Effects of tunicamycin on the biosynthesis of glycosaminoglycans by embryonic chick cornea, J. Biol. Chem. **253**:5795–5801.

Hill, D. L., and Bennett, L. L., Jr., 1969, Purification and properties of 5-phosphoribosyl pyrophosphate aminotransferase from adenocarcinoma 755 cells, Biochemistry **8**:122–130.

Hurmerinta, K., and Thesleff, I., 1982, Diazo-oxo-norleucine (DON)-induced alterations in the extracellular matrix of the mouse tooth germ, Cell Diff. **11**:107–113.

Hurmerinta, K., Thesleff, I., and Saxén, L., 1979, Inhibition of tooth germ differentiation in vitro by diazo-oxo-norleucine (DON), J. Embryol. Exp. Morphol. **50**:99–109.

Ishii, K., and Green, H., 1973, Lethality of adenosine for cultured mammalian cells by interference with pyrimidine biosynthesis, J. Cell Sci. **13**:429–439.

Kallman, F., and Grobstein, C., 1965, Source of collagen at epitheliomesenchymal interfaces during inductive interaction, Dev. Biol. **11**:169–183.

Kallman, F., and Grobstein, C., 1966, Localization of glucosamine-incorporating materials at epithelial surfaces during salivary epitheliomesenchymal interaction in vitro, Dev. Biol. **14**:52–67.

Kim, J. J., and Conrad, H. E., 1974, Effect of D-glucosamine concentration on the kinetics of mucopolysaccharide biosynthesis in cultured chick embryo vertebral cartilage, J. Biol. Chem. **249**:3091–3097.

Lane, J. M., Parkes, L. J., and Prockop, D. J., 1971, Effect of the proline analog azetidine-2-carboxylic acid on collagen synthesis in vivo. II. Morphological and physical properties of collagen containing the analog, Biochim. Biophys. Acta **236**:528–541.

Linsenmayer, T. F., and Kochhar, D. M., 1979, In vitro cartilage formation: Effects of 6-diazo-5-oxo-L-norleucine (DON) on glycosaminoglycan and collagen synthesis, Dev. Biol. **69**:517–528.

Lohmander, S., Madsen, K., and Hinek, A., 1979, Secretion of proteoglycans by chondrocytes: Influence of colchicine, cytochalasin B, and β-D-xyloside, Arch. Biochim. Biophys. **192**:148–157.

Mahoney, W. C., and Duksin, D., 1979, Biological activities of the two major components of tunicamycin, J. Biol. Chem. **254**:6572–6576.

Meier, S., and Hay, E. D., 1974, Control of corneal differentiation by extracellular materials: Collagen as a promoter and stabilizer of epithelial stroma production, Dev. Biol. **38**:249–270.

Pinnell, S. R., and Martin, G. R., 1968, The cross-linking of collagen and elastin: Enzymatic conversion of lysine in peptide linkage to α-aminoadipic-5-semialdehyde by an extract of bone, Proc. Natl. Acad. Sci. USA **61**:708–716.

Pratt, R. M., and Greene, R. M., 1976, Inhibition of palatal epithelial cell death by altered protein synthesis, Dev. Biol. **54**: 135–145.

Sakakura, T., Nishizuka, Y., and Dawe, C. J., 1976, Mesenchyme-dependent morphogenesis and epithelium-specific cytodifferentiation in mouse mammary gland, Science **194**:1439.

Schurer, J. W., Kalicharan, D., Hoedemaeker, J., and Molenaar, I., 1978, The use of polyethyleneimine for demonstration of anionic sites in basement membranes and collagen fibrils, J. Histochem. Cytochem. **26**:688–689.

Schwartz, N. B., 1977, Regulation of chondroitin sulfate synthesis: Effect of β-xylosides on synthesis of chondroitin sulfate chains and core proteins, J. Biol. Chem. **252**:6316–6321.

Schwartz, N. B., Galligani, L., Ho, P. L., and Dorfman, A., 1974, Stimulation of synthesis of free chondroitin sulfate chains by β-D-xylosides in cultured cells, Proc. Natl. Acad. Sci. USA **71**: 4047–4051.

Simionescu, M., Simionescu, N., and Palade, G. E., 1982, Preferential distribution of anionic sites on the basement membrane and the abluminal aspect of the endothelium in fenestrated capillaries, J. Cell Biol. **95**:425–434.

Smith, R. L., and Bernfield, M. R., 1982, Mesenchyme cells degrade epithelial basal lamina proteoglycan, *Dev. Biol.* **94:**378–390.

Spooner, B. S., 1973, Microfilaments, cell shape changes, and morphogenesis of salivary epithelium, *Am. Zool.* **13:** 1007–1022.

Spooner, B. S., 1974, Morphogenesis of vertebrate organs, in: *Concepts of Development* (J. Lash and J. R. Whittaker, Eds.), pp. 213–240, Sinauer Associates, Stanford, Connecticut.

Spooner, B. S., 1975, Microfilaments, microtubules, and extracellular materials in morphogenesis, *BioScience* **25:**440–451.

Spooner, B. S., and Conrad, G. W., 1975, The role of extracellular materials in cell movement. I. Inhibition of mucopolysaccharide synthesis does not stop ruffling membrane activity or cell movement, *J. Cell Biol.* **65:**286–297.

Spooner, B. S., and Faubion, J. M., 1980, Collagen involvement in branching morphogenesis of embryonic lung and salivary gland, *Dev. Biol.* **77:**84–102.

Spooner, B. S., and Wessells, N. K., 1970, Mammalian lung development: Interactions in primordium formation and bronchial morphogenesis, *J. Exp. Zool.* **175:**445–454.

Spooner, B. S., and Wessells, N. K., 1972, An analysis of salivary gland morphogenesis: Role of cytoplasmic microfilaments and microtubules, *Dev. Biol.* **27:**38–54.

Spooner, B. S., Ash, J. F., Wrenn, J. T., Frater, R. B., and Wessells, N. K., 1973, Heavy meromyosin binding to microfilaments involved in cell and morphogenetic movements, *Tissue Cell* **5:**37–46.

Spooner, B. S., Cohen, H. I., and Faubion, J., 1977, Development of the embryonic mammalian pancreas: The relationship between morphogenesis and cytodifferentiation, *Dev. Biol.* **61:**119–130.

Spooner, B. S., Ash, J. F., and Wessells, N. K., 1978, Actin in embryonic organ epithelia, *Exp. Cell Res.* **114:**381–387.

Spooner, B. S., Bassett, K., and Stokes, B., 1985, Sulfated glycosaminoglycan deposition and processing at the basal epithelial surface in branching and β-D-xyloside-inhibited embryonic salivary glands, *Dev. Biol.* **109:**177–183.

Switzer, B. R., and Summer, G. K., 1973, Inhibition of collagen synthesis by α,α'-dipyridyl in human skin fibroblasts in culture, *In Vitro* **9:**160–166.

Taderera, J. V., 1967, Control of lung differentiation *in vitro*, *Dev. Biol.* **16:**489–512.

Takeuchi, T., and Prockop, D. J., 1969, Biosynthesis of abnormal collagens with amino acid analogs. I. Incorporation of L-azetidine-2-carboxylic acid and cis-4-fluoro-L-proline into protocollagen and collagen, *Biochim. Biophys. Acta* **175:**142–155.

Takeuchi, T., Rosenbloom, J., and Prockop, D. J., 1969, Biosynthesis of abnormal collagens with amino acid analogs. II. Inability of cartilage cells to extrude polypeptides containing L-azetidine-2-carboxylic acid or cis-4-fluoro-L-proline, *Biochim. Biophys. Acta* **175:**156–164.

Thompson, H. A., and Spooner, B. S., 1982, Inhibition of branching morphogenesis and alteration of glycosaminoglycan biosynthesis in salivary glands treated with β-D-xyloside, *Dev. Biol.* **89:**417–424.

Thompson, H. A., and Spooner, B. S., 1983, Proteoglycan and glycosaminoglycan synthesis in embryonic mouse salivary glands: Effects of β-D-xyloside, an inhibitor of branching morphogenesis, *J. Cell Biol.* **96:**1443–1450.

Uitto, J., and Prockop, D. J., 1974, Incorporation of proline analogs into collagen polypeptides: Effects on the production of extracellular procollagen and on the stability of the triple-helical structure of the molecule, *Biochim. Biophys. Acta* **336:**234–251.

Uitto, J., Hoffman, H.-P., and Prockop, D. J., 1976, Synthesis of elastin and procollagen by cells from embryonic aorta. Differences in the role of hydroxyproline and the effects of proline analogs on the secretion of the two proteins, *Arch. Biochem. Biophys.* **173:**187–200.

Vaccaro, C. A., and Brody, J. S., 1981, Structural features of alveolar wall basement membrane in the adult rat lung, *J. Cell Biol.* **91:**427–437.

Wessells, N. K., 1970, Mammalian lung development: Interactions in formation and morphogenesis of tracheal buds, *J. Exp. Zool.* **175,**455–466.

Wessells, N. K., and Cohen, J. H., 1968, Effects of collagenase on developing epithelia in vitro: Lung, ureteric bud, and pancreas, *Dev. Biol.* **18:**294–309.

Chapter 13

Cell–Cell Interactions in the Development of *Dictyostelium*

DONNA FONTANA, TIT-YEE WONG, ANNE THEIBERT, and PETER DEVREOTES

1. Life Cycle

Dictyostelium discoideum is a cellular slime mold that, because of its intricate life cycle (Fig. 1), has attracted the attention of developmental biologists. With an adequate supply of food, amoebae of *D. discoideum* will grow and divide as individual cells. When the food supply is depleted (in the laboratory, this is accomplished by removing a nutrient broth or bacteria), the amoebae cease growing and enter into a developmental program. There is a period of protein synthesis and, if the amoebae are on a solid surface, this is followed by the cAMP-mediated aggregation of 10^5–10^6 amoebae into a single mound. This mound becomes encased in slime (hence the name slime mold) and sends up a fingerlike projection that eventually includes most of the amoebae. This projection falls over onto the agar surface and begins to crawl; this is the pseudoplasmodial or slug stage. During this migration, the amoebae differentiate into either prestalk or prespore cells. After a period of time determined by genetic and environmental factors, the prestalk cells that form the anterior section of the slug cease moving, while the prespore cells located in the posterior section of the slug continue migrating until another moundlike structure is formed. The prestalk cells then begin depositing cellulosic walls and push down through the prespore cells to the agar surface, forming a stalk. When the cells are in their proper position and differentiation is complete, the stalk cells die, leaving their walls to support a ball-like structure that contains the mature spores. In the natural setting, the spores are probably dispersed by wind and upon hydration will germinate and the life cycle is begun again. (For a review of the life cycle see Bonner, 1982.)

Thus, the life cycle of this simple eukaryotic organism consists of both

DONNA FONTANA, ANNE THEIBERT, and PETER DEVREOTES • Department of Biological Chemistry, Johns Hopkins University School of Medicine, Baltimore, Maryland 21205.   TIT-YEE WONG • Department of Biology, Johns Hopkins University, Baltimore, Maryland 21218.

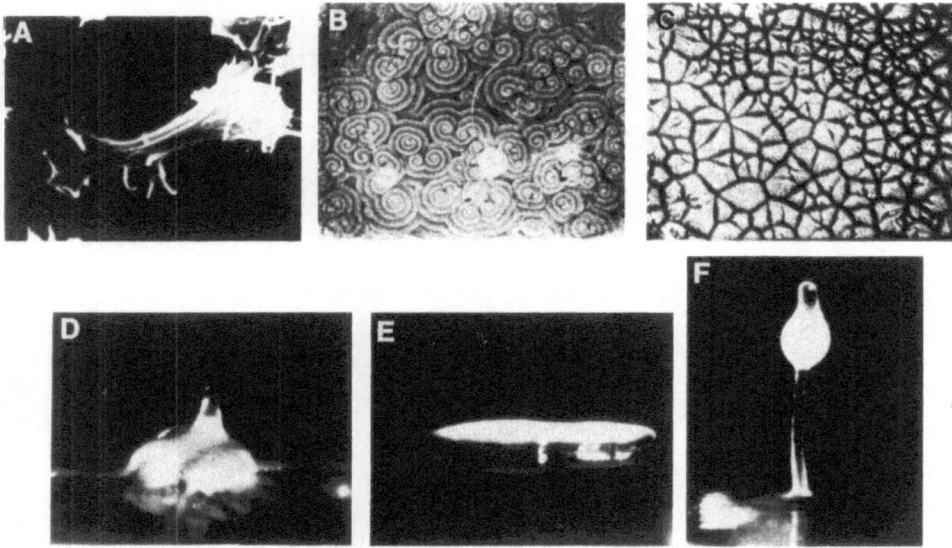

Figure 1. Life cycle of *Dictyostelium discoideum*. Examples of several sequential stages in the developmental cycle are shown (clockwise starting at the top). (a) Scanning electron micrograph of individual amoeba (~8 μm in length); (b) dark-field photograph of early aggregation (~5–7 hr) showing coordinated waves of cell motion (territories are about 1 cm and contain about 10^6 cells); (c) dark-field photograph of streaming patterns in late aggregation (~9–12 hr); (d) "mound" aggregate with apical tip (diameter ~1 mm); (e) migrating slug (length ~1.5 mm); (f) terminal differentiated form consisting of stalk and spore cells (height 1–2 mm).

single and multicellular phases; the developing cells exhibit pattern formation, morphogenesis, differentiation, and specific cell death. The ease of its culture and the ability to generate mutants make it an amenable model that may apply to development in higher systems. This chapter summarizes our studies of the cAMP-mediated cell–cell communication that directs the transition from single cells to a multicellular form.

2. Aggregation

A few hours after starvation, amoebae in a monolayer on an agar surface divide into 1–2-cm territories, each territory encompassing 10^5–10^6 cells. The amoebae move toward the center of these territories in steps, advancing for about 2 min and then stopping for about 5 min. In dark-field photographs, the elongated, moving amoebae appear as white bands that form spirals or concentric circles (Fig. 1B). When these patterns are examined with time-lapse photography, the light bands appear to be initiated every 7–10 min and move out from the territory center. During these movements, the bands maintain a constant width of about 100 cells because as an amoeba at the distal edge begins its inward movement step, one at the proximal edge stops (reviewed in Devreotes, 1982). These organized inward steps, which eventually result in aggregation of the cells into mounds, are mediated by cAMP.

The distribution of cAMP in a monolayer of amoebae was determined by an isotope dilution-fluorographic technique. Regions of high cAMP concentrations form spiral or concentric circular patterns resembling the patterns of cell morphology seen in dark-field photographs (Fig. 2). To determine the relative positions of the cAMP waves and the movement steps, the dark-field photographs were aligned with the fluorographs. The movement steps begin just as the leading edge of a symmetric cAMP wave reaches the amoebae and end as the peak of the wave arrives (Tomchik and Devreotes, 1981). The movement step is a positive chemotactic response to the cAMP gradient formed by the leading edge of the wave. The cAMP wave is the result of the amoebic cAMP signaling response, i.e., when an amoeba senses external cAMP, it synthesizes and secretes cAMP in response. Together these two cAMP responses mediate aggregation.

3. Chemotaxis

The chemotactic response of *D. discoideum* has been studied with the small population assay (Konijn, 1970) in which a drop containing about 100 amoebae is placed on an agar surface and a drop of the test substance is placed nearby. If the amoebae are positively chemotactic with regard to the substance, after about 20 min the amoebae distribute themselves such that most are in the portion of the drop closest to the stimulus. With this assay it was shown that growing amoebae are insensitive to cAMP but are attracted to folic acid (Pan *et al.*, 1972). During development, the amoebae cease being responsive to folic acid and become chemotactically responsive to cAMP.

The small population assay is relatively slow and not useful for determining the kinetics of the chemotactic response. We recently developed another assay for chemotaxis wherein the stimulating cAMP solution is continually perfused past the amoebae to maintain control of the cAMP concentration, minimizing positive feedback effects caused by signaling and degradation of cAMP by phosphodiesterase (PDE) (see Section 4). The perfusion chamber consists of a stainless steel block, the size of a microscope slide, that has a diamond shaped hole cut through it. A glass coverslip is attached to the bottom of the block to form the floor of the chamber. Amoebae are attached to a similar coverslip; then this slip is inverted and placed over the hole, forming the top. Two small needles serve as the fluid inlet and outlet. The needles are attached to tubing and the solutions are pulled through the small diamond-shaped chamber with a pump. With this apparatus, the solution surrounding the amoebae is continually renewed, with the chamber's solution exchanging every 1–2 sec. The amoebae are monitored with a microscope and their actions recorded with a video recorder, a motion picture camera, or a 35-mm camera.

When a buffer solution is perfused past developed amoebae, they move

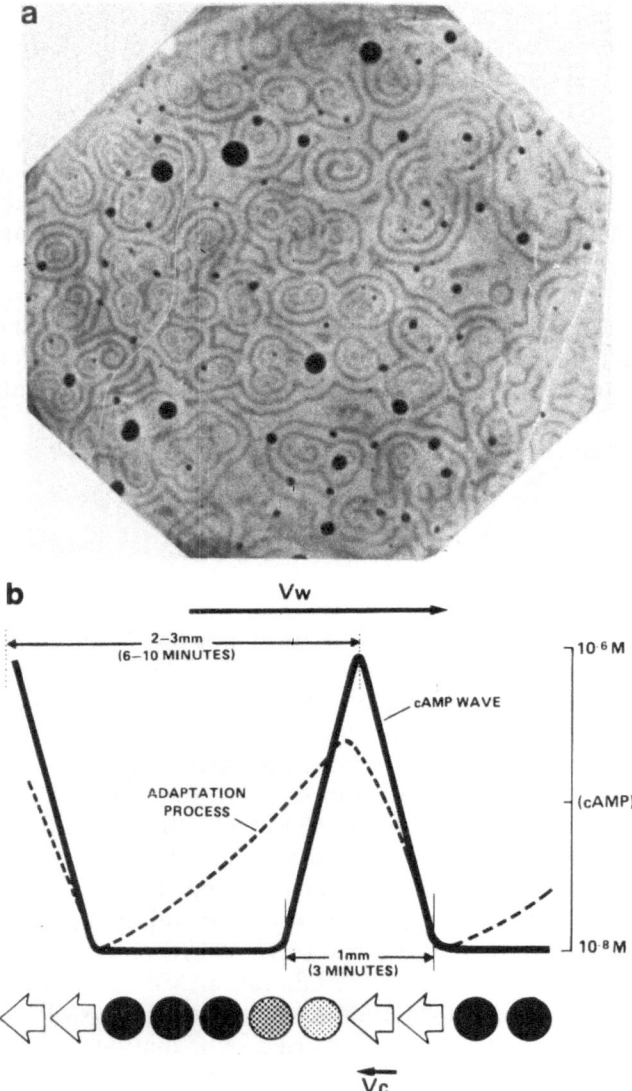

Figure 2. (a) Fluorographic image of cAMP waves. Amoebae were developed to an early aggregation stage (see Fig. 1). Waves were detected by an isotope dilution fluorography technique (Tomchik and Devreotes, 1981). The fluorograph was used as a negative to make this print. (Completely black circles are artifacts caused by air bubbles trapped between the two filters). (b) Dynamics of signal relay and chemotaxis. The heavy line representing the cAMP concentration is drawn from analyses of scans of the optical density of fluorographic images of the cAMP waves. Symbols in lower part of diagram represent a single radial line of cells. Open arrows represent cells moving towards center; shaded circles represent randomly oriented cells. Arrow vectors indicate the speed and direction of motion of the cAMP wave (V_w = 300 μm/min) and the moving cells (V_c = 20 μm/min).

about randomly. When this solution is replaced with cAMP, the amoebae initially retract their pseudopodia. After about 15–30 sec, pseudopodia are extended in all directions and the cells flatten and remain immobilized (Fig. 3). After 2–3 min, the cells begin to move about in a random manner in spite of the continuous presence of cAMP. This flattening response is a reflection of a chemotactic response to a temporal stimulus. The range of cAMP concentrations to which the amoebae exhibit this behavior is the same range in which cAMP elicits a chemotactic response in the small drop assay. If folic acid is perfused past the developed amoebae, they continue their apparent random

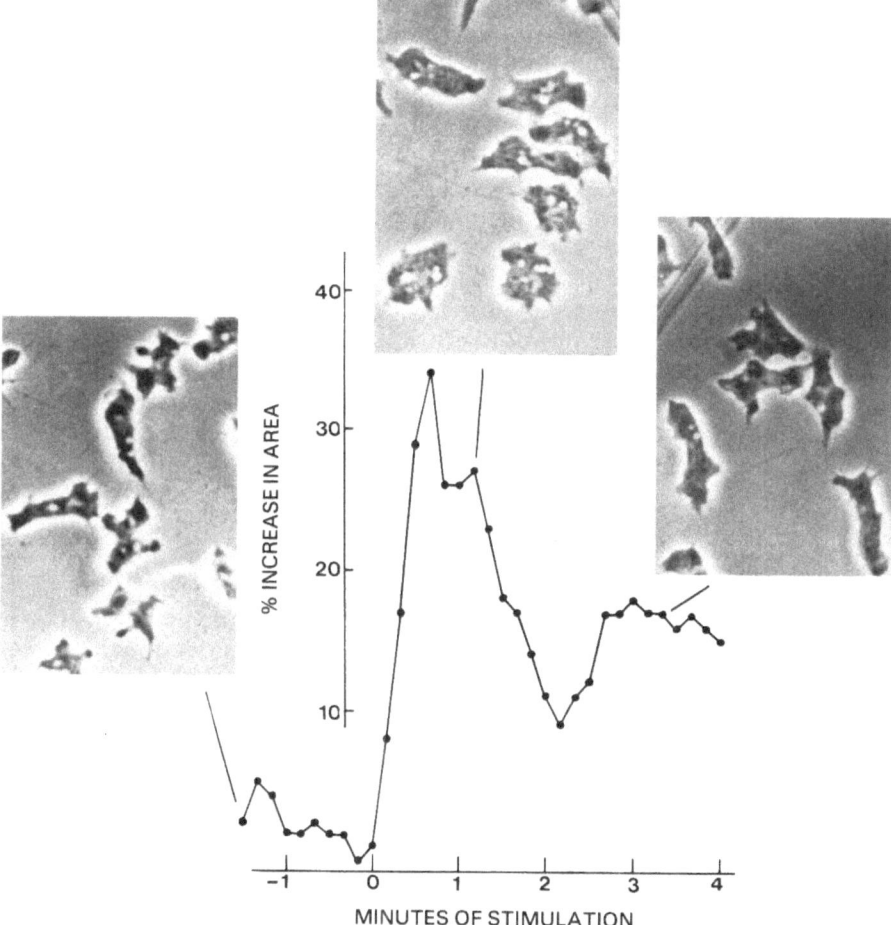

Figure 3. cAMP-elicited shape change. Cells adhering to glass coverslips were perfused rapidly with buffer. The perfusion buffer was rapidly (<5 sec) switched to one containing 10^{-4} M folic acid. Data were collected by tracing areas of cells from 35mm slides taken at 10-sec intervals. (From Dr. Robert Futrelle, University of Illinois, Urbana, Ill.) Solid line indicates average change in area of 17 cells. Photomicrographs show cells 90 sec before and 70 and 200 sec after application of the stimulus.

motion. However, if growing amoebae are placed in the chamber and perfused with folic acid, they respond in the same manner as the developed amoebae responded to cAMP, whereas the growing amoebae do not respond when exposed to cAMP. Therefore, the developmental regulation of their response is identical to the developmental regulation of the chemotactic response. This assay is rather straightforward; it appears to be very sensitive and has yielded interesting qualitative data on the chemotactic response. We are currently collaborating with Dr. Robert Futrelle, University of Illinois, Urbana, in order to quantitate the change in shape and the change in speed of migration.

Since the amoebae resume their amoeboid movement in the continued presence of cAMP (developed amoebae) and folic acid (growing amoebae), it appears that the chemotactic machinery adapts to a continuous presence of chemoattractant. An adaptation process for chemotaxis was also suggested by the observation that cells will orient in a gradient of 10^{-6} M cAMP above a background of 10^{-5} M (van Haastert, 1983a). This adaptation process would explain why, during aggregation, amoebae do not reorient as the cAMP wave passes them and the gradient reverses. The observation that even though amoebae are adapted to a low cAMP concentration they still respond to higher concentrations (seen in both assays) suggests that adaptation is not an all-or-none process, but the degree of adaptation is dependent on the concentration of cAMP to which the amoebae are exposed.

In the perfusion assay, if the stimulating cAMP or folic acid is removed and replaced with buffer for 10–15 min, the amoebae can respond to the chemoattractant when it is reapplied. The recovery of responsiveness is referred to as deadaptation. If amoebae that have been developing for only 3 hr are placed in the perfusion chamber, they will respond to both folic acid and cAMP. With these amoebae, it can be demonstrated that adaptation to folic acid is independent of adaptation to cAMP and that deadaptation to one of the stimuli can occur in the presence of the other. This result was suggested (van Haastert, 1983a) but could not be directly demonstrated with the small population assay. Using the perfusion assay to establish the kinetics of the chemotactic response allows us to determine which biochemical events correlate in time and, it is hoped, how the cAMP chemotactic signal results in directed pseudopodial extension and adaptation.

4. cAMP Signaling

The symmetric cAMP waves, which are seen in monolayers of aggregating amoebae, are the product of the cAMP signaling response. When stimulated by extracellular cAMP, amoebae synthesize and secrete additional cAMP, serving as a signal for more distal amoebae. In a synchronized cell suspension, the response can be elicited by the addition of exogenous cAMP or can occur in response to cAMP secreted by amoebae within the suspension. This response of synchronized populations to cAMP most likely corresponds to that which occurs during aggregation because the concentrations and kinetics of cAMP synthesis are consistent with the dimensions of the cAMP waves detected in monolayers. Also, mutants defective in the generation of cAMP signaling responses do not aggregate normally and specific inhibitors of the response block aggregation (reviewed in Devreotes, 1982).

The cAMP signaling response is triggered by a change in the occupancy of the surface cAMP receptors, which leads to activation of adenylate cyclase. The result is a transient increase in intracellular cAMP and cAMP secretion; the cAMP secretion rate is directly proportional to the intracellular cAMP level (Dinauer *et al.*, 1980*a*). After several minutes, the rate of cAMP synthesis slows down and the cAMP present is degraded by extracellular and membrane-bound phosphodiesterases. The cAMP secretion rate constant or intracellular and extracellular phosphodiesterase activity are not observed to change during the transient response. Thus, the single control point determining the magnitude and duration of the cAMP signaling response appears to be at the level of synthesis of cAMP, i.e., the state of activation of adenylate cyclase (reviewed in Devreotes, 1982).

In cell suspensions (or *in situ* within the monolayer), secreted cAMP binds to surface cAMP receptors, creating a positive feedback loop. Extracellular phosphodiesterase counteracts this feedback loop by decreasing extracellular cAMP. The effect of these two opposing activities results in an excitable system that accounts for both the all-or-none response seen in cell suspensions and the spontaneous oscillations in cAMP levels. The effects of both the positive feedback loop and extracellular phosphodiesterase are minimized when amoebae (labeled with [^{3}H]adenosine) are placed on Millipore filters and stimulated with exogenous cAMP under conditions of rapid perfusion. Secreted cAMP is rapidly removed and degraded cAMP is rapidly replaced. As in the chemotaxis chamber, the filter perfusion clamps the extracellular cAMP concentration at that of the applied stimulus (Devreotes *et al.*, 1979).

Under perfusion conditions, an increment in stimulus concentration elicits a response that subsides after several minutes, even though the stimulating concentration of cAMP is still present (Fig. 4). The extinction of each response by the adjustment of cellular sensitivity to the level of the current stimulus is again referred to as adaptation. An increase in the stimulating cAMP concentration will elicit another response. The sum of the magnitudes of the two responses elicited by the two increments equals the magnitude of the response elicited by a single increment to the highest concentration. The magnitude of the response saturates at 10^{-6} M cAMP, and increments originating at 10^{-6} M cAMP do not elicit another response. These properties suggest that amoebae respond to increases in the fractional occupancy of surface cAMP binding sites (Devreotes and Steck, 1979).

A rapid recovery of responsiveness (deadaptation) occurs after the removal of the cAMP stimulus (Fig. 4). The level of adaptation can be determined by applying an initial stimulus of defined magnitude and duration, and after a variable recovery period, determining the magnitude of the response to a second identical stimulus. Its attenuation is taken as a measure of residual adaptation to the first stimulus. The magnitude of the response to the second stimulus

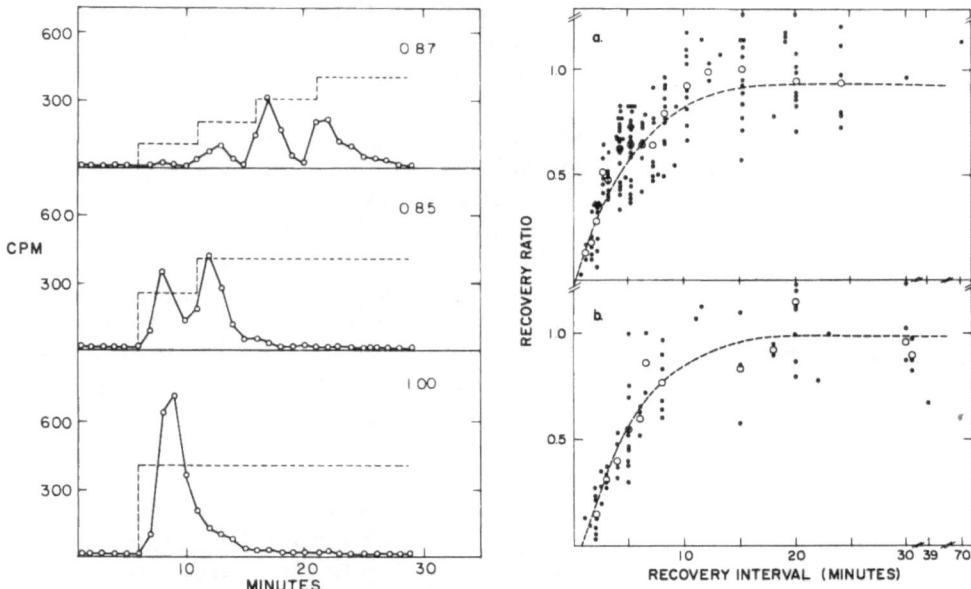

Figure 4. (Left) Response of amoebae to sequential increments in cAMP stimulus concentration. (Bottom) The stimulus was $0-10^{-6}$ M cAMP; (middle) $0-5 \times 10^{-8}-10^{-6}$ M cAMP; (top) 0 to 10^{-9} M to 10^{-8} M to 10^{-7} M to 10^{-6} M cAMP. (- - -) Approximate changes in receptor occupancy. (——) Rate of [3H]cAMP secretion measured by the perfusion technique described in text. Numbers at upper right indicate the total amount of 3H-cAMP secreted in each case, normalized to that secreted in response to the 0 to 10^{-6} M stimulus. (Right) Recovery of the cAMP signaling response after adaptation to cAMP. Two identical stimuli were applied separated by the recovery interval indicated. The magnitude of the second response normalized to that of the first is plotted. (a) 10^{-8} M cAMP; (b) 10^{-5} M cAMP.

increases with the recovery time in a first-order fashion, with a half-life (t½) of 3–4 min for paired stimuli of 10^{-8} or 10^{-5} M cAMP. Recovery is complete in 12–15 min (Dinauer *et al.*, 1980*b*).

Thus far, attempts to activate the adenylate cyclase *in vitro* have failed. This has led to a search for inhibitors of *in vivo* activation, and two have been found. Caffeine does not affect cell viability, intracellular ATP or GTP levels, the cAMP-induced rise in intracellular cGMP, phosphodiesterase activity, binding of cAMP to its surface receptor, or basal adenylate cyclase activity (Brenner and Thoms, 1984). However, both in solution and in perfusion, the presence of caffeine prevents activation of adenylate cyclase on the binding of cAMP to its surface receptors (Theibert and Devreotes, 1983; Brenner and Thoms, 1984). Theibert and Devreotes (1983) showed that this inhibition was instantly and completely reversible and because of this they were able to ask whether adaptation of the signaling response to cAMP occurs in the absence of adenylate cyclase activation. As illustrated in Figure 5, the response to a second increment in the cAMP stimulus concentration is the same whether or not the response to the initial increment is blocked by caffeine, suggesting that caffeine blocks the response but not the adaptation process. Results similar to that shown in Figure 5 were found for initial stimuli of 10^{-9}–10^{-6} M cAMP. In addition, when cells were pretreated with 10^{-6} M cAMP plus caffeine for 1–4 min, sensitivity was lost at the same rate as for treatment with 10^{-6} M cAMP alone. Finally, deadaptation occurred in the same time period whether or not caffeine was present during adaptation. These observations indicate that the adaptation and deadaptation of the signaling response to cAMP occur to the same extent and at the same rate whether or not cAMP synthesis is inhibited.

The second inhibitor that has been shown to block the *in vivo* activation of adenylate cyclase is the lectin, concanavalin A (Con A) (Fontana and De-vreotes, 1984). This inhibition can be blocked by preabsorbing the Con A with α-methylmannoside, a sugar that will block all the binding sites (Fig. 6). The inhibition can also be reversed by high concentrations of α-methylmannoside, indicating that the binding of Con A to the amoebae does not irreversibly harm them. The Con A inhibition is noncompetitive and occurs at a rate consistent with the rate of Con A binding. This time course eliminates Con A-induced capping as the mode of Con A inhibition (Gillette *et al.*, 1974; Condeelis, 1979;

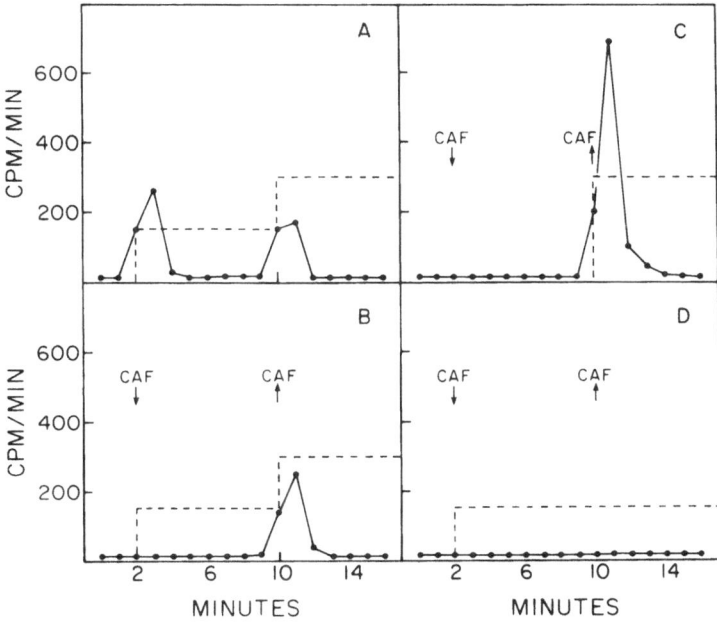

Figure 5. Adaptation is independent of activation of adenylate cyclase. Response to stimulus increment after pretreatment with caffeine and cAMP. The cAMP stimuli (shown by the dashed rectangles) were as follows: (A) 10^{-8}–10^{-6} M cAMP, (B) 10^{-8}–10^{-6} M cAMP, (C) 10^{-6} M cAMP, and (D) 10^{-8} M cAMP. Two mM caffeine was applied and withdrawn as indicated by the arrows. Solid lines indicate rate of [³H]-cAMP secretion measured by the perfusion technique described in text.

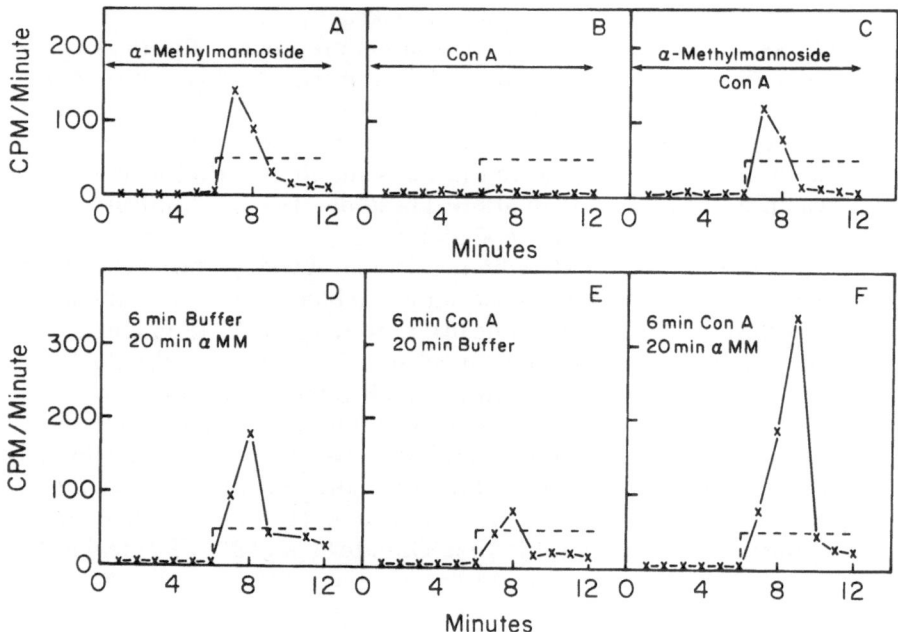

Figure 6. The Con A inhibition is blocked by preabsorption with α-methylmannoside (A,B,C) and can be reversed by application of this sugar (D,E,F). The rates of [³H]-cAMP secretion were measured by the perfusion technique. Plotted is the amount of [³H]-cAMP secreted in the previous minute (x). The dashed lines show where 10^{-6} M cAMP was applied (A–F), with arrows indicating the presence of Con A (10 μg/ml) or α-methylmannoside (50 mM) (A,B,C). The amoebae (D,E,F) were pretreated as indicated with buffer, 2 μg/ml Con A, or 100 mM α-methylmannoside, followed by 9 min with buffer before the cAMP was applied.

Hellio and Ryter, 1980). It was also found that Con A does not alter cAMP binding to its surface receptor, affect basal adenylate cyclase activity, or reduce the ATP level within the amoebae.

In order to examine whether Con A binding to a particular molecule on the surface of the amoebae was inhibiting signaling, other surface binding agents were examined. Wheat germ agglutinin, a polyclonal antibody against a putative adhesion molecule, and four monoclonal antibodies against the surface of *D. discoideum* all inhibited cAMP signaling, suggesting that it is unlikely that binding to a particular component of the cell surface inhibits signaling. A surface-binding agent that was relatively ineffective as an inhibitor was succinylated Con A. This derivative of Con A binds to cells with the same affinity and number as native Con A but has a drastically reduced ability to carry out functions that involve receptor aggregation (Gunther *et al.*, 1973; Wang *et al.*, 1976). This finding suggested that it was the ability of these surface-binding agents to crosslink surface components that resulted in inhibition. To test this, we added a chemical crosslinking agent that binds to primary amines (3,3'-dithiobis(sulfosuccinimidylpropionate) and at 10 mM it inhibited 86%. When the reactive ends of this molecule (not linked) were tested, only 9% inhibition was seen at 20 mM, further suggesting that crosslinking was the cause of inhibition. These results suggest that a membrane rearrangement such as endocytosis or exocytosis or a lateral or transverse movement of molecules within the plasma membrane is necessary for adenylate cyclase activation upon binding of cAMP to its surface receptor.

5. Identification of the Cell Surface cAMP Receptor

The two key responses to cAMP, chemotaxis and signaling, which mediate aggregation are closely integrated. Both occur simultaneously (see Fig. 2b), both are elicited by the same concentrations of cAMP, and adaptation of each response has similar properties and kinetics. Furthermore, the kinetics of the signaling response and wave velocity determine the slope of the cAMP gradient, which is appropriately matched to the chemotactic response of the cells. It seems natural then to ask whether the two responses are mediated by the same surface receptor for cAMP.

One approach to this question is to examine the specificity of the receptor mediating the two responses. The active binding site of the chemotactic receptor has been explored by testing the effectiveness of a series of cAMP analogues, each substituted at a single position on the adenine or ribose rings or within the cyclic phosphate moiety (van Haastert and Kien, 1983). These observations have led to a model for the binding site in which cAMP is held in the anticonformation, with hydrogen bonding at the 6- and 8- and 3'-positions and the adenine moiety bound in a hydrophobic cleft. The 2'-position has little interaction with the receptor.

We have tested similar cAMP analogues for effectiveness in eliciting the cAMP signaling response. For each analogue a dose-response curve was constructed by determining the amount of [3H]-cAMP secreted in response to increasing concentrations of analogue. The analogues tested are listed in Table I in order of effectiveness. Note that substitutions at the 6-, 8-, and 3'-positions greatly reduce effectiveness in eliciting [3H]-cAMP secretion, while the modifications in the 2'- and 5'-positions have little effect.

In Figure 7a, these data for elicitation of the signaling response are compared with the published data for the chemotactic response. Clearly, there is a strong correlation between the two sets of data. This observation indicates that chemotaxis and signaling are both triggered by a receptor of the same affinity and specificity. The simplest interpretation is that the same receptor mediates both responses. Also plotted in Figure 7a are the data for competition by the cAMP analogues of [3H]-cAMP binding to the cell surface and for cGMP accumulation (van Haastert, 1983b). Again a strong correlation with the signaling response data is observed. Taken together, these observations suggest that the receptor detected in the binding assay is that which elicits the signaling, chemotactic, and cGMP responses. There is little correlation of the data for signaling

Table I. Relative Efficiency of cAMP Analogues

| Analogue | R values[a] | | | | | Intracellular receptor |
	Signaling	Binding	Chemotaxis	cGMP	PDE	
a cAMP	0	0	0	0	0	0
b 5′-NH-cAMP	0.47	0.67	0.52	0.71	−1.15	3.35
c 2′-H-cAMP	1.29	.88	1.12	1.08	0.65	4.21
d N′-O-cAMP	1.98	1.61	1.43	1.79	0.45	0.90
e cAMP-(S)	2.46	1.88	1.95	1.88	2.34	0.86
f 6-Cl-cPMP	2.74	2.59	3.00	—	0.40	0.34
g cAMP-N(CH$_3$)$_2$(S)	3.00	2.09	1.99	2.50	1.36	3.35
h 3′-NH-cAMP	3.18	2.69	4.12	2.39	>3.40	2.49
i 8-Br-cAMP	3.19	2.73	2.03	2.45	0.81	−0.50
j cAMP-S(R)	3.67	1.90	—	>4.08	>3.40	2.30
k cAMP-N(CH$_3$)$_2$(R)	4.09	2.32	—	>4.08	2.65	5.22

[a]R is defined as log (concentration of analogue eliciting half-maximal response/concentration of cAMP eliciting half-maximal response).

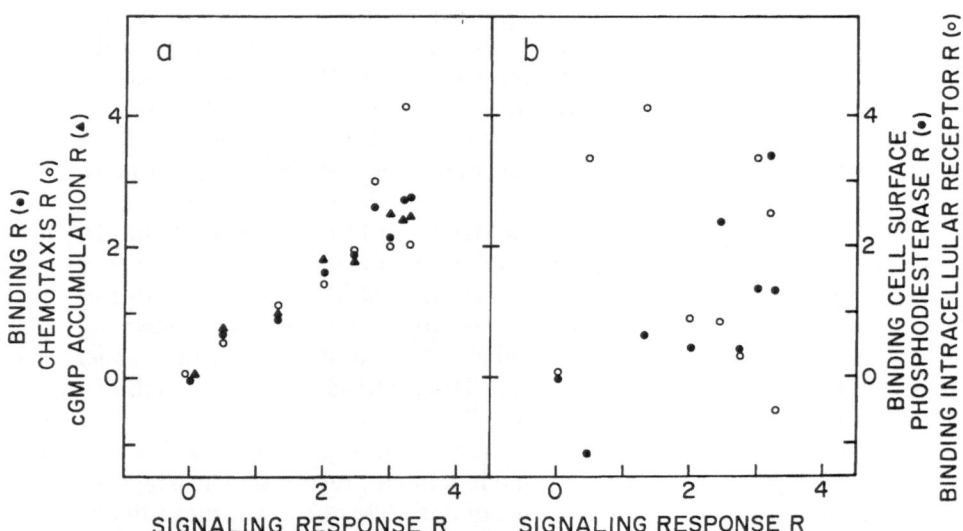

Figure 7. Comparison of cAMP analogue-elicited responses. The concentration of cAMP analogue, which elicited a half-maximal signaling response, was determined (see Fig. 4 and text). R is defined as log (concentration of analogue eliciting half-maximal response/ concentration of cAMP eliciting half-maximal response). (a) R values for signaling response are plotted against R values for [³H]-cAMP binding (●), chemotaxis (○), and cGMP accumulation (▲). (b) R values for signaling response are plotted against R values for membrane phosphodiesterase (●) and intracellular cAMP binding protein (○).

with those for substrate specificity of the cell surface phosphodiesterase or binding to the regulatory subunit of *D. discoideum* protein kinase (Fig. 7b).

The above observations suggest that binding of cAMP to a single surface receptor triggers several key biological responses—chemotaxis, cAMP signaling, and cGMP accumulation. It is of great interest to learn the biochemical identity of this receptor. Previous attempts to identify the surface cAMP receptor by photoaffinity labeling have been thwarted by several fundamental problems: (1) the rapid dissociation rate of the receptor (t ½ ∼ 1–3 sec at 0°) precludes stable association of ligand and receptor; (2) a highly active cell surface phosphodiesterase rapidly degrades exogenous cAMP affinity ligands; and (3) abundant intracellular cAMP binding proteins have confused interpretation of results (Wallace and Frazier, 1979; Juliani and Klein, 1981).

The recent observation that ammonium sulfate stabilizes cell surface $[^3H]$-cAMP binding (van Haastert and Kien, 1983) led us to attempt to identify the surface cAMP receptor by photoaffinity labeling using this technique. $[^{32}P]$-8-N_3-cAMP was bound to surface receptors, and cells were washed in saturated ammonium sulfate. The cells were resuspended in ammonium sulfate and exposed to ultraviolet light. Upon irradiation, the $[^{32}P]$-8-N_3-cAMP was covalently attached to a single protein that runs on sodium dodecyl sulfate–polyacrylamide gel electrophoresis (SDS-PAGE) as a diffuse band of 41,000–45,000 M_r (Fig. 8). The diffuse appearance of the band suggests that it is modified posttranslationally, a possibility we are currently testing. Labeling of this band quantitatively correlates with receptor binding activity in a wide variety of experimental conditions consistent with the known properties of the receptor. Both are competed by cAMP in the range 10^{-8}–10^{-6} M; by 2'-H-cAMP in the range 10^{-7}–10^{-5} M; by 8-Br-cAMP in the range 10^{-6}–10^{-4} M; by N-6-butyryl-cAMP in the range 10^{-5}–10^{-3} M, and by adenosine in the range 10^{-4}–10^{-2} M; neither is competed by 5'-AMP. Both binding of $[^{32}P]$-8-N_3-cAMP and the band on the gel saturate at about 10^{-5} M. Both also have identical time courses of developmental regulation, increasing to a peak activity by 6 hr after starvation and decreasing to low levels at the completion of aggregation by 10–12 hr. These observations strongly suggest that the 41,000–45,000-M_r band contains the binding site of the cell surface receptor. These techniques provide a simple method to identify the receptor on cell surfaces and should permit its purification.

6. Putative GTP-Binding Regulatory Protein in *D. discoideum*

The activation and subsequent reversible adaptation of adenylate cyclase *via* surface cAMP receptors in *D. discoideum* is analogous to the activation–desensitization of the enzyme by receptors for hormones and neurotransmitters in vertebrates. In these systems, the hormone receptor is coupled to the catalytic unit *via* a regulatory component, designated G_s. G_s is a GTP binding protein that exhibits inherent GTPase activity. Persistent activation of adenylate

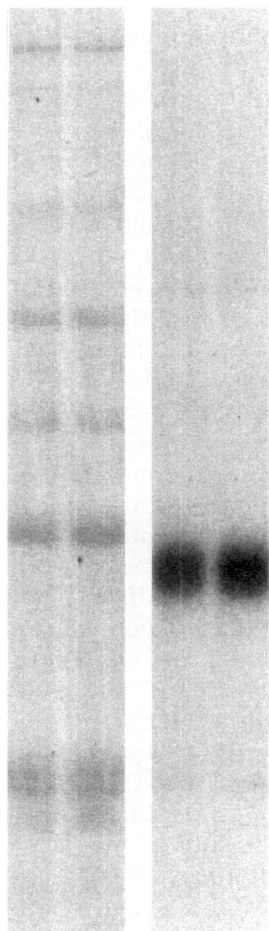

Figure 8. Identification of the surface receptor on *D. discoideum* by photoaffinity labeling. Cells were mixed with [^{32}P]-8-N$_3$-cAMP (5 × 10^{-6} M) washed with saturated (NH$_4$)$_2$ SO$_4$ and the pellet irradiated for 5 min. Cells were lysed by osmotic shock and membranes collected by centrifugation. The pellet was dissolved in SDS sample buffer and analyzed by SDS-PAGE. Gels were stained, dried, and autoradiographed for 15 hr with intensifying screens. Shown are duplicate experiments of Coomassie blue staining (left) and autoradiography (right). The major stained band (left) just above the major radioactive band (right) is *D. discoideum* actin (43,000M$_r$).

cyclase is attained in the presence of nonhydrolyzable GTP analogues (GMP-PNP or GTP-γ-S) or when the GTPase is inactivated by treating membranes with cholera toxin and NAD$^+$ (for a review see Ross and Gilman, 1980). In *D. discoideum*, adenylate cyclase in sonicates prepared from cells at the peak of the cAMP signaling response synthesizes [^{32}P]-cAMP 5–20 times faster than those from unstimulated cells. The activation decays rapidly after cell lysis (t ½ = 20–30 sec at 22°C; t ½ = 5 min at 0°C). Subsequently, the basal activity is relatively stable but unresponsive to cAMP (the hormone in this case), GTP (or GMP-PNP, GTP-S), and cholera toxin–NAD$^+$ (Roos and Gerisch, 1976). Thus, although the response of the cells is very similar to that seen in vertebrates, the *in vitro* system differs; the mechanism of regulation of the enzyme is unknown.

Leichtling *et al.* (1981) reported the occurrence of an apparent G$_s$ protein in *D. discoideum* and cited the following evidence: (1) a protein of 42,000 M$_r$ specifically binds 8-N$_3$-GTP-γ-[^{32}P];(2)a protein of 42,000M$_r$ is ADP-[^{32}P] ribosylated by cholera toxin and NADα-[^{32}P]; and (3) each focuses on a two-dimensional gel in a position similar to the 42,000-M_r subunit of the G_s regulatory protein from vertebrates. We have repeated and extended these observations. In membrane preparations, major bands of 52,000, 44,000, and 38,000 M_r and minor bands at 48,000 and 42,000 M_r are ADP-ribosylated. Labeling is completely dependent on cholera toxin (1–20 μg/ml), GMP-PNP (2–20 μM) and a cytosolic factor from pigeon erythrocytes. On two-dimensional gels, the 44,000-M_r band runs as a series of four spots slightly more acidic and above actin, as Leichtling *et al.* (1981) showed.

In vertebrates, the G_s regulatory protein is reported to be a peripheral membrane protein, although there are reports of soluble G_s in brain, liver, and pigeon erythrocytes, and a homologous protein in rod outer segments is soluble. We have found that the putative G_s polypeptides (44,000 and 42,000 M_r) are also found in supernatants of lysed *D. discoideum*. ADP ribosylation of high-speed supernatants predominantly modifies two bands, at 44,000 and 52,000 M_r (Fig. 9). The former composed about 50–90% of the labeled material (six independent preparations). The reaction is dependent on cholera toxin (1–8 μg/ml), GMP-PNP (1–20 μM) and is abolished by brief heat treatment (65°C, 5') of the extract. A 44,000-M_r polypeptide is also the major band in high-speed supernatants labeled by 4'-azidoanilido-5'-GTP-γ-[^{32}P]. The dependence of the ADP ribosylation reaction of GMP-PNP suggests that the same polypeptide also binds the GTP photolabel. The band at 44,000 M_r labeled by each method was excised from a preparative SDS-slab gel and subjected to limited proteolytic digestion with *Staphylococcus aureus* V8 protease. The two peptide maps are nearly identical, strongly suggesting identity of the two labeled species. By two-dimensional gel analysis the major ADP ribosylated band (44,000 M_r) separates into a series of four spots slightly acidic to and above actin (as does the membrane bound form). The major band (44,000 M_r) labeled with 4'-azidoanilido-5'-GTP-γ-[^{32}P] runs as one major and one minor spot in the same region (Fig. 9).

The structural similarity of the major ADP ribosylated polypeptide (44,000 M_r) in *Dictyostelium* to the ADP-ribosylated G_s regulatory component in human erythrocytes was compared by one-dimensional peptide mapping (Fig. 9). The two polypeptides are not identical, since more peptides are generated by *Staph. aureus* V8 protease cleavage of the human G_s compared with the *D. discoideum* polypeptide. However, there are peptides in common, and the similarity in the overall patterns generated may indicate that the polypeptides are structurally related.

7. Progress in Purification of Adenylate Cyclase and Receptor

Most attempts to study the adenylate cyclase have been hindered by the extreme instability of the enzyme. We have recently developed methods to purify the enzyme about 300-fold and greatly increase its stability (t ½ = 2 hr vs. t ½ = 1 day at 0°C). All the activity is membrane bound. Although extraction

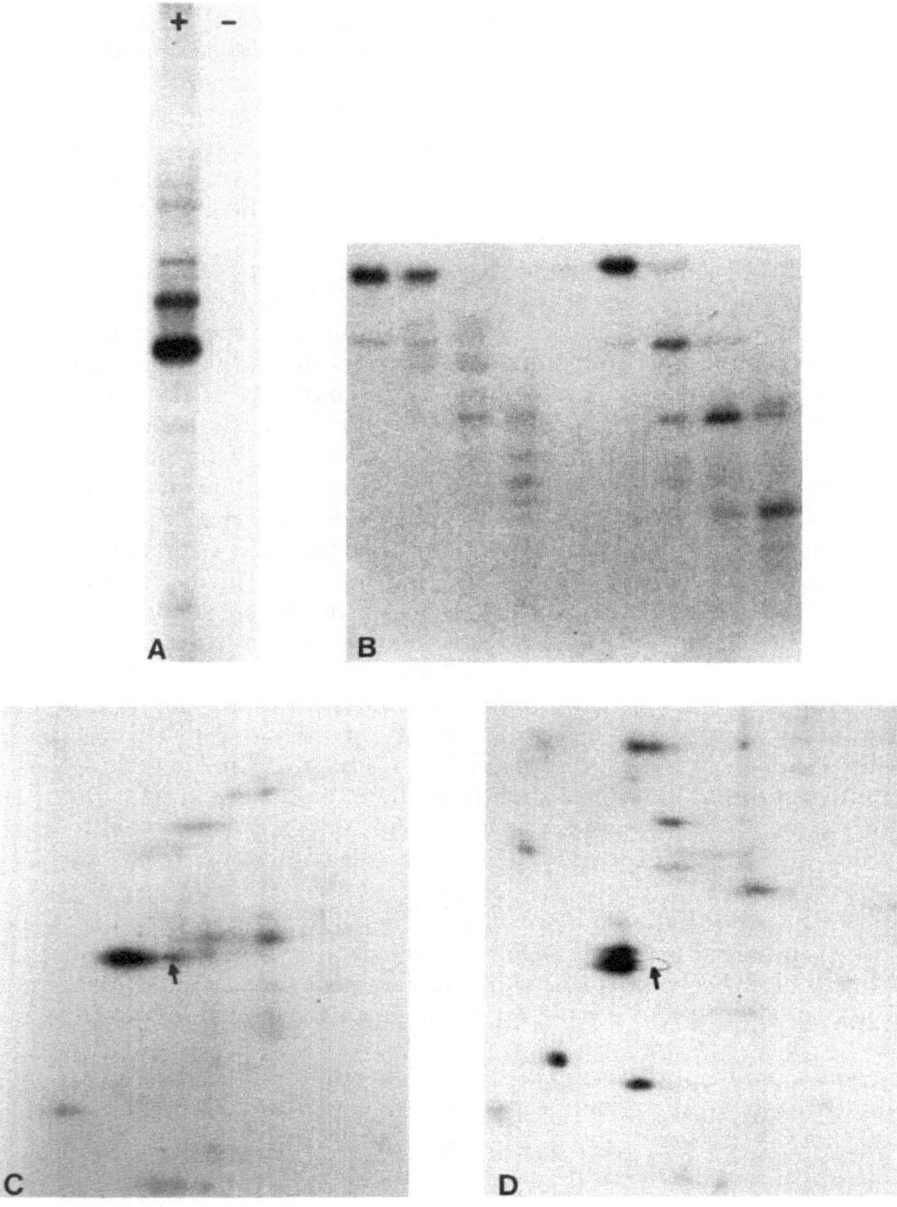

Figure 9. Putative G_s regulatory protein in *D. discoideum*. Cells were lysed and centrifuged at
$50,000 \times g$ for 1 hr. Supernatant was used for all experiments illustrated here. (a) Identification of
band by ADP-ribosylation. Supernatant was incubated with [^{32}P]-NAD$^+$ in the presence (left lane)
or absence (right lane) of 20 µg/ml activated cholera toxin. (b) Peptide maps comparing human
erythrocyte G_s regulatory protein and *D. discoideum* cytosolic protein. Each was labeled with [^{32}P]-
NAD$^+$ and cholera toxin and then eluted from an SDS gel. Aliquots were incubated with 0, 2, 20, or
200 µg/ml *Staph. aureus* V8 protease for 30 min before the SDS-PAGE analysis shown. Left series is

Table II. Purification of Adenylate Cyclase and cAMP Receptor

Step	Protein (% recovery)	Adenylate cyclase		Receptor	
		(% recovery)	(fold purification)	(% recovery)	(fold purification)
1. Cells	100	—	—	100[a]	1
2. Lysate	100	100[b]	1	—	—
3. Membranes[c]	20	80	4	26[d]	1
4. CHAPS pellet[e]	0.6	120	200	19	32
5. Salt washed[f]	0.3	93	310	10	33

[a]Developed cells contain approximately 10^{12} receptors per milligram protein.
[b]Approximately 6 pmol/min per mg protein.
[c]Cells pushed through 5 μM filter and spun at 4500 × g for 20 min.
[d]Low recovery of receptor activity into membrane is the result of a recently discovered technical problem. (At least 60% of photoaffinity labeled receptor is found in membranes.)
[e]Membranes extracted with CHAPS then centrifuged at 30,000 × g for 45 min.
[f]For adenylate cyclase, CHAPS pellet washed with 2 M NaCl, 5 mM pyrophosphate, and 0.5 M urea. For receptor, CHAPS pellet washed with 2 M NaCl, 5 mM pyrophosphate.

of membranes in 16 mM CHAPS solubilizes more than 95% of the membrane protein, nearly all the adenylate cyclase activity remains particulate. The CHAPS extraction followed by washing leads to approximately 200-fold purification; further treatment with CHAPS containing 2 M NaCl, 5 mM PP_i, and 1 M urea increases the fold purification and markedly increases the stability (Table II). These developments should permit identification of the enzyme and greatly facilitate further studies of its regulation.

Electron microscopic examination of this enriched material revealed a homogeneous preparation of vesicles about 0.2 μm in diameter (see Fig. 10). Lipid is probably present, since a definite bilayer is seen. SDS-PAGE shows two major protein bands of 30,000 and 33,000 M_r and about 20 minor bands ranging from 105,000 to 15,000 M_r. The two major bands are enriched over the starting membranes and may maintain the structure of the vesicles. Although the preparations have not been characterized to the same extent, adenylate cyclase behaves similarly in 0.5% Triton X-100, 0.5% Lubrol-PX, 16 mM CHAPSO, 10 mM octylglucoside, and 0.5% sodium cholate. The nearly quantitative retention of activity in the purified vesicle preparation suggests that it is in a specialized membrane domain resistant to detergent extraction. Triton X-100-resistant bilayer structures with similar SDS-PAGE patterns have been previously reported (Luna *et al.*, 1981; Spudich and Spudich, 1982). Recent evidence suggests that the detergent-resistant membrane fragments are plasma membrane derived. Antiserum directed against the vesicles reacts strongly with

human erythrocyte G_s regulatory protein; right series is *D. discoideum* supernatant protein. (c,d) Two-dimensional gel analysis of *D. discoideum* supernatant protein (c) and human erythrocyte G_s-regulatory protein (d). Each was photoaffinity labeled with [^{32}P]azidoanilido-GTP before electrofocusing. Arrows indicate the position of actin run as an internal standard.

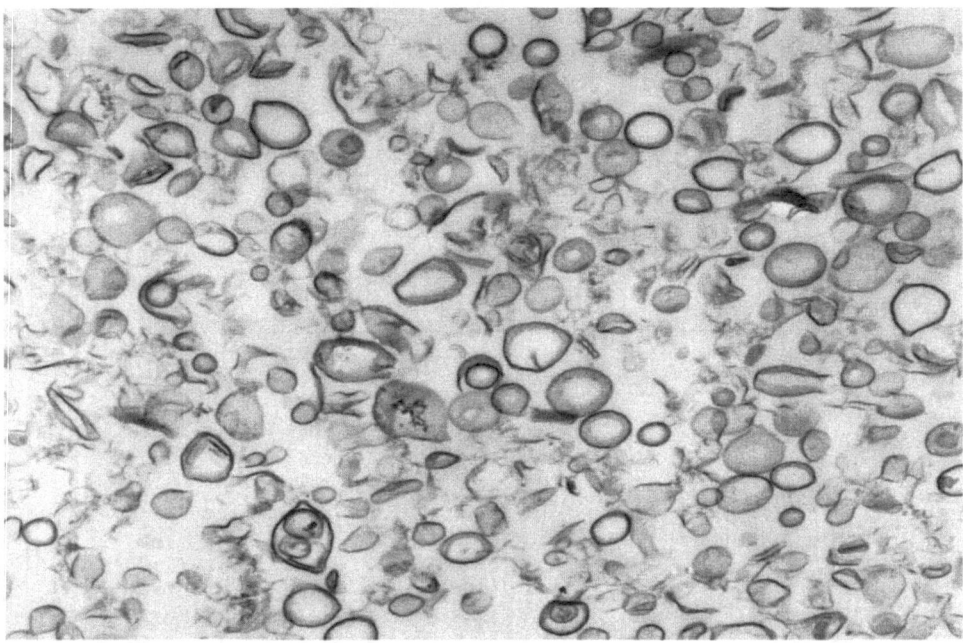

Figure 10. Vesicular adenylate cyclase preparation. Transmission electron micrograph (20,000×) of CHAPS extracted, salt-washed membranes. Fraction exhibiting 300-fold-enriched adenylate cyclase activity.

intact cells. In Western blots, the serum stains about 20 polypeptides. Most of the stained bands disappear when the serum is preabsorbed with intact aggregation-competent cells at 0°C.

We have recently solubilized about 15% of the adenylate cyclase from these detergent-resistant membranes. The solubilization method does not disrupt the vesicles, since the solubilized preparation does not contain the major 31,000- and 33,000-M_r bands. We have begun to examine the chromatographic behavior of the solubilized enzyme. The enzyme adsorbs to DEAE and is eluted by 150 mM NaCl. By gel filtration, the activity behaves as a very large molecule, slightly included on Sepharose CL-4B.

The stabilization of [^{3}H]-cAMP binding to the surface receptor has also allowed for its partial purification in native form. For purification we have devised an assay that takes advantage of the known binding properties of the surface receptor. 5′-AMP does not prevent binding, and the binding is rapid and reversible. [^{3}H]-cAMP binding to subcellular fractions is carried out for 10 sec in the presence of 10^{-4} M 5′-AMP. Nonspecific binding is taken as radioactivity not chased within 30 sec by 10^{-5} M unlabeled cAMP. Specifically bound counts are stabilized and washed in ammonium sulfate. As shown in Figure 11, the receptor detected in membrane fractions using this assay has the same properties as the cell surface receptor. [³H]-cAMP binding to cells or membranes saturates at 10^{-7} M [³H]-cAMP and is effectively competed by 10^{-8}–10^{-6} M cAMP. Using this assay, we have shown that the receptor is highly enriched in the detergent-resistant vesicle fraction (Table II). The recovery of receptor into this fraction is not as high as for adenylate cyclase. However, no receptor activity has been detected in other fractions, and the lower recovery is likely due to insufficient optimization of the receptor assay under all conditions. Photoaffinity labeling of the purified fractions (step 4, Table II) with 8-N₃-cAMP revealed the same 41,000–45,000-M_r band as labeling of intact cells.

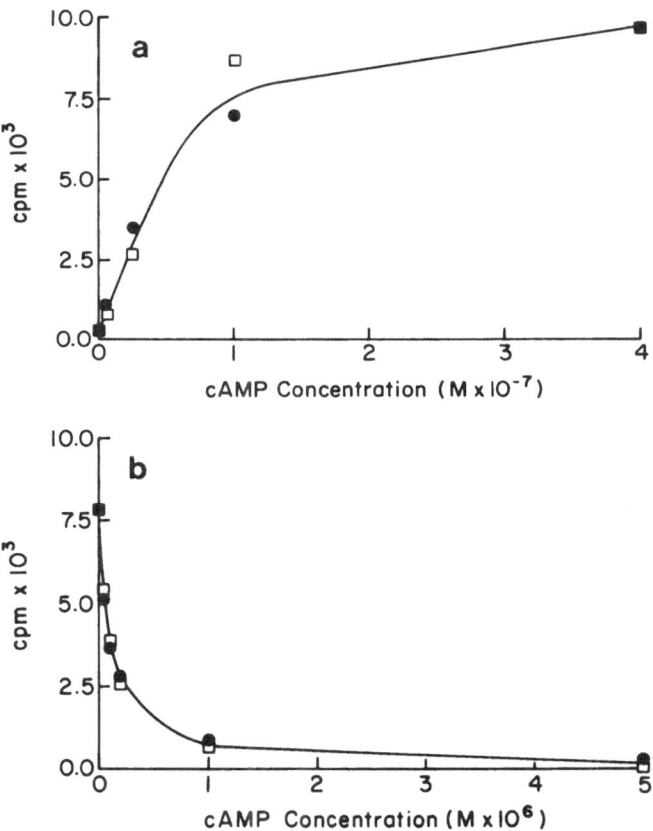

Figure 11. Comparison of [³H]-cAMP binding to intact cells (●) and membranes (□). Fifty-μl cells or membranes were incubated with the indicated concentration of [³H]-cAMP (a) or 10^{-7} M [³H]-cAMP plus the indicated concentration of nonradioactive cAMP (b). Reaction mixture contained 10^{-4} M 5′-AMP and 10 mM DTT to inhibit phosphodiesterase. To stabilize binding, 3 ml saturated ammonium sulfate was added, and samples were washed twice in ammonium sulfate. For total binding, incubation was for 10 sec before addition of ammonium sulfate. For nonspecific binding, 10-sec incubation was followed by 10^{-5} M nonradioactive cAMP for 30 sec before addition of ammonium sulfate.

8. Summary

We have described in *D. discoideum* a highly organized cell aggregation that is mediated by cAMP. After suitable differentiation induced by starvation, the cells develop the capacity to orient in gradients of cAMP and to secrete cAMP in response to cAMP. This signaling response sets up the cell–cell relay of cAMP waves that transiently orients the cells toward the center (see Figs. 1 and 2). Both the signaling response and the chemotactic response, measured in isolated cells, adapt (see Figs. 3 and 4). The kinetics and properties of adaptation of the two responses are similar and may be due to the same mechanism. The mechanism does not involve protein synthesis, a change in the number or affinity of surface receptors, or the activation of adenylate cyclase. Adaptation of signaling is essential for the oscillatory production of cAMP at the aggregation centers and ensures that the cAMP waves move steadily toward the edge of the aggregation territories. Adaptation of the chemotactic response also ensures that cells do not reorient away from the center in the gradient presented by the trailing edge of the wave.

We have demonstrated that both chemotaxis and cAMP signaling are mediated by the same surface receptor (Fig. 7). The polypeptide containing the binding site of the receptor has been identified by photoaffinity labeling with $[^{32}P]$-8-N_3-cAMP as a diffuse band of 41,000–45,000 M_r (Fig. 8). The receptor and adenylate cyclase copurify on a homogeneous class of vesicles resistant to extraction by nonionic detergents (Fig. 10). A GTP-binding protein that is a substrate for cholera toxin-catalyzed ADP ribosylation is found in supernatants and membranes and may be similar to the G_s regulatory protein of adenylate cyclase in higher organisms.

The mechanism of activation of the adenylate cyclase and chemotactic machinery is unknown. We have been able to inhibit the activation of the adenylate cyclase selectively and rapidly with agents acting to crosslink cell surface components, which may give a clue to the activation mechanism.

The elaborate mechanisms of cell–cell communication occurring in *D. discoideum* are without precedent in biological literature, although models of oscillatory wave propagation have been proposed to account for pattern formation. Although it is unlikely that extracellular cAMP would be involved, it is not inconceivable that such mechanisms occur during the development of more evolutionarily advanced organisms. The organized communication system in *D. discoideum* is only apparent when cells are plated uniformly on a flat surface; such organized movements occurring in a three-dimensional structure such as an embryo would be very difficult to discern.

References

Bonner, J. T., 1982, Comparative biology of cellular slime molds, in: *The Development of Dictyostelium discoideum* (W. Loomis, Jr., ed.), pp. 1–33, Academic Press, New York.

Brenner, M., and Thoms, S. D., 1984, Caffeine blocks activation of cyclic AMP synthesis in *Dictyostelium discoideum*, *Dev. Biol.* **101:**136–146.

Condeelis, J., 1979, Isolation of Concanavalin A caps during various stages of formation and their association with actin and myosin, *J. Cell Biol.* **80:**751–758.

Devreotes, P. N., 1982, Chemotaxis, in: *The Development of Dictyostelium discoideum* (W. Loomis, Jr., ed.), pp. 117–168, Academic Press, New York.

Devreotes, P. N., and Steck, T. L., 1979, Cyclic 3′,5′-AMP relay in *Dictyostelium discoideum*. II. Requirements for initiation and termination of the response, *J. Cell Biol.* **80:**300–309.

Devreotes, P. N., Derstine, P. L., and Steck, T. L., 1979, Cyclic AMP relay in *Dictyostelium discoideum*. I. A technique to monitor responses to control stimuli, *J. Cell Biol.* **80:**291–299.

Dinauer, M., MacKay, S., and Devreotes, P., 1980a, Cyclic 3′,5′-AMP relay in *Dictyostelium discoideum*. III. The relationship of cAMP synthesis and secretion during the cAMP signaling response, *J. Cell Biol.* **86:**537–544.

Dinauer, M., Steck, T., and Devreotes, P., 1980b, Cyclic 3′,5′-AMP relay in *Dictyostelium discoideum*. IV. Recovery of the cAMP signaling response after adaptation to cAMP, *J. Cell Biol.* **86:**545–553.

Fontana, D. R., and Devreotes, P. N., 1984, cAMP-stimulated adenylate cyclase activation in *Dictyostelium discoideum* is inhibited by agents acting at the cell surface, *Dev. Biol.* **106:**76–82.

Gillette, M. U., Dengler, R. E., and Filosa, M. F., 1974, The localization and fate of Concanavalin A in amoebae of the cellular slime mold, *Dictyostelium discoideum*, *J. Exp. Zool.* **190:**243–248.

Gunther, G. R., Wang, J. L., Yahara, I., Cunningham, B. A., and Edelman, G. M., 1973, Concanavalin A derivatives with altered biological activities, *Proc. Natl. Acad. Sci. USA* **70:**1012–1016.

Hellio, R., and Ryter, A., 1980, Relationships between anionic sites and lectin receptors in the plasma membrane of *Dictyostelium discoideum* and their role in phagocytosis, *J. Cell Sci.* **41:**89–104.

Juliani, M., and Klein, C., 1981, Photoaffinity labeling of the cell surface adenosine 3′,5′-monophosphate receptor of *Dictyostelium discoideum* and its modification in down-regulated cells, *J. Biol. Chem.* **256:**613–619.

Konijn, T., 1970, Microbiological assay of cyclic 3′,5′-AMP, *Experientia* **26:**367–369.

Leichtling, B. H., Coffman, D. S., Yaeger, E. S., Rickenberg, H. U., Al-Jumaily, W., and Haley, B. E., 1981, Occurrence of the adenylate cyclase "G protein" in membranes of *Dictyostelium discoideum*, *Biochem. Biophys. Res. Commun.* **102:**1187–1195.

Luna, E. J., Fowler, V. M., Swanson, J., Branton, D., and Taylor, D. L., 1981, A membrane cytoskeleton from *Dictyostelium discoideum*. I. Identification and partial characterization of an actin binding activity, *J. Cell Biol.* **88:**396–409.

Pan, P., Hass, E. M., and Bonner, J. T., 1972, Folic acid as second chemotactic substance in the cellular slime molds, *Nature New Biol.* **237:**181–191.

Roos, W., and Gerisch, G., 1976, Receptor mediated adenylate cyclase activation in *Dictyostelium*, *FEBS Lett.* **68:**170–172.

Ross, E. M., and Gilman, A. G., 1980, Biochemical properties of hormone-sensitive adenylate cyclase, *Annu. Rev. Biochem.* **49:**533–564.

Spudich, J. A., and Spudich, A., 1982, Cell motility, in: *The Development of Dictyostelium discoideum* (W. Loomis, Jr., ed.), pp. 169–194, Academic Press, New York.

Theibert, A., and Devreotes, P., 1983, Cyclic 3′,5′-AMP relay in *Dictyostelium discoideum*: Adaptation is independent of activation of adenylate cyclase, *J. Cell Biol.* **97:**173–177.

Tomchik, K. J., and Devreotes, P. N., 1981, cAMP waves in *Dictyostelium discoideum*: Demonstration by a novel isotope dilution fluorography technique, *Science* **212:**443–446.

van Haastert, P., 1983a, Sensory adaptation of *Dictyostelium discoideum* cells to chemotactic signals, *J. Cell Biol.* **96:**1559–1565.

van Haastert, P. 1983b, Binding of cAMP and adenosine derivatives to *Dictyostelium discoideum* cells. Relationships of binding, chemotactic, and antagonistic activities, *J. Biol. Chem.* **258:**9643–9648.

van Haastert, P., and Kien, E., 1983, Binding of cAMP derivatives to *Dictyostelium discoideum* cells. Activation mechanism of the cell surface cAMP receptor, *J. Biol. Chem.* **258:**9636–9642.

Wallace, L., and Frazier, W., 1979, Photoaffinity labeling of cyclic-AMP and AMP-binding proteins of differentiating *Dictyostelium discoideum* cells, *Proc. Natl. Acad. Sci. USA* **76:**4250–4254.

Wang, J. L., Gunther, G. R., and Edelman, G. M., 1976, Properties of dimeric Con A derivatives, in: *Concanavalin A as a Tool* (H. Bittiger and H. P. Schnebli, eds.), pp. 581–595, Wiley, London.

Chapter 14

Growth Cone Guidance and Cell Recognition in Insect Embryos

COREY S. GOODMAN, MICHAEL J. BASTIANI, CHRIS Q. DOE, and SASCHA DULAC

1. Introduction

In contrast to the complex central nervous system of most vertebrates, the insect CNS is relatively simple. The grasshopper CNS, for example, consists of a brain and a chain of segmental ganglia, each of which contains about 1000 neurons. Most of these neurons can be individually identified according to their unique axonal and dendritic morphology and their unique pattern of synaptic connections. This notion of unique identified neurons first arose during the nineteenth century with descriptions of giant axons, which were repeatedly located in particular regions of the nerve cord. With the advent of intracellular dye-injection techniques (e.g., Stretton and Kravitz, 1968; Pitman et al., 1972; Stewart, 1978), many neurons became individually identified according to the location of their axons and dendrites in the neuropil and connectives. The spatial relationship of these processes to one another within the neuropil was then explored (e.g., Tyrer and Gregory, 1982); it was shown that the axons and dendrites of identified neurons run in particular regions of specific tracts and commissures.

For example, the G neuron in the grasshopper mesothoracic (T2) segmental ganglion (Fig. 1A) extends its primary axons anteriorly in a particular region of one specific tract and its symmetrical primary dendrites bilaterally in a different tract (Fig. 1B) (e.g., Bastiani et al., 1984a). These associations of neuronal processes in specific regions of the neuropil appear to influence the choice of synaptic partners profoundly by limiting the neighbors among which each neuron has to choose. How individual axons and dendrites find their way into these neighborhoods, and how specific the association of particular neuronal processes with one another, remained largely unknown until recently. With the advent of studies on the embryonic development of individual neurons in

COREY S. GOODMAN, MICHAEL J. BASTIANI, CHRIS Q. DOE, and SASCHA DULAC • Department of Biological Sciences, Stanford University, Stanford, California 94305.

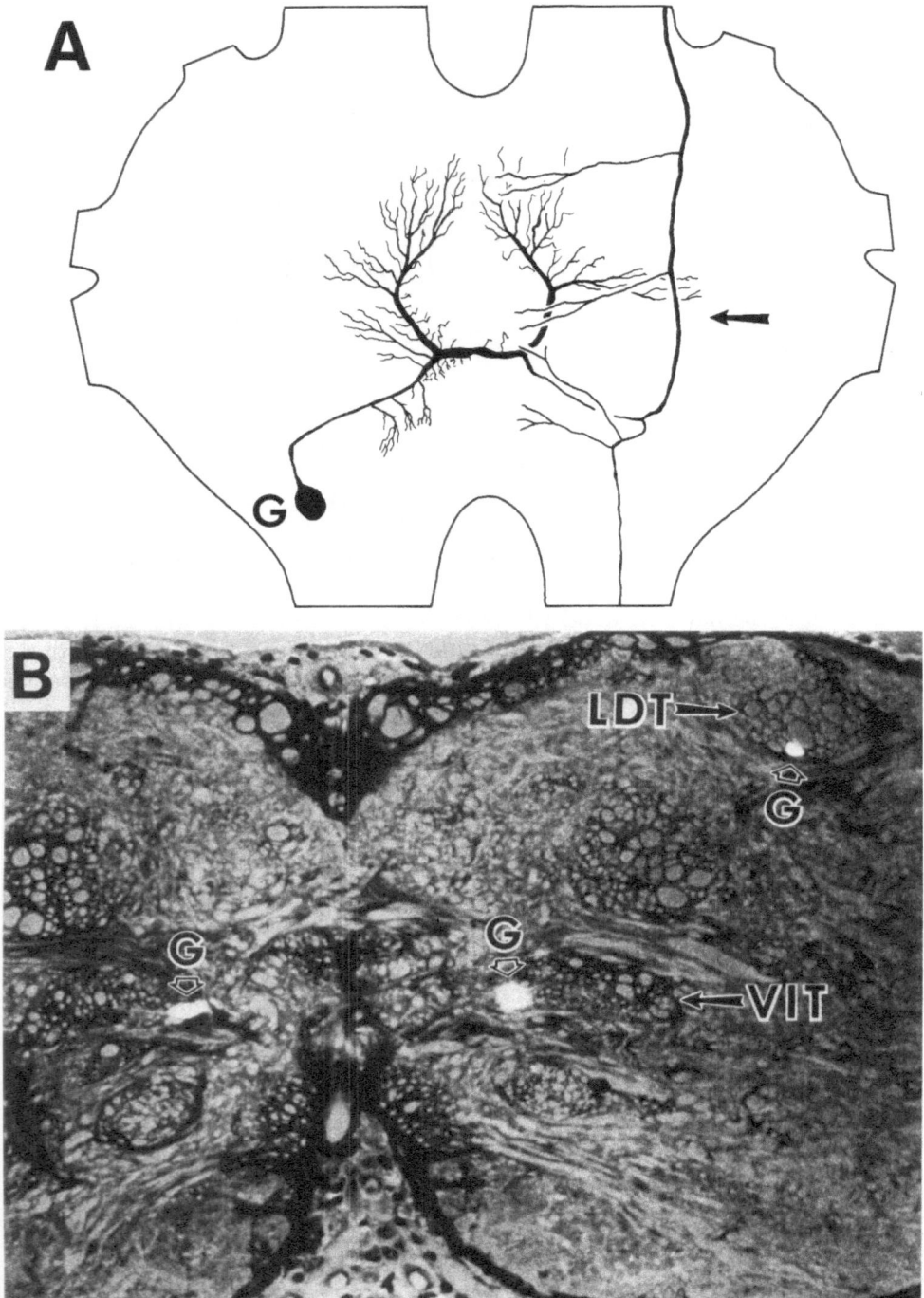

simple nervous systems, one could begin to ask questions about the mechanisms underlying neuronal specificity.

In 1977 we first began studying the development of identified neurons in the grasshopper embryo (Goodman and Spitzer, 1979). Fortunately, the grasshopper embryo is ideally suited to such studies because the individual neurons (and in particular their growth cones) are large and accessible, and the environment in which they make their choices is relatively simply arranged. One of the early observations was that the growth cones of identified neurons make cell-specific stereotyped pathway choices. Confronted with the same environment, different growth cones made very different choices as to which way to grow (e.g., Raper et al., 1983a).

We next learned that in many cases the substrates for these differential choices were the surfaces of other neurons, in particular their axons (Raper et al., 1983b,c; Bastiani et al., 1984b). These observations led to the realization that the tracts and commissures of the adult neuropil were prefigured in the embryo by an orthogonal scaffold of axon bundles (Fig. 2A) (Bastiani et al., 1984a). For example, confronted with the surfaces of about 100 axons organized into 25 different axon fascicles, the G growth cone invariably chooses to fasciculate on a discrete bundle of four axons called the A/P fascicle, and more specifically, on the surface of the P axons within that fascicle (Fig. 3A) (Bastiani et al., 1984b); the A/P fascicle in the embryo gives rise to the LDT tract in the adult.

From these studies arose the notion that many of the earliest events of cell recognition in the developing CNS involve the specific choices made by growth cones as they extend onto particular axonal surfaces. The results suggested that the selective affinities of filopodia guide growth cones and that these differential affinities give rise to the stereotyped patterns of selective fasciculation. From these ideas emerged the labelled pathways hypothesis, which predicts that different neighboring axon fascicles in the embryonic neuropil are differentially labeled by surface-recognition molecules used by growth cones for their selective fasciculation (Goodman et al., 1982; Raper et al., 1983b).

This hypothesis was initially tested experimentally and confirmed for the G growth cone (Raper et al., 1983c, 1984). For example, when the P axons are ablated, the G growth cone behaves abnormally and does not show a high affinity for any of the remaining 100 or so axons, including the A axons. The degree of neuronal specificity observed in these experiments is remarkably exquisite.

Figure 1. Identified neuron in the central nervous system of the grasshopper. The G neuron in the mesothoracic (T2) segmental ganglion of the adult grasshopper. (A) Camera lucida drawings of a Lucifer Yellow filled neuron. (B) Cross section of the neuropil of the T2 segmental ganglion showing the tracts containing the primary axon and two primary dendrites of G. Thick plastic cross section through the mesothoracic ganglion of an adult grasshopper in which G had been filled with Lucifer Yellow; section is taken at level shown by arrow in part A. The G neuron extends its primary axon anteriorly in a particular region of the contralateral lateral dorsal tract (LDT) and its symmetrical dendrites bilaterally in symmetrical regions of the ventral intermediate tracts (VIT). Scale bars: (A) 200 μm; (B) 100 μm. (Courtesy of Keir Pearson.)

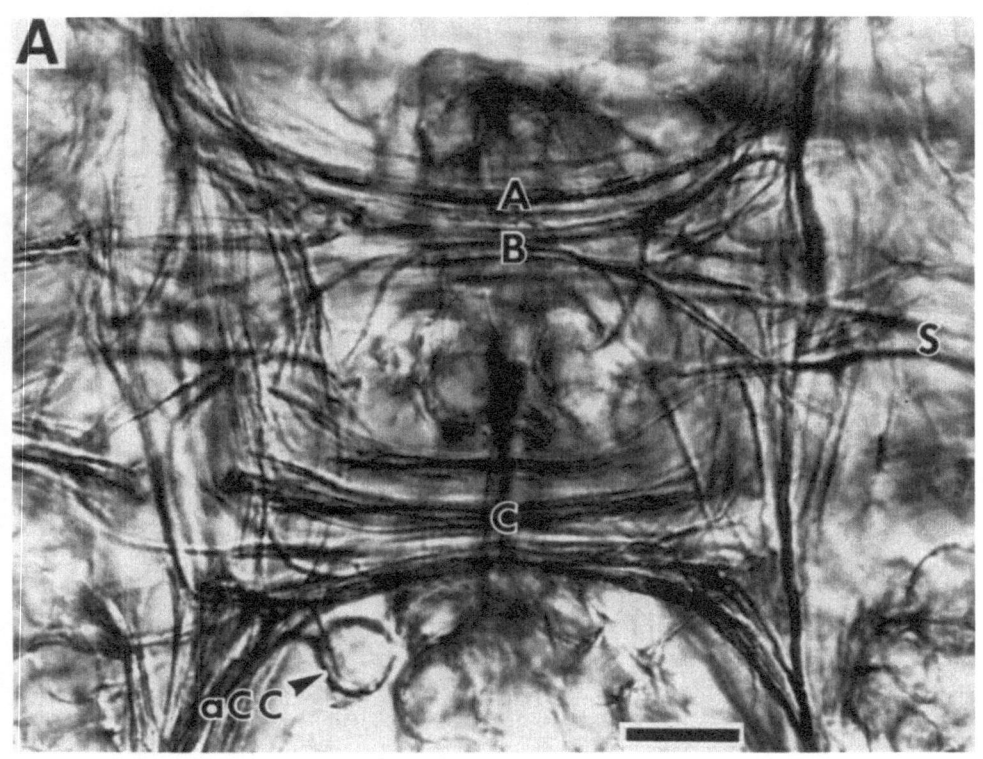

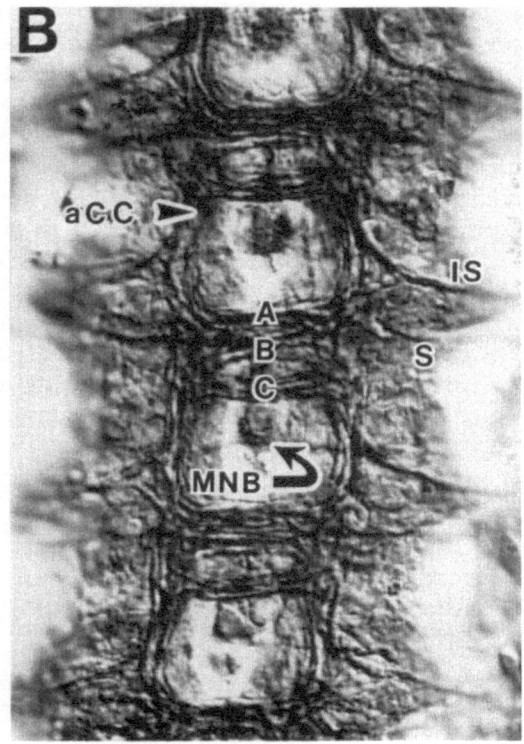

Moreover, the basic observation of selective fasciculation, and the specific affinities of identified growth cones for particular axon surfaces, was soon extended from the grasshopper embryo to the *Drosophila* embryo (Figs. 2B and 3B) (Thomas *et al.*, 1984), making it possible to begin to use these two insect embryos in tandem for a combined cellular and molecular genetic approach (Goodman *et al.*, 1984).

Here we review our recent experimental analysis on selective guidance and recognition by the growth cones of the sibling aCC and pCC neurons in the grasshopper and *Drosophila* embryos (Bastiani *et al.*, 1985). These two growth cones and five others selectively interact with one another and with specific nonneuronal cells and surfaces to form the first three longitudinal axon fascicles and intersegmental nerve in the grasshopper embryo (Fig. 4). Thus, this highly defined and simple network of seven neurons displays the basic three selective affinities characteristic of more complex nervous systems: neuron-to-basement membrane, neuron-to-epidermal (glial) cell, and neuron-to-neuron. In contrast to more complex nervous systems, however, neuron-to-neuron recognition in insect embryos can readily be studied at the level of individual cell affinities.

Our experimental analysis has demonstrated a degree of specificity far beyond what we had initially anticipated: The selective affinities of these growth cones for particular neuronal surfaces are absolute and nonhierarchical, even when temporal and spatial relationships are altered. The results suggest the expression of several different surface specificities on just these first seven neurons.

2. The First Growth Cones in the Insect CNS

The first three longitudinal axon fascicles in the CNS of the grasshopper embryo initially contain the axons of seven identified neurons (Fig. 4). The growth cones of these first seven neurons are able to distinguish one another's surfaces and, by their specific interactions, selectively fasciculate with one another to form three bundles: the MP1/dMP2 fascicle (containing the MP1,

Figure 2. Orthogonal scaffold of axon fascicles in developing neuropil of (A) grasshopper and (B) *Drosophila* embryos. (A) Axon scaffold in a single segment of the grasshopper embryo at 40% of development, as viewed from the dorsal surface of the whole- mount neuroepithelium stained with the I-5 MAb and HRP immunocytochemistry. At this stage of development, the scaffold contains about 100 axons on each side (hemisegment) organized into 25 longitudinal axon fascicles. Each segmental unit also contains three lateral commissures joining the two hemisegments [one posterior (C) and two fused anterior (A,B)] and two peripheral pathways on each side, the intersegmental nerve (IS) at the segment boundary and the segmental nerve (S) at the segment midline [IS not shown in (A)]. (B) Axon scaffold in three segments of the *Drosophila* embryo at 12 hr of development, as viewed from the dorsal surface of the neuroepithelium stained with an anti-tubulin MAb. The same pattern of axon fascicles is seen in the fly as in the grasshopper, albeit on a smaller scale. The cell body of the aCC neuron is identified in both embryos, and that of the median neuroblast (MNB) in the fly embryo. Scale bar: (A) 20 μm; (B) 15 μm.

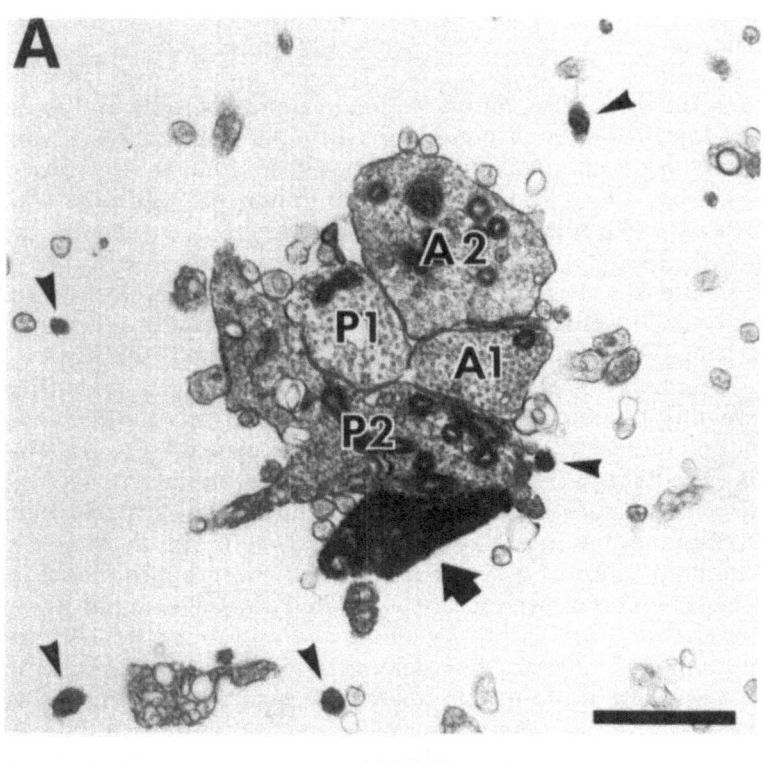

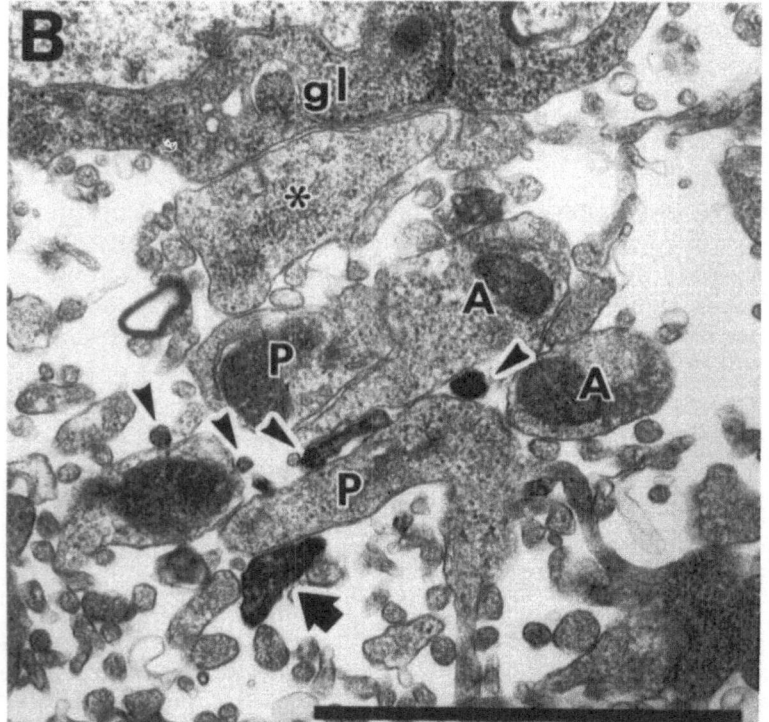

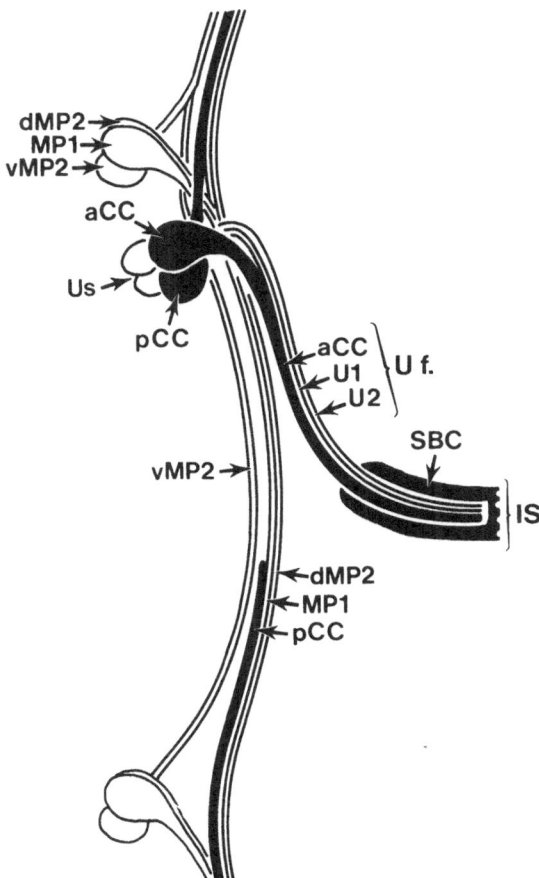

Figure 4. Semischematic diagram showing the first three longitudinal axon fascicles in the CNS of the grasshopper embryo, which initially contain the axons of seven identified neurons. The growth cones of these first seven neurons are able to distinguish one another's surfaces, and by their specific interactions, selectively fasciculate with one another to form these three bundles: the MP1/dMP2 fascicle (containing the MP1, dMP2, and pCC axons), the vMP2 fascicle (initially containing only the vMP2 axon), and the U fascicle (containing the U1, U2, and aCC axons), which turns laterally along the segmental boundary cell (SBC; an identified glial cell) to form the intersegmental nerve (IS).

dMP2, and pCC axons), the vMP2 fascicle (initially containing only the vMP2 axon), and the U fascicle (containing the U1, U2, and aCC axons).

In establishing the locations and directions of these fascicles, the early growth cones also appear to distinguish selectively the dorsal basement membrane, particular epidermal and/or glial cells, and possibly the axial anterior–posterior (A–P) polarity as well. Although our experimental analysis is focused

Figure 3. Selective fasciculation in the grasshopper and *Drosophila* embryos. Transmission electron micrographs of the A/P fascicle and the tip of the G growth cone in a 40% grasshopper embryo (A) and a 12-hr *Drosophila* (B) embryo from semiserial TEM sections of HRP-filled G neurons. In both species, the A/P fascicle contains the A1, A2, P1, and P2 axons, and the G growth cone at this stage. In both species, the tip of the G growth cone (large arrow) is in selective contact with the P axons rather than the A axons. Arrowheads show the filopodia of G. In *Drosophila* the axons are smaller in diameter and the fascicles are closer together than in grasshopper. For example, axon marked by asterisk in B is probably homologous to the D fascicle, which in grasshopper is approximately 10 μm dorsal to the A/P fascicle. Scale bars : 1 μm.

on the selective fasciculation by the aCC and pCC growth cones, we briefly review the selective affinities of all seven growth cones for a variety of non-neuronal and neuronal surfaces.

3. Selective Affinity for Nonneuronal and Neuronal Surfaces

The formation of the first three longitudinal axon fascicles involves a series of cell-specific choices and affinities for both non-neuronal and neuronal surfaces. First, a longitudinal space or channel appears just below the dorsal basement membrane in the neuroepithelium of the grasshopper embryo as epidermal cells retract their processes and end feet; the ganglionic neuropil and longitudinal connectives develop within this expanding space. The MP1, dMP2, and vMP2 neurons send their growth cones up to the dorsal basement membrane. Irrespective of the initial temporal and/or spatial relationship of their growth cones to one another and to the A–P axis, the vMP2 turns anteriorly, while the MP1 and dMP2 turn posteriorly (Bate and Grunewald, 1981; Goodman et al., 1982). The two likely sources of guidance for this choice are (1) polarity information in the form of an A–P gradient on the basement membrane (reflecting the positional information of the epidermal cells that initially secreted the membrane), and (2) positional information in the form of specific "landmark cells" (Taghert et al., 1982). For example, TEM serial section reconstructions show that the MP1 filopodia display a selective affinity for (1) the dorsal basement membrane, and (2) the pCC (as compared with any of the other cells and surfaces in their immediate environment) (Bastiani and Goodman, 1984a,b).

The MP1 and dMP2 growth cones extend posteriorly along the dorsal surface of this channel, i.e., the dorsal basement membrane (their axons however do not adhere to the membrane). The vMP2 growth cone, on the other hand, does not display a selective affinity for the dorsal basement membrane. Rather, it grows along the medioventral surface of the channel, in part extending along an identified glial cell (Fig. 5A,C). When the anteriorly extending vMP2 growth cone (from the next posterior segment) comes within a few microns of the posteriorly extending MP1 and dMP2 growth cones, they do not fasciculate with one another, but rather continue extending along their respective non-neuronal surfaces (Figs. 4 and 5C).

The neurons U1 and U2 send their growth cones up to the dorsal basement membrane. They come within several microns but do not fasciculate with the MP1/dMP2 axons; rather they extend posteriorly along the dorsal basement membrane, several microns lateral to the MP1/dMP2 axons (Figs. 5 and 6) (Bastiani et al., 1985b).

The aCC and pCC neurons arise at the anterior edge of the segment posterior to the one they eventually reside in. [They are the sibling progeny of GMC-1 (ganglion mother cell 1) from NB 1-1 (neuroblast 1-1).] They migrate about 100 μm anteriorly across the segment border to their final location (Fig. 6). The leading edge of the pCC extends around the lateral edge of the aCC and points directly toward the posteriorly extending MP1 and dMP2 growth cones. The pCC growth cone then extends anteriorly, fasciculating with the MP1 and dMP2 axons. When the pCC reaches the anterior extent of MP1/dMP2 cells, it stops; once the MP1/dMP2 growth cones from the next anterior segment come within reach, the pCC extends anteriorly, fasciculating with them (Fig. 4).

While the pCC growth cone fasciculates with the MP1/dMP2 axons, the aCC growth cone remains relatively stationary (Fig. 6). Although the filopodia of the aCC growth cone have access to the MP1/dMP2 axons, the dorsal basement membrane, and the glial cell followed by the vMP2, they do not show a high affinity for any of them. After 10–15 hr, the behavior of the aCC growth cone changes dramatically when the U1 and U2 growth cones appear on the dorsal surface within filopodial grasp. The aCC growth cone extends laterally toward, and then posteriorly along, the U axons.

When the U growth cones reach the segment border, they turn laterally along an identified glial cell called the segment boundary cell (SBC) (Fig. 4) (Goodman *et al.*, 1984; Bastiani *et al.*, 1985). Similarly, when the aCC growth cone reaches the segment border, it too turns laterally. However, although it extends posteriorly by fasciculating with the U axons, it sometimes leaves their surface at the segment border and displays a higher affinity for the SBC. Nevertheless, most growth cones appear uninterested in the SBC. For example, the MP1, dMP2, and vMP2 growth cones extend longitudinally at the segment border through the same channel without displaying any affinity for the SBC.

These observations suggest the following conclusions. First, particular growth cones display cell-specific affinities for particular non-neuronal surfaces. For example, the MP1 and dMP2 growth cones display an affinity for the dorsal basement membrane, the vMP2 growth cone for the longitudinal glial cell, and the U1, U2, and aCC growth cones for the intersegmental nerve glial cell called the SBC.

Second, neuronal growth cones do not simply fasciculate with other axons, nor do they indiscriminately prefer neuronal surfaces over non-neuronal surfaces. For example, the vMP2 does not fasciculate with the MP1 and dMP2 axons, the U axons do not fasciculate with MP1 and dMP2 axons, and the aCC sometimes leaves the U axons to grow separately on the SBC.

Third, particular neuronal growth cones display selective affinities for specific neuronal surfaces. For example, the MP1 and dMP2 growth cones display an affinity for each other's axons, the U1 and U2 growth cones for each other, the pCC growth cone for the MP1 and dMP2 axons, and the aCC growth cone for the U1 and U2 axons. Although all three types of selective affinities are important early in development for the establishment of axonal pathways, once the axon scaffold is erected, the third form predominates; later in development, growth cones in the CNS are guided largely by their selective fasciculation with specific axons. It is this highly specific neuron-to-neuron recognition that led to the labeled pathways hypothesis and that forms the basis of our experimental analysis.

In general, the same neurons display the same selective affinities for one another in the *Drosophila* embryo (Fig. 7B), albeit on a smaller temporal and spatial scale (Goodman *et al.*, 1984; Thomas *et al.*, 1984; Bastiani *et al.*, 1985). For example, the dMP2, MP1, and pCC axons join the MP1/dMP2 fascicle, whereas the aCC, U1, and U2 axons join the U fascicle. This is a fortunate stroke of evolutionary conservation, because the experimental analysis conducted in the grasshopper embryo would not be so easy in the fly. Moreover, the similarities extend to non-neuronal cells; at the segment border, the axons in the U fascicle extend laterally along the segment boundary cell in fly as in grasshopper.

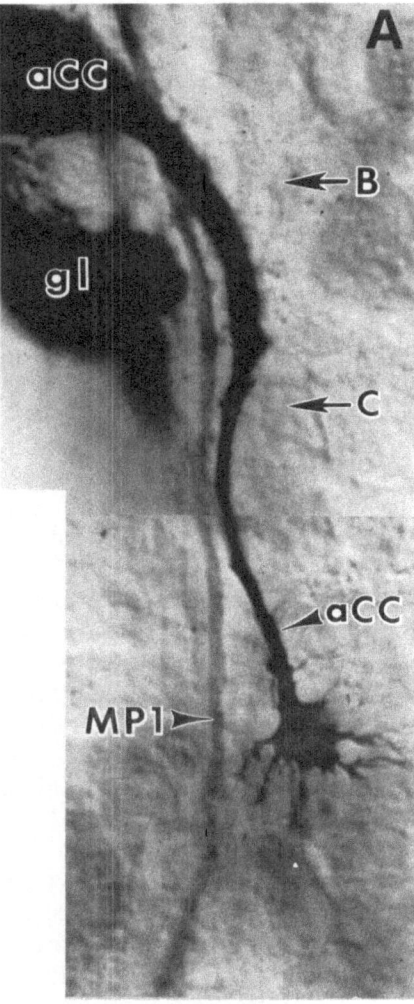

Figure 5. Selective affinity of the aCC growth cone for the U fascicle in the grasshopper embryo. Light (A) and electron (B,C) micrographs of HRP-filled neurons. (A) Dorsal light micrograph of whole-mount preparation (arrows note location of sections shown in B and C). At this stage (35%), the aCC growth cone has just reached the segment border by extending posteriorly along the U fascicle; at the segment border, the aCC growth cone is about to turn laterally along a large glial cell called the segment boundary cell (SBC; not shown). (B,C) TEM sections of the same preparation at

However, there are some interesting differences as well. A large channel with broad expanses of exposed basement membrane does not develop in the fly; nevertheless, the MP1/dMP2 and U fascicles develop in the appropriate places. TEM serial section reconstruction of the *Drosophila* embryo at hour 10:15 (Fig. 7B), a stage comparable to 35% in the grasshopper embryo (Fig. 7A), reveals some interesting temporal differences: (1) the vMP2 fascicle has not yet

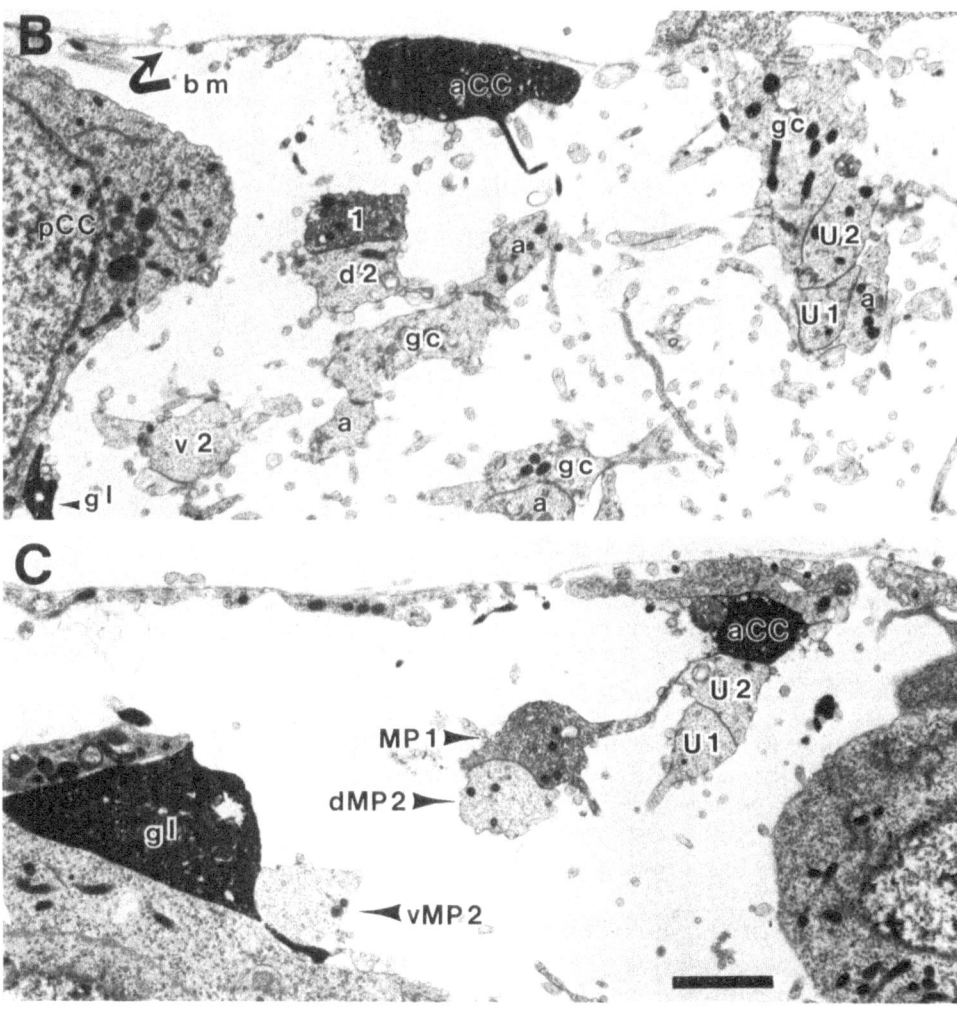

the locations marked in A. The pCC axon (not shown) extends anteriorly along the MP1/dMP2 fascicle and is not present in this plane of section. (C) The first three fascicles contain the axons of seven neurons, only six of which are shown in this plane of section (aCC and MP1 have been filled with HRP). gc, growth cone; a, axon; gl, glia of the connective; bm, basement membrane. Scale bar: (A) 12 μm; (B,C) 2 μm.

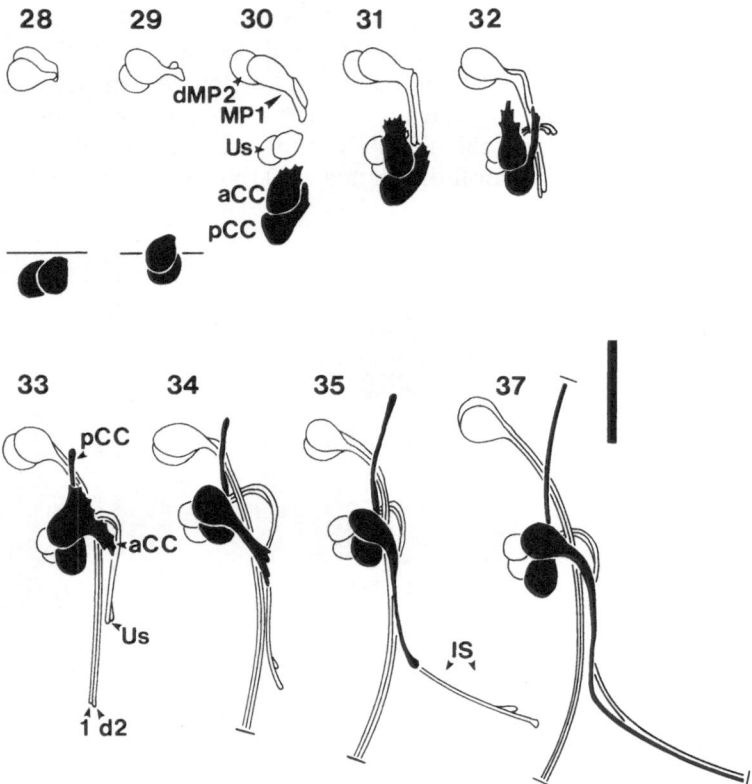

Figure 6. Temporal development of the pCC and aCC neurons, and the MP1/dMP2 and U fascicles, in the grasshopper embryo. Camera lucida drawings of the pCC, aCC, MP1, dMP2, U1, and U2 neurons (vMP2 not shown for simplicity) from whole-mount embryos stained with the I-5 MAb and HRP immunocytochemistry between 28% and 37% of embryonic development. IS, intersegmental nerve; horizontal line in 28 and 29, segment border over which aCC and pCC neurons migrate. Scale bar: 50 μm.

formed, (2) the MP1/dMP2 and U fascicles already contain several additional axons, and (3) the aCC is the leader rather than the follower in the U fascicle (Goodman *et al.*, 1984; Bastiani *et al.*, 1985).

4. Cell-Ablation Experiments

The labeled pathways hypothesis predicts that the first three fascicles are differentially labeled and that the aCC and pCC growth cones are differentially determined to follow specific labeled pathways. To test this hypothesis and further examine the specificity of these interactions, we ablated individual neurons *in vitro* (S. duLac, M. J. Bastiani, and C. S. Goodman, in preparation) and individual neuronal precursor cells *in ovo* with a laser microbeam (C. Q. Doe, M. J. Bastiani, and C. S. Goodman, in preparation).

Experiment #1: When the U1 and U2 neurons were ablated, the aCC did not extend along any other axon pathway (Fig. 7C).

Experiment #2: when the MP1 and dMP2 neurons were ablated, the pCC did not extend along any other axon pathway (Fig. 7D).

Experiment #3: When the MP1 and dMP2 neurons from the next anterior segment were ablated, the pCC extended anteriorly and then stopped where it reached the anterior extent of the MP1/dMP2 from its own segment (Fig. 7E). The pCC did not extend along any other axon pathway in the middle of the developing neuropil (S. duLac, M. J. Bastiani, and C. S. Goodman, in preparation).

The results of experiments #1, #2, and #3 argue against (1) the simple location of axons, (2) subtle timing mechanisms, or (3) simple quantitative differences in the expression of a common surface label being the major determinant of the aCC and pCC selective affinities. Rather, the aCC growth cone appears to display an absolute preference for the U axons, and the pCC growth cone an absolute preference for the MP1/dMP2 axons. No hierarchy in preferences is implicated by these results. These experiments support the notion that the surfaces of the U axons, MP1/dMP2 axons, and vMP2 axon have special recognition labels that allow the aCC and pCC growth cones to distinguish amongst them and other axons that develop within the neuropil.

We were interested in testing the temporal and spatial specificity of these affinities. How important is precise timing? Do the U axons continue to express the surface label that guides the aCC? Once many other axons and fascicles have developed, can the aCC growth cone still selectively recognize the U axons? If the aCC growth cone extends into other regions of the developing neuropil, will it follow other fascicles? And finally, if the aCC growth cone contacts the U axons in a different location and setting, will it still cue on them?

To achieve such temporal and spatial transplant experiments, we ablated neuroblast 1-1 (NB 1-1, the precursor cell that generates aCC and pCC) with a laser microbeam *in ovo*. We knew that neighboring epidermal cells would regulate and replace the ablated NB 1-1, generating the aCC and pCC neurons with a temporal delay of 10–15 hr, depending on the experimental paradigm (Doe and Goodman, 1985). Although delayed, in most cases the aCC and pCC neurons still migrate anteriorly to their normal location.

Experiment #4: We ablated NB 1-1 before its first division and in so doing delayed the aCC and pCC neurons by about 10 hours (2%) (Fig. 7G). Although confronted with many additional axons because of the temporal delay, the pCC growth cone extended anteriorly along the MP1/dMP2 fascicle while the aCC growth cone extended posteriorly and then laterally along the U fascicle (C. Q. Doe, M. J. Bastiani, and C. S. Goodman, in preparation).

Experiment #5: We ablated NB 1-1 and its first ganglion mother cell after its first division and in so doing delayed the aCC and pCC neurons by about 15 hr (3%) (Figs. 7H and 8). Although confronted with even more axons and fascicles, the pCC growth cone always extended anteriorly along the MP1/dMP2 fascicle. However, in one-half of the experiments, the aCC growth cone did not extend posteriorly as normal along the U fascicle, possibly because the U fascicle had become inaccessible due to increasing distance and density of other fascicles. Instead, the aCC growth cone wandered anteriorly along the dorsal basement membrane, just above and within reach of the developing axon scaffold of the ganglionic neuropil. Although confronted with many new and different axon surfaces, it did not selectively fasciculate with any of them. Rather, once it had extended far enough anteriorly to contact the axons of the U fascicle from the next anterior segment, it turned laterally and fasciculated with them (Fig. 8) C. Q. Doe, M. J. Bastiani, and C. S. Goodman, in preparation).

The results of experiments #4 and #5 further demonstrate that the selective affinities of the aCC and pCC growth cones are absolute and not hierarchical. Furthermore, they suggest that precise timing is not important; the

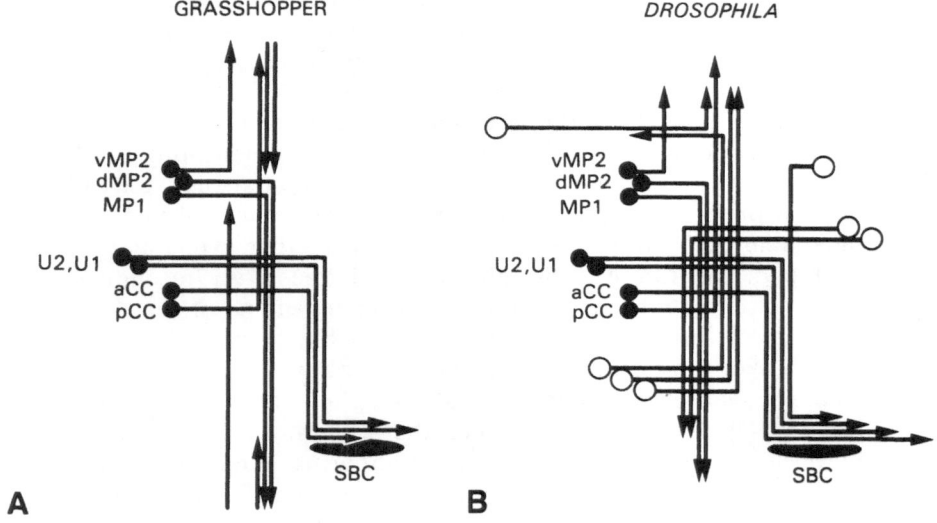

Figure 7. Schematic diagrams of the first longitudinal axon fascicles in the grasshopper (A) and *Drosophila* (B) embryos, and the results of five different cell ablation experiments (C–H). (A,B) TEM serial section reconstructions reveal the first longitudinal fascicles in the grasshopper embryo at day 7 (7/20 days) and in the *Drosophila* embryo at hour 10:15 (10:15/22 hr). (SBC is the identified glial cell called the segment boundary cell.) Note the similar specificity of the aCC and pCC growth cones. The major differences involve timing. In the grasshopper, the U axons pioneer the U fascicle, whereas in the fly, the aCC does. In the grasshopper, the vMP2 fascicle has already formed, whereas in the fly, it has not. In the fly, several other axons have already joined the MP1/dMP2 and U fascicles. (C–E) *In vitro* experiments (1–3 in text) in which the U neurons (C), MP1 and dMP2 (D), and MP1 and dMP2 in the next anterior segment (E) have been ablated. (G,H) *In ovo* experiments (4–5 in text) in which the precursor of the aCC and pCC neurons, neuroblast (NB) 1-1, has been ablated. A neural epidermal cell takes the place of the ablated NB 1-1, generating the aCC and pCC neurons with a temporal delay. When NB 1-1 is ablated before its first division, the aCC and pCC neurons are delayed by about 10% of development; when NB 1-1 and its first ganglion mother cell are ablated after its first division, the aCC and pCC neurons are delayed by about 15% of development. These five different ablation experiments demonstrate that the selective affinities of the aCC and pCC growth cones are absolute and not hierarchical; furthermore, they suggest that precise timing is not important. See text for details and discussion.

aCC growth cone was able to recognize selectively the U fascicle and/or its surrounding glial cells (such as the SBC) after a delay of 10–15 hr. Moreover, although the aCC appears uninterested in any of the other axons within the developing ganglionic neuropil (up to 100 different axons in 25 fascicles), it still joins the U fascicle from the next anterior segment even though it contacts this fascicle in a different time, location, and spatial orientation.

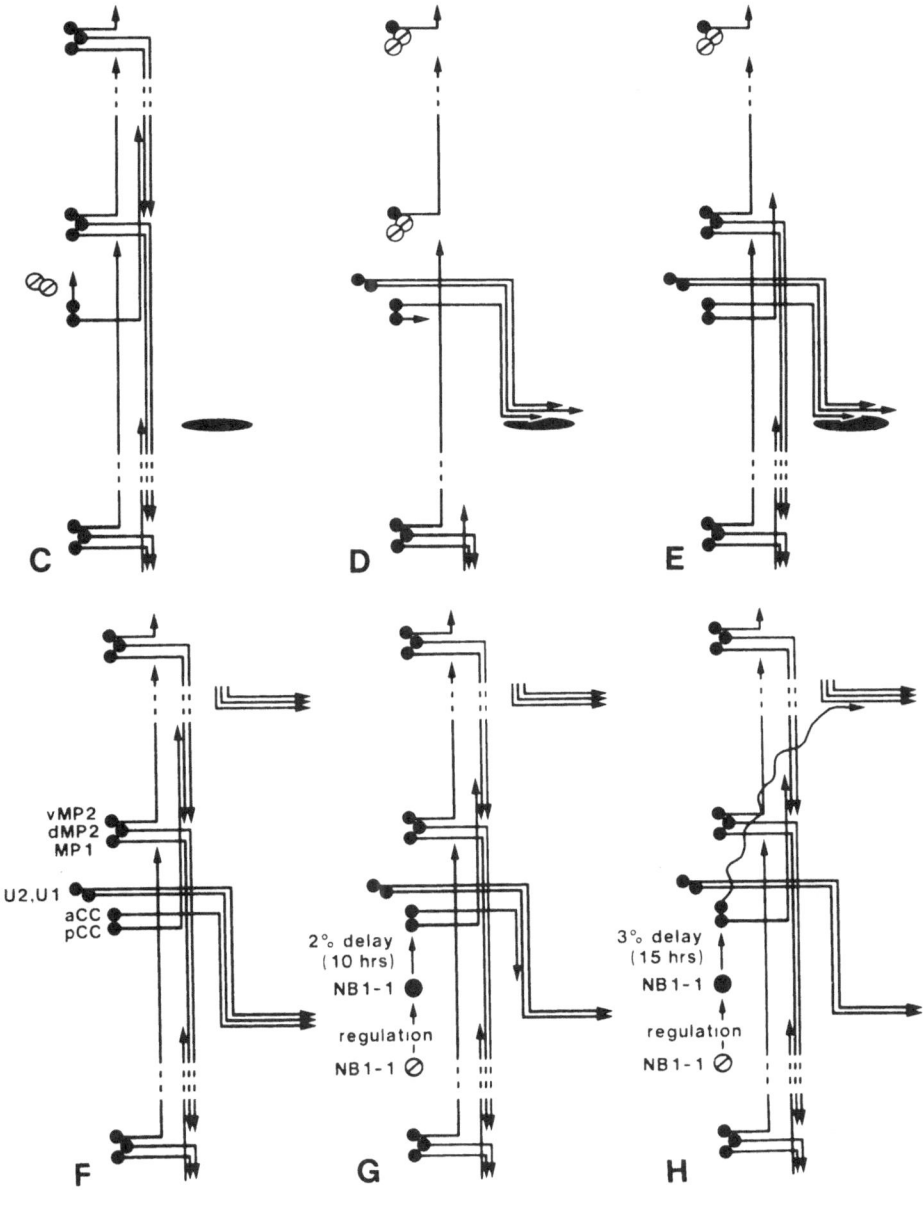

Figure 7. (Continued)

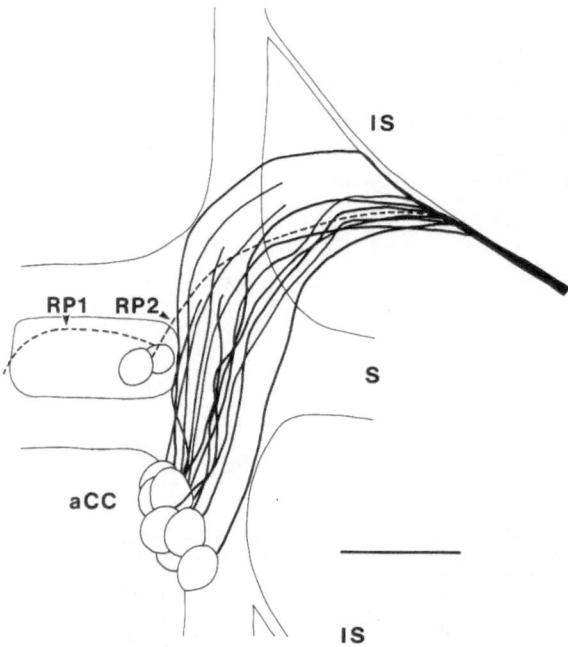

Figure 8. Behavior of aCC growth cone in temporal delay experiment 5. We ablated NB 1-1 and its first ganglion mother cell after its first division and in so doing delayed the aCC and pCC neurons by about 15 hr (3%). Although confronted with even more axons and fascicles, the pCC growth cone always extended anteriorly along the MP1/dMP2 fascicle. However, in one-half the experiments, the aCC growth cone did not extend posteriorly as normal along the U fascicle, possibly because the U fascicle had become inaccessible due to increasing distance and density of other fascicles. Instead, the aCC growth cone wandered anteriorly along the dorsal basement membrane, just above and within reach of the developing axon scaffold of the ganglionic neuropil. Although confronted with many new and different axon surfaces, it did not fasciculate selectively with any of them. Rather, once it had extended far enough anteriorly to contact the axons of the U fascicle from the next anterior segment, it turned laterally and fasciculated with them. The axon pathway of the aCC neuron is shown in 13 superimposed examples here. Scale bar: 50 μm.

5. Conclusions

The cellular analysis of growth-cone guidance in the grasshopper embryo reviewed here indicates that surface-recognition molecules are likely to be expressed on subsets of fasciculating axons early in development. How many recognition molecules remains an open question. In a very general sense, however, Sperry may well have been right (Sperry, 1963). Certainly every neuron does not have its own unique chemical label (interestingly, Sperry never postulated this, even though the notion is often attributed to him). Rather, correctly wiring a nervous system is likely to require molecular heterogeneity in the form of surface-recognition molecules that enable neurons to be guided to and identify their correct targets. Passive and spatiotemporal constraints no doubt help reduce the number of molecules needed, but neuronal growth cones make specific and divergent choices that implicate different molecules in pathway and target recognition.

Although general cell and substrate adhesion molecules are of utmost importance in neuronal development, we think they are likely to be only part of the story. Our results encourage us to predict in addition a family or families of less abundant and more specifically expressed surface-recognition molecules. Our recent immunocytochemical and biochemical data with monoclonal antibodies directed against glycoproteins on the surfaces of axons in both grasshopper and *Drosophila* embryos support this prediction (M. J. Bastiani, A. L. Harrelson, N. H. Patel, and P. M. Snow, unpublished results).

Needless to say, this is not a universally held opinion. The two most commonly asked questions are as follows: Are insects truly a good model system for vertebrates? Do such recognition molecules really exist in insects? Our results suggest that such molecules exist, and evolutionary arguments suggest that such mechanisms underlie neuronal development in all organisms. At this juncture, however, it is difficult to answer these two questions with precision, since sufficient biochemical and functional data are not yet at hand for any organism. Fortunately, a detailed understanding of the molecular basis of cell recognition during neuronal development in insect embryos may be within reach over the next decade. The highly accessible embryonic neurons of grasshopper, the powerful molecular genetics of *Drosophila*, and the ability to apply information learned from one to the other make it appealing to use these two insect embryos in tandem for a combined cellular and molecular genetic approach. In a few years, we may be in a better position to understand how neuronal specificity is generated during the development of the simple nervous system of insects and to ask whether these mechanisms apply to more complex nervous systems as well.

ACKNOWLEDGMENTS. Supported by Scholars Award from F.E.S.N. Foundation to M.J.B., NICHHD Developmental Biology Traineeship to C.Q.D., NIMH Neurobiology Traineeship to S.d.L., and grants and awards from the NIH, NSF, March of Dimes, McKnight Foundation, and Weingart Foundation to C.S.G.

References

Bastiani, M. J., and Goodman, C. S., 1984a, Neuronal growth cones: Specific interactions mediated by filopodial insertion and induction of coated vesicles, *Proc. Natl. Acad. Sci. USA* **81**:1849–1853.

Bastiani, M. J., and Goodman, C. S., 1984b, The first growth cones in the central nervous system of the grasshopper embryo, pp. 63–84, in: *Cellular and Molecular Approaches to Neuronal Development* (I. Black, ed.), Plenum Press, New York.

Bastiani, M. J., Pearson, K. G., and Goodman, C. S., 1984a, From embryonic fascicles to adult tracts: Organization of neuropil from a developmental perspective, *J. Exp. Biol.* **112**:45–64.

Bastiani, M. J., Raper, J. A., and Goodman, C. S., 1984b, Pathfinding by neuronal growth cones in grasshopper embryos. III. Selective affinity of the G growth cone for the P cells within the A/P fascicle, *J. Neurosci.* **4**:2311–2328.

Bastiani, M. J., Doe, C. Q., Helfand, S. L., and Goodman, C. S., 1985a, Neuronal specificity and growth cone guidance in grasshopper and Drosophila embryos, *Trends Neurosci.* **8**:257–266.

Bastiani, M. J., duLac, S., and Goodman, C. S., 1985b, The first growth cones in insect embryos: Model system for studying the development of neuronal specificity, pp. 149–174, in: *Model Neural Networks and Behavior* [A. Selverston, ed.], Plenum Press, New York.

Bate, C. M., and Grunewald, E. B., 1981, Embryogenesis of an insect nervous system. II. A second class of precursor cells and the origin of the intersegmental connectives, *J. Embryol. Exp. Morphol.* **61**:317–330.

Doe, C. Q., and Goodman, C. S., 1985, Early events in insect neurogenesis: II. The role of cell interactions and cell lineage in the determination of neuronal precursor cells, *Dev. Biol.* **111**:206–219.

Goodman, C. S., and Spitzer, N. C., 1979, Embryonic development of identified neurones: Differentiation from neuroblast to neurone, *Nature (Lond.)* **280**:208–214.

Goodman, C. S., Raper, J. A., Ho, R., and Chang, S., 1982, Pathfinding by neuronal growth cones in grasshopper embryos, *Symp. Soc. Dev. Biol.* **40**:275–316.

Goodman, C. S., Bastiani, M. J., Doe, C. Q., and duLac, S., Helfand, S. L., Kuwada, K. Y., and Thomas, J. B., 1984, Cell recognition during neuronal development, *Science* **225**:1271–1279.

Pitman, R. M., Tweedle, C. D., and Cohen, M. J., 1972, Branching of central neurons: Intracellular cobalt injection for light and electron microscopy, *Science* **176**:412–414.

Raper, J. A., Bastiani, M. J., and Goodman, C. S., 1983a, Pathfinding by neuronal growth cones in grasshopper embryos: I. Divergent choices made by the growth cones of sibling neurons, *J. Neurosci.* **3**:20–30.

Raper, J. A., Bastiani, M. J., and Goodman, C. S., 1983b, Pathfinding by neuronal growth cones in grasshopper embryos: II. Selective fasciculation onto specific axonal pathways, *J. Neurosci.* **3**:31–41.

Raper, J. A., Bastiani, M. J., and Goodman, C. S., 1983c, Guidance of neuronal growth cones: Selective fasciculation in the grasshopper embryo, *Cold Spring Harbor Symp. Quant. Biol. Mol. Neurobiol.* **48**:587–598.

Raper, J. A., Bastiani, M. J., and Goodman, C. S., 1984, Pathfinding by neuronal growth cones in grasshopper embryos. IV. The effects of ablating the A and P axons upon the behavior of the G growth cone, *J. Neurosci.* **4**:2329–2345.

Sperry, R. W., 1963, Chemoaffinity in the orderly growth of nerve fiber patterns and connections, *Proc. Natl. Acad. Sci. USA* **50**:703–710.

Stewart, W., 1978, Functional connections between cells revealed by dye-coupling with highly fluorescent naphthalamide tracer, *Cell* **14**:741–759.

Stretton, A. O. W., and Kravitz, E. A., 1968, Neuronal geometry: Determination with a technique of intracellular dye injection, *Science* **162**:132–134.

Taghert, P., Bastiani, M., Ho, R. K., and Goodman, C. S., 1982, Guidance of pioneer growth cones: Filopodial contacts and coupling revealed with an antibody to Lucifer Yellow, *Dev. Biol.* **94**:391–399.

Thomas, J. B., Bastiani, M. J., Bate, C. M., and Goodman, C. S. (1984) From grasshopper to *Drosophila*: A common plan for neuronal development, *Nature (Lond.)* **310**:203–207.

Tyrer, N. M., and Gregory, G. E., 1982, A guide to the neuroanatomy of locust suboesophageal and thoracic ganglia, *Philos. Trans. R. Soc. Lond. B* **297**:91–123.

Chapter 15

Guidance of Neural Crest Migration
Latex Beads as Probes of Surface–Substratum Interactions

MARIANNE BRONNER-FRASER

1. Neural Crest Migratory Pathways

The vertebrate neural crest is first recognizable at the onset of neurulation as the neural folds appose to form the neural tube. Neural crest cells emigrate from the dorsal neural tube shortly after tube closure. From this site of origin, they migrate extensively and differentiate into a wide variety of derivatives. Neural crest-derived cell types include neurons and supportive cells of the peripheral nervous system, melanocytes, several neurosecretory cell types, bone and cartilage of the face, and adrenal chromaffin cells. Because of the migratory ability of neural crest cells and the diversity of crest derivatives, this transient embryonic structure provides an important model system for studying interactions involved in cell movement and differentiation.

In avian embryos, neurulation begins in the head region and moves progressively tailward. Similarly, initiation of neural crest migration progresses as a rostral to caudal wave down the length of the embryo. Along the neural axis, there are several distinct populations of neural crest cells (Weston, 1963; LeDouarin, 1973; LeDouarin and Teillet, 1974; Noden, 1975), which differ both in their migratory pathways and in their range of potential derivatives. In the head region, the cranial neural crest cells undergo complex migrations and distribute according to their brain level of origin (Noden, 1975). The cranial crest gives rise to the skeletal and connective tissue of the face as well as to neuronal and supportive cells of the cranial ganglia. The vagal neural crest cells, which arise between the first and seventh somites, undergo some of the most extensive migrations of any embryonic cell type. These cells enter the gut in the anterior regions, migrate caudally as far as the umbilicus and rectum, and

MARIANNE BRONNER-FRASER • Developmental Biology Center, University of California—Irvine, Irvine, California 92715.

differentiate into the parasympathetic ganglia of the gut (LeDouarin and Teillet, 1973; Allan and Newgreen, 1980).

The most extensively studied and best understood neural crest migratory routes are those in the "trunk" region of the embryo, extending from somites 8 to 28. After departure from the dorsal neural tube, trunk neural crest cells migrate along two major routes (Fig. 1)—either dorsally beneath the ectoderm or ventrally between the neural tube and the somites (Weston, 1963). Those cells following the *dorsal* route eventually give rise to melanocytes. The cells following the *ventral* route settle in three locations: (1) adjacent to the neural tube, where they form the dorsal root ganglia; (2) above the dorsal aorta, where they condense into sympathetic ganglia; and (3) lateral to the dorsal aorta, where they form the chromaffin cells of the adrenal medulla.

During their initial migratory phase, neural crest cells move through a narrow cell-free space that contains extracellular matrix (ECM) molecules. The primary components of this matrix are glycosaminoglycans (GAGs) and fibro-

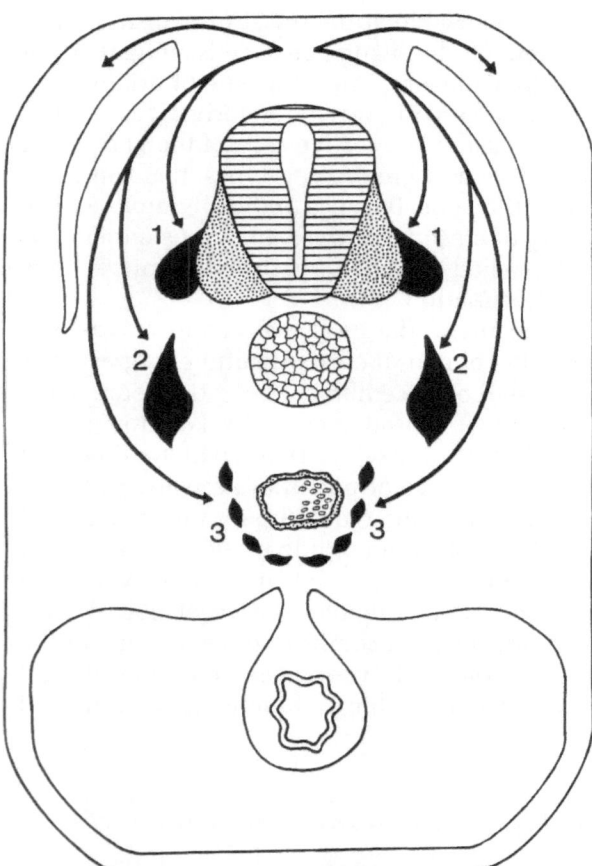

Figure 1. Diagram illustrating the normal routes of neural crest migration in the trunk region of the avian embryo. Cells choosing the ventral pathway localize in three main areas: (1) the sensory ganglia; (2) the primary sympathetic chain; and (3) the adrenal gland, aortic plexuses, and some cells of the metanephric mesenchyme. Crest cells following the dorsolateral pathway migrate under the ectoderm and become skin melanocytes.

nectin (FN). Hyaluronic acid is the major GAG present on both the dorsal and ventral trunk neural crest pathways at the onset of crest migration (Derby, 1978; Pintar, 1978). Another glycosaminoglycan, chondroitin sulfate, appears to be present predominantly on the dorsal route. Because entry of neural crest cells onto the dorsal route is significantly delayed over initiation of migration along the ventral pathway in the quail/chick chimera (Teillet and LeDouarin, 1970), it has been suggested that chondroitin sulfate may retard neural crest migration (Newgreen et al., 1982). Fibronectin is another ECM molecule present in high concentrations along both the dorsal and ventral migratory routes (Newgreen and Thiery, 1980; Mayer et al., 1981). FN is a cell-surface and ECM-associated glycoprotein with important functions in cell adhesion and motility. In addition to GAGs and FN, the extracellular spaces through which crest cells migrate also contain fibrillar ECM components that can be seen by electron microscopy (Cohen and Hay, 1971; Bancroft and Bellairs, 1976; Ebendal, 1977; Tosney, 1978); these may be composed predominantly of collagen (Cohen and Hay, 1971).

The composition of the ECM changes during embryogenesis. For example, with increasing embryonic age, the glycosaminoglycan composition of the neural crest pathways changes such that the concentration of hyaluronate decreases and the concentration of chondroitin sulfate increases (Derby, 1978; Pintar, 1978). The levels of FN along neural crest routes are also reduced after neural crest migration and during gangliogenesis (Newgreen and Thiery, 1980; Thiery et al., 1982b). Since fibronectin is thought to promote cell migration (Ali and Hynes, 1978), the loss of this adhesive glycoprotein from the ECM may help shift the balance of adhesions to favor crest cell aggregation. These dynamic properties of the ECM may be important for cessation of neural crest migration and final cell localization.

Temporal changes in cell surface molecules also appear to contribute to changes in adhesiveness along neural crest migratory routes. For example, neural cell adhesion molecule (N-CAM) can be identified by immunohistofluorescent staining in the dorsal neural tube before neural crest migration (Thiery et al., 1982a). During active crest migration, no N-CAM staining is observed; however, staining returns in neural crest sites of localization after the cells condense to form ganglia. Thus, alterations in both cell surface and ECM components along neural crest pathways may cause changes in adhesiveness that contribute to neural crest cell localization.

The matrix lining neural crest pathways appears to be synthesized and modified by both embryonic tissues and by neural crest cells themselves. The neural tube produces collagen, which appears to contribute to the fibrillar material of the ECM (Cohen and Hay, 1981). In addition, the somites, ectoderm, and neural tube synthesize fibronectin (Newgreen and Thiery, 1980) and glycosaminoglycans (Loring et al., 1977; Manasek and Cohen, 1977; Pintar, 1978; Solursh et al., 1979). In tissue culture, neural crest cells have also been shown to synthesize glycosaminoglycans and glycoproteins (Manasek and Cohen, 1977; Pintar, 1978), which are predominantly associated with the cell surface and the substratum. The primary glycosaminoglycan produced by neural crest cells is hyaluronic acid. These correlations between spatial and temporal changes in ECM molecules and timing of neural crest morphogenesis sum to indicate a potentially important role for cell surface–ECM interactions in neural crest cell movement and localization.

2. Microinjection Technique

Because it is difficult to distinguish neural crest cells from the tissues through which they migrate, it has become necessary to mark neural crest cells during their migration. For avian embryos, the most useful method for cell marking combines quail and chick tissues (LeDouarin, 1973); these two species are developmentally similar, but quail cells contain a heterochromatin marker that permits their distinction from chick cells. Cell marking techniques in combination with neural tube transplantations have made it possible to study the migration and differentiation of neural crest cell populations (Weston, 1963; Johnston, 1966; LeDouarin, 1973; LeDouarin and Teillet, 1974; Noden, 1975). Recently, several new techniques have emerged for placing labeled cells into the embryo (LeDouarin et al., 1978; Bronner and Cohen, 1979; Erickson et al., 1980). These methods enable implantation of small numbers of cells, in contrast to the large cell numbers emerging from grafted neural tubes, and reduce some of the surgical trauma incurred during neural tube transplantation.

An injection technique for implanting cells into defined regions of the embryo has been developed in our laboratory (Bronner and Cohen, 1979; Bronner-Fraser and Cohen, 1980; Bronner-Fraser et al., 1980). This method entails the use of suspensions of selected cell types or Latex beads. The suspension is placed into a micropipette (with tip of about 30 μm) mounted onto a micromanipulator and connected to a pressure-injection apparatus. The micromanipulator is used to accurately maneuver the injection micropipette to the surface of the embryo, to puncture the ectoderm, and to lower the pipette carefully to the desired region. The contents in the tip of the micropipette are then expelled into the embryo with a pulse of pressure.

For studies of cell migration, the injected cells or beads must be placed in a site that is identifiable, reproducible, and localized. Therefore, for most experiments, the embryonic somites were chosen as an injection site. At the beginning of neural crest migration, the somites in the trunk region are epithelial sacs with a cavity suitable for initially retaining the injected cell or bead suspension. During normal somitic development, the sclerotomal cells of the ventromedial wall of the somite disperse. The cells or latex beads within the somitic cavity are deposited onto the ventral neural crest pathway adjacent to the ventromedial aspect of the neural tube. Because this method produces no incision adjacent to the neural tube, the ECM material along the ventral pathway remains intact and undisturbed. An additional advantage of this injection site results from the presence of the dorsal aorta immediately below the somite at the time of injection. Embryonic bleeding therefore serves as a useful assay for an improperly placed implant. By utilizing the somitic injection site, all the requirements for a proper implantation site are fulfilled; the cells are initially contained in a reproducible location and are subsequently exposed to a normal neural crest migratory route. The injection technique can be used to implant small numbers of cells or particles into the embryo and serves to characterize properties either of the embryonic pathways or of the injected cells themselves.

In one experimental series, latex beads were injected adjacent to the neural tube in order to control for the somitic injection site. The injections were performed as described above, except that the micropipette was placed just lateral to the neural tube instead of inside the somite. It is possible to deposit a small number of beads in this location by careful pressure injection.

3. Translocation of Embryonic Cells on the Ventral Neural Crest Pathway

Some useful aspects of cell microinjection techniques include the ability to place cells into regions of the embryo they would not normally encounter or to implant already differentiated cells into younger regions of the embryo. Initial experiments using the injection technique employed pigment cells that were clonally derived from one single neural crest cell in tissue culture. Because avian neural crest cells become nonmotile after expression of pigmentation, these cells are past their migratory phase. The effects of the embryonic environment on these already differentiated neural crest cells could be examined by injecting them into young embryos. The cloned melanocytes (derived from a quail neural crest cell) were injected into the posterior somites of 2½-day-old chicken embryos; crest migration in this region was just beginning at the time of injection. At various times postinjection, the host embryos were fixed, serially sectioned, stained, and examined. The implanted cells were originally contained in the cavity of the somite (Fig. 2A). After release from the somitic lumen, the melanocytes distributed along the ventral neural crest pathway. This route is normally followed by endogenous precursors to neuronal and supportive cells (see Fig. 1). The injected melanocytes were found in progressively more ventral sites along the pathway with increasing time (Fig. 2B–E). By 2 and 3 days postinjection, the pigment cells were found in regions where endogenous trunk neural crest cells localize, such as around the dorsal aorta where the adrenal chromaffin cells and aortic plexuses differentiate. In some cases, the pigment cells invaded the gut and were found in the region in which parasympathetic ganglia (derived from vagal neural crest) normally differentiate (Bronner and Cohen, 1979). These results demonstrated that the neural crest-derived pigment cells were capable of ventral translocation when introduced into the proper embryonic milieu.

Neural crest cells actively migrate in tissue culture (Davis, 1980) and *in vivo*, but the motility of avian neural crest cells ceases upon expression of pigmentation. The observed resumption of movement in postmigratory pigment cells after intraembryonic injection may specifically be caused by implan-

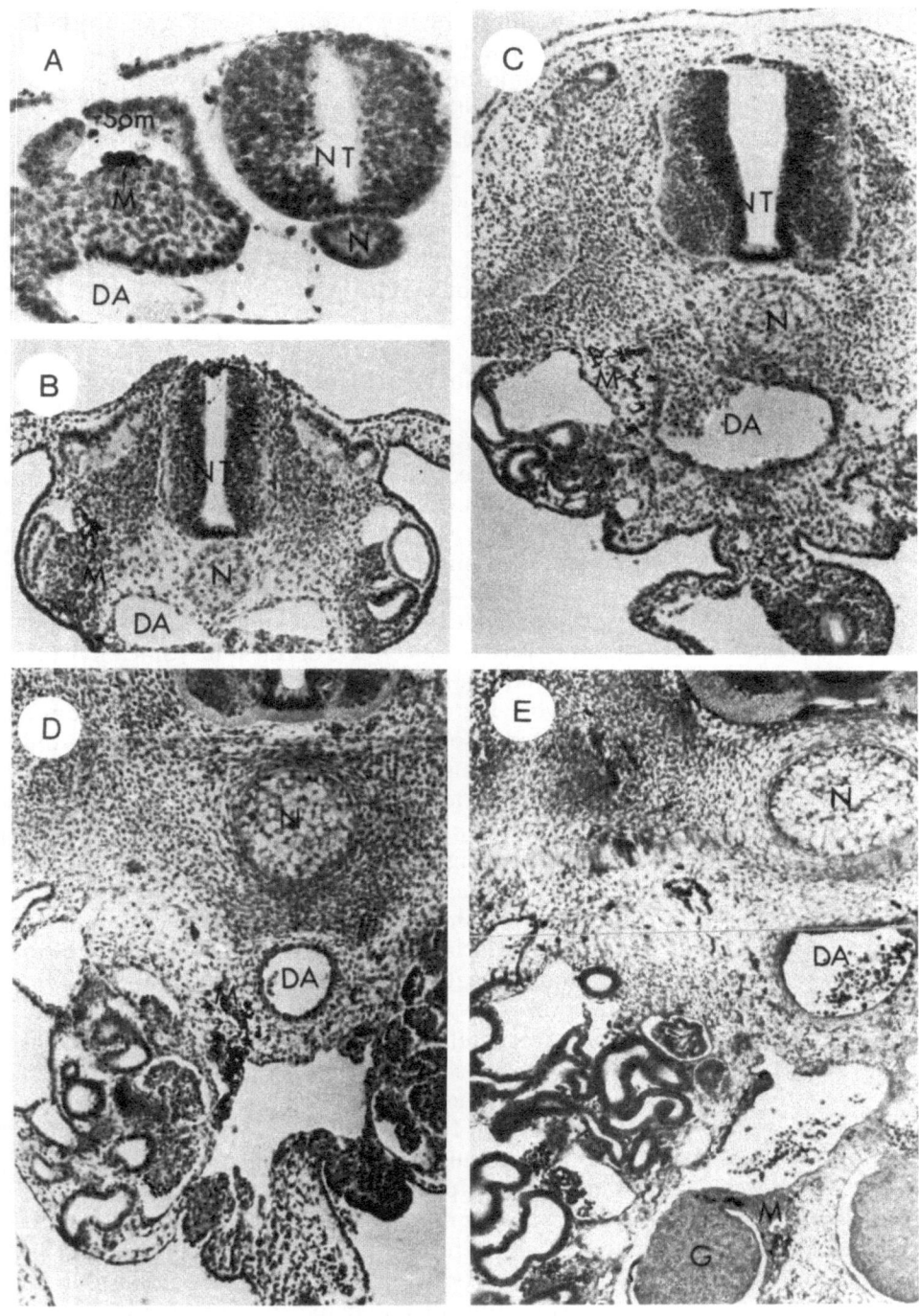

tation onto the neural crest pathway. Alternatively, the translocation of these cells may result from confounding effects such as (1) potential species incompatibilities between the implanted quail cells and the chick host; or (2) the potential effects of long-term culture on the melanocytes. Although it is often assumed that the quail and the chick are developmentally homologous, the two species do differ in embryonic size and gestation time. In addition, quail cells appear to be more adhesive and invasive than do chick cells under some culture conditions (Bellairs *et al.*, 1981). We therefore examined the possibility that chick and quail cells may differ by injecting cloned chicken melanocytes into the trunk somites. Because the host embryos were devoid of pigment cells at the times considered, the melanin in the cytoplasm of the pigment cells permitted the distinction of the injected cells from host cells. Both chicken and quail pigment cells translocated identically after injection, indicating that species incompatibilities did not account for movement of the injected cells (Bronner-Fraser and Cohen, 1980). A second factor that may contribute to the ventral distribution of the quail melanocytes might be the long-term culture conditions to which the cells were exposed. To control for this possibility, pigment cells were removed from quail embryonic skin and injected in the same manner as the cultured melanocytes used previously. The movement of these skin-derived melanocytes (which are also past their migratory phase *in vivo*) was indistinguishable from that of the cloned melanocytes described previously (Bronner-Fraser and Cohen, 1980). Therefore, prolonged time *in vitro* does not account for the movement of pigment cells after injection into young embryos.

In tissue culture, neural crest cells give rise to other derivatives in addition to pigment cells. Some neural crest cells remain unpigmented and do not differentiate into any identifiable cell types, whereas others become neuroblasts, which produce catecholamines, acetylcholine (ACh), or serotonin (Cohen, 1977; Greenberg and Schrier, 1977; Kahn *et al.*, 1980; Maxwell *et al.*, 1982; Sieber-Blum *et al.*, 1983). To examine the migration of other neural crest phenotypes, unpigmented quail neural crest cells were injected into the trunk somites of chicken hosts. Donor cells were distinguished from host chicken cells on the basis of their heterochromatin marker, which is unique to the quail (LeDouarin, 1973). Like the pigmented cells, unpigmented neural crest cells translocated along the ventral pathway and settled within sympathetic ganglia

---

Figure 2. Light photomicrographs of transverse sections demonstrating the ventral migration of cloned quail melanocytes (M) injected into the newly formed somites (Som) of 2.5-day-old chick embryos and fixed at progressively older stages of development. (A) Immediately after injection, melanocytes are restricted to the somitic cavity. (B) One day after injection, melanocytes have migrated to the region of the sensory ganglion. Note the single melanocyte adjacent to the dorsal aorta. (C) Two days after injection, melanocytes are found further ventrally in the region of the sympathetic ganglia and lateral to the dorsal aorta. (D) Three days after injection, melanocytes have migrated beyond the dorsal aorta and are seen next to the developing adrenal gland, metanephric mesenchyme, and aortic plexuses. (E) Four days after injection, melanocytes are present further ventrally in the vicinity of the gonads (G) and mesentery of the gut; neural tube (NT), notochord (N), and dorsal aorta (DA). (From Bronner-Fraser and Cohen, 1980.)

or in adrenal medullary sites (Bronner-Fraser et al., 1980). In these locations, some of the injected cells became catecholamine-containing neuroblasts.

Is this phenomenon of ventral migration unique to cells of neural crest origin or will any cell type placed onto this pathway distribute similarly? Non-neural crest cells were injected into the posterior somites in order to determine whether any cell type would migrate ventrally when exposed to the somitic milieu. Somite cells and fibroblast cells were chosen for these experiments because of the inherent migratory properties of these two cell types. The somite cells on the ventromedial wall of the somite disperse during the course of normal somitic development and are therefore motile during a portion of their development. The migratory behavior of fibroblast cells in vitro has been the object of much study and probably reflects their in situ involvement in wound healing. Suspensions of quail somite or fibroblast cells were injected into the trunk region of 2½-day-old chicken embryos. Unlike the crest-derived cells described above, neither the fibroblasts nor the somitic cells moved ventrally along the neural crest pathway (Bronner-Fraser and Cohen, 1980). The quail somite cells were found to be associated with the somitic mesenchyme of the host (Fig. 3A,B). The quail fibroblasts were usually found in the ectodermal or dermamyotomal region of the host tissue (Fig. 3C–E). Thus, certain cell types such as somitic and fibroblast cells were apparently excluded from the pathway, whereas neural crest-derived melanocytes distributed along the ventral route similarly to endogenous crest cells. These findings suggest that the embryonic environment along the ventral migratory route is selective.

In an effort to examine the behavior of a nonmotile cell type after implantation onto the ventral pathway, retinal pigment epithelial (RPE) cells were injected into chick embryos during endogenous neural crest migration. The RPE cells, derived from the optic cup, are the only pigmented cells of the body not derived from the neural crest. Although neural crest melanocytes and RPE cells have similar metabolic pathways for melanin production, these two cell types are morphologically quite different. The RPE cells are cuboidal in shape and lack the dendritic branching pattern characteristic of the neural crest-derived melanocytes. After injection into the trunk somites, RPE cells distributed along the ventral neural crest pathway similarly to the neural crest-derived melanocytes (Fig. 4). Although the RPE cells retained their cuboidal morphology and did not extend pseudopods, they settled in the vicinity of endogenous trunk neural crest sites. The neural crest ventral pathway therefore supports movement of other cell types as well as neural crest-derived cells. The pathway is somewhat selective (excluding fibroblast and somitic cells) but does not exclude all non-neural crest cells. Those cells that translocate ventrally tend to localize near endogenous neural crest sites. Therefore, the embryonic environment appears to (1) influence the selection of cell types that translocate along the neural crest pathways, and (2) play some role in determining the pattern of neural crest cell distribution.

4. Latex Beads as Probes of the Neural Crest Ventral Pathway

Cell implantation methods have shown that the embryonic milieu in re-

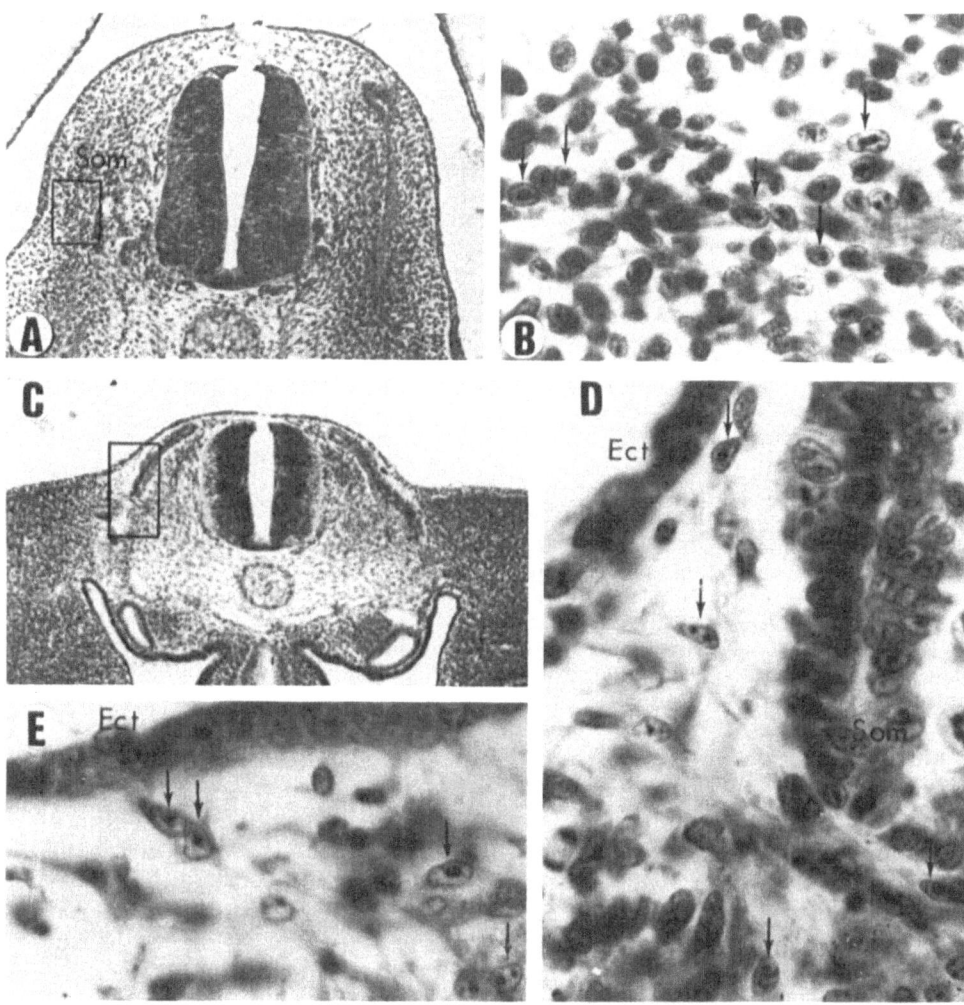

Figure 3. Light photomicrographs of transverse sections through a chick embryo fixed after injection of quail somite or fibroblast cells into the posterior somites. Quail cells are identified by a Feulgen-positive nucleolar mass. Arrows indicate cells in which the quail marker is most visible. (A) Chicken host 2 days after quail somite cells were injected into the somite (Som). (B) At higher magnification of (A) inset, the quail somite cells remained associated with the host somitic mesenchyme. (C) Embryo fixed 1 day after injection of quail fibroblast cells into the posterior somites. (D) At higher magnification of (C) inset, the quail fibroblasts are found at the base of the somite (Som) and between the somite and the ectoderm (Ect). (E) Another embryo 1 day after injection with quail fibroblasts located under the ectoderm and closer to the neural tube. (From Bronner-Fraser and Cohen, 1980.)

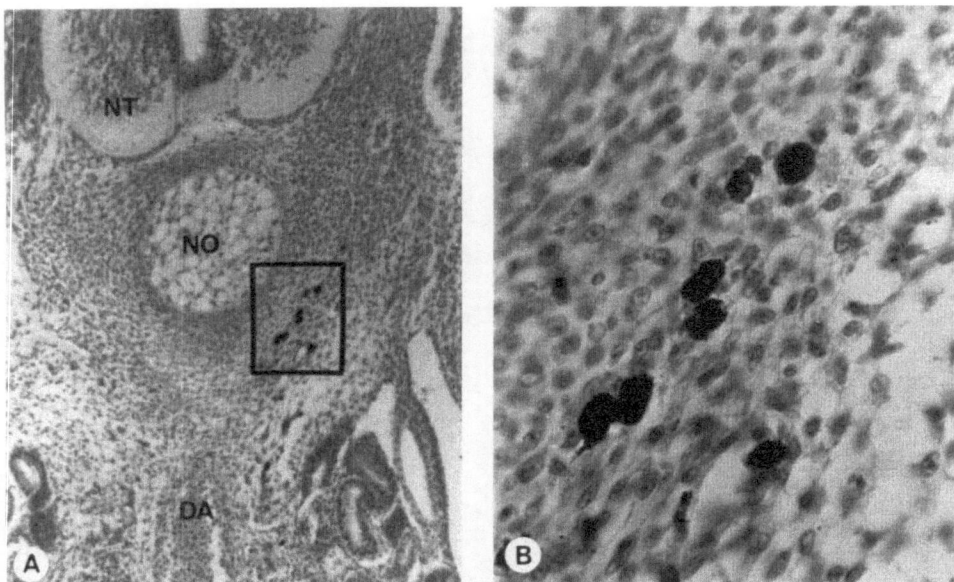

Figure 4. Light photomicrographs of a transverse section through a chick embryo 3 days after retinal pigment epithelial (RPE) cells were injected into the posterior somites. (A) At lower magnification, the RPE cells are found on the ventral pathway below the notochord and extending adjacent to the dorsal aorta. RPE cells in adjacent sections are predominantly adjacent to the aorta. (B) High magnification of (A) inset demonstrates that the RPE cells maintain their cuboidal morphology after injection. NT, neural tube; NO, notochord; DA, dorsal aorta. (From Bronner-Fraser, 1982.)

gions of active crest migration can promote movement of differentiated, postmigratory neural crest cells (LeDouarin *et al.*, 1978; Bronner and Cohen, 1979; Bronner-Fraser and Cohen, 1980; LeLievre *et al.*, 1980). Some other cell types also appear to be capable of neural crestlike translocation (Erickson *et al.*, 1980; Bronner-Fraser, 1982). However, there is no evidence for a correlation between migratory ability and the ability to distribute along the ventral route. Both fibroblast and somite cells (which did not translocate ventrally) are inherently migratory, whereas retinal pigment epithelial cells (which did translocate ventrally) are not normally migratory. Factors other than active cell motility may therefore be important for movement along neural crest pathways.

4.1. Uncoated Latex Beads Injected into the Somites

In order to examine events independent of cell motility that may be involved in crest cell distribution and localization, latex polystyrene beads were placed onto the ventral neural crest pathway as probes of the embryonic environment. The particles, which are inert and nonmotile, were chosen to approximate the dimensions of neural crest cells during active migration. Beads spanning a 10-fold range of diameters yielded similar results (Bronner-Fraser,

1982). Using the injection technique, the bead suspension was implanted onto the ventral neural crest pathway. The latex beads were injected into the posterior somites of chicken embryos and were initially contained within the somitic cavity (Fig. 5). In embryos fixed and sectioned at progressively older stages after injection, the latex beads distributed ventrally along the neural crest pathway with a time course similar to that of endogenous neural crest cells. Figure 6 illustrates cross sections through representative embryos fixed 2 and 3 days after injection, where the beads have localized ventrally adjacent to the dorsal aorta. Figure 7A is a schematic representation of the distribution of beads in all embryos fixed 2 and 3 days after somitic injection of latex beads. In these embryos, 97% of the beads settled in ventral sites along the neural crest pathway. The beads localized in the vicinity of host neural crest sites, i.e., near the sympathetic ganglion, ventral nerve cord, or adjacent to the dorsal aorta in the region in which the adrenal chromaffin cells coalesce. Thus, even inert particles can translocate along the ventral neural crest pathway after injection into the posterior somites.

4.2. Lateral Injection of Uncoated Beads

After somitic injection, the beads were initially localized in the lumen of the somite and subsequently released onto the ventral neural crest pathway as

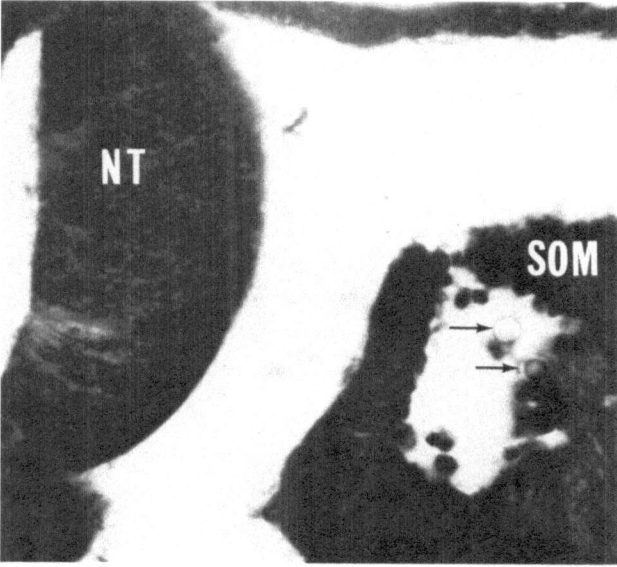

Figure 5. Transverse section through an embryo immediately after injection of beads into the somite (magnification: ×570). NT, neural tube; SOM, somite. Arrows indicate the position of the beads. (From Bronner- Fraser, 1984.)

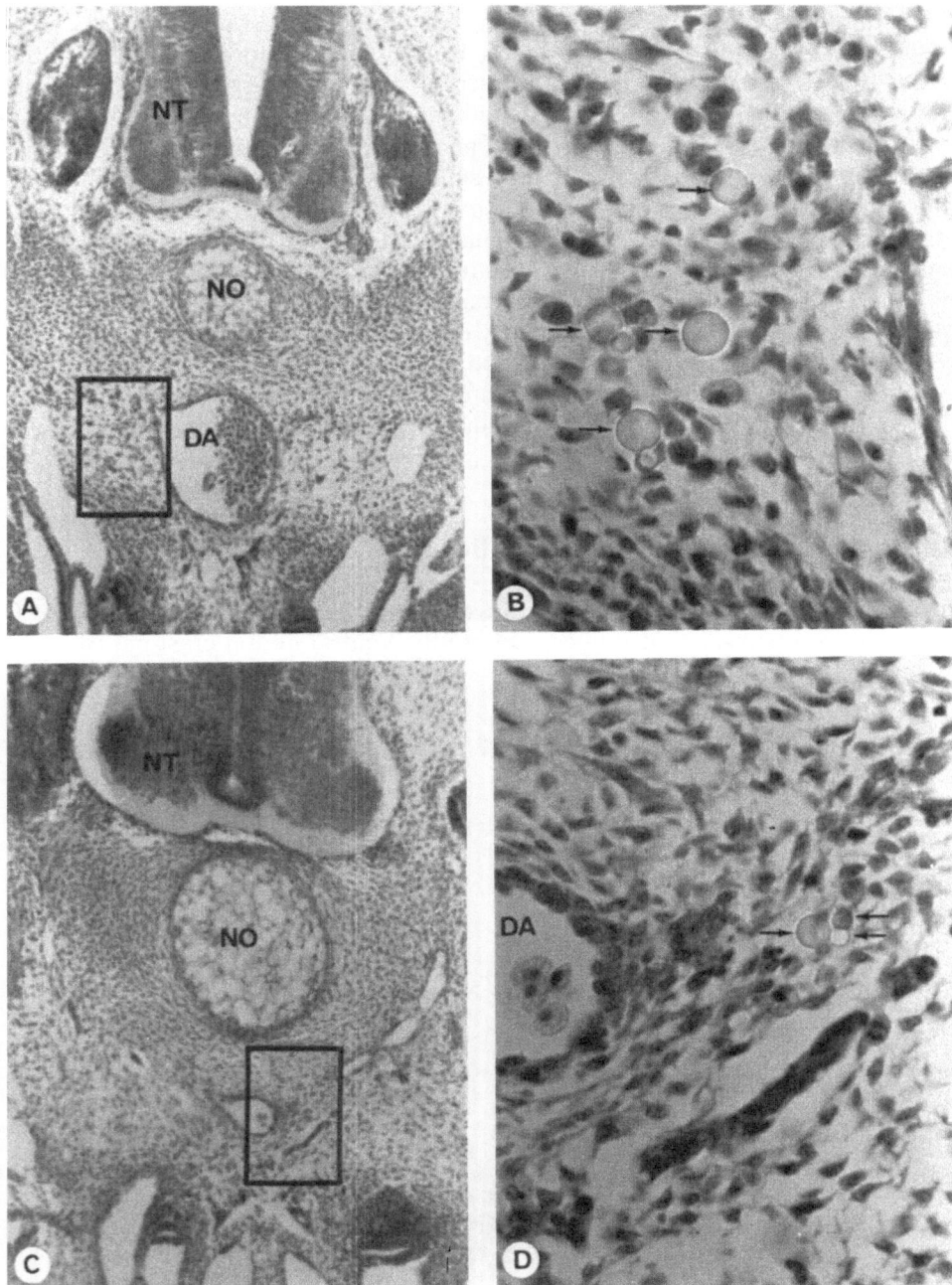

Figure 6. Transverse sections through chick embryos after uncoated beads were injected into the posterior somites. (A) An embryo fixed 2 days after injection. (B) At higher magnification of (A) inset, the latex particles are clearly visible adjacent to the dorsal aorta in close proximity to

the sclerotomal cells dispersed. Upon release, they presumably intermingled with neural crest cells (Weston, 1963). Endogenous neural crest cells initiate their migration from the dorsal neural tube and may migrate ventrally between the neural tube and somites or between the somites (Thiery et al., 1982b). Therefore, it seems important to consider the effects of the somitic implantation site on the localization of injected beads. To control for possible effects related to the injection site, latex beads were also injected "laterally," i.e., placed adjacent to the neural tube as crest migration was beginning. This injection site also places beads onto the ventral pathway but is more difficult than the somitic injection site because no natural cavity exists for initially containing the beads. The final distribution of beads was indistinguishable for both the "lateral" and the "somitic" injection sites. Figure 8 illustrates a cross section through an embryo fixed 3 days after injection of latex beads lateral to the neural tube. The beads translocated ventrally and were found around the dorsal aorta adjacent to adrenomedullary cells. Figure 7B is a schematic representation of the distribution of uncoated beads after lateral injection in those embryos fixed 2 or 3 days postinjection. The distribution pattern is similar to that observed after somitic injection (Fig. 7A); 95% of the beads settled in ventral sites. The fact that similar results were obtained for both injection sites lends credence to the idea that beads localize according to their surface properties and not as a function of their implantation site. The somitic injection site remains preferred because the injected particles can be initially contained within the somitic cavity in a reproducible manner.

4.3. BSA-Coated Beads

The translocation of the latex particles could potentially be caused by nonspecific adhesion of the beads to host neural crest cells. To minimize this adhesion, the beads were coated with bovine serum albumin (BSA) because, in another cell system, BSA-coated beads exhibited reduced binding to cells compared to uncoated latex beads (Grinnell, 1980). By using a protein such as BSA which is not normally present along the crest pathway, it was also possible to assess whether coating the bead surface with any protein would alter the distribution of the latex beads after injection. In embryos monitored one to three days following injection, the BSA-coated beads translocated along the ventral pathway and settled near the sympathetic ganglia, ventral nerve cord, and dorsal aorta. Figure 9 is a schematic representation of the distribution pattern of BSA beads in embryos fixed 2 and 3 days postinjection. Coating the bead with this protein apparently does not alter the translocation process, since uncoated and BSA-coated beads distribute analogously.

---

endogenous crest cells. (C) An embryo fixed 3 days after injection. (D) At higher magnification of (C) inset, the beads can be seen adjacent to the dorsal aorta and close to adrenomedullary crest cells. Arrows indicate the position of the beads. NT, neural tube; NO, notochord; DA, dorsal aorta. (From Bronner-Fraser, 1982.)

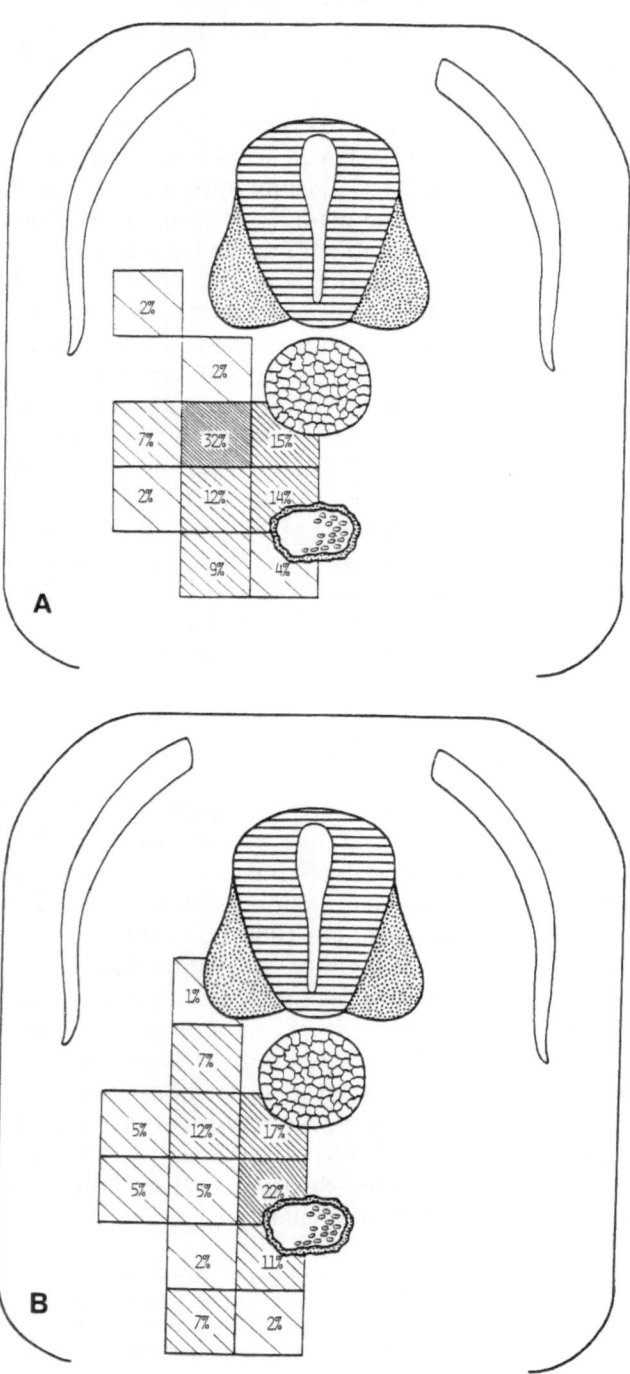

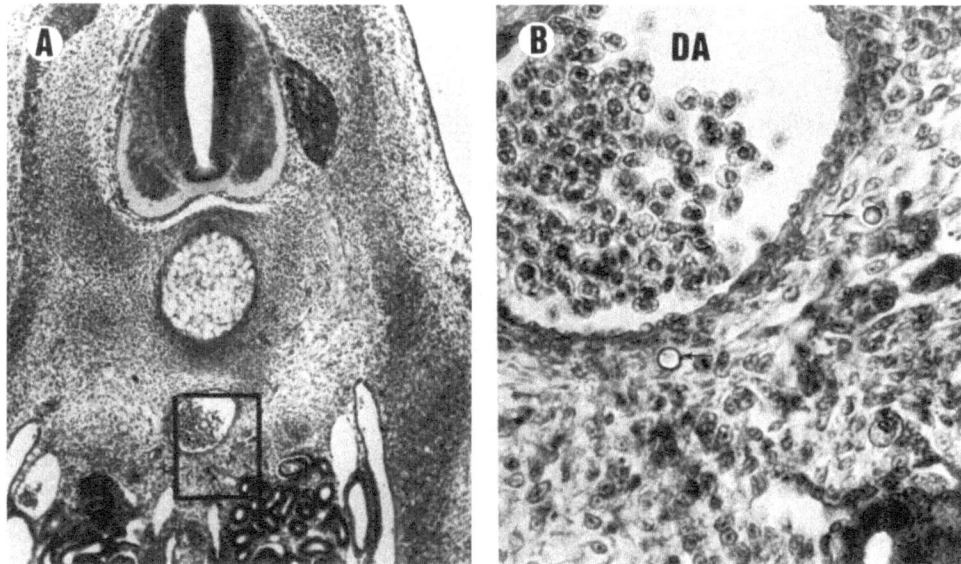

Figure 8. Transverse section through chicken embryo 3 days after injection of beads "lateral" to the neural tube. The latex beads were found in ventral sites around the dorsal aorta. (a) Low-power view. (b) Higher-power of (A) inset.

4.4. Fibronectin-Coated Beads

Using the injection technique for placing cells or latex beads into young embryos, the experiments described above have shown that RPE cells are capable of neural crestlike translocation, whereas somite cells and fibroblasts are not. Because inert latex particles also translocate ventrally, these findings suggest that the selection of cell types that move along the ventral route may be independent of inherent cell motility. A possible distinction between those cells that are restricted from movement versus those capable of ventral translocation may be differential cell surface properties. Fibronectin is one known surface component that somite and fibroblast cells produce (Engvall and Ruoslahti, 1977; Loring *et al.*, 1977), but neural crest cells (Newgreen and Thiery, 1980) and RPE cells (Kurkinen *et al.*, 1979) do not.

Figure 7. Schematic representation of the distribution of uncoated beads in embryos fixed 2 and 3 days postinjection. The localization of every bead in all embryos was plotted by camera lucida on a two-dimensional grid of an idealized embryo in cross section. In each embryo, the percentage of beads in a given quadrant was determined and then weight averaged for all embryos. The data are expressed as percentage of beads in a quadrant, rounded off to the nearest integer; values of less than 1% are not shown. The density of lines in each quadrant is proportional to the percentage of beads localized in that region. (A) uncoated beads after injection into the somites; N = 9 embryos. (B) Uncoated beads after lateral injection; N = 13 embryos.

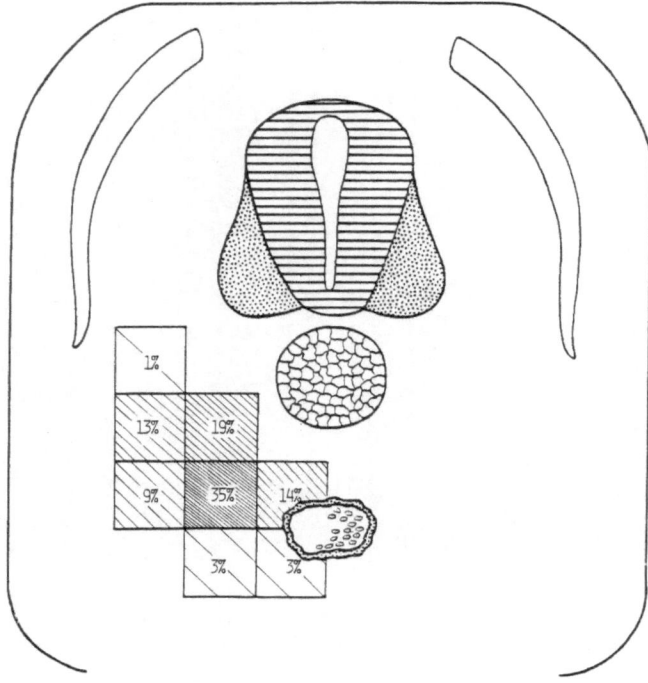

Figure 9. Schematic representation of the distribution of bovine serum albumin (BSA)-coated beads in embryos fixed and analyzed 2 and 3 days postinjection. Data were analyzed as described in Figure 7; N = 6 embryos.

To test whether FN might have some role in restriction along the ventral pathway, latex beads coated with FN were injected into the trunk somites of host embryos. The FN-coated beads (FN beads) behaved quite differently from either the uncoated or the BSA-coated beads. The FN beads remained predominantly at the injection site (associated with the dermamyotomal or sclerotomal remnant of the somite) and did not distribute ventrally along the neural crest pathway (Fig. 10). Occasionally, the FN beads were associated with the ectoderm. Figure 11 is a schematic representation of the distribution of FN beads in embryos fixed 2 or 3 days postinjection. The distribution pattern demonstrates that 95% of beads in all embryos remained in dorsal sites, generally adhering to cells of the dermamyotome. Only 5% of the beads were found in ventral sites. This is a contrast to the results obtained with uncoated and BSA-coated beads, where 97% of the beads were found ventrally along the neural crest pathway. Therefore, the FN beads were excluded from the ventral neural crest pathway, as were the somite and fibroblast cells.

The presence of cell surface fibronectin correlates with the inability to move along the ventral route. Embryonic fibroblasts and somite cells both synthesize FN (Engvall and Ruoslahti, 1977; Loring et al., 1977; Yamada and Olden, 1978), and both cell types fail to translocate after intraembryonic implantation. However, the cell and bead types capable of translocation along the ventral pathway lack surface FN. Melanocytes and other crest-derived cells, sarcoma 180 cells (a transformed fibroblastic cell line that has lost FN), and retinal pigment epithelial cells all are deficient in cell surface FN. Yet they distributed along the ventral pathway after injection (Bronner-Fraser and Cohen, 1980; Erickson et al., 1980). Cell surface FN may therefore function to restrict certain embryonic cell types from neural crest sites. Experiments using FN beads to produce an "artificial cell" with a single surface component offered a simple system to test the correlation between surface fibronectin and the inability to translocate. Because FN beads behaved similarly to certain cell types with surface FN, the results strengthen the correlation. These studies demonstrate the utility of latex beads as a tool for probing embryonic interactions. One can better characterize the nature of neural crest migratory pathways and their role in guiding cell movement and localization by observing the effects of surface macromolecules on bead distribution.

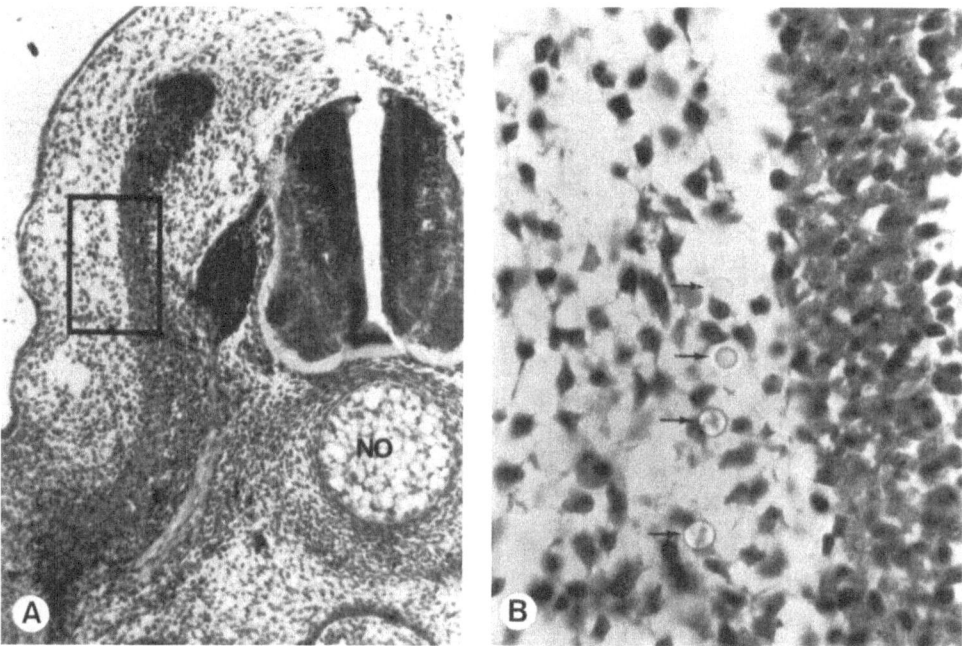

Figure 10. Transverse sections through chick embryos after fibronectin FN-coated beads were injected into the posterior somite. (A) An embryo fixed 3 days after injection with FN beads. (B) At (A) inset, the FN beads are found associated with the dermatome. Arrows indicate the positions of the FN beads. NO, neural tube; NT, notochord. (From Bronner-Fraser, 1982.)

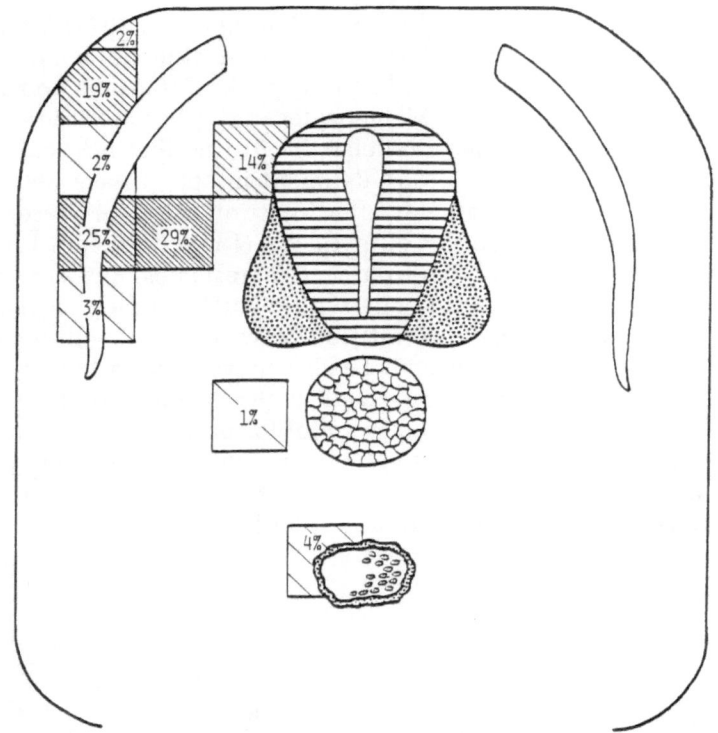

Figure 11. Schematic representation of the distribution of fibronectin-(FN)-coated beads in embryos fixed and analyzed 2 and 3 days postinjection. Data were analyzed as described in Figure 7; N = 7 embryos.

4.5. Beads Coated with Fibronectin Fragments

Fibronectin is a large dimeric glycoprotein. Each polypeptide chain has a molecular weight on the order of 220,000 (Mosesson *et al.*, 1975). FN has a wide range of functional activities such as binding to collagens (Engvall and Ruoslahti, 1977; Dessau *et al.*, 1978; Jilek and Hormann, 1979), glycosaminoglycans (Stathakis and Mosesson, 1977; Jilek and Hormann, 1979), actin (Keski-Oja *et al.*, 1980), FN, and fibrin (Ruoslahti and Vaheri, 1975; Stemberger and Hormann, 1976) and to cell surfaces (Klebe, 1974; Pearlstein, 1976). These binding activities appear to be localized in distinct domains of the molecule. By proteolytic digestion of the FN molecule, fragments of fibronectin containing one or more of the functional binding activities have been isolated (Balian *et al.*, 1979; Gold *et al.*, 1979; Hahn and Yamada, 1979a,b; Rubin *et al.*, 1979; Ruoslahti and Hayman, 1979; Ruoslahti *et al.*, 1979; Sekiguchi and Hakomori, 1980; Yamada *et al.*, 1980).

In order to examine which domains of FN may cause restriction of FN-coated beads from the ventral neural crest pathway, latex beads coated with distinct fragments of FN were injected into chicken embryos during endogenous neural crest migration (Bronner-Fraser, 1985). Three fragments were used for these studies: (1) a 66,000-M_r fragment that contains the collagen-binding domain (FN collagen-binding fragment); (2) a mixture of 35,000-M_r and 42,000-M_r fragments that contains heparin-binding activity (FN heparin-binding fragment); and (3) a 120,000-M_r fragment that contains the cell-binding site (FN cell-binding fragment).

After injection of beads coated with the FN fragments, embryos were monitored for 1–3 days postinjection. Beads coated with the FN collagen-binding fragment translocated along the ventral neural crest pathway and settled adjacent to sympathetic and adrenomedullary sites. Similarly, beads coated with the FN heparin-binding fragment translocated ventrally and settled around the dorsal aorta and sympathetic ganglia. No distinct differences in distribution between beads coated with the FN collagen or heparin-binding fragments and uncoated latex beads were found. Figures 12 and 13 are schematic representations of the distribution of beads coated with FN collagen-binding fragment and the FN heparin-binding fragment, respectively, illustrating that

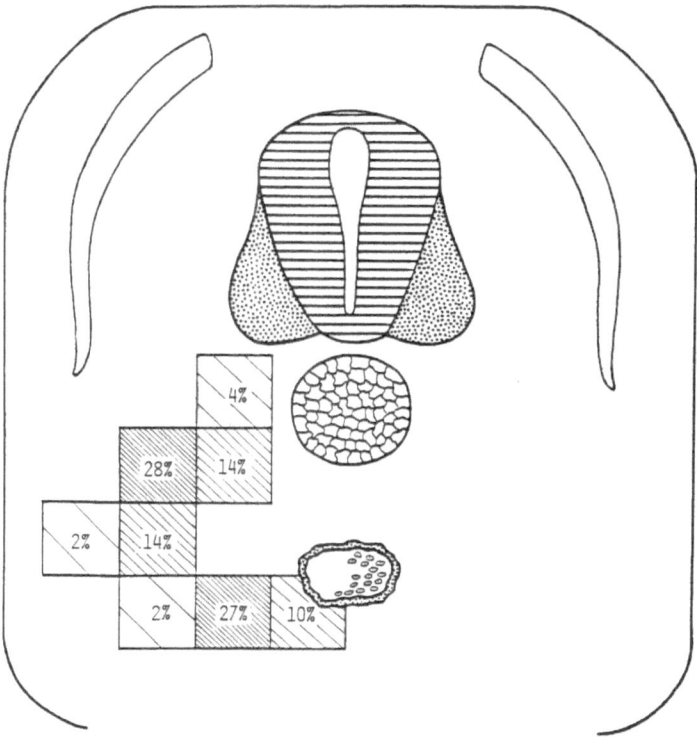

Figure 12. Schematic representation of the distribution of beads coated with the fibronectin (FN) collagen-binding fragment after injection. Embryos were fixed and analyzed 2 and 3 days postinjection. Data were analyzed as described in Figure 7; N = 5 embryos.

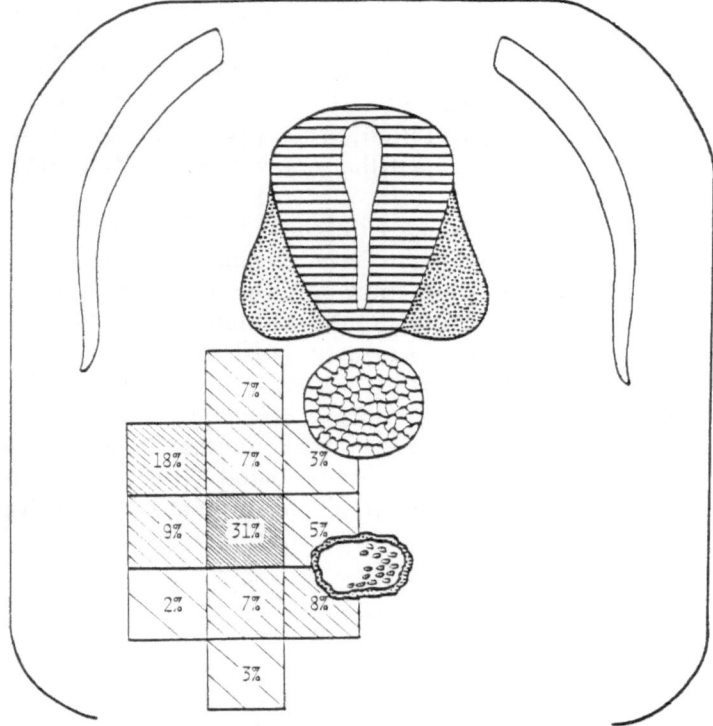

Figure 13. Schematic representation of the distribution of beads coated with the fibronectin (FN) heparin-binding fragment in embryos fixed 2 and 3 days postinjection. Data were analyzed as described in Figure 7; N = 11 embryos.

the beads have localized in ventral sites. By contrast, most beads coated with the FN cell-binding fragment did not translocate ventrally but remained associated with the dermamyotomal cells at the implantation site. Figure 14 is a schematic representation of the distribution of beads coated with the FN cell-binding site for embryos fixed 2 and 3 days postinjection. Eighty percent of the beads remained dorsally in these embryos, associated with the dermamyotome or sclerotome. These findings indicate that it is predominantly the cell-binding domain of the FN molecule that accounts for the adhesion of FN beads to cells at the somitic implantation site.

4.6. Beads Coated with Other ECM Molecules

Coating beads with FN changes their distribution pattern within the embryo. Perhaps other molecules on the bead surface would affect bead localization as well. By coating beads with other ECM molecules, it is possible to examine (1) whether restriction of FN beads from the ventral route is specific

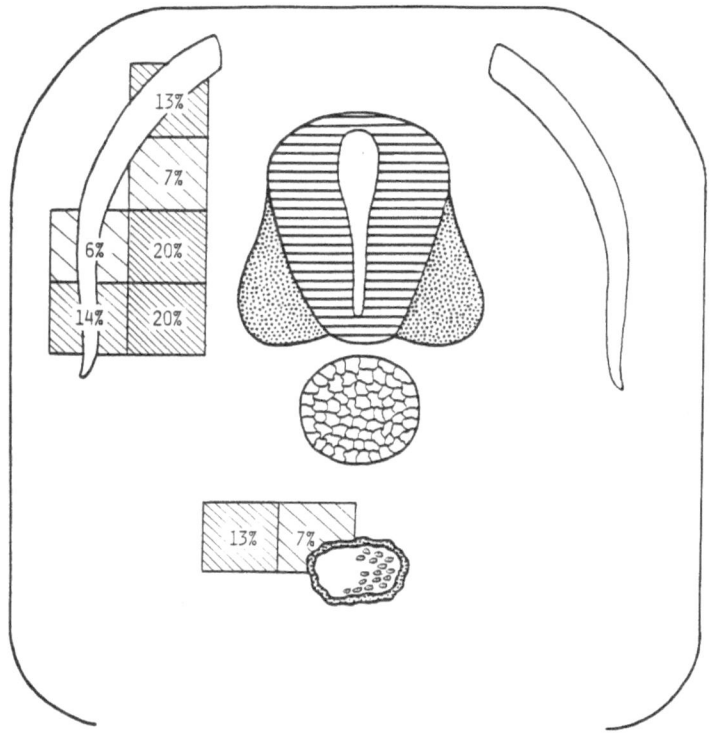

Figure 14. Schematic representation of the distribution of beads coated with the fibronectin (FN) cell-binding fragment in embryos fixed 2 and 3 days postinjection. Data were analyzed as described in Figure 7; N = 5 embryos.

for fibronectin; or (2) whether other adhesive molecules will cause similar restriction of the latex beads. The first possibility would indicate that a selective binding event at the implantation site recognizes cell surface FN and retains FN-coated cells at that location. The latter possibility suggests a less specific mechanism whereby general adhesive interactions could cause immobilization of some bead types.

Other ECM molecules that may be involved in cell migration and guidance include collagens and laminin. Type I collagen, produced by the epithelial cells of the neural tube lying adjacent to the ventral neural crest pathway (Cohen and Hay, 1971), appears to be a major component of the fibrillar material found in the ECM during crest migration (Tosney, 1978). In order to probe the possible surface function of this molecule, latex beads were coated with type I collagen and subsequently injected into the trunk somites. Scanning electron micrographs (SEMs) of the bead surface after coating suggest that the collagen was in fibrillar form (Fig. 15). After injection into the somites, beads coated with type I collagen translocated ventrally along the pathway taken by endogenous crest cells (Bronner-Fraser, 1984). The collagen beads settled adja-

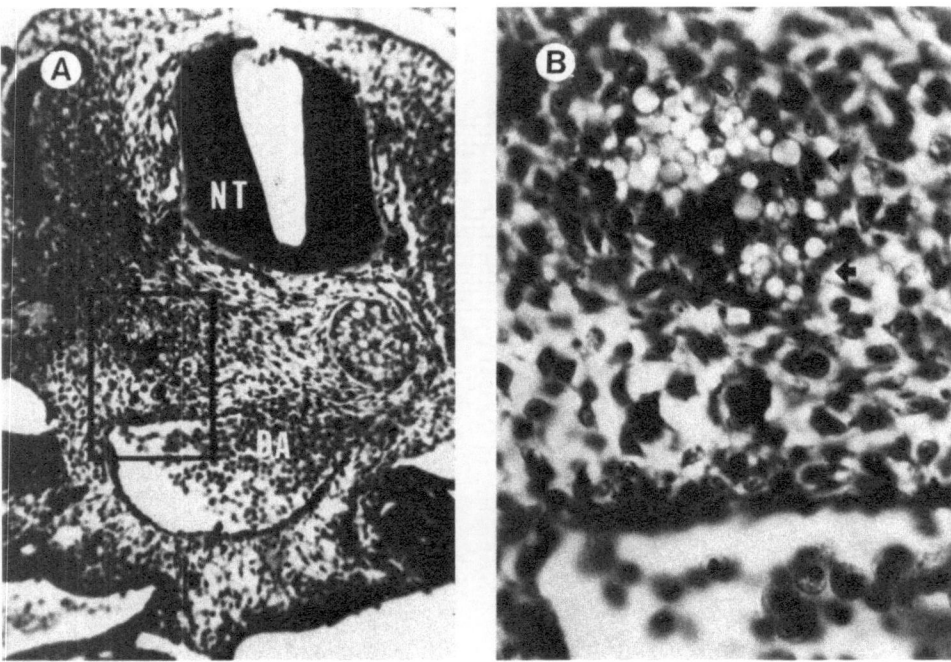

Figure 15. Light photomicrographs of an embryo 2 days after injection of collagen-coated beads into the somite. (A) Low magnification; (b) higher magnification of (A) inset shows the collagen beads surrounding and within the primary sympathetic ganglion, just above the dorsal aorta. NT, neural tube; DA, dorsal aorta. Arrows mark the location of the beads. (From Bronner-Fraser, 1984.)

cent to sympathetic ganglia or adrenomedullary sites. In fact, 30% of the embryos fixed at 2 days postinjection had collagen beads localized within the primary sympathetic ganglia (Fig. 16). Figure 17 is a schematic representation of the distribution of collagen-coated beads in embryos fixed 2 and 3 days postinjection. Ninety percent of the beads were found in ventral sites at these times. Thus, collagen beads moved ventrally along the neural crest pathway in much the same way as uncoated latex beads.

Another adhesive glycoprotein that may be important for the adhesion of some epithelial cell types is laminin. Laminin, which has been found in a wide variety of basement membranes (Chung *et al.*, 1979; Rohde *et al.*, 1979; Timpl *et al.*, 1979; Ekblom *et al.*, 1980; Foidart *et al.*, 1980; Wartiovaara *et al.*, 1980), is distinct from FN by many criteria, including amino acid composition and absence of immunological crossreactivity. It has been proposed to mediate the attachment to basement membranes of certain epithelial cell types that cannot use FN (Terranova *et al.*, 1980; Vlodasky and Gospodarowicz, 1980). In order to further explore surface adhesion effects that might be involved in restriction from the ventral pathway, Latex beads were coated with laminin. Laminin-coated beads (LM beads) were injected into the trunk somites at the beginning of host crest migration. In embryos examined 1–3 days postinjection, the LM beads were found associated with the dermamyotomal remnant of the somite (Fig. 18). Thus the LM beads, like FN-coated beads, did not translocate along the ventral neural crest pathway but remained associated with cells at the dorsal implantation site (Bronner-Fraser, 1984). Figure 19 is a schematic representation of the distribution of laminin beads in embryos fixed 2 and 3 days postinjection. Ninety-seven percent of the beads were found in dorsal sites, associated with the dermamyotome, ectoderm, and occasionally dorsolateral to the neural tube. These results indicate that the immobilization of FN beads does not necessarily reflect binding specific for FN alone; rather, a more general adhesion mechanism is likely to be responsible for restriction of some bead or cell types from the ventral route.

4.7. Effects of Surface Charge on Bead Translocation

Two distinct patterns of bead localization have been observed after injection of beads with differing surface properties. Beads coated with FN or laminin do not translocate along the ventral neural crest pathway, whereas uncoated, BSA-coated, or collagen-coated beads undergo ventral translocation. The differences in distribution pattern along the ventral neural crest pathway

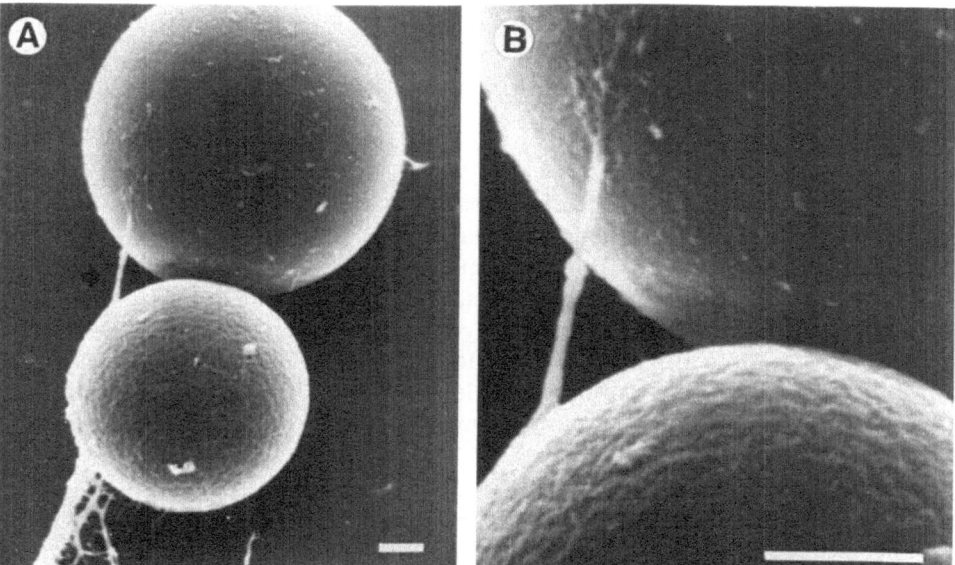

Figure 16. Scanning electron micrograph of collagen-coated beads. The collagen-bead surface generally had an irregular texture. Fibrils can be observed coming off the surface of the beads as well as connecting adjacent beads. (A) the arrow indicates collagen fibril shown at higher magnification in (B). Scale bar: 1 μm.

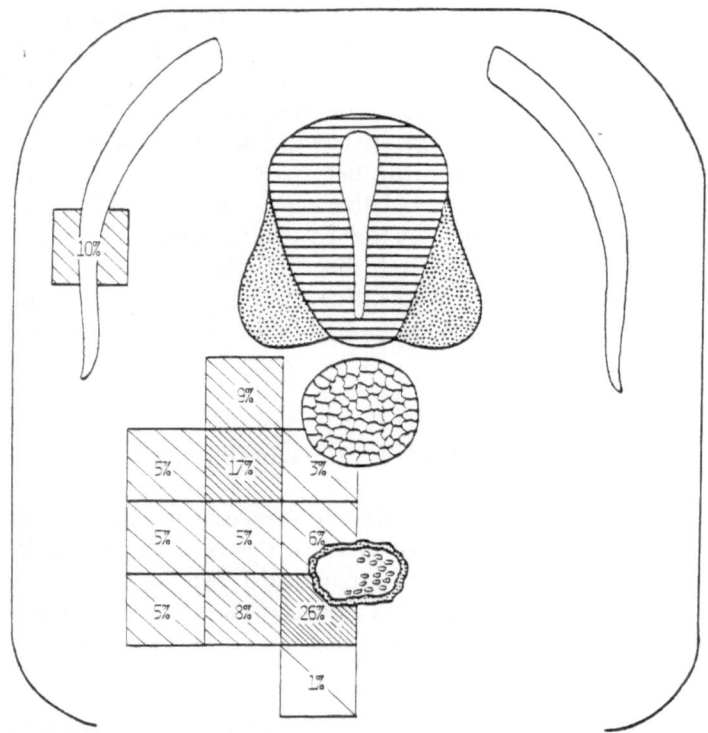

Figure 17. Schematic representation of the distribution of collagen-coated beads in embryos fixed 2 and 3 days after injection into somites. Data were analyzed as described in Figure 7; N = 11 embryos.

appear to be related to the macromolecule adsorbed onto the bead surface. Several possible factors could account for the observed differences in distribution. These include the possibilities that (1) the specific macromolecular coat confers adhesive properties to the surface of the beads so that they associate with the cells at the implantation site by some specific adhesion mechanisms; or (2) the macromolecule may change the surface charge on the beads, thereby altering their electrostatic properties. Because latex beads carry an inherent negative charge, the latter possibility introduces a confounding variable that

Figure 18. Light photomicrographs of tranverse sections through embryos several days after injection of laminin beads into the somites. (A) An embryo 2 days postinjection shown at low magnification. (B) Higher magnification of (A) inset demonstrates that the laminin beads remain associated with the dermamyotomal remnant of the somite. (C) Another embryo fixed 3 days postinjection. (D) Higher magnification of (C) inset shows that the laminin beads are associated with cells of the dermamyotome. NT, neural tube; NO, notochord; DA, dorsal aorta; DM, dermamyotome. Arrows indicate the position of the beads. (From Bronner-Fraser, 1984.)

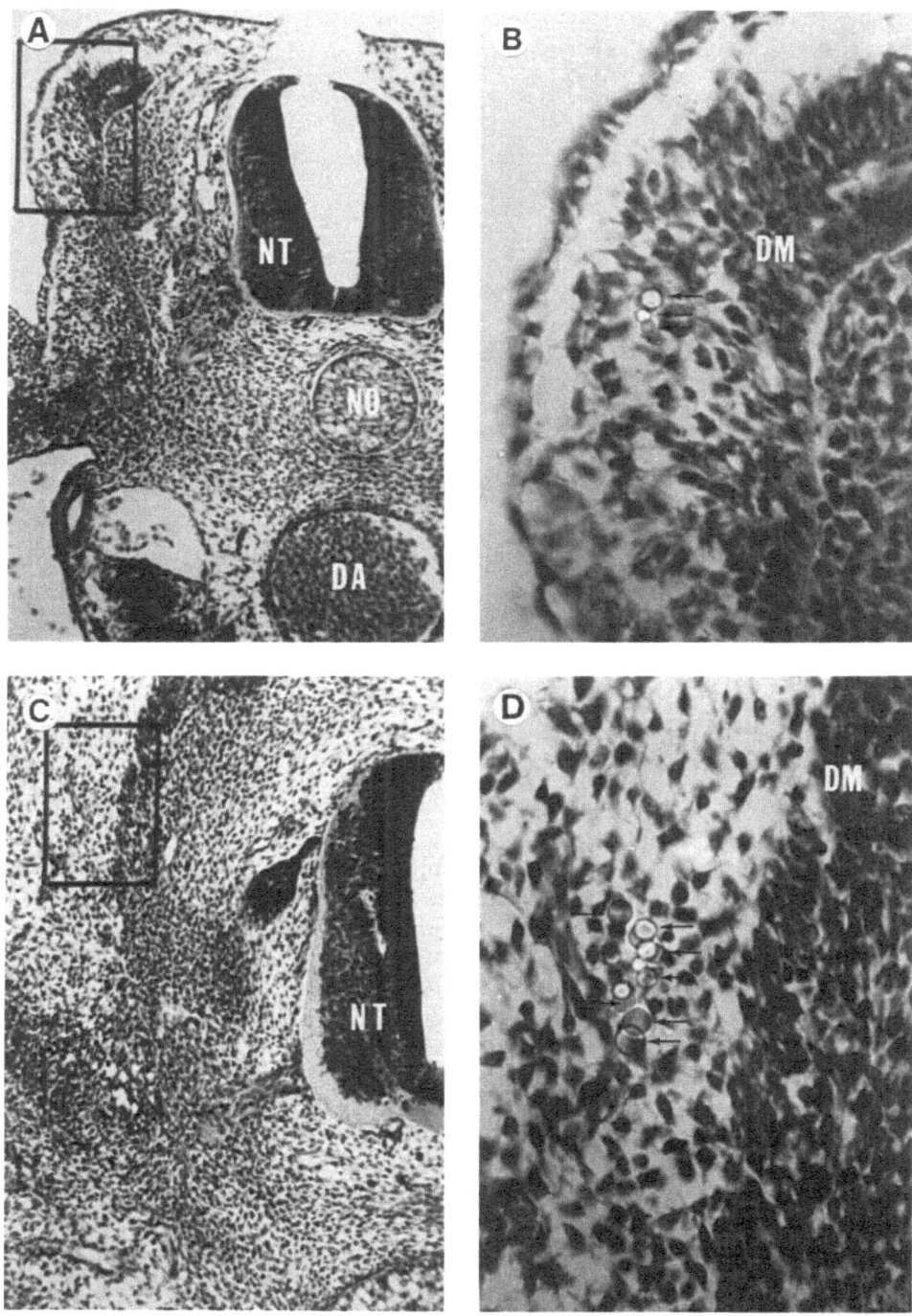

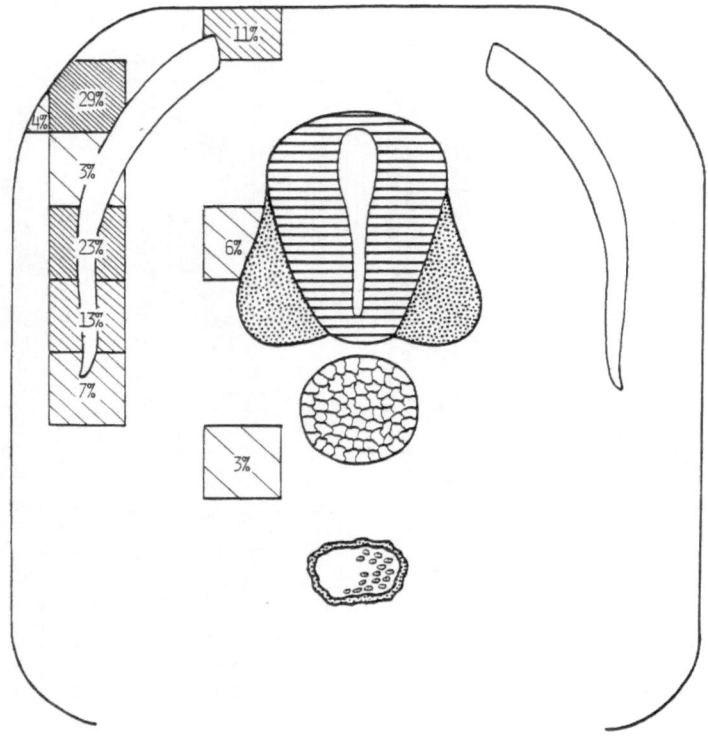

Figure 19. Schematic representation of the distribution of laminin-coated beads in embryos fixed 2 and 3 days postinjection into the somites. Data were analyzed as described in Figure 7; N = 6 embryos.

may affect the results in a nonbiologically relevant way. Therefore, electrostatic effects on bead movement were examined by coating the latex beads with amino acid polymers in order to alter predictably the initial surface charge.

The latex beads were coated with polytyrosine in order to produce a bead with neutral surface charge. At physiological pH, tyrosine is a neutrally charged amino acid. The polytyrosine-coated beads (PT beads) were injected into the trunk somites at the onset of host neural crest migration. The PT beads translocated along the ventral neural crest pathway with a final pattern of distribution very similar to that of uncoated latex beads and of endogenous crest cells. Figure 20 is a schematic representation of the distribution of PT beads in embryos fixed 2 and 3 days postinjection. Ninety-five percent of the beads localized in ventral sites, typically above the dorsal aorta, adjacent to sympathetic ganglion, or around the notochord (Bronner-Fraser, 1984).

Beads with a positive surface charge were produced by coating the latex particles with polylysine (PL). Lysine is a basic amino acid at physiological pH. After injection into the trunk somites, the PL beads behaved differently from either uncoated latex beads or the PT beads. Instead of translocating along the

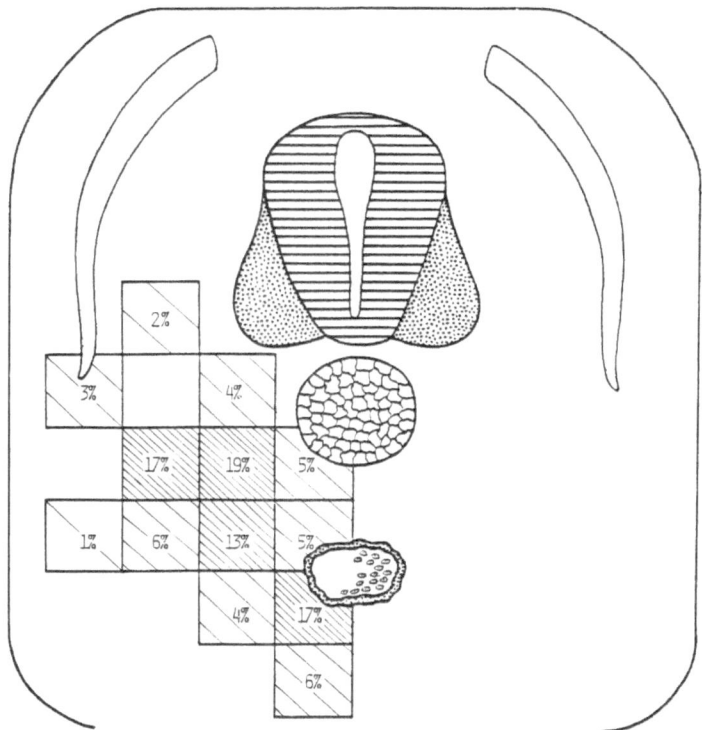

Figure 20. Schematic representation of the distribution of polytyrosine (PT)-coated beads in embryos fixed 2 and 3 days postinjection into the somites. Data were analyzed as described in Figure 7; N = 9 embryos.

ventral neural crest pathway, the PL beads remained at the implantation site. Figure 21 is a schematic representation of the distribution of PL-coated beads in embryos fixed 2 and 3 days postinjection; 88% of the beads were found in dorsal sites associated with the dermamyotome, sclerotome, or adjacent to the ectoderm. Thus, PL on the bead surface caused restriction of beads from the ventral neural crest pathway (Bronner-Fraser, 1984). These results are similar to those obtained for beads coated with the adhesive glycoproteins FN and laminin. The adhesion of PL-coated beads to the cells at the implantation site is not surprising, since this molecule is well known to adhere to embryonic cells by means of strong electrostatic interactions.

These results demonstrate that the positively charged beads exhibit a pattern of distribution similar to that of beads coated with FN or laminin and remain at the implantation site after injection onto a neural crest pathway. Is the immobilization of laminin and fibronectin beads caused by some specific adhesion mechanism? Or, do FN and laminin beads also carry a positive charge and, like PL beads, remain at the implantation site because of electrostatic interactions? To distinguish between these possibilities, the surface charge of

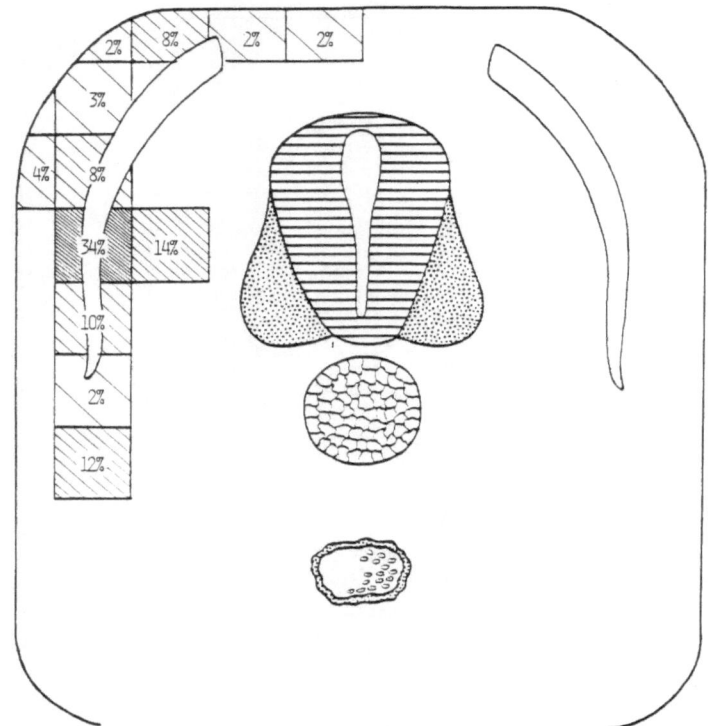

Figure 21. Schematic representation of the distribution of polylysine (PL)-coated beads in embryos fixed 2 and 3 days postinjection into the somites. Data were analyzed as described in Figure 7; N = 7 embryos.

the beads coated with various macromolecules was determined before injection and was compared with their subsequent distribution patterns after injection.

The surface charge can be determined by measuring the electrophoretic mobility of beads coated with a variety of macromolecules in an electrophoresis chamber. From this information, the surface potential and the surface charge can in turn be calculated. Uncoated latex particles and beads coated with the following types of molecules were tested: PL, PT, BSA, type I collagen, FN, and laminin. The data, summarized in Table I, demonstrate that PL-coated beads were positively charged and the PT beads were neutrally charged as predicted. Uncoated beads, beads coated with collagen, BSA, FN, and laminin were all negatively charged.

After injection into embryos at the time of host neural crest migration, beads coated with PL, FN, and laminin were all restricted from the ventral pathway. Yet the surface charge of the PL beads is positive, whereas that of both FN and laminin is negative. Laminin beads were, in fact, the most negatively charged of all the coated beads produced. Because both positively charged beads and some negatively charged beads fail to translocate along the ventral pathway while other negatively charged beads do translocate ventrally, no correlation exists between the sign of the surface charge and subsequent bead localization. These findings suggest that an adhesion mechanism independent of charge effects may account for restriction of beads coated with FN or laminin from the ventral neural crest pathway. PL, which is known to interact electrostatically with embryonic cell types, can mimic the more specific binding of these biologically important macromolecules.

4.8. Modifications to the Bead Surface after Intraembryonic Injection

The microinjection method for implanting cells or inert substances into embryos makes it possible to explore some aspects of cell surface–ECM interactions during development. By using latex beads, the system can be reduced to simpler components because the beads lack motility and metabolic activity and, unlike living cells, do not turn over their surface molecules. A limitation of this approach, however, is that it is only possible to ascertain the surface properties of the bead before injection. Because the distribution of latex beads depends on surface properties, it seems likely that the bead surface interacts with the embryonic milieu. Within the embryo, subsequent modifications may occur such that additional macromolecules are absorbed onto the bead surface or portions of the surface coat are enzymatically cleaved. Both the initial surface coat and possible modifications incurred by interactions with embryonic tissues may be important for final localization of latex beads after injection.

Because it is possible that the surface coat might be modified by interactions within the embryo, we examined (1) whether the surface coat remains on the beads after injection; and (2) whether additional substances are added to the surface of the beads within the embryo. Using immunofluorescence, the presence of various proteins and glycoproteins can be detected on the bead surface (Bronner-Fraser, 1982). In order to determine whether the surface coat remains on the bead after injection *in vivo*, the surface of beads coated with laminin or collagen was examined for the presence of the homologous sub-

Table I. Surface Properties of Coated Latex Beads

Latex bead surface coat	Surface charge (charges/$Å^2 \times 10^{-5}$)	Electrophoretic mobility (μm/sec/V/cm)	Ventral translocation
Polylysine	+3.1	+1.4	No
Polytyrosine	0.0	0.0	Yes
Collagen	−1.6	−0.8	Yes
Bovine serum albumin	−2.0	−1.0	Yes
Uncoated	−2.3	−1.1	Yes
Fibronectin	−2.5	−1.2	No
Laminin	−5.9	−2.6	No

stance before and after injection. Before injection, all beads appeared to be well coated. When laminin- or collagen-coated beads were recovered after 2 days *in ovo*, 70% of the beads retained their surface coat as identified by immunofluorescence. For those beads that were not antigenic, it was not clear whether the surface coat was enzymatically removed or masked by the binding of other molecules. The results suggest that most beads, however, do retain their surface coat. To determine whether the beads accumulated additional molecules from the embryonic milieu after intraembryonic injection, uncoated latex beads or beads coated with collagen, laminin, PT, or PL were processed for the detection of laminin or FN. FN and laminin were detectable by immunofluorescence on the surface of a small percentage (15% and 4%, respectively) of beads coated with collagen. For other bead types, however, no accumulation of FN or laminin was observed at levels detectable by immunofluorescence (Bronner-Fraser, 1984).

5. Laser Ablation of the Avian Neural Crest

The mechanisms that account for movement of latex beads along neural crest pathways remain unclear. One possible explanation is that the latex beads are "carried" to ventral sites by nonspecifically adhering to migrating neural crest cells. Under tissue culture conditions, latex beads and BSA-coated beads do not stick avidly to cultured neural crest cells (Bronner-Fraser, 1982). This may not, however, be an accurate reflection of events in the intact embryo.

One can more directly address the possibility that latex beads translocate by adhering to neural crest cells by ablating the host neural crest. Surgical ablations have been used in the past to study neural crest derivatives (Yntema and Hammond, 1945). Because these methods require precise microsurgical manipulations, they are often difficult to perform accurately. In addition, surgical procedures tend to be traumatic and may lead to deformities that complicate interpretation of the results.

Many of the problems incurred by surgical ablation can be avoided by using a laser microbeam to ablate the neural crest. Neural crest cells can be reproducibly removed from embryos by means of laser irradiation without disrupting the integrity of the embryo (Coulombe and Bronner-Fraser, 1982, 1984). The embryos are stained with the vital dye neutral red and subsequently exposed to laser irradiation. The Q-switched, frequency-doubled output at 532 nm from a YAG : Nd laser is used for these experiments. By means of a motorized stage the movement of which is coupled to the firing of the laser, the specimen is moved through the laser microbeam.

In each embryo, six or more somite lengths of neural tube are ablated. Changes in the dorsal neural tube are evident by a few seconds after irradiation. In embryos fixed at these times, necrosis of the cells localized on the dorsal aspect of the neural tube was observed. In order to examine the effectiveness of the ablation, the presence of neural crest derivatives (e.g., dorsal root or sympathetic ganglia) was observed in embryos fixed 2 or 3 days after injection.

These structures are easily detectable in fixed and stained histological sections and serve as convenient markers for the effectiveness of the laser irradiation. By these criteria, irradiation with the laser microbeam caused complete or partial removal of neural crest structures in the majority of irradiated embryos. No severe malformations were observed in the ablated embryos, with the exception of slight neural tube deformities.

In order to examine the translocation of latex beads in the absence of neural crest cells, we have injected beads into the somites of embryos whose endogenous neural crest was removed using a laser microbeam. Embryos were monitored 1 and 2 days postinjection. The latex beads in these embryos translocated ventrally even in the absence of host neural crest cells (Fig. 22). Figure 23 is a schematic representation that illustrates the distribution of beads in embryos following removal of host neural crest cells with laser irradiation. The distribution pattern is similar to that of latex beads injected into embryos whose neural crest was not ablated (see Fig. 7A). Eighty-nine percent of the beads localized in ventral sites, with most settling around the dorsal aorta. In

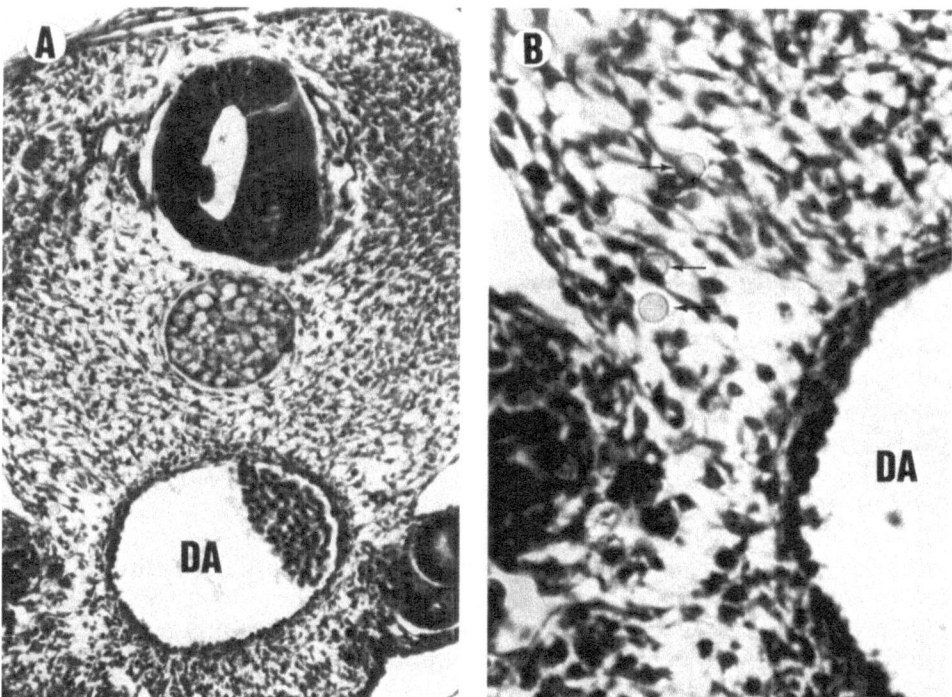

Figure 22. Transverse section through an embryo fixed 2 days postinjection of latex into the embryonic somites of chick embryo whose neural crest cells were removed with a laser microbeam. The beads translocate ventrally even without the endogenous neural crest, localizing adjacent to the dorsal aorta. The embryos appear normal, with the exception of some neural tube anomalies. (A) Low power, (B) Higher power of (A) inset.

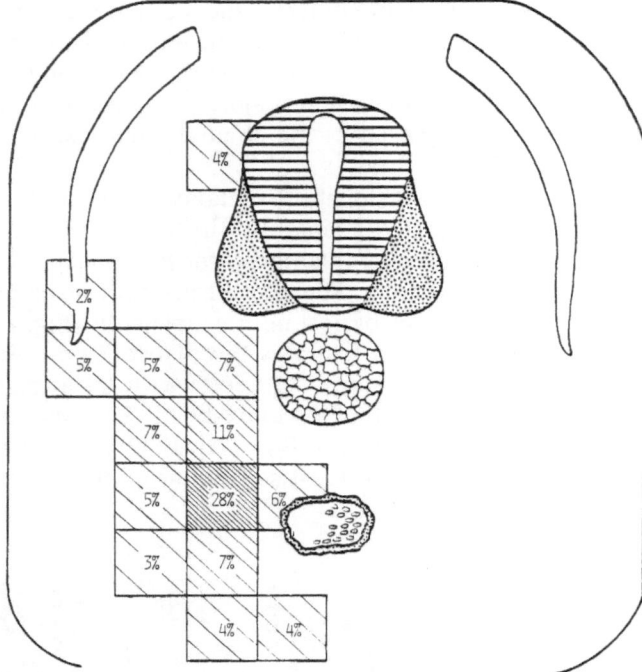

Figure 23. Schematic representation of the distribution of uncoated Latex beads in embryos fixed 2 days after laser ablation of the neural crest. Data were analyzed as described in Figure 7; N = 8 embryos.

these ablated embryos, no neural crest-derived dorsal root or sympathetic ganglia were present in the vicinity of the injected beads. The findings indicate that Latex beads do not translocate by adhering to already migrating crest cells. Rather, properties of the bead surface may cause sorting out from other tissues, which results in ventral translocation.

6. Summary

The experiments reviewed in this chapter examine the translocation of various cell types and latex beads on a neural crest pathway. The cells and beads are implanted into the embryo *via* an injection technique that can be used to characterize the embryonic pathways or the injected cells themselves. The results demonstrate that postmigratory neural crest cells, undifferentiated neural crest cells, and retinal pigment epithelial cells will translocate to ventral sites after implantation. By contrast, somitic and fibroblastic cells fail to translocate. No correlation was found between the inherent motile ability of a cell and the ability to move along the ventral route. Therefore, the role of cell surface molecules in movement along the neural crest pathway was examined.

Latex beads, which lack inherent motility, were used as probes of the neural crest pathway. Uncoated beads as well as latex beads coated with a variety of ECM molecules and polyamino acids were injected into embryos in order to explore interactions between the cell surface and the embryonic substrata that might be involved in neural crest localization. The distribution pattern of the latex beads was altered by the nature of the surface properties of the beads. Two distinct patterns of localization were observed. Those beads coated with FN, cell-binding fragment of FN, laminin, or PL remained primarily associated with the dermamyotomal cells of the implantation site. By contrast, uncoated beads or beads coated with BSA, collagen, or PT translocated to ventral sites, usually around the sympathetic ganglia or dorsal aorta.

In order to analyze mechanisms that may be involved in translocation of latex beads along neural crest pathways, we examined the possible effects of (1) bead surface charge; and (2) the removal of endogenous neural crest cells. To examine the effects of electrostatic interactions in bead translocation or restriction, the initial surface charge of beads coated with various macromolecules was measured and compared with their subsequent ability to translocate along the ventral pathway. No correlation was observed between the sign of the surface charge and subsequent distribution of beads, suggesting that initial surface charge properties alone cannot account for the restriction or translocation. To dissect the role of endogenous neural crest cells in bead movement, the host neural crest was ablated using a laser microbeam. After injection of latex beads into ablated embryos, the latex beads translocated ventrally even in the absence of the neural crest. Thus, latex beads are not merely carried ventrally by adhering to migrating neural crest cells.

The findings reviewed in this chapter suggest that although neural crest cells are inherently migratory, many events in neural crest distribution may involve cell surface properties rather than the motile apparatus of the cell. Adhesive molecules such as FN on the cell surface, for example, may restrict classes of cells from ventral translocation along neural crest pathways. When latex beads are implanted into embryonic tissues, they also seem to localize according to surface adhesive properties. Beads with highly adhesive surfaces appear to adhere to cells at the implantation site. Less adhesive beads translocate and may be mixed by the movement of embryonic cells around them, enabling them to respond to their surface properties. Therefore, the surface properties on both the latex bead and the cell surface may account for restriction from or distribution along the neural crest ventral pathway. By implanting latex beads into embryos during active neural crest migration, we have discovered that the mechanisms governing cell localization needn't be complex or highly discriminatory. They may not, in fact, completely distinguish cells from some inert particles. During morphogenesis, perhaps only recognition of large surface differences between embryonic cells is necessary for proper tissue organization and pattern formation.

ACKNOWLEDGMENTS. I would like to thank Virginia Bayer for illustrations. Parts of the research reviewed in this article were supported by U.S. Public

Health Service Grant HD-15527-01 and by Basil O'Connor Starter Research Grant 5-312 from the March of Dimes Birth Defects Foundation.

References

Ali, I. U., and Hynes, R. O., 1978, The effects of LETS glycoprotein on cell motility, *Cell* **14:**439–446.

Allan, I. J., and Newgreen, D. F., 1980, The origin and differentiation of enteric neurons of the intestines of the fowl embryo, *Am. J. Anat.* **157:**137–154.

Balian, G., Click, E. M., Crouch, E., Davidson, J., and Bornstein, P., 1979, Isolation of a collagen-binding fragment for fibronectin and cold-soluble globulin, *J. Biol. Chem.* **254:**1429–1432.

Bancroft, M., and Belliars, R., 1976, The neural crest cells of the trunk region of the chick embryo studied by SEM and TEM, *Zoon* **4:**73–85.

Bellairs, R., Ireland, G. W., Sanders, E. J., and Stern, C. D., 1981, The behavior of embryonic chick and quail tissues in culture, *J. Embryol. Exp. Morphol.* **61:**15–33.

Bronner, M. E., and Cohen, A. M., 1979, Migratory patterns of cloned neural crest melanocytes injected into host chicken embryos, *Proc. Natl. Acad. Sci. USA* **76:**1843–1848.

Bronner-Fraser, M. E., 1982, Distribution of latex beads and retinal pigment epithelial cells along the ventral neural crest pathway, *Dev. Biol.* **91:**50–63.

Bronner-Fraser, M. E., 1984, Latex beads as probes of a neural crest pathway: the effects of laminin, collagen, and surface charge on bead translocation, *J. Cell Biol.* **98:**1947–1960.

Bronner-Fraser, M. E., 1985, Effects of different fragments of the fibronectin molecule on latex bead translocation along a neural crest pathway, *Dev. Biol.* **108:**(in press).

Bronner-Fraser, M. E., and Cohen, A. M., 1980, Analysis of the neural crest ventral pathway using injected tracer cells, *Dev. Biol.* **77:**130–141.

Bronner-Fraser, M. E., Sieber-Blum, M., and Cohen, A. M., 1980, Clonal analysis of the avian neural crest: Migration and maturation of mixed neural crest clones injected into host chicken embryos, *J. Comp. Neurol.* **193:**423–434.

Chung, A. E., Jaffe, R., Freeman, I. L., Vergnes, J. P., Bragninski, J. E., and Carlin, B., 1979, Properties of a basement membrane-related glycoprotein synthesized in culture by a mouse embryonal carcinoma-derived cell line, *Cell* **16:**277–287.

Cohen, A. M., 1977, Independent expression of the adrenergic phenotype by neural crest *in vitro*, *Proc. Natl. Acad. Sci. USA* **74:**2899–2903.

Cohen, A. M., and Hay, E. D., 1971, Secretion of collagen by embryonic neuroepithelium at the time of spinal cord–somite interactions, *Dev. Biol.* **26:**578–605.

Coulombe, J. N., and Bronner-Fraser, M. E., 1982, Laser ablation of the avian neural crest, *Soc. Neurosci.* **8(2):**320.

Coulombe, J. N., and Bronner-Fraser, M. E., 1984, Laser ablation of the avian neural crest, *Dev. Biol.* **106:**121–134.

Davis, E. M., 1980, Translocation of neural crest cells within a hydrated collagen lattice, *J. Embryol. Exp. Morphol.* **55:**17–31.

Derby, M. A., 1978, Analysis of glycosaminoglycans within the extracellular environments encountered by migrating neural crest cells, *Dev. Biol.* **66:**321–336.

Dessau, W., Adelmann, B. C., Timple, R., and Martin, G. R., 1978, Identification of the sites in collagen alpha-chains that bind serum anti-gelatin factor (cold-insoluble globulin), *Biochem. J.* **169:**55–59.

Ebendal, T., 1977, Extracellular matrix fibrils and cell contacts in the chick embryo. Possible roles in orientation of cell migration and axon extension, *Cell Tissue Res.* **175:**439–458.

Ekblom, P., Alitalo, L., Vaheri, A., Timpl, R., and Saxen, L., 1980, Induction of a basement membrane glycoprotein in embryonic kidney: Possible role of laminin in morphogenesis, *Proc. Natl. Acad. Sci. USA* **77:**485–489.

Engvall, E., and Ruoslahti, E., 1977, Binding of soluble form of fibroblast surface protein, fibronectin to collagen, *Int. J. Cancer* **20:**1–5.

Erickson, C. A., Tosney, K. W., and Weston, J. A., 1980, Analysis of migratory behavior of neural crest and fibroblastic cells in embryonic tissues, Dev. Biol. **77**:142–156.

Foidart, J.-M., Bere, E. W., Yaar, M., Rennard, S. I., Gullino, M., Martin, G. R., and Katz, S. I., 1980, Distribution and immunoelectron microscopic localization of laminin, a noncollagenous basement membrane glycoprotein, Lab. Invest. **42**:336–342.

Gold, L. I., Garcia-Pardo, A., Frangione, B., Franklin, E. C., and Pearlstein, E., 1979, Subtilisin and cyanogen bromide cleavage products of fibronectin that retain gelatin-binding activity, Proc. Natl. Acad. Sci. USA **76**:4803–4807.

Greenberg, J. H., and Schrier, B. K. 1977, Development of choline acetyltransferase activity in chick cranial neural crest cells in culture, Dev. Biol. **61**:86–103.

Grinnell, F., 1980, Fibroblast receptors for cell-substratum adhesion: Studies on the interaction of baby hamster kidney cells with latex beads coated by cold insoluble globulin (plasma fibronectin), J. Cell Biol. **86**:104–112.

Hahn, L.-H. E., and Yamada, K., 1979a, Identification and isolation of a collagen-binding fragment of the adhesive glycoprotein fibronectin, Proc. Natl. Acad. Sci. USA **76**:1160–1163.

Hahn, L.-H. E., and Yamada, K., 1979b, Isolation and biological characterization of active fragments of the adhesive glycoprotein fibronectin, Cell **18**:1043–1051.

Jilek, F., and Hormann, H., 1979, Fibronectin (Cold Insoluble Globulin) VI. Influence of heparin and hyaluronic acid on the binding of native collagen, Hoppe-Seylers Z. Physiol. Chem. **360**:597–603.

Johnston, M. C., 1966, A radioautographic study of migration and fate of cranial neural crest cells in the chick embryo, Anat. Rec. **156**:143–156.

Kahn, C. R., Coyle, J. T., and Cohen, A. M., 1980, Head and trunk neural crest in vitro: Autonomic neuron differentiation, Dev. Biol. **77**:340–348.

Keski-Oja, J., Sen. A., and Todaro, G. J., 1980, Direct association of fibronectin and actin molecules in vitro, J. Cell Biol. **85**:527–533.

Klebe, R. J., 1974, Isolation of a collagen dependent cell attachment factor, Nature (Lond.) **250**:248–251.

Kurkinen, M., Alitalo, K., Vaheri, A., Stenman, S., and Saxen, L., 1979, Fibronectin in the development of the embryonic chick eye, Dev. Biol. **30**:217–222.

LeDouarin, N. M., 1973, A biological cell labelling technique and its use in experimental embryology, Dev. Biol. **30**:217–222.

LeDouarin, N. M., and Teillet, M. A., 1973, The migration of neural crest cells to the walls of the digestive tract in avian embryos, J. Embryol. Exp. Morphol. **30**:31–48.

LeDouarin, N. M., and Teillet, M. A., 1974, Experimental analysis of the migration and differentiation of neuroblasts of the autonomic nervous system and of neuroectodermal mesenchymal derivatives using a biological cell marking technique, Dev. Biol. **41**:162–184.

LeDouarin, N. M., Teillet, M. A., Ziller, C., and Smith, J., 1978, Adrenergic differentiation of cells of the cholinergic ciliary and Remak's ganglia in avian embryos after in vivo transplantation, Proc. Natl. Acad. Sci. USA **75**:2030–2034.

LeLievre, C. S., Schweizer, G. G., Ziller, C. M., and LeDouarin, N. M., 1980, Restrictions of developmental capabilities in neural crest derivatives as tested by in vivo transplantation, Dev. Biol. **77**:362–378.

Loring, J., Erickson, C. A., and Weston, J. A., 1977, Surface proteins of neural crest, crest-derived, and somite cells in vivo, J. Cell Biol. **75**:71a.

Manasek, F. J., and Cohen, A. M., 1977, Anionic glycopeptides and glycosaminoglycans synthesized by embryonic neural tube and neural crest, Proc. Natl. Acad. Sci. USA **74**:1057–1061.

Maxwell, G., Seitz, P. D., and Rafford, C. E., 1982, Synthesis and accumulation of putative neurotransmitters by cultured neural crest cells, J. Neurosci. **2**:879–888.

Mayer, B. W., Hay, E. D., and Hynes, R. O., 1981, Immunocytochemical localization of fibronectin in embryonic chick trunk and area vasculosa, Dev. Biol. **82**:267–286.

Mosesson, M. W., Chen, A. B., and Huseby, R. M., 1975, The cold insoluble globulin of human plasma: Studies of its essential structural features, Biochem. Biophys. Acta **386**:509–524.

Newgreen, D. F., and Thiery, J.-P., 1980, Fibronectin in early avian embryos: Synthesis and distribution along the migration pathways of neural crest cells, Cell Tissue Res. **211**:269–291.

Newgreen, D. F., Gibbins, I. L., Sauter, J., Wallenfels, B., and Wutz, R., 1982, Ultrastructural and tissue-culture studies on the role of fibronectin, collagen and glycosaminoglycans in the migration of neural crest cells in the fowl embryo, *Cell Tissue Res.* **221:**521–549.

Noden, D. M., 1975, An analysis of the migratory behavior of avian cephalic neural crest cells, *Dev. Biol.* **42:**106–130.

Pearlstein, E., 1976, Plasma membrane glycoprotein which mediates adhesion of fibroblast to collagen, *Nature (Lond.)* **262:**407–500.

Pintar, J., 1978, Distribution and synthesis of glycosaminoglycans during quail neural crest morphogenesis, *Dev. Biol.* **67:**444–464.

Rohde, H., Wick, G., and Timple, R., 1979, Immunochemical characterization of the membrane glycoprotein laminin, *Eur. J. Biochem.* **102:**195–201.

Rubin, K., Johansson, S., Pettersson, I., Ocklind, C., Obrink, B., and Hook, M., 1979, Attachment of rat hepatocytes to collagen and fibronectin: A study using antibodies directed against cell surface components, *Biochem. Biophys. Res. Commun.* **91:**86–94.

Ruoslahti, E., and Hayman, E. G., 1979, Two active sites with different characteristics in fibronectin, *FEBS Lett.* **97:**221–224.

Ruoslahti, E., and Vaheri, A., 1975, Interaction of soluble fibroblast surface antigen with fibrinogen and fibrin, *J. Exp. Med.* **141:**497–501.

Ruoslahti, E., Hayman, E. G., Kuusela, P. Shively, J. E., and Engvall, E., 1979, Isolation of a tryptic fragment containing the collagen-binding site of plasma fibronectin, *J. Biol. Chem.* **254:**6054–6059.

Sekiguchi, K., and Hakomori, S., 1980, Functional domain structure of fibronectin, *Proc. Natl. Acad. Sci. USA* **77:**2661–2665.

Sieber-Blum, M., Reed, W., and Lidov, H., 1983, Serotonergic differentiation of quail neural crest cells *in vitro*, *Dev. Biol.* **99:**352–359.

Solursh, M., Fisher, M., Meier, S., and Singley, C. T., 1979, The synthesis of hyaluronic acid by ectoderm during early organogenesis in the chick embryo, *Differentiation* **14:**77–85.

Stathakis, N. E., and Mosesson, M. W., 1977, Interactions among heparin, cold-insoluble globulin, and fibronectin in formation of heparin-precipitable fraction of plasma, *J. Clin. Invest.* **60:**855–865.

Stemberger, A., and Hormann, H., 1976, Affinity chromatography on immobilized fibrinogen and fibrin monomer. II. The behavior of cold-insoluble globulin, *Hoppe-Seylers Z. Physiol. Chem.* **357:**1003–1005.

Teillet, M. A., and LeDouarin, N. M., 1970, La migration de cellules pigmentaires etudiée par la methode des greffes heterospecifiques de tube nerveux chex l'embryon d'oiseau, *C. R. Acad. Sci. Ser. D* **270:**3095–3098.

Terranova, V. P., Rohrbach, D. H., Murray, J. C., Martin, G. R., and Yuspa, S. H., 1980, The role of laminin in epidermal cell attachment to basement membrane collagen, in: *Cold Spring Harbor Symposium on the Biology of the Vascular Cell*, Cold Spring Harbor Press, Cold Spring Harbor, New York.

Thiery, J. P., Duband, J. L., Rutishauser, U., and Edelman, G. M., 1982a, Cell adhesion molecules in early chick embryogenesis, *Proc. Natl. Acad. Sci. USA* **79:**6737–6741.

Thiery, J. P., Duband, J. L., and Delouvee, A., 1982b, Pathways and mechanisms of avian trunk neural crest cell migration and localization, *Dev. Biol.* **93:**324–343.

Timpl, R., Rohde, H., Robye, P. G., Rennard, S. I., Foidart, J. M., and Martin, G. R., 1979, Laminin— A glycoprotein for basement membranes, *J. Biol. Chem.* **254:**9933–9937.

Tosney, K., 1978, The early migration of neural crest cells in the trunk region of the avian embryo: An electron microscopic study, *Dev. Biol.* **62:**317–333.

Vlodasky, I., and Gospodarowicz, D., 1980, Respective involvement of laminin and fibronectin in the adhesion of human carcinoma and sarcoma cells, *Nature (Lond.)* **289:**304–306.

Wartiovaara, J., Leivo, I., and Vaheri, A., 1980, Matrix glycoproteins in early mouse development and in differentiation of teratocarcinoma cells, in: *The Cell Surface: Mediator of Cellular and Developmental Events* (S. Subtelny and N. Wessells, eds.), pp. 305–324, Academic Press, New York.

Weston, J. A., 1963, A radioautographic analysis of the migration and localization of trunk neural crest cells in the chick, *Dev. Biol.* **6:**279–310.

Yamada, K. M., and Olden, K., 1978, Fibronectin—Adhesive glycoproteins of cell surface and blood, *Nature (Lond.)* **275:**179–184.

Yamada, K. M., Kennedy, D. W., Kimata, K., and Pratt, R. M., 1980, Characterization of fibronectin interactions with glycosaminoglycans and identification of active proteolytic fragments, *J. Cell Biol.* **255:**6055–6063.

Yntema, C. L., and Hammond, W. S., 1945, Depletions and abnormalities in the cervical sympathetic system of the chick following extirpation of the neural crest, *J. Exp. Zool.* **100:**237–263.

Chapter 16

Cell Traction in Relationship to Morphogenesis and Malignancy

ALBERT K. HARRIS

1. Introduction

If we explant a fragment of embryonic tissue into a suitable culture medium on a reasonably adhesive substratum such as glass, the component cells of the tissue will crawl actively over this substratum, almost as if they were so many amoebae. Concerning this familiar phenomenon, three questions arise: How do the cells propel themselves? What functions does their propulsive mechanism serve in the development and maintenance of the body? Why doesn't the locomotion of cells disrupt and disperse the proper histological arrangements of cells inside the body?

Unfortunately, it is premature to give complete and definite answers to these questions. Nevertheless, recent attempts to answer the first of these three questions, using flexible elastic substrata to make visible the forces exerted by tissue cells, have suggested some rather unexpected answers to all three. This chapter describes these results and some of the apparent implications, leading to the following conclusions:

1. Tissue cells propel themselves by exerting shearing forces tangentially through adhering areas of their plasma membranes. Structural cells, especially fibroblasts, exert shearing forces several orders of magnitude stronger than necessary for their own propulsion, or than are exerted by leukocytes, which are far more mobile than they are.

2. In addition to cell locomotion, the exertion of these traction forces by tissue cells serves several other morphogenetic and physiological functions involving the displacement relative to cells of extracellular matrix materials, especially collagen. There are reasons to believe that this *tractional structuring* mechanism, as it has been called, is the principal means of connective tissue morphogenesis during embryonic develop-

ALBERT K. HARRIS • Department of Biology, University of North Carolina, Chapel Hill, North Carolina 27514.

ment and remains an effective mechanism for adjusting the tensile
stress of collagen fibers during life.

3. As to the reason why the capacity of structural cells for autonomous
   locomotion does not disrupt normal anatomical relationships, such disruption is an important part of what actually does occur in cancer.
   Studies of traction in certain transformed cells suggest that perhaps the
   invasiveness of cancerous cells reflects some sort of switch or misap-
   plication of traction exertion from normal matrix organization functions, such as capsule formation, to the production of unnecessary cell
   locomotion.

2. Traction and Its Detection: Observations using Elastic Substrata

2.1. Plasma Clots as Substrata

The exertion of microscopic forces by and on individual cells is the central
issue in the study of cell locomotion. Unfortunately, the small absolute magnitude of these forces makes them difficult to measure or even detect directly.
Micromanipulation and similar methods, including the application of small
resistance forces, have been successfully applied to various sorts of amebae but
have been really effective in the case of tissue and embryonic cells only in cases
in which the collective or additive behavior of groups of cells was being measured (James and Taylor, 1969; Phillips and Davis, 1978; Phillips and Steinberg, 1978). Ideally, one would hope for a method that (1) would reveal the
forces exerted even by individual cells; (2) would tell something about the
geometric arrangement of these forces (e.g., where they were exerted, as well as
in what directions; and (3) would provide this information over as long as
possible a period of time with the minimum disturbance to the normal behavior
of the cell. It was with this set of goals in mind that the attempts to develop a
suitable elastic substratum were begun.

My own initial attempts to deduce the directions and relative strengths of
cellular forces employed films of clotted plasma as the elastic substratum.
Clotted plasma was the first successful tissue culture substratum (used by Ross
Harrison); it long remained a standard material for culturing cells through
about four decades, when tissue culture continued to be more an art than a
standard technique. Indeed, it may be that the practical difficulties of taking
unclotted (but still clottable) plasma from an animal, not to mention its unpre-
dictable behavior when isolated, was a part of what kept tissue culture an art. In
any event, it is somewhat ironic that the physical distorting effects of cell
traction on plasma clots had not previously been noted or used as a way to
study the forces responsible for cell locomotion. Paul Weiss did in fact make a
close study of some of these distorting effects on plasma clots. But he interpreted these distortions as consequences of the shrinkage of the clot fibers
themselves, probably caused by a hypothetical "dehydration" due to the up-

take of water from the medium by growing cells. Weiss observed that this distortion effect was greater around neural explants containing extensive glia, as opposed to those containing primarily neurons; he proposed that this was because the glial cells grew more and took up more water from the medium. Proponents of even slightly divergent explanations were long met with quite vigorous criticisms (Katzberg, 1951; Weiss, 1952), which might explain why these phenomena were not studied for so long. Certainly, Weiss's ideas about clot and matrix dehydration and shrinkage became widely accepted, taught, and written into textbooks. Concerning this dehydration hypothesis, Stopak has now been able to show not only that the difference in distorting effect between nerves and glia is actually attributable to the greater contractile strength of the latter, but also that local foci of dehydration do not in fact have the supposed shrinking effect—not, at least, in the case of collagen gels (Stopak and Harris, 1982).

My reinvestigation of the clot distortion phenomenon began in the laboratory of Professor Michael Abercrombie, to whom is due the credit for proposing that Weiss's "two-center effects" and "one-center effects" were entirely attributable to mechanical side effects of cellular propulsion.

Clottable plasma, taken from the carotid artery of adult chickens, kept cool in waxed test tubes to minimize (delay) clotting and centrifuged to remove blood cells, can easily be spread to form a thin, even film on the surface of a slide, coverslip, or petri dish. One simply uses the standard technique to make blood smears for microscopic examination. A small drop of plasma is put onto the surface, and the end of a slide is pulled quickly across the surface laterally. By adjusting the pressure and angle of the end of the slide, it is not difficult to make reproducible clot films only 15 μm thick. The standard technique for clotting plasma is to mix it with embryo extract; therefore, in earlier attempts a drop of embryo extract was simply placed on the surface next to the drop of plasma so that the lateral sweep of the slide accomplished both mixing and spreading simultaneously. However, one quickly learns that almost anything will cause plasma to clot, including the spreading process itself or simple warming. Mixing the plasma with various proportions of embryo extract (or merely with tissue culture medium) before clotting is a satisfactory way to produce films that are even more distortable than are ordinary clot films, affording detection of traction forces exerted by even weaker cells. To permit visualization of small distortions of the plasma clot substrata, small particles of carbon black were suspended in the plasma before spreading and clotting.

Cells of a variety of differentiated types were plated out onto these plasma clot films, primarily cells from chick embryos but also mouse peritoneal macrophages as well as polymorphonuclear leukocytes from both rabbit peritoneum and human blood. Time-lapse cinemicrography was then used to detect and record the displacements of the suspended carbon particles. Because small distortions in the plasma clot, unlike the wrinkles in silicone rubber substrata, are not themselves inherently visible, embedded marker particles and time-lapse recording were utilized.

All the cells studied by this method were able to make at least some distortions in the plasma clot films, as judged by the displacement of embedded carbon particles near their surfaces. In all cases, these distortions were centripetal relative to the cells producing them and rearward relative to any locomotion or spreading of these cells. A fair description would be to say that motile tissue cells pull their substratum rearward past themselves as they move, pulling it toward their centers as they spread. On a rigid substratum, such as glass or polystyrene, the only appreciable effect of this pulling force is the movement of the cell or the outward spreading of its margins. But on these films of clotted plasma, the effect of the force was partly to pull the cells "forward" and partly to pull the clot and carbon particles "backward." The relative amount of these two movements, the cell versus the substratum, depended on the contractile strength of the given cell relative to the elastic resistance of the layer of clotted plasma.

The different cell types turned out to differ enormously with respect to their capacity to distort these substrata. The fibroblasts seemed to be the strongest by far, and the leukocytes the weakest. The polymorphonuclear leukocytes would merely give the adjacent plasma clot a sort of twitch as they surged rapidly along, while the macrophages distorted it centripetally along their leading lamellae as they spread. In this respect they resembled miniature fibroblasts, except that in order to permit detection of the small forces exerted by macrophages, the clot films had to be made so weak (thin and made from maximally diluted plasma) that fibroblasts cultured on the same film would simply shred it. On such a weak clot film, the effect of the strong contractility of a fibroblast is simply to tear the film and compress the shreds onto the surface of the cell, which itself receives insufficient elastic resistance to be able to spread. Therefore, it is necessary to use undiluted plasma for fibroblasts. Part of this difference in contractile or "tractional" strength can be attributed to the small size of leukocytes, including macrophages, which, despite the misleading etymology of their name, are comparatively tiny cells. But even apart from size, the difference in strength appears to be enormous.

A number of time-lapse films were made of the behavior of 3T3 cells and of explanted fibroblasts on these fibrin clots. These substrata have a notable effect on the morphology of these cells, which is to narrow the width of their cytoplasmic extentions ("leading lamellae") and increase their length. Cells that would ordinarily have been shaped like pancakes on a glass or plastic surface become long and thin, "spindly," or spiderlike. As these cell processes elongate, the clot fibers near their surfaces, both in front of their leading edges and along their sides, are pulled rearward relative to the direction of extension. One sees this from the displacements of the carbon particles in the time-lapse films. These distortions of the clot are elastic in the sense that the clot fibers and carbon particles spring back to their original positions whenever a cell margin happens to detach and retract or when a cell rounds up to enter mitosis. When the distortion becomes too great, for example when fibroblasts actually tear the weaker clots or when groups of cells draw in masses of clot fibers around themselves, the distortion becomes irreversible.

The directions of clot fiber (and carbon particle) displacement are, as stated, opposite to the direction of cellular extension. This makes the distortions centripetal in the case of bipolar or tripolar cells, which were in the majority. In these cases, the carbon particles near the center of the cell, toward which all this pulling was going on, could often be seen to move slightly (1 or 2 μm) outward laterally away from the cell body, as if being pushed out by the clot material being pulled centripetally along the leading lamellae. This slight outward movement as well as the directionality of the clot distortion parallel but opposite to the direction of extension are both strong indications that one is observing a simple mechanical effect of cellular contractility upon the clot. If some chemical activity of the cell (Weiss's postulated dehydration, for example) were causing the clot itself to shrink or contract somehow, the centripetal displacement of the particles would be more or less radially symmetrical around the cell and greatest near the central, nuclear region of the cell, exactly where this outward movement occurs. The immediate elastic retraction of the clot when cells detach or round up would also not be expected if a chemically induced shrinkage were responsible.

Cells of several lines of transformed cells were also cultured on these plasma clot films, including L-929 cells and Hep-2 cells. The contractile strength of these cells proved to be much weaker than that of the untransformed fibroblasts, although not so weak as that of leukocytes. The basic pattern of centripetal distortion seemed not to be different, however—just a much weaker version of the same thing.

After this and a few parallel efforts at measurement of forces, this approach was abandoned until it could be proved that clot shrinkage could not be responsible for these phenomena. The use of plasma clot substrata to study cell traction still has many advantages, however, even in comparison with silicone rubber. For one thing, plasma clots can be weak enough to be distorted appreciably even by macrophages and leukocytes, something that has not yet been achieved with rubber.

Good use of this same basic approach has more recently been made by Steinberg *et al.* (1980) using gels of reprecipitated collagen. In this case the measure of relative contractile strength was the minimum population density of cells needed to produce a certain degree of matrix contraction in a given time. Again, the overall finding was that transformed cells are generally weaker than their untransformed counterparts. There were some apparent contradictions to this rule, however.

2.2. Cell Spreading on Surface Films and Liquids

As a related side issue, it was frequently noticed in these studies that fully spread macrophages could be found on the inner surface of the liquid–air interface. Observations of this kind, involving both leukocytes and nerve growth cones, are apparently the source of the opinion sometimes encountered that tissue cells do not actually require substrata in order to spread but can extend themselves at any interface without regard to its mechanical strength. I

believe that this is a mistake. Protein films form spontaneously at such interfaces by denaturation of serum proteins. These protein films themselves provide the mechanical equivalent of a "solid" substratum, which resists and
supports the contractility of the cells. These meniscus films sometimes appear
surprisingly rigid, certainly more so than the weakest plasma clot films. The
stronger the traction exerted by the cell, the stronger the mechanical resistance
of the substratum must be in order to support it.

It is intriguing to consider the fluidity of the plasma membrane itself,
either when it is exerting traction or in the case of another cell exerting traction
onto the surface of a neighbor. It might also be worth exploring the micro-
physics of these situations. Does the fluidity of the plasma membrane limit the
strength of the traction forces that can be exerted through it?

2.3. Strain-Induced Birefringence

Another possible way of detecting and potentially even measuring the
forces exerted by cells would be to observe the birefringence produced in
susceptible substrata by the forces of cellular traction. Engineers have long
made good use of this approach ("photoelasticity") to make visible the spatial
distribution of stresses and strains within models (usually plastic) of load-
bearing machinery too complex in shape for the stresses to be calculated analyt-
ically (Jessop and Harris, 1960). Plastics and other materials vary widely in
their susceptibility to this effect, measured as the "strain optical coefficient."
But if a material could be found or produced with a sufficiently great strain
optical coefficient, this material could simply be used as a culture substratum
in conjunction with an ordinary polarizing microscope to measure the birefringence induced by each cell's traction and thereby calculate the force exerted.
The difficulty lies in finding such a material. Even the most sensitive glasses
and plastics are not remotely sensitive enough and, although it might seem
helpful to make an extremely thin layer of the material so that cells will be
better able to strain it, this is actually of no help because the effect of strain is to
increase the birefringence per unit thickness of the material. Thus, the net
birefringence should remain unchanged for a given imposed force.

Ironically, when one pursues the photoelasticity literature to learn what is
the most sensitive of the materials used by engineers in this technique, the
answer turns out to be none other than gelatin—that is, collagen. I have recently tried to extend this approach by using liquid crystalline substances.
Liquid crystals, particularly the nematic ones, represent virtually the ultimate
in susceptibility to alignment in that they will align spontaneously. Unfortunately, the available nematic substances that display this behavior in the necessary temperature range turn out to hydrolyze when exposed to water; cho-
lesteric liquid crystals have not produced useful results either.

The most successful efforts to use birefringence for the detection and measurement of microscopic biological forces have been those of J. K. Harris (1978),
the near identity of whose initials and name to my own have unfairly concealed

his achievements. He is an independent and unrelated scientist, not a misprint. What J. K. Harris did was to use gelatin as a substratum, both for cultured cells and for walking insects, and observed between crossed polarizing filters the pattern of birefringence showing the forces exerted and their changes with time. His efforts were very successful in the case of walking insects but less so in the case of cultured cells. Gelatin begins to melt at body temperatures, and its strain optical coefficient is not really high enough for the purpose. It should also be realized that photoelastic birefringence is essentially only an indirect means of measuring distortion, which it may sometimes be preferable to measure directly.

3. Silicone Rubber Substrata

3.1. The Substrata Themselves

In an attempt to find an inert substitute for plasma clot substrata, a great many different plastic and rubberlike materials were examined. Some were toxic, others insufficiently transparent or flexible, and for many no suitable way was found to form the material into layers thin enough to be appreciably distorted by the traction of individual cells. Eventually, success was achieved with silicone rubber made by crosslinking dimethylpolysiloxane (ordinarily silicone fluid) by brief exposure to a flame. Thin films of rubber (as little as 1 μm thick) formed on the surface of still uncrosslinked liquid silicone fluid turn out to be nearly ideal for the purpose. They are totally inert to biological activity and are nontoxic and completely transparent. An unanticipated feature that has proved useful is that these rubber films develop pronounced wrinkles when stressed; thus, even static pictures can reveal the patterns of imposed forces without the need to trace the paths of marker particles, as is necessary with plasma clots.

Virtually the only drawback is that the relative crudeness of the method of crosslinking (by flaming the surface in a Bunsen burner for about 1½ sec) limits one's ability to prepare substrata that have exactly the same elasticity each time. Equally thin layers of silicone rubber can also be prepared by chemical catalysis with organic peroxide catalysts. When the catalyst phase and the silicone fluid are mutually insoluble, the rubber will form only along their interface; with careful control of catalyst concentration and duration of exposure, it should be possible to prepare elastic substrata that are both extremely thin and highly reproducible in their mechanical properties. In my own attempts in this direction, the layers formed proved toxic to cells. Furthermore, the peroxide catalysts happen to be explosive. Polysiloxanes are available in which the side groups are vinyl rather than methyl groups, so that crosslinking to form a rubber occurs spontaneously. This is the material called Silastic, but since the whole mass of fluid crosslinks, instead of just the outermost surface, it does not form sufficiently flexible layers for individual cells to distort.

3.2. Substratum Distortion by Cellular Traction

Cell traction produces two classes of wrinkles in the silicone rubber substrata (Fig. 1). Compression wrinkles are formed directly beneath the cell bodies and are oriented perpendicular to the direction in which the traction force is maximal. For example, in the case of a bipolar fibroblast, the compression wrinkles would be transverse to the long axis of the cell. Tension wrinkles, on the other hand, form beyond the advancing margins of the cell, from which they radiate outward along the axis of maximum stretching of the substratum. As the traction force increases, the compression wrinkles grow larger and more numerous, while the tension wrinkles grow longer. Particularly in crowded cultures, the distinction between these two kinds of wrinkles becomes obscured as the tension wrinkles from one cell merge into the compression wrinkles of its neighbor (Fig. 2).

The locations of compression wrinkles provide information about the locations on the cell surface where the traction forces are exerted. These wrinkles always form an appreciable distance ($\geq$ 5–10 μm) behind the advancing margin of the cell, implying that traction is not exerted at the outermost margin. The lamellipodia or ruffles do not appear to participate directly in the exertion

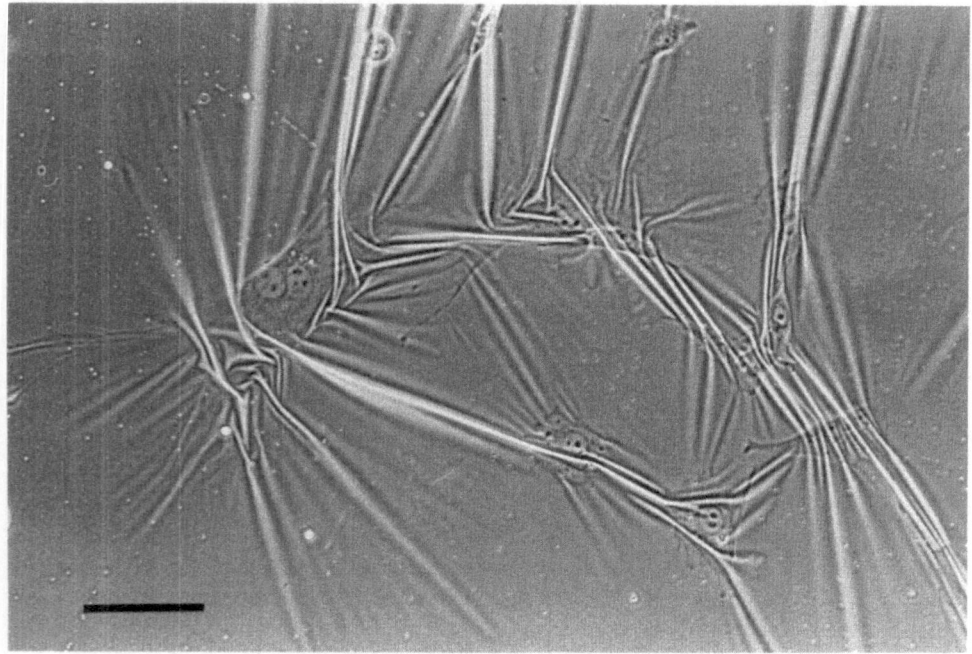

Figure 1. Fibroblast cells from embryonic chicken heart growing in tissue culture on a thin sheet of silicone rubber. Notice the patterns of wrinkles produced in this rubber sheet by the traction forces which the cells exert as they spread and crawl. Scale bar: 100 μm.

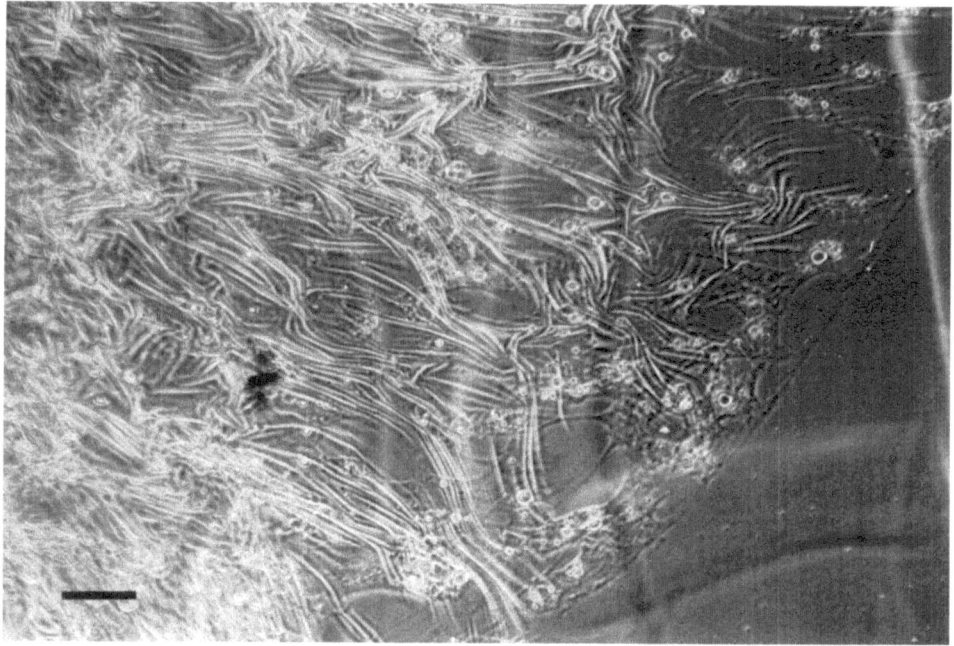

Figure 2. A more crowded culture of embryonic chick heart fibroblasts crawling on a thin sheet of silicone rubber, the edge of which has been torn loose along the lower right corner. The cumulative effect of the cells' forces is folding the rubber sheet into more and more complex patterns of compression wrinkles and drawing it toward the left. Eventually the entire rubber sheet will be contracted into a single tight ball of cells and rubber within about 24 hr more in culture. Scale bar: 100 μm.

of the force, and there is no indication of any simple process of reaching out, grabbing, and pulling backward by cytoplasmic protrusions of any kind. Instead, traction is exerted steadily by a broad area of the lower cell surface behind the leading margin. The only relationship that seems to exist between traction and the ruffling movements of the cell margin is that the exertion of this force tends to be concentrated behind those parts of the cell margin where ruffling is most active and that the direction of the ruffling movements corresponds to the direction of the traction force. The formation of substratum wrinkles accompanied cell spreading and elongation and tended to be progressive—the more a cell spreads, the larger and more numerous its wrinkles become. It is not a matter of a cell extending first and then contracting. These descriptions refer specifically to fibroblasts but also apply to the other cell types studied.

As was the case with the plasma clot substrata, cell types differed widely in the amount of distortion produced in the silicone rubber substrata (Fig. 3). A number of additional cell types have been tried; of these, blood platelets deserve special mention. These cellular fragments serve in blood clotting and

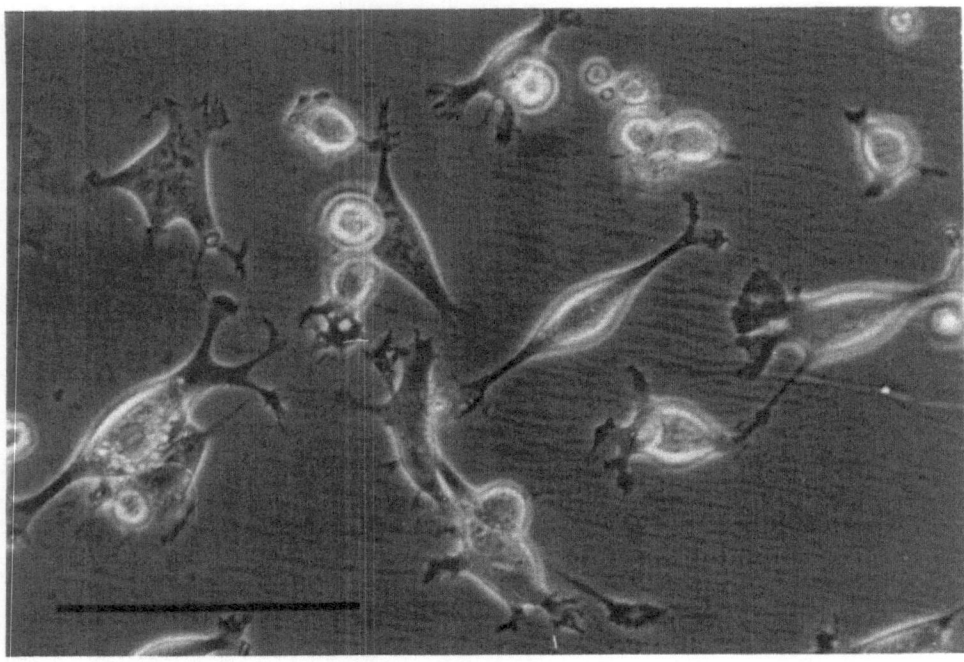

Figure 3. L cells, a neoplastically transformed line of fibroblastic cells originally derived from mice, cultured on the surface of a thin sheet of silicone rubber. Although this rubber sheet has been made as thin as those shown in Figures 1 and 2, the traction forces exerted by these transformed fibroblasts are so much weaker than those exerted by untransformed cells that they produce little or no wrinkling of the rubber surface. Scale bar: 100 μm.

those from human blood had approximately the same power to wrinkle rubber substrata as did fibroblasts. That is, an individual 3–4-μm platelet had a strength comparable to that of a whole fibroblast. The comparison is actually not easy to make directly, because platelets generate only tension wrinkles; apparently their size is less than the "wavelength" of compression wrinkles. Certainly platelets are very strong. Incidentally, their traction is also exerted as a shear force, and it is those platelets with the flattened pancake shape that are strongest, rather than those with filopodia.

Nerve growth cones were found to be so weak that no wrinkling or other distortions of the substrata by these cell processes could ever be detected. All rubber distortion in neural explants could be traced to glial cells or fibroblasts. Neuroblastoma cells were also extremely weak and produced only small and occasional wrinkles. No wrinkles at all were seen with polymorphonuclear leukocytes, again from human blood. Macrophages, specifically Kupffer cells from embryonic chick liver, were barely able to produce compression wrinkles and produced no tension wrinkles. These seem to represent the weakest cells whose traction can successfully be detected by this method. Epithelial cells, such as pigmented retina, liver, and epidermis from chick embryos, were found to exert moderate amounts of tension, intermediate in strength between macrophages and fibroblasts. Fibroblasts themselves, whether from embryonic heart, dermis, or other tissues, were found to increase their already strong traction in a sharp jump about 70 hr after explantation. We believe that this hypertrophy of cellular traction is a delayed response to trauma and is equivalent to that which serves normally to constrict wounds (see Fig. 4). Those transformed cells that we have examined were always much weaker in exerting traction than were fibroblasts from explants; the changes in cells of the Chinese hamster ovary (CHO) line are described more fully in the next section.

4. Collagen Substrata and the Tractional Structuring Phenomenon

Cellular traction also produces distortions in gels made of collagen. These gels can be easily prepared by dissolving collagen from rat tail tendons and then reprecipitating it (Elsdale and Bard, 1972). The amount of distortion produced in collagen gels by different kinds of cells varies in the same way observed with the silicone rubber substrata and is apparently a straightforward mechanical effect. Our studies concentrated on fibroblasts, which produce by far the greatest effects.

Around an individual fibroblast, the collagen fibers become reoriented into a sort of a swirl, reminiscent of the arrangement of iron filings around a magnet. Collagen becomes somewhat compacted around the cell body, and its fibers become oriented along the axes of the major cytoplasmic extensions, from the

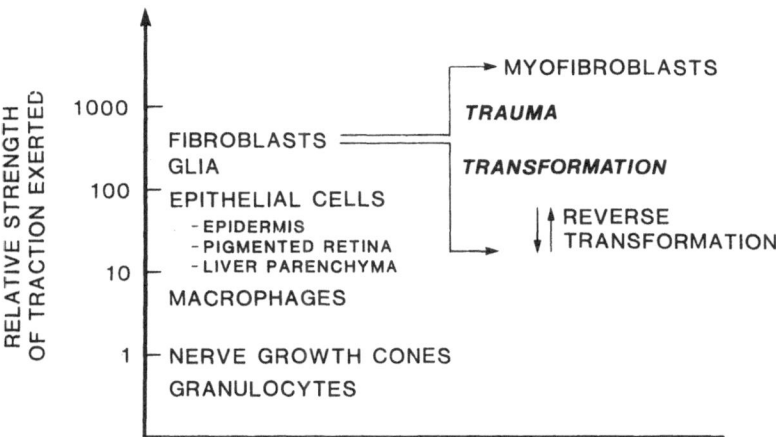

Figure 4. Diagramatic summary of the relative strengths of tractional forces exerted by different cell types, as well as the consequences of trauma, neoplastic transformation, and reverse transformation. The units of relative strength indicated as a logarithmic scale on the vertical axis represent a subjective estimate based on the sizes and numbers of wrinkles produced by cells on silicone rubber substrata and the population densities of the different kinds of cells required to produce a given degree of substratum distortion.

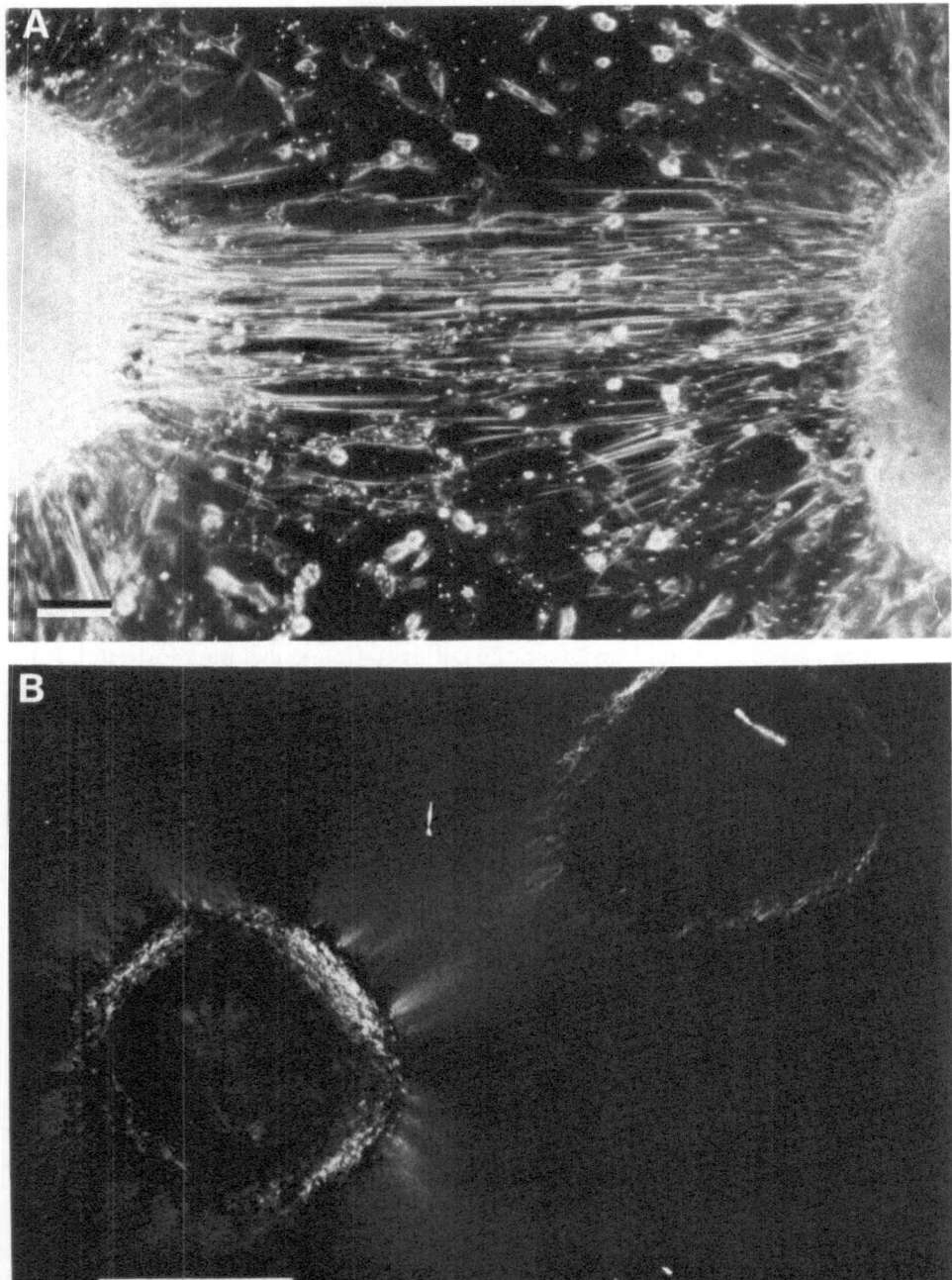

Figure 5. (A) Pattern of tensile stresses produced in a silicone rubber sheet between two explants of embryonic chick heart. Scale bar: 100 μm. (B) Pattern of stress realignment produced in an initially isotropic collagen gel between and surrounding two explants of embryonic chick heart. The culture has been photographed between crossed polarizing filters and the bright areas result from the

ends of which they splay outward radially, much like the tension wrinkles formed on rubber. The net effect is as if the cell had adhered to various points along the collagen fibers and then "capped" these points of attachment, pulling them centripetally across the cell surface, with the fibers becoming rearranged and reoriented accordingly. The cytoplasmic extensions or "leading lamellae" of fibroblasts, which have such a flattened form on planar substrata, instead take on an elongate conical shape in collagen gels.

Groups of fibroblasts, and in particular explanted tissue fragments containing fibroblasts, produce large-scale distortions in collagen gels (Fig. 5). These distortions are both cumulative and permanent (not elastic). Unlike the silicone rubber or even the clotted plasma, collagen has a strong tendency to "set up" or lock into the arrangements produced in it by cellular traction. This relative permanence of the distorted state seems to be explainable as the result of increased crosslinking between collagen fibers when the imposed forces bring them into parallel alignment. Once aligned, they tend to stay that way.

The fibroblasts themselves are strongly aligned by the collagen fibers. The alignment of cells parallel to fibers is an example of the phenomenon called contact guidance, discovered by Paul Weiss. Because the cells exert most of their traction parallel to their own long axes, the combination of fiber alignment by traction and cell alignment by fibers constitutes a positive feedback cycle. This cycle, in combination with the rheological properties of the collagen itself, results in some surprisingly long-range effects. Tracts of laterally compacted and highly aligned collagen form between fibroblast-containing explants as much as 40 mm apart. On the surfaces of such explants, collagen gels are pulled into dense compacted sheets with the fibers running circumferentially.

The aligned tracts of collagen are much like ligaments and tendons, while the compacted sheets of collagen have the same basic structure as organ capsules or the dermis. These similarities suggested to David Stopak and myself the hypothesis that these anatomical structures are actually formed by the same mechanical process as one observes when tissues are explanted into collagen gels. The basic idea is that the traction force exerted by fibroblasts is specialized and intended to rearrange and align fibers of collagen, and not just those collagen fibers secreted by the cells themselves. In other words, a macrophage exerts traction in order to move itself, but a fibroblast exerts traction in order to move collagen; this is the reason why the force exerted by fibroblasts is

birefringence of the collagen gel where its fibers have been aligned by cell traction. The collagen has been aligned under tension along the axis connecting the two explants. This alignment corresponds to the tension wrinkles seen on the silicone rubber substrata as well as perhaps to the mechanism by which ligaments and tendons are normally formed in the embryo. Other parts of the collagen have been aligned circumferentially around the surfaces of these explants; this circumferential wrapping of collagen corresponds to the formation of the compression wrinkles formed in the equivalent positions on silicone rubber substrata and may also correspond to the normal mechanism by which collagen fibers are formed into perichondria and organ capsules. Scale bar: 1 mm.

so much stronger than that exerted by a macrophage, even though a macrophage is more invasive. A useful analogy might be made to the uses of cilia. In *Paramecium*, the purpose of the cilia is to move the cell. But in the case of a ciliated epithelium in human lungs, say, the function is to move water relative to the cells. We would suggest that much of the locomotion undergone by fibroblasts in culture represents what amounts to an attempt by the cells to rearrange their substrata—an attempt that is both thwarted and concealed by the rigidity of glass and polystyrene. Our proposed morphogenetic mechanism, which we have termed *tractional structuring*, has interesting implications for the mechanical nature of the changes in malignant cells that render them invasive.

5. Reduced Tractional Strength and Its Relationship to Invasiveness

Intuitively, one would probably have expected that the correlation between invasiveness and the strength of the forces used by cells to propel themselves would be a positive one—in other words, that more invasive cells should be stronger. However, to judge both from leukocytes (including macrophages) and those cells from transformed lines that have been examined, the actual correlation is a negative one; the more invasive cells exert the weakest traction as they spread and crawl. The eventual solution to this paradox may help us to understand how malignant cells move through the body, as well as what mechanical changes in cells render them malignant. But it remains most unclear how to study the relationship directly.

5.1. Reverse Transformation of CHO Cells

In an effort to clarify the relationship among cell transformation, the strength of cell traction, adhesiveness, and other factors, my former students Mark Leader and David Stopak and I carried out a study of the reversibly transformed cell line (CHO) (Leader *et al.*, 1983). These cells, a transformed line originally derived from explants of Chinese hamster ovaries, were shown by Hsie and Puck (1971) as well as by Johnson *et al.* (1971) to be subject to a marked morphological change back toward an untransformed appearance when treated with cAMP, an effect that is potentiated by testosterone. The cells spread more fully and extend flatter lamellae in response to treatment with these hormones, either the two together or cAMP alone at a higher concentration. It has also been shown by Bloom and Lockwood (1980) that the cytoskeleton, especially fibers staining with antibodies against myosin, becomes better organized as a result of this reverse transformation. All these changes occur within a few hours and are reversible upon removal of the hormones. We therefore undertook a study of the accompanying changes in both contractility and adhesiveness, using the silicone rubber substrata as well as the special optical technique known as interference reflection microscopy by which the locations and gap distances of cell-to-substratum contacts can be seen (Izzard and Lochner, 1980).

As anticipated, we found that the CHO cells gradually increase the amount by which they wrinkle silicone rubber substrata, in parallel with the morphological changes of reverse transformation. The increased contractile strength is apparent after 2 hr, reaching a maximum after 5 hr of exposure to cAMP and testosterone. The reversal process has virtually this same time course. CHO cells turned out to be only very weakly contractile, however, even in the reverse transformed state, so that cells had to be plated at high (indeed confluent) population densities in order for these changes to be seen in substratum wrinkling. (It was important to have some wrinkling even before the reverse transformation began.) Thus, with this cell line it proved impossible to study the changes in traction at the individual cell level.

The changes in adhesion to the substratum were carried out at a considerably lower, subconfluent, density. The interference reflection images of the CHO cells in their ordinary transformed state showed that their contacts to the substratum were of the gray-appearing 30-nm gap-distance type, as well as that these contacts extended over most or all of the cells' lower surfaces, including the area beneath the nucleus, which in many cells does not adhere to the substratum. As these cells underwent morphological reverse transformation in response to the hormones, they developed the black-appearing, 10-nm adhesions to the glass substratum. These focal adhesions occupied only a small fraction of the cells' lower surfaces. They became concentrated near the cells' margins and the central areas near the nuclei lost contact with the substratum altogether, so that they appeared white in the interference reflection image. In other words, the reverse-transformed CHO cell gradually developed the adhesion pattern of a normal, untransformed fibroblast, both as regards to the types of contacts formed to the substratum and in respect to the spatial distribution of these contacts.

To our surprise, these changes in the interference reflection images, as well as in the spread areas of individual CHO cells, required considerably longer times (18–25 hr) to reach completion or to be completely reversed after the return of the cells to medium lacking hormone. We believe this must have been because of the much lower population density used in the interference reflection studies as compared with those on rubber substrata. Incidentally, neither in the transformed nor in the reverse-transformed state did the CHO cells undergo appreciable net locomotion. They merely spread in place.

5.2. Malignancy and Capsule Formation

It is one of the characteristics of malignant tumors that they lack capsules. Having previously proposed that cellular traction exerted on the extracellular matrix is the mechanism by which organ capsules are normally formed, finding unusually weak traction exerted by cells regarded as the tissue culture equivalents of malignant cells caused us to wonder whether weakened traction might underlie a malignant tumor's failure to form capsules. This is an attractive possibility and may well be true. There are some difficulties, however. One problem is that most human malignancies are carcinomas, that is, epithelial cancers, whereas our expectation was that fibroblasts (mesenchymal cells) exerted the traction responsible for arranging collagen into capsules. A parallel problem is that it is not at all clear why the malignant cells should prevent their normal neighbors, whether epithelial or mesenchymal, from forming a capsule.

A pilot study was undertaken to determine whether aggregates of CHO cells would rearrange collagen gels into capsulelike sheets and how reverse transformation of the cells with cAMP would affect the ability to form capsules. Unfortunately, these cells proved too weak, even in the reverse transformed state, to produce much in the way of encapsulation. Collagen fibers were pulled

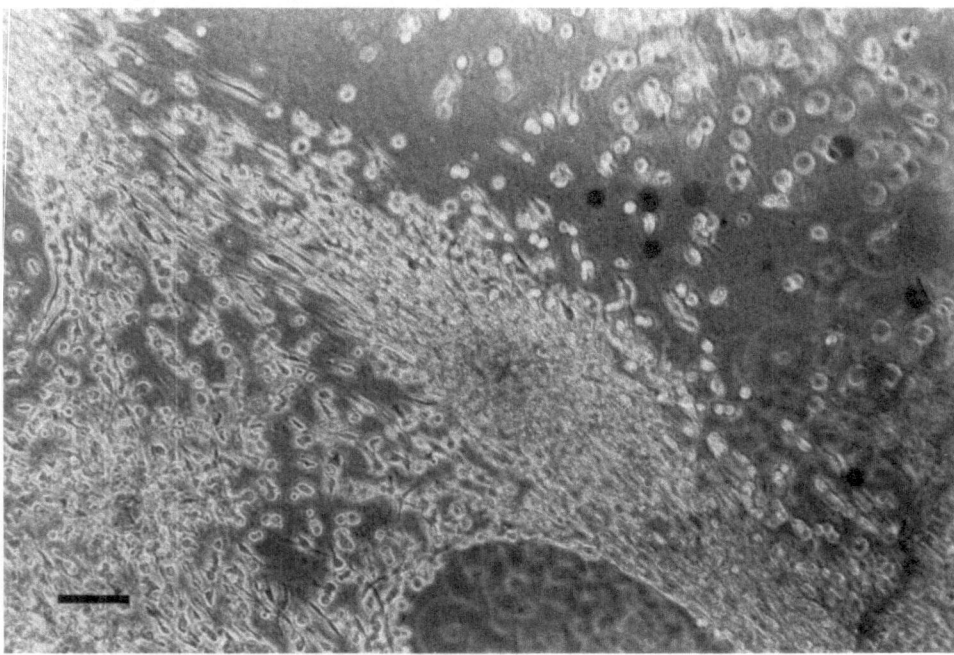

Figure 6. CHO cells (a neoplastically transformed line of fibroblastic cells originally derived from the ovaries of a Chinese hamster), cultured as clumps of cells in a gel of reprecipitated rat tail tendon collagen. The traction forces exerted by these cells have become stronger in response to treatment (reverse transformation) with dibutyryl cyclic adenosine monophosphate. The traction exerted by the reverse transformed cells has produced some degree of mechanical reorganization of the initially homogeneous and isotropic collagen gel into ligamentlike and capsulelike regions. This reorganization was considerably greater than that which occurred when these transformed cells were not treated with the cyclic AMP, but still falls short of the degree of collagen gel reorganization produced by normal untransformed fibroblasts. A characteristic trait of malignant tumors is that they do not form capsules; benign tumors in contrast are encapsulated. Scale bar: 100 μm.

in and accumulated around clusters of the CHO cells; this accumulation was increased by reverse transformation, but even then, nothing like the tight wrappings of fibers that form around fibroblastic explants could be seen (Fig. 6). Another system of conditionally transformed cells might yield better results.

5.3. Malignancy and the Mechanical Properties of Leukocytes

Macrophages and other leukocytes possess, among their normal differentiated characters, the capacity for tissue invasion, including blood vessel penetration and escape, the development of which in epithelial and mesenchymal cells renders these cells malignant. But are carcinoma cells invasive for the same reasons as macrophages? Are the mechanical properties that render them invasive the same in both cases? These questions cannot be answered until the mechanical properties responsible have been identified. But it is worth noting that both in their weak traction and in their broad distribution of the gray-appearing 30-nm substratum contacts (instead of localized tight adhesions), normal macrophages and CHO in the transformed state are very similar indeed. These and other transformed cells could reasonably be described as fibroblasts that have somehow acquired the traction and adhesive properties of macrophages.

Although it makes intuitive sense that reduced adhesiveness should permit cells to migrate more invasively, it remains a puzzle why weaker traction should be associated with invasiveness. This might merely be a by-product of reduced adhesiveness—weak adhesions being unable to transmit strong tension to the substratum—but it certainly does not appear that the reason macrophages, say, or nerve growth cones do not exert stronger traction is because their adhesions are pulling away from the substratum. They spread as well as or better than normal fibroblasts on silicone rubber substrata; they just produce little or no wrinkling. Although it is true that one can make 3T3 cells or chick heart fibroblasts look very much like transformed cells by culturing them on a substratum like cellulose acetate, teflon, or untreated polystyrene (to which they cannot form sufficiently strong adhesions to support full spreading), it is also a fact that macrophages, L cells, sarcoma-180 cells, and other transformed lines will spread on these less adhesive surfaces just as well as they do on glass. What mechanical factors limit the spreading ability of transformed cells is far from clear.

Their appearance in time-lapse films tempts one to visualize spreading cells as if they were themselves so many little bits of rubber, being stretched this way and that by tractorlike pulling forces distributed along their margins. Such a mechanical analogy implies that the degree of spreading achieved should result from the balance between the traction forces and the (imagined) elastic resistance to spreading on the part of the stretched cell body. When this elastic stretching force, pulling inward on the adhesions of the cell, becomes strong enough either to break the adhesions themselves or to match in strength the outward-pulling traction, spreading cannot proceed further. In that case, the reduced spreading of transformed cells could be explained as a direct result of their reduced traction.

One trouble with this analogy is that it separates the overall contractility of the cell body from the traction being exerted through its plasma membrane (although both forces are presumably created by the same cytoplasmic actomyosins). Another problem is that the resistance of the cell to being stretched out flat is merely assumed to increase monotonically with length or spread area. A piece of rubber or other elastic material will resist stretching more and more strongly the more it is stretched and, to judge from the increasing degree of elastic distortion cells produce in flexible substrata as they themselves spread, much the same seems to be the case for cells. But perhaps we should be surprised by this. Why should it necessarily be true of an active, actomyosin-based contraction? What is the restoring force that pulls back on leading lamellae? Does it pull harder the more the cell spreads? Can it remain strong while the traction force pulling against it is weakened? Is this what happens in transformed cells?

5.4. The Paradox of Weaker Traction and Greater Invasiveness

A macrophage spreads at the same speed as a fibroblast while exerting perhaps one-hundredth the propulsive force; a polymorphonuclear leukocyte crawls at 50 times the speed of a fibroblast while exerting even weaker traction than a macrophage. How can this be, and what can it tell us about the mechanical changes in cancerous cells that render them malignant? The root of the question may lie in the use of traction for two such different functions as cellular propulsion and the structural rearrangement of collagen fibers. We do not know what factors besides simple relative mass determine whether traction exerted by a cell on a collagen fiber will cause the displacement of the cell or the rearrangement of the collagen. One possibility is that the difference lies in the balance or imbalance in the forces exerted in opposed directions by a given cell. If the forces are all or mostly in one direction, cellular displacement will be the result even though the total force is quite small. On the other hand, if the forces are very large but equally balanced against each other, distortion of the substratum will be the primary result of traction. To follow this line of reasoning one step further, invasiveness might be expected to result from cellular changes that interfere with the balance of cellular forces, even though this is achieved by weakening some or all of them.

In closing, it may be worth raising one last question. Malignant tissues are morphologically distinguishable from their untransformed equivalents. The morphological criteria are empirical, based on the accumulated experience of pathologists, but the criteria do seem to be reliable, and texts can be found illustrating the lore of this topic with great numbers of photographic examples (Gompel, 1978). Certainly a great deal does depend upon the pathologist's accuracy, though his criteria are not based upon the cell biologist's theories about what a cancer cell "should" look like. Now a change in the mechanical properties of a cell, the viscosity of its cytoplasm or membranes, the adhesiveness of its surface, the strength of its contractility, or other properties can be expected to produce a morphological change in the cell—not that it is necessarily a straightforward task to deduce the nature of this change from its morphological consequences. The converse also seems true—any morphological change should ultimately be traceable to a mechanical cause. A changed distribution of matter implies a changed distribution in the forces acting on this matter, and it is not too large a logical jump to suppose that the possession by malignant cells of consistent morphological peculiarities implies that their mechanical properties must also differ from equivalent normal cells.

Current cancer chemotherapy seems to be based almost entirely on poisoning cell growth and depending on the rapid growth rate of malignant cells for whatever selectivity the drugs achieve in killing cancerous but not normal cells. How to poison cells selectively on the basis of whatever makes their nuclei asymmetrical, or causes their chromatin to clump against the inner surface of their nuclear membranes, or weakens their traction is, most unfortunately, not obvious.

References

Elsdale, T., and Bard, J., 1972, Collagen substrata for studies on cell behavior, *J. Cell Biol.* **54**:626–637.

Gompel, C., 1978, *An Atlas of Diagnostic Cytology*, Wiley, New York.

Harris, J. K., 1978, A photoelastic substrate technique for dynamic measurements of forces exerted by moving organisms, *J. Microsc.* **114**:219–228.

Hsie, A. W., and Puck, T. T., 1971, Morphological transformation of Chinese hamster cells by dibutyryl adenosine cyclic 3′ : 5′-monophosphate and testosterone, *Proc. Natl. Acad. Sci. USA* **68**:358–361.

Izzard, C. S., and Lochner, L. R., 1980, Formation of cell to substrate contacts during fibroblast motility: An interference reflexion study, *J. Cell Sci.* **42**:81–116.

James, D. W., and Taylor, J. F., 1969, The stress developed by sheets of chick fibroblasts *in vitro*, *Exp. Cell Res.* **54**:107–110.

Jessop, H. T., and Harris, F. C., 1960, *Photoelasticity, Principles and Methods*, Dover, New York.

Johnson, G. S., Friedman, R. M., and Pastan, I., 1971, Restoration of several morphological characteristics of normal fibroblasts in sarcoma cells treated with adenosine-3′ : 5′-cyclic monophosphate and its derivatives, *Proc. Natl. Acad. Sci. USA* **68**:425–429.

Katzberg, A. A., 1951, Distance as a factor in the development of attraction fields between growing tissues in culture, *Science* **114**:431–432.

Leader, W. M., Stopak, D., and Harris, A. K., 1983, Increased contractile strength and tightened adhesions to the substratum result from reverse transformation of CHO cells by dibutyryl cyclic adenosine monophosphate, *J. Cell Sci.* **64**:1–11.

Phillips, H. M., and Davis, G. S., 1978, Liquid-tissue mechanics in amphibian gastrulation. Germ layer assembly in *Rana pipiens*, *Am. Zool.* **18**:81–93.

Phillips, H. M., and Steinberg, M. S., 1978, Embryonic tissues as elasticoviscous liquids. I. Rapid and slow shape changes in centrifuged cell aggregates, *J. Cell Sci.* **30**:1–20.

Steinberg, B. M., Smith, K., Colozzo, M., and Pollack, R., 1980, Establishment and transformation diminish the ability of fibroblasts to contract a native collagen gel, *J. Cell Biol.* **87**:304–308.

Stopak, D., and Harris, A. K., 1982, Connective tissue morphogenesis by fibroblast traction. I. Tissue culture observations, *Dev. Biol.* **90**:383–398.

Weiss, P., 1952, "Attraction fields" between growing tissue cultures, *Science* **115**:293–295.

Index